# THE BOOK OF R

# THE BOOK OF R

## A First Course in Programming and Statistics

by Tilman M. Davies

**no starch press**

San Francisco

Printed in USA

Second printing

20 19 18 17          2 3 4 5 6 7 8 9

ISBN-10: 1-59327-651-6
ISBN-13: 978-1-59327-651-5

Publisher: William Pollock
Production Editor: Riley Hoffman
Cover Illustration: Josh Ellingson
Interior Design: Octopod Studios
Developmental Editor: Liz Chadwick
Technical Reviewer: Debbie Leader
Copyeditor: Kim Wimpsett
Compositor: Riley Hoffman
Proofreader: Paula Fleming
Indexer: BIM Creatives, LLC

For information on distribution, translations, or bulk sales, please contact No Starch Press, Inc. directly:

No Starch Press, Inc.
245 8th Street, San Francisco, CA 94103
phone: 1.415.863.9900; info@nostarch.com
www.nostarch.com

*Library of Congress Cataloging-in-Publication Data*

Names: Davies, Tilman M., author.
Title: The book of R : a first course in programming and statistics / by
    Tilman M. Davies.
Description: San Francisco : No Starch Press, [2016] | Includes
    bibliographical references and index.
Identifiers: LCCN 2015035305| ISBN 9781593276515 | ISBN 1593276516
Subjects: LCSH: R (Computer program language) | Computer programming. |
    Statistics--Data processing.
Classification: LCC QA76.73.R3 D38 2016 | DDC 519.50285--dc23
LC record available at http://lccn.loc.gov/2015035305

Prof. Dr. Dr. Jochen Weniger, 1925–2015
Ein Wissenschaftler. Mein Opa.

# BRIEF CONTENTS

## PART V: ADVANCED GRAPHICS

*Plates follow page 736*

# CONTENTS IN DETAIL

## PART I
## THE LANGUAGE

## 1
## GETTING STARTED                                                          3

# 2
# NUMERICS, ARITHMETIC, ASSIGNMENT, AND VECTORS    17

# 3
# MATRICES AND ARRAYS    39

# 4
# NON-NUMERIC VALUES    59

# 5
# LISTS AND DATA FRAMES 89

# 6
# SPECIAL VALUES, CLASSES, AND COERCION 103

# 7
# BASIC PLOTTING 127

# 8
# READING AND WRITING FILES 147

# PART II
# PROGRAMMING

# 9
# CALLING FUNCTIONS 165

## 22
## LINEAR MODEL SELECTION AND DIAGNOSTICS     527

# PART V
# ADVANCED GRAPHICS

# 25
# DEFINING COLORS AND PLOTTING IN HIGHER DIMENSIONS 631

# 26
# INTERACTIVE 3D PLOTS         691

*Plates follow page 736*

# PREFACE

The aim of *The Book of R: A First Course in Programming and Statistics* is to provide a relatively gentle yet informative exposure to the statistical software environment R, alongside some common statistical analyses, so that readers may have a solid foundation from which to eventually become experts in their own right. Learning to use and program in a computing language is much the same as learning a new spoken language. At the beginning, it is often difficult and may even be daunting—but total immersion in and active use of the language is the best and most effective way to become fluent.

Many beginner-style texts that focus on R can generally be allocated to one of two categories: those concerned with computational aspects (that is, syntax and general programming tools) and those with statistical modeling and analysis in mind, often one particular type. In my experience, these texts are extremely well written and contain a wealth of useful information but better suit those individuals wanting to pursue fairly specific goals from the outset. This text seeks to combine the best of both worlds, by first focusing on *only* an appreciation and understanding of the language and its style and subsequently using these skills to fully introduce, conduct, and interpret some common statistical practices. The target audience is, quite simply, anyone who wants to gain a foothold in R as a first computing language, perhaps with the ultimate goal of completing their own statistical analyses. This includes but is certainly not limited to undergraduates, postgraduates, academic researchers, and practitioners in the applied sciences with little or no experience in programming or statistics in general. A basic understanding of elementary mathematical behavior (for example, the order of operations) and associated operators (for example, the summation symbol $\Sigma$) is desirable, however.

In view of this, *The Book of R* can be used purely as a programming text to learn the language or as an introductory statistical methods book with accompanying instruction in R. Though it is not intended to represent an exhaustive dictionary of the language, the aim is to provide readers with a comfortable learning tool that eliminates the kind of foreboding many have voiced to me when they have considered learning R from scratch. The fact remains that there are usually many different ways to go about any given task—something that holds true for most so-called high-level computer languages. What this text presents reflects my own way of thinking about learning and programming in R, which I approach less as a computer scientist and more as an applied data analyst.

In part, I aim to provide a precursor and supplement to the work in *The Art of R Programming: A Tour of Statistical Software Design*, the other R text published by No Starch Press (2011), written by Professor Norman Matloff (University of California, Davis). In his detailed and well-received book, Professor Matloff comes at R from a computer science angle, that is, treating it as a programming language in its own right. As such, *The Art of R Programming* provides some of the best descriptions of R's computational features I've yet to come across (for example, running external code such as C from R programs, handling and manipulating R's memory allocations, and formal debugging strategies). Noteworthy, however, is the fact that some previous experience and knowledge of programming in general goes a long way to appreciating some of these more advanced features. It is my hope that my text will not only provide this experience but do so in R itself at a comfortable pace, with statistical analyses as the supplementary motivation.

This text, which serves as a "traveler's guide" as we backpack our way through R country, was born out of a three-day introductory R workshop I began teaching at the University of Otago in New Zealand. The emphasis is on active use of the software, with each chapter containing a number of code examples and practice exercises to encourage interaction. For those readers not part of a workshop, just fire up your computer, grab a drink and a comfy chair, and start with Chapter 1.

Tilman M. Davies
Dunedin, New Zealand

# ACKNOWLEDGMENTS

First, I must of course acknowledge the creators and maintainers of R, both past and present. Whether or not any prominent contributor ever picks up this book (perhaps as a handy stabilizer for a wonky piece of furniture), they should be reminded of the appreciation that many, including myself, have for the incredible role R plays as a data analysis tool worldwide and a glowing example of what can be achieved with free software. R has transformed how academic and industry statisticians, and many others, make contributions to research and data analysis. This doesn't refer just to their respective pursuits but also to the ability to make computational functionality readily available so that others may perform the same analyses. I'm just one voice of many to echo this, and *The Book of R* essentially represents my own professional love letter to all things R. The best way to profess this sentiment is to encourage and educate both students and professionals in the analytic freedom R provides and hence perpetuate the enthusiasm many already hold for the software.

The team at No Starch Press is undoubtedly an absolute pleasure to work with. If "informal professionalism" is actually a thing, it's exactly how I like to work and exactly how they guided the progress of what turned out to be a large, time-consuming project. Thanks to Bill Pollock and Tyler Ortman for taking on this project. Thanks especially to those at the coalface of my drafts, specifically Liz Chadwick and Seph Kramer for picking through and streamlining the language and exposition in numerous TeX files, and to those who turned the collection of these files into a cohesive, polished book, headed by Riley Hoffman. Dr. Debbie Leader, a senior university tutor in statistics at Massey University (Palmerston North, New Zealand), can't escape my appreciation for the scrutiny she applied to the R code and statistical content, improving the clarity and interpretation of the core focus of the book. Comments from other reviewers and early readers were also instrumental in shaping the finished product.

A special mention must be made of Professor Martin Hazelton at Massey University. Once my undergraduate lecturer during my freshman year at the University of Western Australia; then my graduate and doctoral supervisor in New Zealand, where I followed him; and now my ongoing collaborator, mentor, and friend, Martin is pretty much solely responsible for my interest and career in statistics. Although my scribblings mightn't include all the big fancy Oxford words he'd use, it's only with this education that I'm now in a position to write this book. With it, I hope I can inspire my own students and any others wanting to learn more about statistical programming in the same way he inspired me.

Last, but certainly not least, my immediate family in Perth and my extended family in Germany must be thanked for their ongoing and unwavering love and support through many years. My New Zealand family must also be thanked, especially Andrea, who would often put up with a glowing Apple logo instead of my face on the other end of the couch as I continued tapping away on the manuscript after work, only pausing to reach for a beer (or if something juicy was happening on *Shortland Street*—Google it). The cat, Sigma, was always nearby at these moments too, ensuring her insertion in a couple of exercises in this book.

I suppose all this could be summarized in one word, though:

Cheers.

# INTRODUCTION

R plays a key role in a wide variety of research and data analysis projects because it makes many modern statistical methods, both simple and advanced, readily available and easy to use. It's true, however, that a beginner to R is often new to programming in general. As a beginner, you must not only learn to use R for your specific data analysis goals but also learn to think like a programmer. This is partly why R has a bit of a reputation for being "hard"—but rest assured, that really isn't the case.

## A Brief History of R

R is based heavily on the S language, first developed in the 1960s and 1970s by researchers at Bell Laboratories in New Jersey (for an overview, see, for example, Becker et al., 1988). With a view to embracing open source software, R's developers—Ross Ihaka and Robert Gentleman at the University of Auckland in New Zealand—released it in the early 1990s under the

GNU public license. (The software was named for Ross and Robert's shared first initial.) Since then, the popularity of R has grown in leaps and bounds because of its unrivaled flexibility for data analysis and powerful graphical tools, all available for the princely sum of *nothing*. Perhaps the most appealing feature of R is that any researcher can contribute code in the form of *packages* (or *libraries*), so the rest of the world has fast access to developments in statistics and data science (see Section A.2).

Today, the main source code archives are maintained by a dedicated group known as the *R Core Team*, and R is a collaborative effort. You can find the names of the most prominent contributors at *http://www.r-project.org/*; these individuals deserve thanks for their ongoing efforts, which keep R alive and at the forefront of statistical computing!

The team issues updated versions of R relatively frequently. There have been substantial changes to the software over time, though neighboring versions are typically similar to one another. In this book, I've employed versions 3.0.1–3.2.2. You can find out what's new in the latest version by following the NEWS link on the relevant download page (see Appendix A).

# About This Book

*The Book of R* is intended as a resource to help you get comfortable with R as a first programming language and with the statistical thought that underpins much of its use. The goal is to lay an introductory yet comprehensive foundation for understanding the computational nature of modern data science.

The structure of the book seeks to progress naturally in content, first focusing on R as a computational and programming tool and then shifting gears to discuss using R for probability, statistics, and data exploration and modeling. You'll build your knowledge up progressively, and at the end of each chapter, you'll find a section summarizing the important code as a quick reference.

## Part I: The Language

Part I, which covers the fundamental syntax and object types used across all aspects of R programming, is essential for beginners. Chapters 2 through 5 introduce the basics of simple arithmetic, assignment, and important object types such as vectors, matrices, lists, and data frames. In Chapter 6, I'll discuss the way R represents missing data values and distinguishes among different object types. You're given a primer on plotting in Chapter 7, using both built-in and contributed functionality (via the ggplot2 package—see Wickham, 2009); this chapter lays the groundwork for graphical design discussed later in the book. In Chapter 8, I'll cover the fundamentals of reading data in from external files, essential for analysis of your own collected data.

## Part II: Programming

Part II focuses on getting you familiar with common R programming mechanisms. First, I'll discuss functions and how they work in R in Chapter 9. Then, in Chapter 10, I'll cover loops and conditional statements, which are used to control the flow, repetition, and execution of your code, before teaching you how to write your own executable R functions in Chapter 11. The examples in these two chapters are designed primarily to help you understand the behavior of these mechanisms rather than to present real-world analyses. I'll also cover some additional topics, such as error handling and measuring function execution time, in Chapter 12.

## Part III: Statistics and Probability

With a firm handle on R as a language, you'll shift your attention to statistical thinking in Part III. In Chapter 13, you'll look at important terminology used to describe variables; elementary summary statistics such as the mean, variance, quantiles, and correlation; and how these statistics are implemented in R. Turning again to plotting, Chapter 14 covers how to visually explore your data (with both built-in and ggplot2 functionality) by using and customizing common statistical plots such as histograms and box-and-whisker plots. Chapter 15 gives an overview of the concepts of probability and random variables, and then you'll look at the R implementation and statistical interpretation of some common probability distributions in Chapter 16.

## Part IV: Statistical Testing and Modeling

In Part IV, you're introduced to statistical hypothesis testing and linear regression models. Chapter 17 introduces sampling distributions and confidence intervals. Chapter 18 details hypothesis testing and $p$-values and demonstrates implementation and interpretation using R; the common ANOVA procedure is then discussed in Chapter 19. In Chapters 20 and 21, you'll explore linear regression modeling in detail, including model fitting and dealing with different types of predictor variables, inferring and predicting, and dealing with variable transformation and interactive effects. Rounding off Part IV, Chapter 22 discusses methods for selecting an appropriate linear model and assessing the validity of that model with various diagnostic tools.

Linear regression represents just one class of parametric models and is a natural starting point for learning about statistical regression. Similarly, the R syntax and output used to fit, summarize, predict from, and diagnose linear models of this kind are much the same for other regression models—so once you're comfortable with these chapters, you'll be ready to tackle the R implementation of more complicated models covered in more advanced texts with relative ease.

Parts III and IV represent much of what you'd expect to see in first- and second-year college statistics courses. My aim is to keep mathematical details to a minimum and focus on implementation and interpretation. I'll provide references to other resources where necessary if you're interested in looking more closely at the underlying theory.

### Part V: Advanced Graphics

The final part looks at some more advanced graphing skills. Chapter 23 shows you how to customize traditional R graphics, from handling the graphics devices themselves to controlling the finer aspects of your plot's appearance. In Chapter 24, you'll study the popular ggplot2 package further, looking at more advanced features such as adding smooth scatterplot trends and producing multiple plots via faceting. The final two chapters concentrate on higher dimensional plotting in R. Chapter 25 covers color handling and 3D surface preparation before discussing contour plots, perspective plots, and pixel images with the aid of multiple examples. Chapter 26 then focuses on interactive plots and includes some simple instructions for plotting multivariate parametric equations.

Though not strictly necessary, it's helpful to have some familiarity with the linear regression methods discussed in Part IV before tackling Part V, since some of the examples in this last part use fitted linear models.

## For Students

Like many, I first started becoming proficient in R programming and the associated implementation of various statistical methods when I began my graduate studies (at Massey University in Palmerston North, New Zealand). Building on little more than the odd line or two of code I'd encountered during my undergraduate years in Australia, being "thrown in the deep end" had both benefits and drawbacks. While the immersion accelerated my progress, not knowing what to do when things don't work properly is of course frustrating.

*The Book of R* thus represents the introduction to the language that I wish I'd had when I began exploring R, combined with the first-year fundamentals of statistics as a discipline, implemented in R. With this book, you'll be able to build a well-rounded foundation for using R, both as a programming language and as a tool for statistical analyses.

This book was written to be read cover to cover, like a story (albeit with no plot twists!). Ideas are built up progressively within each part of the book, so you can choose to begin either right at the start or where you feel your level of knowledge currently stands. With that in mind, I offer the following recommendation to students of R:

- Try not to be afraid of R. It will do exactly what you tell it to—nothing more, nothing less. When something doesn't work as expected or an error occurs, this literal behavior works in your favor. Look carefully at the commands line by line and try to narrow down the instructions that caused the fault.

- Attempt the practice exercises in this book and check your responses using the suggested solutions—these are all available as R script files on the book's website, *https://www.nostarch.com/bookofr/*. Download the *.zip* file and extract the *.R* files, one for each part of the book. Open these in your R session, and you can run the lines like you would any R code to see the output. The short practice exercises are intended to be exactly that—practice—as opposed to being hard or insurmountable challenges. Everything you need to know to complete them will be contained in the preceding sections of that chapter.

- Especially in your early stages of learning, when you're away from this book, try to use R for everything, even for very simple tasks or calculations you might usually do elsewhere. This will force your mind to switch to "R mode" more often, and it'll get you comfortable with the environment quickly.

## For Instructors

This book was designed from a three-day workshop, Introduction to R, that I run at my current institution—the Department of Mathematics & Statistics at the University of Otago in New Zealand—as part of our Statistics Workshops for Postgraduates and Staff (SWoPS). Succeeded by the SWoPS class Statistical Modelling 1 run by two of my colleagues, the aim of Introduction to R is, as the title suggests, to address the programming side of things. Your coverage will naturally depend on your target audience.

Here I provide some recommendations for using the content in *The Book of R* for workshops of similar length to our SWoPS series. Particular chapters can be added or dropped depending on your target workshop duration and students' existing knowledge.

- **Programming Introduction**: Parts I and II. Selected material from Part V, especially Chapter 23 (Advanced Plot Customization), might also suit the scope of such a course.

- **Statistics Introduction**: Parts III and IV. If a brief introduction to R is warranted beforehand, consider dropping, for example, Chapter 13 from Part III and Chapters 17 through 19 in Part IV and building an initial foundation from content in Part I.

- **Intermediate Programming and Statistics**: Parts II and IV. Consider dropping Chapters 17 through 19 from Part IV to include Part V if the audience is interested in developing plotting skills.

- **R Graphics**: Parts I and V. Depending on audience knowledge, material from Part I may be dropped so that Chapter 14 in Part II can be included (Basic Data Visualization).

If you're planning to go even further and structure a longer course around this book, the practice exercises make particularly good lecture-specific homework to keep students abreast of the skills in R and statistics as they're developed. The main points of the sections making up each chapter are relatively easy to translate into slides that can be initially structured with help from the Contents in Detail.

# PART I

**THE LANGUAGE**

# 1

## GETTING STARTED

R provides a wonderfully flexible programming environment favored by the many researchers who do some form of data analysis as part of their work. In this chapter, I'll lay the groundwork for learning and using R, and I'll cover the basics of installing R and certain other things useful to know before you begin.

## 1.1   Obtaining and Installing R from CRAN

R is available for Windows, OS X, and Linux/Unix platforms. You can find the main collection of R resources online at the Comprehensive R Archive Network (CRAN). If you go to the R project website at *http://www.r-project.org/*, you can navigate to your local CRAN mirror and download the installer relevant to your operating system. Section A.1 provides step-by-step instructions for installing the base distribution of R.

## 1.2   Opening R for the First Time

R is an interpreted language that's strictly case- and character-sensitive, which means that you enter instructions that follow the specific syntactic rules of the language into a console or command-line interface. The software then interprets and executes your code and returns any results.

**NOTE**    *R is what's known as a* high-level *programming language. Level refers to the level of abstraction away from the fundamental details of computer execution. That is, a low-level language will require you to do things such as manually manage the machine's memory allotments, but with a high-level language like R, you're fortunately spared these technicalities.*

When you open the base R application, you're presented with the R console; Figure 1-1 shows a Windows instance, and the left image of Figure 1-2 shows an example in OS X. This represents R's naturally incorporated *graphical user interface (GUI)* and is the typical way base R is used.

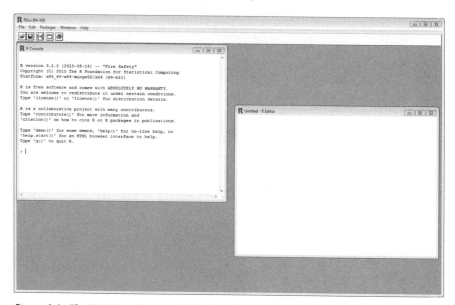

*Figure 1-1: The R GUI application (default configuration) in Windows*

The functional, "no-frills" appearance of the interpreter, which in my experience has struck fear into the heart of many an undergraduate, stays true to the very nature of the software—a blank statistical canvas that can be used for any number of tasks. Note that OS X versions use separate windows for the console and editor, though the default behavior in Windows is to contain these panes in one overall R window (you can change this in the GUI preferences if desired; see Section 1.2.1).

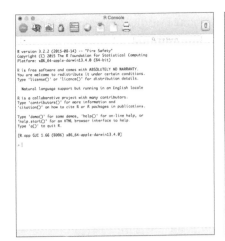

Figure 1-2: The base R GUI console pane (left) and a newly opened instance of the built-in editor (right) in OS X

*As I've just done, in some parts of the book I'll refer specifically to the R GUI functionality in Windows and OS X, given these are the two platforms most often used by beginners. As well as Linux/Unix implementations, it's possible to run R from a terminal or shell or, indeed, in the alternative* batch *mode. The vast majority of the code in this book is functional in all settings.*

### 1.2.1   Console and Editor Panes

There are two main window types used for programming R code and viewing output. The console or command-line interpreter that you've just seen is where all execution takes place and where all textual and numeric output is provided. You may use the R console directly for calculations or plotting. You would typically use the console directly only for short, one-line commands.

By default, the R *prompt* that indicates R is ready and awaiting a command is a > symbol, after which a text cursor appears. To avoid confusion with the mathematical symbol for "greater than," >, some authors (including me) prefer to modify this. A typical choice is R>, which you can set as follows:

```
> options(prompt="R> ")
R>
```

With the cursor placed at the prompt, you can use the keyboard up arrow (↑) and down arrow (↓) to scroll through any previously executed commands; this is useful when making small tweaks to earlier commands.

For longer chunks of code and function authoring, it's more convenient to first write your commands in an *editor* and execute them in the console only when you're done. There is a built-in R code editor for this purpose. The R *scripts* you write in the code editor are essentially just plain-text files with a *.R* extension.

You can open a new instance of the editor using the R GUI menus (for example, File → New script in Windows or File → New Document in OS X).

The built-in editor features useful keystroke shortcuts (for example, CTRL-R in Windows or ⌘-RETURN in OS X), which automatically send lines to the console. You can send the line upon which the cursor sits, a highlighted line, a highlighted part of a line, or a highlighted chunk of code. It's common to have multiple editor panes open at once when working with multiple R script files; keystroke code submissions simply operate with respect to the currently selected editor.

Aesthetics such as coloring and character spacing of both the console and editor can be tailored to a certain extent depending on operating system; you simply need to access the relevant GUI preferences. Figure 1-3 shows the R GUI preferences in Windows (Edit → GUI preferences...) and OS X (R → Preferences...). A nice feature of the OS X version of R in particular is the code-coloring and bracket-matching features of the editor, which can improve the authoring and readability of large sections of code.

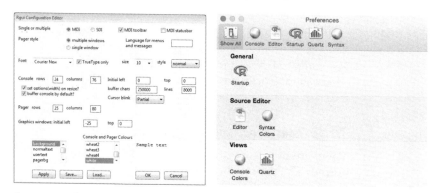

Figure 1-3: The R GUI preferences in Windows (left) and OS X (right)

## 1.2.2   Comments

In R, you can annotate your code with *comments*. Just preface the line with a hash mark (#), and anything that comes thereafter will be ignored by the interpreter. For example, executing the following in the console does nothing but return you to the prompt:

```
R> # This is a comment in R...
```

Comments can also appear after valid commands.

```
R> 1+1 # This works out the result of one plus one!
[1] 2
```

If you're writing large or complicated chunks of code in the editor, this kind of annotation can be helpful to others (and indeed yourself!) who want to understand what your code is doing.

## 1.2.3    Working Directory

An active R session always has a *working directory* associated with it. Unless you explicitly specify a file path when saving or importing data files, R will use this working directory by default. To check the location of the working directory, use the getwd function.

```
R> getwd()
[1] "/Users/tdavies"
```

File paths are always enclosed in double quotation marks, and R uses forward slashes, not backslashes, when specifying folder locations.

You can change the default working directory using the function setwd as follows:

```
R> setwd("/folder1/folder2/folder3/")
```

You may provide your file path relative to the current working directory or fully (in other words, from a system root drive). Either way, it's important to remember the case-sensitive nature of R; you must match the naming and punctuation of any folder names exactly or an error will be thrown.

That said, if you're happy specifying a full and correct file path each time you read or write a file (there are further details in Chapter 8), then the files of interest can reside anywhere on your computer.

## 1.2.4    Installing and Loading R Packages

The base installation of R comes ready with a plethora of built-in commands for numeric calculations, common statistical analyses, and plotting and visualization. These commands can be used right from the outset and needn't be loaded or imported in any way. I'll refer to these functions as *built-in* or *ready-to-use* in this text.

Slightly more specialized techniques and data sets are contained within *packages* (also referred to as *libraries*) of code. Using contributed packages is common, and you'll be doing so throughout this book, so it's important to get comfortable with installing and loading the required libraries.

Section A.2 covers the relevant details concerning package download and installation from CRAN, but I'll provide a brief overview here.

### Loading Packages

There are a small number of recommended packages that are included with the base distribution of R (listed in Section A.2.2). They don't need to be installed separately, but to use them, you do need to load them by calling library. One package you'll use in this book is named MASS (Venables and Ripley, 2002). To load it (or any other installed package) and gain access to its functions and data sets, simply execute library at the prompt as follows:

```
R> library("MASS")
```

Note that calling library provides access to a package's functionality only for the running R session. When you close R and reopen a fresh instance, you'll need to reload any packages you want to use.

### Installing Packages

There are thousands of contributed packages not included with the typical R installation; to make them loadable in R, you must first download and install them from a repository (usually CRAN). The easiest way to do this is by using the install.packages function directly at the R prompt (for this you need an Internet connection).

For example, one such package is ks (Duong, 2007), which you'll use in Chapter 26. Executing the following will attempt to connect to your local CRAN mirror and download and install ks, as well as several packages upon which it relies (called *dependencies*):

```
R> install.packages("ks")
```

The console will show running output as the procedure completes.

You need to install a package only once; thereafter it will be available for your R installation. You can then load your installed package (like ks) in any newly opened instance of R with a call to library, just as you did for MASS.

Section A.2.3 offers more detail on package installation.

### Updating Packages

The maintainers of contributed packages periodically provide version updates to fix bugs and add functionality. Every so often, you might want to check for updates to your collection of installed packages.

From the R prompt, a simple execution of the following will attempt to connect to your set package repository (defaulting to CRAN), looking for versions of all your installed packages that are later than those you currently have.

```
R> update.packages()
```

Section A.3 offers more details about updating packages and Section A.4 discusses alternate CRAN mirrors and repositories.

## 1.2.5   Help Files and Function Documentation

R comes with a suite of *help files* that you can use to search for particular functionality, to seek information on precisely how to use a given function and specify its *arguments* (in other words, the values or objects you supply to the function when you execute it), to clarify the role of arguments in the operations, to learn about the form of any returned objects, to provide possible examples of using the function, and to get details on how you may cite any software or data sets.

To access the help file for a given command or other object, use the `help` function at the console prompt or use the convenient shortcut ?. For example, consider the ready-to-use arithmetic mean function, `mean`.

```
R> ?mean
```

This brings up the file in the top image of Figure 1-4.

Figure 1-4: The R help file for the function mean (top) and the results of a help search for the string "mean" (bottom) in OS X

If you're unsure of the precise name of the desired function, you can search the documentation across all installed packages using a character string (a statement in double quotes) passed to `help.search`, or you can use `??` as a shortcut:

```
R> ??"mean"
```

This search brings up a list of functions, with their host packages and descriptions, whose help files contain the string of interest, as shown in the bottom image of Figure 1-4 (the highlighted entry is that of the arithmetic mean).

All help files follow the general format shown in the top image of Figure 1-4; the length and level of detail in the file typically reflect the complexity of the operations carried out by the function. Most help files include the first three items listed here; the others are common but optional:

- The *Description* section provides a short statement about the operations carried out.

- The *Usage* section specifies the form of the function in terms of how it should be passed to the R console, including the natural order of the arguments and any default values (these are the arguments that are shown being set using =).

- In the *Arguments* section, more detail is given about what each argument does as well as the possible values that they're allowed to take on.

- The nature of the object that's returned by the function (if anything) is specified under *Value*.

- The *References* section provides relevant citations for the command or the methodology behind the function.

- The help files for related functions are linked under *See Also*.

- *Examples* provides executable code that you can copy and paste into the console, demonstrating the function in action.

There are several more possible fields in a help file—functions with longer explanations often contain a *Details* section after the *Arguments* section. Common traps or mistakes made when calling the function are usually placed in a *Warnings* section, and additional information can be placed in *Notes*.

Although they might seem quite technical when you're first starting out, I encourage you to keep looking at help files—even if you already know how a function works, getting comfortable with the layout and interpretation of function documentation is an important part of becoming a skilled R user.

### 1.2.6    Third-Party Editors

The popularity of R has led to the development of several third-party code editors, or compatible plug-ins for existing code-editing software, which can enhance the experience of coding in R.

One noteworthy contribution is RStudio (RStudio Team, 2015). This is an integrated development environment (IDE) available free for Windows, OS X, and Linux/Unix platforms at *http://www.rstudio.com/*.

RStudio includes a direct-submission code editor; separate point-and-click panes for things such as file, object, and project management; and the creation of markup documents incorporating R code. Appendix B discusses RStudio and its capabilities in more detail.

Use of any third-party editor, including RStudio, is by and large a personal choice. In this book, I simply assume use of the typical base R GUI application.

## 1.3    Saving Work and Exiting R

So, you've spent a few hours coding in R, and it's time to go home? When saving work in R, you need to pay attention to two things: any R objects that have been created (and stored) in the active session and any R script files written in an editor.

### 1.3.1    Workspaces

You can use the GUI menu items (for example, under File in Windows and under Workspace in OS X) to save and load *workspace image* files. An R workspace image contains all the information held in the R session at the time of exit and is saved as a *.RData* file. This will include all objects you've created and stored (in other words, *assigned*) within the session (you'll see how to do this in Chapter 2), including those that may have been loaded from a previous workspace file.

Essentially, loading a stored *.RData* file allows you to "pick up from where you left off." At any point in an R session, you can execute ls() at the prompt, which lists all objects, variables, and user-defined functions currently present in the active workspace.

Alternatively, you can use the R commands save.image and load at the console for handling workspace *.RData* files—both of these functions contain a file argument to which you pass the folder location and name of the target *.RData* file (see the corresponding help files ?save.image and ?load for further information on the use of these).

Note that saving a workspace image in this way doesn't retain the functionality of any contributed packages that were loaded in the previously active R session. As mentioned in Section 1.2.4, you'll need to use library to load any packages required for your work for each new instance of R.

The quickest way to exit the software is to enter q() at the prompt:

```
R> q()
```

Simply exiting the console will bring up a dialog asking if you'd like to save the workspace image. In this case, choosing to save doesn't open a file browser to name your file but creates (or overwrites) a "no-name" file as one with a *.RData* extension in your working directory (refer to Section 1.2.3).

If an unnamed *.RData* file exists in the default working directory when a new instance of R is opened, the program will automatically load that default workspace—if that has happened, you'll be notified in the console's welcoming text.

**NOTE** *Alongside the .RData file, R will automatically save a file containing a line-by-line history of all the commands executed in the console for the associated workspace in the same directory. It's this history file that allows you to scroll through the previously executed commands using the keyboard directional arrows, as noted earlier.*

### 1.3.2   Scripts

For tasks requiring anything more than a handful of commands, you'll usually want to work in the built-in code editor. Saving your R scripts is therefore at least as important as saving a workspace, if not more so.

You save editor scripts as plain-text files with a *.R* extension (noted in Section 1.2.1); this allows your operating system to associate these files with the R software by default. To save a script from the built-in editor, ensure the editor is selected and navigate to File → Save (or press CTRL-S in Windows or ⌘-S in OS X). To open a previously saved script, select File → Open script... (CTRL-O) in Windows or File → Open Document... (⌘-O) in OS X.

Often, you won't really need to save a workspace *.RData* file if your script files are saved. Once any required commands in a saved script are reexecuted in a new R console, the objects created previously (in other words, those contained within a saved *.RData* file) are simply created once more. This can be useful if you're working on multiple problems at one time because it can be easy to mistakenly overwrite an object when relying solely on the stand-alone default workspace. Keeping your collection of R scripts separate is therefore a simple way to separate several projects without needing to worry about overwriting anything important that may have been stored previously.

R also provides a number of ways to write individual objects, such as data sets and image files of plots, to disk, which you'll look at in Chapter 8.

## 1.4   Conventions

There are a few conventions that I'll follow in the book in terms of the presentation of code and math.

## 1.4.1 Coding

As mentioned, when you code with R, you execute the code in the console, possibly after writing the script in the editor first. The following points are important to note:

- R code that's entered directly into the console for execution is shown preceded by the R> prompt and followed by any output displayed in the console. For example, this simple division of 14 by 6 from Section 2.1.1 looks like this:

```
R> 14/6
[1] 2.333333
```

  If you want to copy and paste console-executed code directly from the text of the book, you'll need to omit the R> prompt.

- For code that should be written in the editor before it's executed in the console, I'll indicate as such in the text, and the code will be presented *without* the prompt. The following example comes from Section 10.2.1:

```
for(myitem in 5:7){
    cat("--BRACED AREA BEGINS--\n")
    cat("the current item is",myitem,"\n")
    cat("--BRACED AREA ENDS--\n\n")
}
```

  My preferred coding style for actually arranging and indenting chunks like this will become clearer as you progress through Part II.

- There will occasionally be long lines of code (either executed directly in the console or written in the editor), which, for the sake of print, will be split and indented at an appropriate place to fit on the page. For example, take this line from Section 6.2.2:

```
R> ordfac.vec <- factor(x=c("Small","Large","Large","Regular","Small"),
                        levels=c("Small","Regular","Large"),
                        ordered=TRUE)
```

  Although this can be written out as a single line when using R, you can also break the line at a comma (in this case, the comma splits the arguments to the factor function). The broken line will be indented to the level of the opening parenthesis of the relevant command. Both forms—single line or split—will work as is when executed in R.

- Lastly, in a couple of places when the console output is lengthy and not essential to your understanding of the immediate content, it'll be suppressed for the sake of print. I'll say as much in the text, and you'll see the designation *--snip--* in the affected chunk of code.

### 1.4.2  Math and Equation References

Mathematics and equations that appear in this book (mainly in Parts III and IV) will be kept to a minimum, but in certain sections it's sometimes necessary to go into a little mathematical detail.

Important equations will be presented on their own lines as follows:

$$y = 4x \tag{1.1}$$

Equations will be numbered in parentheses, and references to equations in the text will use these parenthesized numbers and may or may not be preceded by *Equation*. For example, you'll see equations referred to in both of the following ways:

- As per Equation (1.1), $y = 8$ when $x = 2$.
- Inversion of (1.1) yields $x = y/4$.

When numeric results are rounded to a certain level, they'll be noted as such according to the number of *decimal places*, abbreviated to *d.p.* Here are some examples:

- The famous geometric value pi is given as $\pi = 3.1416$ (rounded to 4 d.p.).
- Setting $x = 1.467$ in (1.1) results in $y = 5.87$ (2 d.p.).

### 1.4.3  Exercises

Exercise questions in the chapters appear in a rounded box:

---

### Exercise 1.1

   a.   Say the word *cat* aloud.

   b.   Using nothing but your brain, find the solution to $1 + 1$.

---

These exercises are optional. If you choose to tackle them, they are intended to be completed as and when they appear in the text to help you practice and understand the specific content and code in the sections that immediately precede them.

All the data sets you'll use in this book for coding and plotting examples are available either as built-in R objects or as part of one of the contributed packages you'll install. These packages will be noted in the relevant text (for a short list of them, see Section A.2.3).

For your convenience, all code examples in this book, as well as complete suggested solutions to *all* practice exercises, are freely available as runnable *.R* script files on the book's web page at *https://www.nostarch.com/bookofr/*.

You should think of these solutions (and any accompanying commentary) as "suggested" because there are often multiple ways to perform a certain task in R, which may not necessarily be any better or worse than those supplied.

## Important Code in This Chapter

| Function/operator | Brief description | First occurrence |
|---|---|---|
| options | Set various R options | Section 1.2.1, p. 5 |
| # | A comment (ignored by interpreter) | Section 1.2.2, p. 6 |
| getwd | Print current working directory | Section 1.2.3, p. 7 |
| setwd | Set current working directory | Section 1.2.3, p. 7 |
| library | Load an installed package | Section 1.2.4, p. 7 |
| install.packages | Download and install package | Section 1.2.4, p. 8 |
| update.packages | Update installed packages | Section 1.2.4, p. 8 |
| help or ? | Function/object help file | Section 1.2.5, p. 9 |
| help.search or ?? | Search help files | Section 1.2.5, p. 10 |
| q | Quit R | Section 1.3.1, p. 12 |

# 2

## NUMERICS, ARITHMETIC, ASSIGNMENT, AND VECTORS

In its simplest role, R can function as a mere desktop calculator. In this chapter, I'll discuss how to use the software for arithmetic. I'll also show how to store results so you can use them later in other calculations. Then, you'll learn about vectors, which let you handle multiple values at once. Vectors are an essential tool in R, and much of R's functionality was designed with vector operations in mind. You'll examine some common and useful ways to manipulate vectors and take advantage of vector-oriented behavior.

## 2.1   R for Basic Math

All common arithmetic operations and mathematical functionality are ready to use at the console prompt. You can perform addition, subtraction, multiplication, and division with the symbols +, -, *, and /, respectively. You can create exponents (also referred to as *powers* or *indices*) using ^, and you control the order of the calculations in a single command using parentheses, ().

## 2.1.1 Arithmetic

In R, standard mathematical rules apply throughout and follow the usual left-to-right order of operations: parentheses, exponents, multiplication, division, addition, subtraction (PEMDAS). Here's an example in the console:

```
R> 2+3
[1] 5
R> 14/6
[1] 2.333333
R> 14/6+5
[1] 7.333333
R> 14/(6+5)
[1] 1.272727
R> 3^2
[1] 9
R> 2^3
[1] 8
```

You can find the square root of any non-negative number with the sqrt function. You simply provide the desired number to x as shown here:

```
R> sqrt(x=9)
[1] 3
R> sqrt(x=5.311)
[1] 2.304561
```

When using R, you'll often find that you need to translate a complicated arithmetic formula into code for evaluation (for example, when replicating a calculation from a textbook or research paper). The next examples provide a mathematically expressed calculation, followed by its execution in R:

$$10^2 + \frac{3 \times 60}{8} - 3$$

```
R> 10^2+3*60/8-3
[1] 119.5
```

$$\frac{5^3 \times (6-2)}{61-3+4}$$

```
R> 5^3*(6-2)/(61-3+4)
[1] 8.064516
```

$$2^{2+1} - 4 + 64^{-2^{2.25-\frac{1}{4}}}$$

```
R> 2^(2+1)-4+64^((-2)^(2.25-1/4))
[1] 16777220
```

$$\left(\frac{0.44 \times (1-0.44)}{34}\right)^{\frac{1}{2}}$$

```
R> (0.44*(1-0.44)/34)^(1/2)
[1] 0.08512966
```

Note that some R expressions require extra parentheses that aren't present in the mathematical expressions. Missing or misplaced parentheses are common causes of arithmetic errors in R, especially when dealing with exponents. If the exponent is itself an arithmetic calculation, it must always appear in parentheses. For example, in the third expression, you need parentheses around 2.25-1/4. You also need to use parentheses if the number being raised to some power is a calculation, such as the expression $2^{2+1}$ in the third example. Note that R considers a negative number a calculation because it interprets, for example, -2 as -1*2. This is why you also need the parentheses around -2 in that same expression. It's important to highlight these issues early because they can easily be overlooked in large chunks of code.

### 2.1.2 Logarithms and Exponentials

You'll often see or read about researchers performing a *log transformation* on certain data. This refers to rescaling numbers according to the *logarithm*. When supplied a given number *x* and a value referred to as a *base*, the logarithm calculates the power to which you must raise the base to get to *x*. For example, the logarithm of $x = 243$ to base 3 (written mathematically as $\log_3 243$) is 5, because $3^5 = 243$. In R, the log transformation is achieved with the log function. You supply log with the number to transform, assigned to the value x, and the base, assigned to base, as follows:

```
R> log(x=243,base=3)
[1] 5
```

Here are some things to consider:

- Both *x* and the base must be positive.
- The log of any number *x* when the base is equal to *x* is 1.
- The log of $x = 1$ is always 0, regardless of the base.

There's a particular kind of log transformation often used in mathematics called the *natural log*, which fixes the base at a special mathematical number—*Euler's number*. This is conventionally written as *e* and is approximately equal to 2.718.

Euler's number gives rise to the *exponential function*, defined as *e* raised to the power of *x*, where *x* can be any number (negative, zero, or positive). The exponential function, $f(x) = e^x$, is often written as exp(*x*) and represents the *inverse* of the natural log such that $\exp(\log_e x) = \log_e \exp(x) = x$. The R command for the exponential function is exp:

```
R> exp(x=3)
[1] 20.08554
```

The default behavior of log is to assume the natural log:

```
R> log(x=20.08554)
[1] 3
```

You must provide the value of base yourself if you want to use a value other than *e*. The logarithm and exponential functions are mentioned here because they become important later on in the book—many statistical methods use them because of their various helpful mathematical properties.

### 2.1.3   E-Notation

When R prints large or small numbers beyond a certain threshold of significant figures, set at 7 by default, the numbers are displayed using the classic scientific e-notation. The e-notation is typical to most programming languages—and even many desktop calculators—to allow easier interpretation of extreme values. In e-notation, any number $x$ can be expressed as $x$e$y$, which represents exactly $x \times 10^y$. Consider the number $2,342,151,012,900$. It could, for example, be represented as follows:

- $2.3421510129$e$12$, which is equivalent to writing $2.3421510129 \times 10^{12}$

- $234.21510129$e$10$, which is equivalent to writing $234.21510129 \times 10^{10}$

You could use any value for the power of $y$, but standard e-notation uses the power that places a decimal just after the first significant digit. Put simply, for a *positive* power $+y$, the e-notation can be interpreted as "move the decimal point $y$ positions to the *right*." For a *negative* power $-y$, the interpretation is "move the decimal point $y$ positions to the *left*." This is exactly how R presents e-notation:

```
R> 2342151012900
[1] 2.342151e+12
R> 0.0000002533
[1] 2.533e-07
```

In the first example, R shows only the first seven significant digits and hides the rest. Note that no information is lost in any calculations even if R hides digits; the e-notation is purely for ease of readability by the user, and the extra digits are still stored by R, even though they aren't shown.

Finally, note that R must impose constraints on how extreme a number can be before it is treated as either infinity (for large numbers) or zero (for small numbers). These constraints depend on your individual system, and I'll discuss the technical details a bit more in Section 6.1.1. However, any modern desktop system can be trusted to be precise enough by default for most computational and statistical endeavors in R.

## Exercise 2.1

a. Using R, verify that

$$\frac{6a + 42}{3^{4.2 - 3.62}} = 29.50556$$

when $a = 2.3$.

b. Which of the following squares negative 4 and adds 2 to the result?
   i.   (-4)^2+2
   ii.  -4^2+2
   iii. (-4)^(2+2)
   iv.  -4^(2+2)

c. Using R, how would you calculate the square root of half of the average of the numbers 25.2, 15, 16.44, 15.3, and 18.6?

d. Find $\log_e 0.3$.

e. Compute the exponential transform of your answer to (d).

f. Identify R's representation of −0.00000000423546322 when printing this number to the console.

## 2.2   Assigning Objects

So far, R has simply displayed the results of the example calculations by printing them to the console. If you want to save the results and perform further operations, you need to be able to *assign* the results of a given computation to an *object* in the current workspace. Put simply, this amounts to storing some item or result under a given name so it can be accessed later, without having to write out that calculation again. In this book, I will use the terms *assign* and *store* interchangeably. Note that some programming books refer to a stored object as a *variable* because of the ability to easily overwrite that object and change it to something different, meaning that what it represents can vary throughout a session. However, I'll use the term *object* throughout this book because we'll discuss variables in Part III as a distinctly different statistical concept.

You can specify an assignment in R in two ways: using arrow notation (<-) and using a single equal sign (=). Both methods are shown here:

```
R> x <- -5
R> x
[1] -5
```

```
R> x = x + 1  # this overwrites the previous value of x
R> x
[1] -4

R> mynumber = 45.2

R> y <- mynumber*x
R> y
[1] -180.8

R> ls()
[1] "mynumber" "x"          "y"
```

As you can see from these examples, R will display the value assigned to an object when you enter the name of the object into the console. When you use the object in subsequent operations, R will substitute the value you assigned to it. Finally, if you use the ls command (which you saw in Section 1.3.1) to examine the contents of the current workspace, it will reveal the names of the objects in alphabetical order (along with any other previously created items).

Although = and <- do the same thing, it is wise (for the neatness of code if nothing else) to be consistent. Many users choose to stick with the <-, however, because of the potential for confusion in using the = (for example, I clearly didn't mean that $x$ is *mathematically* equal to $x + 1$ earlier). In this book, I'll do the same and reserve = for setting function arguments, which begins in Section 2.3.2. So far you've used only numeric values, but note that the procedure for assignment is universal for all types and classes of objects, which you'll examine in the coming chapters.

Objects can be named almost anything as long as the name begins with a letter (in other words, not a number), avoids symbols (though underscores and periods are fine), and avoids the handful of "reserved" words such as those used for defining special values (see Section 6.1) or for controlling code flow (see Chapter 10). You can find a useful summary of these naming rules in Section 9.1.2.

## Exercise 2.2

a.  Create an object that stores the value $3^2 \times 4^{1/8}$.

b.  Overwrite your object in (a) by itself divided by 2.33. Print the result to the console.

c.  Create a new object with the value $-8.2 \times 10^{-13}$.

d.  Print directly to the console the result of multiplying (b) by (c).

## 2.3 Vectors

Often you'll want to perform the same calculations or comparisons upon multiple entities, for example if you're rescaling measurements in a data set. You could do this type of operation one entry at a time, though this is clearly not ideal, especially if you have a large number of items. R provides a far more efficient solution to this problem with *vectors*.

For the moment, to keep things simple, you'll continue to work with numeric entries only, though many of the utility functions discussed here may also be applied to structures containing non-numeric values. You'll start looking at these other kinds of data in Chapter 4.

### 2.3.1  Creating a Vector

The vector is the essential building block for handling multiple items in R. In a numeric sense, you can think of a vector as a collection of observations or measurements concerning a single variable, for example, the heights of 50 people or the number of coffees you drink daily. More complicated data structures may consist of several vectors. The function for creating a vector is the single letter c, with the desired entries in parentheses separated by commas.

```
R> myvec <- c(1,3,1,42)
R> myvec
[1]  1  3  1 42
```

Vector entries can be calculations or previously stored items (including vectors themselves).

```
R> foo <- 32.1
R> myvec2 <- c(3,-3,2,3.45,1e+03,64^0.5,2+(3-1.1)/9.44,foo)
R> myvec2
[1]    3.000000   -3.000000    2.000000    3.450000 1000.000000    8.000000
[7]    2.201271   32.100000
```

This code created a new vector assigned to the object myvec2. Some of the entries are defined as arithmetic expressions, and it's the result of the expression that's stored in the vector. The last element, foo, is an existing numeric object defined as 32.1.

Let's look at another example.

```
R> myvec3 <- c(myvec,myvec2)
R> myvec3
 [1]    1.000000    3.000000    1.000000   42.000000    3.000000   -3.000000
 [7]    2.000000    3.450000 1000.000000    8.000000    2.201271   32.100000
```

This code creates and stores yet another vector, myvec3, which contains the entries of myvec and myvec2 appended together in that order.

## 2.3.2 Sequences, Repetition, Sorting, and Lengths

Here I'll discuss some common and useful functions associated with R vectors: seq, rep, sort, and length.

Let's create an equally spaced sequence of increasing or decreasing numeric values. This is something you'll need often, for example when programming loops (see Chapter 10) or when plotting data points (see Chapter 7). The easiest way to create such a sequence, with numeric values separated by intervals of 1, is to use the colon operator.

```
R> 3:27
 [1]  3  4  5  6  7  8  9 10 11 12 13 14 15 16 17 18 19 20 21 22 23 24 25 26 27
```

The example 3:27 should be read as "from 3 to 27 (by 1)." The result is a numeric vector just as if you had listed each number manually in parentheses with c. As always, you can also provide either a previously stored value or a (strictly parenthesized) calculation when using the colon operator:

```
R> foo <- 5.3
R> bar <- foo:(-47+1.5)
R> bar
 [1]   5.3   4.3   3.3   2.3   1.3   0.3  -0.7  -1.7  -2.7  -3.7  -4.7
[12]  -5.7  -6.7  -7.7  -8.7  -9.7 -10.7 -11.7 -12.7 -13.7 -14.7 -15.7
[23] -16.7 -17.7 -18.7 -19.7 -20.7 -21.7 -22.7 -23.7 -24.7 -25.7 -26.7
[34] -27.7 -28.7 -29.7 -30.7 -31.7 -32.7 -33.7 -34.7 -35.7 -36.7 -37.7
[45] -38.7 -39.7 -40.7 -41.7 -42.7 -43.7 -44.7
```

### Sequences with seq

You can also use the seq command, which allows for more flexible creations of sequences. This ready-to-use function takes in a from value, a to value, and a by value, and it returns the corresponding sequence as a numeric vector.

```
R> seq(from=3,to=27,by=3)
[1]  3  6  9 12 15 18 21 24 27
```

This gives you a sequence with intervals of 3 rather than 1. Note that these kinds of sequences will always start at the from number but will not always include the to number, depending on what you are asking R to increase (or decrease) them by. For example, if you are increasing (or decreasing) by even numbers and your sequence ends in an odd number, the final number won't be included. Instead of providing a by value, however, you can specify a length.out value to produce a vector with that many numbers, evenly spaced between the from and to values.

```
R> seq(from=3,to=27,length.out=40)
 [1]  3.000000  3.615385  4.230769  4.846154  5.461538  6.076923  6.692308
 [8]  7.307692  7.923077  8.538462  9.153846  9.769231 10.384615 11.000000
[15] 11.615385 12.230769 12.846154 13.461538 14.076923 14.692308 15.307692
```

```
[22] 15.923077 16.538462 17.153846 17.769231 18.384615 19.000000 19.615385
[29] 20.230769 20.846154 21.461538 22.076923 22.692308 23.307692 23.923077
[36] 24.538462 25.153846 25.769231 26.384615 27.000000
```

By setting length.out to 40, you make the program print exactly 40 evenly spaced numbers from 3 to 27.

For decreasing sequences, the use of by must be negative. Here's an example:

```
R> foo <- 5.3
R> myseq <- seq(from=foo,to=(-47+1.5),by=-2.4)
R> myseq
 [1]   5.3   2.9   0.5  -1.9  -4.3  -6.7  -9.1 -11.5 -13.9 -16.3 -18.7 -21.1
[13] -23.5 -25.9 -28.3 -30.7 -33.1 -35.5 -37.9 -40.3 -42.7 -45.1
```

This code uses the previously stored object foo as the value for from and uses the parenthesized calculation (-47+1.5) as the to value. Given those values (that is, with foo being greater than (-47+1.5)), the sequence can progress only in negative steps; directly above, we set by to be -2.4. The use of length.out to create decreasing sequences, however, remains the same (it would make no sense to specify a "negative length"). For the same from and to values, you can create a decreasing sequence of length 5 easily, as shown here:

```
R> myseq2 <- seq(from=foo,to=(-47+1.5),length.out=5)
R> myseq2
[1]   5.3  -7.4 -20.1 -32.8 -45.5
```

There are shorthand ways of calling these functions, which you'll learn about in Chapter 9, but in these early stages I'll stick with the explicit usage.

## Repetition with rep

Sequences are extremely useful, but sometimes you may want simply to repeat a certain value. You do this using rep.

```
R> rep(x=1,times=4)
[1] 1 1 1 1
R> rep(x=c(3,62,8.3),times=3)
[1]  3.0 62.0  8.3  3.0 62.0  8.3  3.0 62.0  8.3
R> rep(x=c(3,62,8.3),each=2)
[1]  3.0  3.0 62.0 62.0  8.3  8.3
R> rep(x=c(3,62,8.3),times=3,each=2)
 [1]  3.0  3.0 62.0 62.0  8.3  8.3  3.0  3.0 62.0 62.0  8.3  8.3  3.0  3.0 62.0
[16] 62.0  8.3  8.3
```

The rep function is given a single value or a vector of values as its argument x, as well as a value for the arguments times and each. The value for times provides the number of times to repeat x, and each provides the number of times to repeat each element of x. In the first line directly above, you simply repeat a single value four times. The other examples first use rep and times on a vector to repeat the entire vector, then use each to repeat each member of the vector, and finally use both times and each to do both at once.

If neither times nor each is specified, R's default is to treat the values of times and each as 1 so that a call of rep(x=c(3,62,8.3)) will just return the originally supplied x with no changes.

As with seq, you can include the result of rep in a vector of the same data type, as shown in the following example:

```
R> foo <- 4
R> c(3,8.3,rep(x=32,times=foo),seq(from=-2,to=1,length.out=foo+1))
[1]   3.00   8.30 32.00 32.00 32.00 32.00 -2.00 -1.25 -0.50  0.25  1.00
```

Here, I've constructed a vector where the third to sixth entries (inclusive) are governed by the evaluation of a rep command—the single value 32 repeated foo times (where foo is stored as 4). The last five entries are the result of an evaluation of seq, namely a sequence from −2 to 1 of length foo+1 (5).

## Sorting with sort

Sorting a vector in increasing or decreasing order of its elements is another simple operation that crops up in everyday tasks. The conveniently named sort function does just that.

```
R> sort(x=c(2.5,-1,-10,3.44),decreasing=FALSE)
[1] -10.00  -1.00   2.50   3.44

R> sort(x=c(2.5,-1,-10,3.44),decreasing=TRUE)
[1]   3.44   2.50  -1.00 -10.00

R> foo <- seq(from=4.3,to=5.5,length.out=8)
R> foo
[1] 4.300000 4.471429 4.642857 4.814286 4.985714 5.157143 5.328571 5.500000
R> bar <- sort(x=foo,decreasing=TRUE)
R> bar
[1] 5.500000 5.328571 5.157143 4.985714 4.814286 4.642857 4.471429 4.300000

R> sort(x=c(foo,bar),decreasing=FALSE)
 [1] 4.300000 4.300000 4.471429 4.471429 4.642857 4.642857 4.814286 4.814286
 [9] 4.985714 4.985714 5.157143 5.157143 5.328571 5.328571 5.500000 5.500000
```

The sort function is pretty straightforward. You supply a vector to the function as the argument x, and a second argument, decreasing, indicates the order in which you want to sort. This argument takes a type of value you have not yet met: one of the all-important *logical* values. A logical value can be only one of two specific, case-sensitive values: TRUE or FALSE. Generally speaking, logicals are used to indicate the satisfaction or failure of a certain *condition*, and they form an integral part of all programming languages. You'll investigate logical values in R in greater detail in Section 4.1. For now, in regards to sort, you set decreasing=FALSE to sort from smallest to largest, and decreasing=TRUE sorts from largest to smallest.

### Finding a Vector Length with length

I'll round off this section with the length function, which determines how many entries exist in a vector given as the argument x.

```
R> length(x=c(3,2,8,1))
[1] 4

R> length(x=5:13)
[1] 9

R> foo <- 4
R> bar <- c(3,8.3,rep(x=32,times=foo),seq(from=-2,to=1,length.out=foo+1))
R> length(x=bar)
[1] 11
```

Note that if you include entries that depend on the evaluation of other functions (in this case, calls to rep and seq), length tells you the number of entries *after* those inner functions have been executed.

---

## Exercise 2.3

a. Create and store a sequence of values from 5 to −11 that progresses in steps of 0.3.

b. Overwrite the object from (a) using the same sequence with the order reversed.

c. Repeat the vector c(-1,3,-5,7,-9) twice, with each element repeated 10 times, and store the result. Display the result sorted from largest to smallest.

d. Create and store a vector that contains, in any configuration, the following:
   i. A sequence of integers from 6 to 12 (inclusive)
   ii. A threefold repetition of the value 5.3
   iii. The number −3
   iv. A sequence of nine values starting at 102 and ending at the number that is the total length of the vector created in (c)

e. Confirm that the length of the vector created in (d) is 20.

---

### 2.3.3   Subsetting and Element Extraction

In all the results you have seen printed to the console screen so far, you may have noticed a curious feature. Immediately to the left of the output there is a square-bracketed [1]. When the output is a long vector that spans the width of the console and wraps onto the following line, another square-bracketed number appears to the left of the new line. These numbers represent the *index* of the entry directly to the right. Quite simply, the index corresponds to the *position* of a value within a vector, and that's precisely why the first value always has a [1] next to it (even if it's the only value and not part of a larger vector).

These indexes allow you to retrieve specific elements from a vector, which is known as *subsetting*. Suppose you have a vector called myvec in your workspace. Then there will be exactly length(x=myvec) entries in myvec, with each entry having a specific position: 1 or 2 or 3, all the way up to length(x=myvec). You can access individual elements by asking R to return the values of myvec at specific locations, done by entering the name of the vector followed by the position in square brackets.

```
R> myvec <- c(5,-2.3,4,4,4,6,8,10,40221,-8)
R> length(x=myvec)
[1] 10
R> myvec[1]
[1] 5

R> foo <- myvec[2]
R> foo
[1] -2.3

R> myvec[length(x=myvec)]
[1] -8
```

Because length(x=myvec) results in the final index of the vector (in this case, 10), entering this phrase in the square brackets extracts the final element, -8. Similarly, you could extract the second-to-last element by subtracting 1 from the length; let's try that, and also assign the result to a new object:

```
R> myvec.len <- length(x=myvec)
R> bar <- myvec[myvec.len-1]
R> bar
[1] 40221
```

As these examples show, the index may be an arithmetic function of other numbers or previously stored values. You can assign the result to a new object in your workspace in the usual way with the <- notation. Using your knowledge of sequences, you can use the colon notation with the length of the specific vector to obtain all possible indexes for extracting a particular element in the vector:

```
R> 1:myvec.len
[1]  1  2  3  4  5  6  7  8  9 10
```

You can also delete individual elements by using *negative* versions of the indexes supplied in the square brackets. Continuing with the objects myvec, foo, bar, and myvec.len as defined earlier, consider the following operations:

```
R> myvec[-1]
[1]    -2.3     4.0     4.0     4.0     6.0     8.0    10.0 40221.0    -8.0
```

This line produces the contents of myvec without the first element. Similarly, the following code assigns to the object baz the contents of myvec without its second element:

```
R> baz <- myvec[-2]
R> baz
[1]     5     4     4     4     6     8    10 40221    -8
```

Again, the index in the square brackets can be the result of an appropriate calculation, like so:

```
R> qux <- myvec[-(myvec.len-1)]
R> qux
[1]  5.0 -2.3  4.0  4.0  4.0  6.0  8.0 10.0 -8.0
```

Using the square-bracket operator to extract or delete values from a vector does not change the original vector you are subsetting *unless* you explicitly overwrite the vector with the subsetted version. For instance, in this example, qux is a new vector defined as myvec without its second-to-last entry, but in your workspace, myvec itself *remains unchanged*. In other words, subsetting vectors in this way simply returns the requested elements, which can be assigned to a new object if you want, but doesn't alter the original object in the workspace.

Now, suppose you want to piece myvec back together from qux and bar. You can call something like this:

```
R> c(qux[-length(x=qux)],bar,qux[length(x=qux)])
 [1]     5.0    -2.3     4.0     4.0     4.0     6.0     8.0    10.0 40221.0
[10]    -8.0
```

As you can see, this line uses c to reconstruct the vector in three parts: qux[-length(x=qux)], the object bar defined earlier, and qux[length(x=qux)]. For clarity, let's examine each part in turn.

- qux[-length(x=qux)]

This piece of code returns the values of qux except for its last element.

```
R> length(x=qux)
[1] 9
R> qux[-length(x=qux)]
[1]  5.0 -2.3  4.0  4.0  4.0  6.0  8.0 10.0
```

Now you have a vector that's the same as the first eight entries of myvec.

- bar

Earlier, you had stored bar as the following:

```
R> bar <- myvec[myvec.len-1]
R> bar
[1] 40221
```

This is precisely the second-to-last element of myvec that qux is missing. So, you'll slot this value in after qux[-length(x=qux)].

- qux[length(x=qux)]

Finally, you just need the last element of qux that matches the last element of myvec. This is extracted from qux (not deleted as earlier) using length.

```
R> qux[length(x=qux)]
[1] -8
```

Now it should be clear how calling these three parts of code together, in this order, is one way to reconstruct myvec.

As with most operations in R, you are not restricted to doing things one by one. You can also subset objects using *vectors of indexes*, rather than individual indexes. Using myvec again from earlier, you get the following:

```
R> myvec[c(1,3,5)]
[1] 5 4 4
```

This returns the first, third, and fifth elements of myvec in one go. Another common and convenient subsetting tool is the colon operator (discussed in Section 2.3.2), which creates a sequence of indexes. Here's an example:

```
R> 1:4
[1] 1 2 3 4
R> foo <- myvec[1:4]
R> foo
[1]  5.0 -2.3  4.0  4.0
```

This provides the first four elements of myvec (recall that the colon operator returns a numeric vector, so there is no need to explicitly wrap this using c).

The order of the returned elements depends entirely upon the index vector supplied in the square brackets. For example, using foo again, consider the order of the indexes and the resulting extractions, shown here:

```
R> length(x=foo):2
[1] 4 3 2
R> foo[length(foo):2]
[1]  4.0  4.0 -2.3
```

Here you extracted elements starting at the end of the vector, working backward. You can also use rep to repeat an index, as shown here:

```
R> indexes <- c(4,rep(x=2,times=3),1,1,2,3:1)
R> indexes
 [1] 4 2 2 2 1 1 2 3 2 1
R> foo[indexes]
 [1]  4.0 -2.3 -2.3 -2.3  5.0  5.0 -2.3  4.0 -2.3  5.0
```

This is now something a little more general than strictly "subsetting"—by using an index vector, you can create an entirely new vector of any length consisting of some or all of the elements in the original vector. As shown earlier, this index vector can contain the desired element positions in any order and can repeat indexes.

You can also return the elements of a vector after deleting more than one element. For example, to create a vector after removing the first and third elements of foo, you can execute the following:

```
R> foo[-c(1,3)]
[1] -2.3  4.0
```

Note that it is not possible to mix positive and negative indexes in a single index vector.

Sometimes you'll need to overwrite certain elements in an existing vector with new values. In this situation, you first specify the elements you want to overwrite using square brackets and then use the assignment operator to assign the new values. Here's an example:

```
R> bar <- c(3,2,4,4,1,2,4,1,0,0,5)
R> bar
 [1] 3 2 4 4 1 2 4 1 0 0 5
R> bar[1] <- 6
R> bar
 [1] 6 2 4 4 1 2 4 1 0 0 5
```

This overwrites the first element of bar, which was originally 3, with a new value, 6. When selecting multiple elements, you can specify a single value to replace them all or enter a vector of values that's equal in length to the number of elements selected to replace them one for one. Let's try this with the same bar vector from earlier.

```
R> bar[c(2,4,6)] <- c(-2,-0.5,-1)
R> bar
 [1]  6.0 -2.0  4.0 -0.5  1.0 -1.0  4.0  1.0  0.0  0.0  5.0
```

Here you overwrite the second, fourth, and sixth elements with -2, -0.5, and -1, respectively; all else remains the same. By contrast, the following code overwrites elements 7 to 10 (inclusive), replacing them all with 100:

```
R> bar[7:10] <- 100
R> bar
 [1]   6.0  -2.0   4.0  -0.5   1.0  -1.0 100.0 100.0 100.0 100.0   5.0
```

Finally, it's important to mention that this section has focused on just one of the two main methods, or "flavors," of vector element extraction in R. You'll look at the alternative method, using logical flags, in Section 4.1.5.

## Exercise 2.4

a. Create and store a vector that contains the following, in this order:
   - A sequence of length 5 from 3 to 6 (inclusive)
   - A twofold repetition of the vector c(2,-5.1,-33)
   - The value $\frac{7}{42} + 2$

b. Extract the first and last elements of your vector from (a), storing them as a new object.

c. Store as a third object the values returned by omitting the first and last values of your vector from (a).

d. Use only (b) and (c) to reconstruct (a).

e. Overwrite (a) with the same values sorted from smallest to largest.

f. Use the colon operator as an index vector to reverse the order of (e), and confirm this is identical to using sort on (e) with decreasing=TRUE.

g. Create a vector from (c) that repeats the third element of (c) three times, the sixth element four times, and the last element once.

h. Create a new vector as a copy of (e) by assigning (e) as is to a newly named object. Using this new copy of (e), overwrite the first, the fifth to the seventh (inclusive), and the last element with the values 99 to 95 (inclusive), respectively.

### 2.3.4 Vector-Oriented Behavior

Vectors are so useful because they allow R to carry out operations on multiple elements simultaneously with speed and efficiency. This *vector-oriented*, *vectorized*, or *element-wise* behavior is a key feature of the language, one that you will briefly examine here through some examples of rescaling measurements.

Let's start with this simple example:

```
R> foo <- 5.5:0.5
R> foo
[1] 5.5 4.5 3.5 2.5 1.5 0.5
R> foo-c(2,4,6,8,10,12)
[1]   3.5   0.5  -2.5  -5.5  -8.5 -11.5
```

This code creates a sequence of six values between 5.5 and 0.5, in increments of 1. From this vector, you subtract another vector containing 2, 4, 6, 8, 10, and 12. What does this do? Well, quite simply, R matches up the elements according to their respective positions and performs the operation on each corresponding pair of elements. The resulting vector is obtained by subtracting the first element of c(2,4,6,8,10,12) from the first element of foo (5.5 − 2 = 3.5), then by subtracting the second element of c(2,4,6,8,10,12) from the second element of foo (4.5 − 4 = 0.5), and so on. Thus, rather than inelegantly cycling through each element in turn (as you could do by hand or by explicitly using a loop), R permits a fast and efficient alternative using vector-oriented behavior. Figure 2-1 illustrates how you can understand this type of calculation and highlights the fact that the positions of the elements are crucial in terms of the final result; elements in differing positions have no effect on one another.

The situation is made more complicated when using vectors of different lengths, which can happen in two distinct ways. The first is when the length of the longer vector can be evenly divided by the length of the shorter vector. The second is when the length of the longer vector *cannot* be divided by the length of the shorter vector—this is usually unintentional on the user's part. In both of these situations, R essentially attempts to replicate, or *recycle*, the shorter vector by as many times as needed to match the length of the longer vector, before completing the specified operation. As an example, suppose you wanted to alternate the entries of foo shown earlier as negative

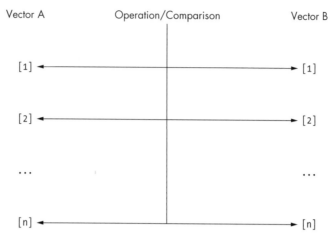

| Vector A | Operation/Comparison | Vector B |

Figure 2-1: A conceptual diagram of the element-wise behavior of a comparison or operation carried out on two vectors of equal length in R. Note that the operation is performed by matching up the element positions.

and positive. You could explicitly multiply foo by c(1,-1,1,-1,1,-1), but you don't need to write out the full latter vector. Instead, you can write the following:

```
R> bar <- c(1,-1)
R> foo*bar
[1]  5.5 -4.5  3.5 -2.5  1.5 -0.5
```

Here bar has been applied repeatedly throughout the length of foo until completion. The left plot of Figure 2-2 illustrates this particular example. Now let's see what happens when the vector lengths are not evenly divisible.

```
R> baz <- c(1,-1,0.5,-0.5)
R> foo*baz
[1]  5.50 -4.50  1.75 -1.25  1.50 -0.50
Warning message:
In foo * baz :
  longer object length is not a multiple of shorter object length
```

Here you see that R has matched the first four elements of foo with the entirety of baz, but it's not able to fully repeat the vector again. The repetition has been attempted, with the first two elements of baz being matched with the last two of the longer foo, though not without a protest from R, which notifies the user of the unevenly divisible lengths (you'll look at warnings in more detail in Section 12.1). The plot on the right in Figure 2-2 illustrates this example.

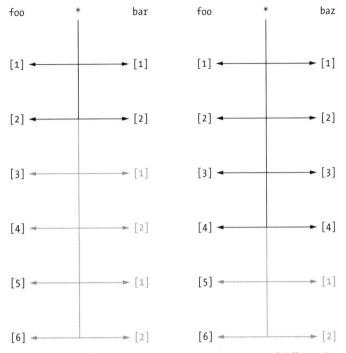

*Figure 2-2: An element-wise operation on two vectors of differing lengths. Left:* foo *multiplied by* bar; *lengths are evenly divisible. Right:* foo *multiplied by* baz; *lengths are not evenly divisible, and a warning is issued.*

As I noted in Section 2.3.3, you can consider single values to be vectors of length 1, so you can use a single value to repeat an operation on all the values of a vector of any length. Here's an example, using the same vector foo:

```
R> qux <- 3
R> foo+qux
[1] 8.5 7.5 6.5 5.5 4.5 3.5
```

This is far easier than executing foo+c(3,3,3,3,3,3) or the more general foo+rep(x=3,times=length(x=foo)). Operating on vectors using a single value in this fashion is quite common, such as if you want to rescale or translate a set of measurements by some constant amount.

Another benefit of vector-oriented behavior is that you can use vectorized functions to complete potentially laborious tasks. For example, if you want to sum or multiply all the entries in a numeric vector, you can just use a built-in function.

Recall foo, shown earlier:

```
R> foo
[1] 5.5 4.5 3.5 2.5 1.5 0.5
```

You can find the sum of these six elements with

```
R> sum(foo)
[1] 18
```

and their product with

```
R> prod(foo)
[1] 162.4219
```

Far from being just convenient, vectorized functions are faster and more efficient than an explicitly coded iterative approach like a loop. The main takeaway from these examples is that much of R's functionality is designed specifically for certain data structures, ensuring neatness of code as well as optimization of performance.

Lastly, as mentioned earlier, this vector-oriented behavior applies in the same way to overwriting multiple elements. Again using foo, examine the following:

```
R> foo
[1] 5.5 4.5 3.5 2.5 1.5 0.5
R> foo[c(1,3,5,6)] <- c(-99,99)
R> foo
[1] -99.0   4.5  99.0   2.5 -99.0  99.0
```

You see four specific elements being overwritten by a vector of length 2, which is recycled in the same fashion you're familiar with. Again, the length of the vector of replacements must evenly divide the number of elements being overwritten, or else a warning similar to the one shown earlier will be issued when R cannot complete a full-length recycle.

## Exercise 2.5

a. Convert the vector c(2,0.5,1,2,0.5,1,2,0.5,1) to a vector of only 1s, using a vector of length 3.

b. The conversion from a temperature measurement in degrees Fahrenheit $F$ to Celsius $C$ is performed using the following equation:

$$C = \frac{5}{9}(F - 32)$$

Use vector-oriented behavior in R to convert the temperatures 45, 77, 20, 19, 101, 120, and 212 in degrees Fahrenheit to degrees Celsius.

c.  Use the vector c(2,4,6) and the vector c(1,2) in conjunction with rep and * to produce the vector c(2,4,6,4,8,12).

d.  Overwrite the middle four elements of the resulting vector from (c) with the two recycled values -0.1 and -100, in that order.

## Important Code in This Chapter

| Function/operator | Brief description | First occurrence |
|---|---|---|
| +, *, -, /, ^ | Arithmetic | Section 2.1, p. 17 |
| sqrt | Square root | Section 2.1.1, p. 18 |
| log | Logarithm | Section 2.1.2, p. 19 |
| exp | Exponential | Section 2.1.2, p. 19 |
| <-, = | Object assignment | Section 2.2, p. 21 |
| c | Vector creation | Section 2.3.1, p. 23 |
| :, seq | Sequence creation | Section 2.3.2, p. 24 |
| rep | Value/vector repetition | Section 2.3.2, p. 25 |
| sort | Vector sorting | Section 2.3.2, p. 26 |
| length | Determine vector length | Section 2.3.2, p. 27 |
| [ ] | Vector subsetting/extraction | Section 2.3.3, p. 28 |
| sum | Sum all vector elements | Section 2.3.4, p. 36 |
| prod | Multiply all vector elements | Section 2.3.4, p. 36 |

# 3

## MATRICES AND ARRAYS

By now, you have a solid handle on using vectors in R. A *matrix* is simply several vectors stored together. Whereas the size of a vector is described by its length, the size of a matrix is specified by a number of rows and a number of columns. You can also create higher-dimensional structures that are referred to as *arrays*. In this chapter, we'll begin by looking at how to work with matrices before increasing the dimension to form arrays.

## 3.1 Defining a Matrix

The matrix is an important mathematical construct, and it's essential to many statistical methods. You typically describe a matrix $A$ as an $m \times n$ matrix; that is, $A$ will have exactly $m$ rows and $n$ columns. This means $A$ will have a total of $mn$ entries, with each entry $a_{i,j}$ having a unique position given by its specific row ($i = 1, 2, \ldots, m$) and column ($j = 1, 2, \ldots, n$).

You can therefore express a matrix as follows:

$$A = \begin{bmatrix} a_{1,1} & a_{1,2} & \cdots & a_{1,n} \\ a_{2,1} & a_{2,2} & \cdots & a_{2,n} \\ \vdots & \ddots & \ddots & \vdots \\ a_{m,1} & a_{m,2} & \cdots & a_{m,n} \end{bmatrix}$$

To create a matrix in R, use the aptly named `matrix` command, providing the entries of the matrix to the data argument as a vector:

```
R> A <- matrix(data=c(-3,2,893,0.17),nrow=2,ncol=2)
R> A
     [,1]   [,2]
[1,]  -3 893.00
[2,]   2   0.17
```

You must make sure that the length of this vector matches exactly with the number of desired rows (nrow) and columns (ncol). You can elect not to supply nrow and ncol when calling `matrix`, in which case R's default behavior is to return a single-column matrix of the entries in data. For example, `matrix(data=c(-3,2,893,0.17))` would be identical to `matrix(data=c(-3,2,893,0.17),nrow=4,ncol=1)`.

### 3.1.1   Filling Direction

It's important to be aware of how R fills up the matrix using the entries from data. Looking at the previous example, you can see that the $2 \times 2$ matrix A has been filled in a *column-by-column* fashion when reading the data entries from left to right. You can control how R fills in data using the argument byrow, as shown in the following examples:

```
R> matrix(data=c(1,2,3,4,5,6),nrow=2,ncol=3,byrow=FALSE)
     [,1] [,2] [,3]
[1,]    1    3    5
[2,]    2    4    6
```

Here, I've instructed R to provide a $2 \times 3$ matrix containing the digits 1 through 6. By using the optional argument byrow and setting it to FALSE, you explicitly tell R to fill this $2 \times 3$ structure in a column-wise fashion, by filling each column before moving to the next, reading the data argument vector from left to right. This is R's default handling of the `matrix` function, so if the byrow argument isn't supplied, the software will assume byrow=FALSE. Figure 3-1 illustrates this behavior.

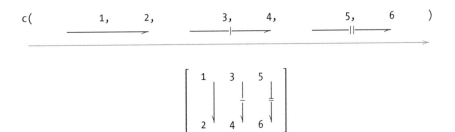

Figure 3-1: Filling a 2 × 3 matrix in a column-wise fashion with byrow=FALSE (R default)

Now, let's repeat the same line of code but set byrow=TRUE.

```
R> matrix(data=c(1,2,3,4,5,6),nrow=2,ncol=3,byrow=TRUE)
     [,1] [,2] [,3]
[1,]    1    2    3
[2,]    4    5    6
```

The resulting 2 × 3 structure has now been filled in a row-wise fashion, as shown in Figure 3-2.

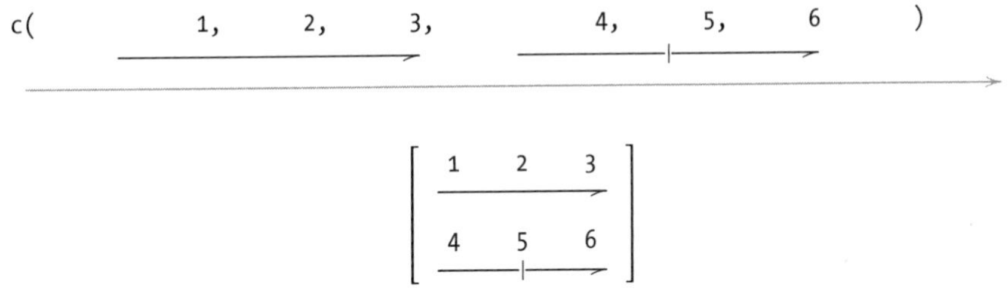

c(            1,      2,      3,            4,      5,      6      )

$$\begin{bmatrix} 1 & 2 & 3 \\ 4 & 5 & 6 \end{bmatrix}$$

Figure 3-2: Filling a 2 × 3 matrix in a row-wise fashion with byrow=TRUE

### 3.1.2   Row and Column Bindings

If you have multiple vectors of equal length, you can quickly build a matrix by binding together these vectors using the built-in R functions, rbind and cbind. You can either treat each vector as a row (by using the command rbind) or treat each vector as a column (using the command cbind). Say you have the two vectors 1:3 and 4:6. You can reconstruct the 2 × 3 matrix in Figure 3-2 using rbind as follows:

```
R> rbind(1:3,4:6)
     [,1] [,2] [,3]
[1,]    1    2    3
[2,]    4    5    6
```

Here, rbind has bound together the vectors as two rows of a matrix, with the top-to-bottom order of the rows matching the order of the vectors supplied to rbind. The same matrix could be constructed as follows, using cbind:

```
R> cbind(c(1,4),c(2,5),c(3,6))
     [,1] [,2] [,3]
[1,]    1    2    3
[2,]    4    5    6
```

Here, you have three vectors each of length 2. You use cbind to glue together these three vectors in the order they were supplied, and each vector becomes a column of the resulting matrix.

### 3.1.3  Matrix Dimensions

Another useful function, dim, provides the dimensions of a matrix stored in your workspace.

```
R> mymat <- rbind(c(1,3,4),5:3,c(100,20,90),11:13)
R> mymat
      [,1] [,2] [,3]
[1,]    1    3    4
[2,]    5    4    3
[3,]  100   20   90
[4,]   11   12   13

R> dim(mymat)
[1] 4 3
R> nrow(mymat)
[1] 4
R> ncol(mymat)
[1] 3
R> dim(mymat)[2]
[1] 3
```

Having defined a matrix mymat using rbind, you can confirm its dimensions with dim, which returns a vector of length 2; dim always supplies the number of rows first, followed by the number of columns. You can also use two related functions: nrow (which provides the number of rows only) and ncol (which provides the number of columns only). In the last command shown, you use dim and your knowledge of vector subsetting to extract the same result that ncol would give you.

## 3.2  Subsetting

Extracting and subsetting elements from matrices in R is much like extracting elements from vectors. The only complication is that you now have an additional dimension. Element extraction still uses the square-bracket operator, but now it must be performed with both a row *and* a column position, given strictly in the order of [*row*,*column*]. Let's start by creating a 3 × 3 matrix, which I'll use for the examples in this section.

```
R> A <- matrix(c(0.3,4.5,55.3,91,0.1,105.5,-4.2,8.2,27.9),nrow=3,ncol=3)
R> A
      [,1]  [,2] [,3]
[1,]   0.3  91.0 -4.2
[2,]   4.5   0.1  8.2
[3,]  55.3 105.5 27.9
```

To tell R to "look at the third row of A and give me the element from the second column," you execute the following:

```
R> A[3,2]
[1] 105.5
```

As expected, you're given the element at position [3,2].

### 3.2.1   Row, Column, and Diagonal Extractions

To extract an entire row or column from a matrix, you simply specify the desired row or column number and leave the other value blank. It's important to note that *you must still include* the comma that separates the row and column numbers—this is how R distinguishes between a request for a row and a request for a column. The following returns the second column of A:

```
R> A[,2]
[1]  91.0   0.1 105.5
```

The following examines the first row:

```
R> A[1,]
[1]  0.3 91.0 -4.2
```

Note that whenever an extraction (or deletion, covered in a moment) results in a single value, single row, or single column, R will always return stand-alone vectors comprised of the requested values. You can also perform more complicated extractions, for example requesting whole rows or columns, or multiples rows or columns, where the result must be returned as a new matrix of the appropriate dimensions. Consider the following subsets:

```
R> A[2:3,]
      [,1]  [,2] [,3]
[1,]   4.5   0.1  8.2
[2,]  55.3 105.5 27.9

R> A[,c(3,1)]
      [,1] [,2]
```

```
[1,] -4.2  0.3
[2,]  8.2  4.5
[3,] 27.9 55.3

R> A[c(3,1),2:3]
      [,1] [,2]
[1,] 105.5 27.9
[2,]  91.0 -4.2
```

The first command returns the second and third rows of A, and the second command returns the third and first columns of A. The last command accesses the third and first rows of A, in that order, and from those rows it returns the second and third column elements.

You can also identify the values along the diagonal of a square matrix (that is, a matrix with an equal number of rows and columns) using the diag command.

```
R> diag(x=A)
[1]  0.3  0.1 27.9
```

This returns a vector with the elements along the diagonal of A, starting at A[1,1].

### 3.2.2  Omitting and Overwriting

To delete or omit elements from a matrix, you again use square brackets, but this time with negative indexes. The following provides A without its second column:

```
R> A[,-2]
      [,1] [,2]
[1,]  0.3 -4.2
[2,]  4.5  8.2
[3,] 55.3 27.9
```

The following removes the first row from A and retrieves the third and second column values, in that order, from the remaining two rows:

```
R> A[-1,3:2]
      [,1]  [,2]
[1,]  8.2   0.1
[2,] 27.9 105.5
```

The following produces A without its first row and second column:

```
R> A[-1,-2]
      [,1] [,2]
[1,]  4.5  8.2
[2,] 55.3 27.9
```

Lastly, this deletes the first row and then deletes the second and third columns from the result:

```
R> A[-1,-c(2,3)]
[1]  4.5 55.3
```

Note that this final operation leaves you with only the last two elements of the first column of A, so this result is returned as a stand-alone vector rather than a matrix.

To overwrite particular elements, or entire rows or columns, you identify the elements to be replaced and then assign the new values, as you did with vectors in Section 2.3.3. The new elements can be a single value, a vector of the same length as the number of elements to be replaced, or a vector whose length evenly divides the number of elements to be replaced. To illustrate this, let's first create a copy of A and call it B.

```
R> B <- A
R> B
      [,1]  [,2] [,3]
[1,]  0.3  91.0 -4.2
[2,]  4.5   0.1  8.2
[3,] 55.3 105.5 27.9
```

The following overwrites the second row of B with the sequence 1, 2, and 3:

```
R> B[2,] <- 1:3
R> B
      [,1]  [,2] [,3]
[1,]  0.3  91.0 -4.2
[2,]  1.0   2.0  3.0
[3,] 55.3 105.5 27.9
```

The following overwrites the second column elements of the first and third rows with 900:

```
R> B[c(1,3),2] <- 900
R> B
      [,1] [,2] [,3]
[1,]  0.3  900 -4.2
[2,]  1.0    2  3.0
[3,] 55.3  900 27.9
```

Next, you replace the third column of B with the values in the third *row* of B.

```
R> B[,3] <- B[3,]
R> B
      [,1] [,2]  [,3]
```

```
[1,]  0.3  900  55.3
[2,]  1.0    2 900.0
[3,] 55.3  900  27.9
```

To try R's vector recycling, let's now overwrite the first and third column elements of rows 1 and 3 (a total of four elements) with the two values -7 and 7.

```
R> B[c(1,3),c(1,3)] <- c(-7,7)
R> B
     [,1] [,2] [,3]
[1,]   -7  900   -7
[2,]    1    2  900
[3,]    7  900    7
```

The vector of length 2 has replaced the four elements *in a column-wise fashion*. The replacement vector c(-7,7) overwrites the elements at positions $(1,1)$ and $(3,1)$, in that order, and is then repeated to overwrite $(1,3)$ and $(3,3)$, in that order.

To highlight the role of index order on matrix element replacement, consider the following example:

```
R> B[c(1,3),2:1] <- c(65,-65,88,-88)
R> B
     [,1] [,2] [,3]
[1,]   88   65   -7
[2,]    1    2  900
[3,]  -88  -65    7
```

The four values in the replacement vector have overwritten the four specified elements, again in a column-wise fashion. In this case, because I specified the first and second columns in reverse order, the overwriting proceeded accordingly, filling the second column before moving to the first. Position $(1,2)$ is matched with 65, followed by $(3,2)$ with -65; then $(1,1)$ becomes 88, and $(3,1)$ becomes -88.

If you just want to replace the diagonal of a square matrix, you can avoid explicit indexes and directly overwrite the values using the diag command.

```
R> diag(x=B) <- rep(x=0,times=3)
R> B
     [,1] [,2] [,3]
[1,]    0   65   -7
[2,]    1    0  900
[3,]  -88  -65    0
```

## Exercise 3.1

a.  Construct and store a $4 \times 2$ matrix that's filled row-wise with the values 4.3, 3.1, 8.2, 8.2, 3.2, 0.9, 1.6, and 6.5, in that order.

b.  Confirm the dimensions of the matrix from (a) are $3 \times 2$ if you remove any one row.

c.  Overwrite the second column of the matrix from (a) with that same column sorted from smallest to largest.

d.  What does R return if you delete the fourth row and the first column from (c)? Use matrix to ensure the result is a single-column matrix, rather than a vector.

e.  Store the bottom four elements of (c) as a new $2 \times 2$ matrix.

f.  Overwrite, in this order, the elements of (c) at positions $(4, 2)$, $(1, 2)$, $(4, 1)$, and $(1, 1)$ with $-\frac{1}{2}$ of the two values on the diagonal of (e).

## 3.3 Matrix Operations and Algebra

You can think of matrices in R from two perspectives. First, you can use these structures purely as a computational tool in programming to store and operate on results, as you've seen so far. Alternatively, you can use matrices for their mathematical properties in relevant calculations, such as the use of matrix multiplication for expressing regression model equations. This distinction is important because the mathematical behavior of matrices is not always the same as the more generic data handling behavior. Here I'll briefly describe some special matrices, as well as some of the most common mathematical operations involving matrices, and the corresponding functionality in R. If the mathematical behavior of matrices isn't of interest to you, you can skip this section for now and refer to it later as needed.

### 3.3.1 Matrix Transpose

For any $m \times n$ matrix $A$, its *transpose*, $A^{\top}$, is the $n \times m$ matrix obtained by writing either its columns as rows or its rows as columns.

Here's an example:

$$\text{If } A = \begin{bmatrix} 2 & 5 & 2 \\ 6 & 1 & 4 \end{bmatrix}, \text{ then } A^{\top} = \begin{bmatrix} 2 & 6 \\ 5 & 1 \\ 2 & 4 \end{bmatrix}.$$

In R, the transpose of a matrix is found with the function t. Let's create a new matrix and then transpose it.

```
R> A <- rbind(c(2,5,2),c(6,1,4))
R> A
     [,1] [,2] [,3]
[1,]    2    5    2
[2,]    6    1    4
R> t(A)
     [,1] [,2]
[1,]    2    6
[2,]    5    1
[3,]    2    4
```

If you "transpose the transpose" of A, you'll recover the original matrix.

```
R> t(t(A))
     [,1] [,2] [,3]
[1,]    2    5    2
[2,]    6    1    4
```

### 3.3.2   Identity Matrix

The *identity matrix* written as $I_m$ is a particular kind of matrix used in mathematics. It's a square $m \times m$ matrix with ones on the diagonal and zeros elsewhere.

Here's an example:

$$I_3 = \begin{bmatrix} 1 & 0 & 0 \\ 0 & 1 & 0 \\ 0 & 0 & 1 \end{bmatrix}$$

You can create an identity matrix of any dimension using the standard matrix function, but there's a quicker approach using diag. Earlier, I used diag on an existing matrix to extract or overwrite its diagonal elements. You can also use it as follows:

```
R> A <- diag(x=3)
R> A
     [,1] [,2] [,3]
[1,]    1    0    0
[2,]    0    1    0
[3,]    0    0    1
```

Here you see diag can be used to easily produce an identity matrix. To clarify, the behavior of diag depends on what you supply to it as its argument x. If, as earlier, x is a matrix, diag will retrieve the diagonal elements of the matrix. If x is a single positive integer, as is the case here, then diag will produce the identity matrix of the corresponding dimension. You can find more uses of diag on its help page.

### 3.3.3 Scalar Multiple of a Matrix

A scalar value is just a single, univariate value. Multiplication of any matrix $A$ by a scalar value $a$ results in a matrix in which every individual element is multiplied by $a$.

Here's an example:

$$2 \times \begin{bmatrix} 2 & 5 & 2 \\ 6 & 1 & 4 \end{bmatrix} = \begin{bmatrix} 4 & 10 & 4 \\ 12 & 2 & 8 \end{bmatrix}$$

R will perform this multiplication in an element-wise manner, as you might expect. Scalar multiplication of a matrix is carried out using the standard arithmetic $*$ operator.

```
R> A <- rbind(c(2,5,2),c(6,1,4))
R> a <- 2
R> a*A
     [,1] [,2] [,3]
[1,]    4   10    4
[2,]   12    2    8
```

### 3.3.4 Matrix Addition and Subtraction

Addition or subtraction of two matrices of equal size is also performed in an element-wise fashion. Corresponding elements are added or subtracted from one another, depending on the operation.

Here's an example:

$$\begin{bmatrix} 2 & 6 \\ 5 & 1 \\ 2 & 4 \end{bmatrix} - \begin{bmatrix} -2 & 8.1 \\ 3 & 8.2 \\ 6 & -9.8 \end{bmatrix} = \begin{bmatrix} 4 & -2.1 \\ 2 & -7.2 \\ -4 & 13.8 \end{bmatrix}$$

You can add or subtract any two equally sized matrices with the standard + and - symbols.

```
R> A <- cbind(c(2,5,2),c(6,1,4))
R> A
     [,1] [,2]
[1,]    2    6
[2,]    5    1
[3,]    2    4
R> B <- cbind(c(-2,3,6),c(8.1,8.2,-9.8))
R> B
     [,1] [,2]
[1,]   -2  8.1
[2,]    3  8.2
[3,]    6 -9.8
R> A-B
```

```
      [,1] [,2]
[1,]    4 -2.1
[2,]    2 -7.2
[3,]   -4 13.8
```

### 3.3.5  Matrix Multiplication

In order to *multiply* two matrices $A$ and $B$ of size $m \times n$ and $p \times q$, it must be true that $n = p$. The resulting matrix $A \cdot B$ will have the size $m \times q$. The elements of the product are computed in a row-by-column fashion, where the value at position $(AB)_{i,j}$ is computed by element-wise multiplication of the entries in row $i$ of $A$ by the entries in column $j$ of $B$, summing the result. Here's an example:

$$\begin{bmatrix} 2 & 5 & 2 \\ 6 & 1 & 4 \end{bmatrix} \cdot \begin{bmatrix} 3 & -3 \\ -1 & 1 \\ 1 & 5 \end{bmatrix}$$
$$= \begin{bmatrix} 2{\times}3 + 5{\times}(-1) + 2{\times}1 & 2{\times}(-3) + 5{\times}(1) + 2{\times}5 \\ 6{\times}3 + 1{\times}(-1) + 4{\times}1 & 6{\times}(-3) + 1{\times}(1) + 4{\times}5 \end{bmatrix}$$
$$= \begin{bmatrix} 3 & 9 \\ 21 & 3 \end{bmatrix}$$

Note that, in general, multiplication of appropriately sized matrices (denoted, say, with $C$ and $D$) is not commutative; that is, $CD \neq DC$.

Unlike addition, subtraction, and scalar multiplication, matrix multiplication is not a simple element-wise calculation, and the standard * operator cannot be used. Instead, you must use R's matrix product operator, written with percent symbols as %*%. Before you try this operator, let's first store the two example matrices and check to make sure the number of columns in the first matrix matches the number of rows in the second matrix using dim.

```
R> A <- rbind(c(2,5,2),c(6,1,4))
R> dim(A)
[1] 2 3
R> B <- cbind(c(3,-1,1),c(-3,1,5))
R> dim(B)
[1] 3 2
```

This confirms the two matrices are compatible for multiplication, so you can proceed.

```
R> A%*%B
      [,1] [,2]
[1,]    3    9
[2,]   21    3
```

You can show that matrix multiplication is noncommutative using the same two matrices. Switching the order of multiplication gives you an entirely different result.

```
R> B%*%A
     [,1] [,2] [,3]
[1,]  -12   12   -6
[2,]    4   -4    2
[3,]   32   10   22
```

### 3.3.6 Matrix Inversion

Some square matrices can be *inverted*. The inverse of a matrix $A$ is denoted $A^{-1}$. An invertible matrix satisfies the following equation:

$$AA^{-1} = I_m$$

Here's an example of a matrix and its inverse:

$$\begin{bmatrix} 3 & 1 \\ 4 & 2 \end{bmatrix}^{-1} = \begin{bmatrix} 1 & -0.5 \\ -2 & 1.5 \end{bmatrix}$$

Matrices that are not invertible are referred to as *singular*. Inverting a matrix is often necessary when solving equations with matrices and has important practical ramifications. There are several different approaches to matrix inversion, and these calculations can become extremely computationally expensive as you increase the size of a matrix. We won't go into too much detail here, but if you're interested, see Golub and Van Loan (1989) for formal discussions.

For now, I'll just show you the R function solve as one option for inverting a matrix.

```
R> A <- matrix(data=c(3,4,1,2),nrow=2,ncol=2)
R> A
     [,1] [,2]
[1,]    3    1
[2,]    4    2
R> solve(A)
     [,1] [,2]
[1,]    1 -0.5
[2,]   -2  1.5
```

You can also verify that the product of these two matrices (using matrix multiplication rules) results in the $2 \times 2$ identity matrix.

```
R> A%*%solve(A)
      [,1] [,2]
[1,]    1    0
[2,]    0    1
```

### Exercise 3.2

a.  Calculate the following:

$$\frac{2}{7}\left(\begin{bmatrix} 1 & 2 \\ 2 & 4 \\ 7 & 6 \end{bmatrix} - \begin{bmatrix} 10 & 20 \\ 30 & 40 \\ 50 & 60 \end{bmatrix}\right)$$

b.  Store these two matrices:

$$A = \begin{bmatrix} 1 \\ 2 \\ 7 \end{bmatrix} \qquad B = \begin{bmatrix} 3 \\ 4 \\ 8 \end{bmatrix}$$

Which of the following multiplications are possible? For those that are, compute the result.

  i.   $A \cdot B$
  ii.  $A^\mathsf{T} \cdot B$
  iii. $B^\mathsf{T} \cdot (A \cdot A^\mathsf{T})$
  iv.  $(A \cdot A^\mathsf{T}) \cdot B^\mathsf{T}$
  v.   $[(B \cdot B^\mathsf{T}) + (A \cdot A^\mathsf{T}) - 100I_3]^{-1}$

c.  For

$$A = \begin{bmatrix} 2 & 0 & 0 & 0 \\ 0 & 3 & 0 & 0 \\ 0 & 0 & 5 & 0 \\ 0 & 0 & 0 & -1 \end{bmatrix},$$

confirm that $A^{-1} \cdot A - I_4$ provides a $4 \times 4$ matrix of zeros.

## 3.4  Multidimensional Arrays

Just as a matrix (a "rectangle" of elements) is the result of increasing the dimension of a vector (a "line" of elements), the dimension of a matrix can be increased to get more complex data structures. In R, vectors and matrices can be considered special cases of the more general *array*, which is how I'll refer to these types of structures when they have more than two dimensions.

So, what's the next step up from a matrix? Well, just as a matrix is considered to be a collection of vectors of equal length, a three-dimensional array can be considered to be a collection of equally dimensioned matrices, providing you with a rectangular prism of elements. You still have a fixed number of rows and a fixed number of columns, as well as a new third dimension called a *layer*. Figure 3-3 illustrates a three-row, four-column, two-layer ($3 \times 4 \times 2$) array.

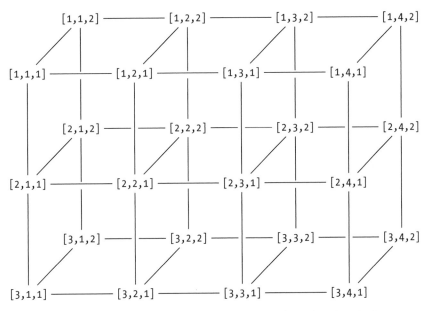

*Figure 3-3: A conceptual diagram of a 3 × 4 × 2 array. The index of each element is given at the corresponding position. These indexes are provided in the strict order of [row,column,layer].*

### 3.4.1 Definition

To create these data structures in R, use the array function and specify the individual elements in the data argument as a vector. Then specify size in the dim argument as another vector with a length corresponding to the number of dimensions. Note that array fills the entries of each layer with the elements in data in a strict column-wise fashion, starting with the first layer. Consider the following example:

```
R> AR <- array(data=1:24,dim=c(3,4,2))
R> AR
, , 1

     [,1] [,2] [,3] [,4]
[1,]    1    4    7   10
[2,]    2    5    8   11
[3,]    3    6    9   12

, , 2
```

```
     [,1] [,2] [,3] [,4]
[1,]   13   16   19   22
[2,]   14   17   20   23
[3,]   15   18   21   24
```

This gives you an array of the same size as in Figure 3-3—each of the two layers constitutes a 3 × 4 matrix. In this example, note the order of the dimensions supplied to dim: c(*rows*,*columns*,*layers*). Just like a single matrix, the product of the dimension sizes of an array will yield the total number of elements. As you increase the dimension further, the dim vector must be extended accordingly. For example, a four-dimensional array is the next step up and can be thought of as *blocks* of three-dimensional arrays. Suppose you had a four-dimensional array comprised of three copies of AR, the three-dimensional array just defined. This new array can be stored in R as follows (once again, the array is filled column-wise):

```
R> BR <- array(data=rep(1:24,times=3),dim=c(3,4,2,3))
R> BR
, , 1, 1

     [,1] [,2] [,3] [,4]
[1,]    1    4    7   10
[2,]    2    5    8   11
[3,]    3    6    9   12

, , 2, 1

     [,1] [,2] [,3] [,4]
[1,]   13   16   19   22
[2,]   14   17   20   23
[3,]   15   18   21   24

, , 1, 2

     [,1] [,2] [,3] [,4]
[1,]    1    4    7   10
[2,]    2    5    8   11
[3,]    3    6    9   12

, , 2, 2

     [,1] [,2] [,3] [,4]
[1,]   13   16   19   22
[2,]   14   17   20   23
[3,]   15   18   21   24

, , 1, 3
```

```
       [,1] [,2] [,3] [,4]
[1,]    1    4    7   10
[2,]    2    5    8   11
[3,]    3    6    9   12

, , 2, 3

       [,1] [,2] [,3] [,4]
[1,]   13   16   19   22
[2,]   14   17   20   23
[3,]   15   18   21   24
```

With BR you now have three copies of AR. Each of these copies is split into its two layers so R can print the object to the screen. As before, the rows are indexed by the first digit, the columns by the second digit, and the layers by the third digit. The new fourth digit indexes the blocks.

### 3.4.2  Subsets, Extractions, and Replacements

Even though high-dimensional objects can be difficult to conceptualize, R indexes them consistently. This makes extracting elements from these structures straightforward now that you know how to subset matrices—you just have to keep using commas in the square brackets as separators of the dimensions being accessed. This is highlighted in the examples that follow.

Suppose you want the second row of the second layer of the previously created array AR. You just enter these exact dimensional locations of AR in square brackets.

```
R> AR[2,,2]
[1] 14 17 20 23
```

The desired elements have been extracted as a vector of length 4. If you want specific elements from this vector, say the third and first, in that order, you can call the following:

```
R> AR[2,c(3,1),2]
[1] 20 14
```

Again, this literal method of subsetting makes dealing with even high-dimensional objects in R manageable.

An extraction that results in multiple vectors will be presented as columns in the returned matrix. For example, to extract the first rows of both layers of AR, you enter this:

```
R> AR[1,,]
       [,1] [,2]
[1,]    1   13
```

```
[2,]    4   16
[3,]    7   19
[4,]   10   22
```

The returned object has the first rows of each of the two matrix layers. However, it has returned each of these vectors as a *column* of the single returned matrix. As this example shows, when multiple vectors are extracted from an array, they will be returned as columns by default. This means extracted rows will not necessarily be returned as rows.

Turning to the object BR, the following gives you the single element of the second row and first column of the matrix in the first layer of the three-dimensional array located in the third block.

```
R> BR[2,1,1,3]
[1] 2
```

Again, you just need to look at the position of the index in the square brackets to know which values you are asking R to return from the array. The following examples highlight this:

```
R> BR[1,,,1]
     [,1] [,2]
[1,]    1   13
[2,]    4   16
[3,]    7   19
[4,]   10   22
```

This returns all the values in the first row of the first block. Since I left the column and layer indexes blank in this subset [1,,,1], the command has returned values for all four columns and both layers in that block of BR.

Next, the following line returns all the values in the second layer of the array BR, composed of three matrices:

```
R> BR[,,2,]
, , 1

     [,1] [,2] [,3] [,4]
[1,]   13   16   19   22
[2,]   14   17   20   23
[3,]   15   18   21   24

, , 2

     [,1] [,2] [,3] [,4]
[1,]   13   16   19   22
[2,]   14   17   20   23
[3,]   15   18   21   24
```

```
, , 3

     [,1] [,2] [,3] [,4]
[1,]  13   16   19   22
[2,]  14   17   20   23
[3,]  15   18   21   24
```

This last example highlights a feature noted earlier, where multiple vectors from AR were returned as a matrix. Broadly speaking, if you have an extraction that results in multiple *d*-dimensional arrays, the result will be an array of the next-highest dimension, *d* + 1. In the last example, you extracted multiple (two-dimensional) matrices, and they were returned as a three-dimensional array. This is demonstrated again in the next example:

```
R> BR[3:2,4,,]
, , 1

     [,1] [,2]
[1,]  12   24
[2,]  11   23


, , 2

     [,1] [,2]
[1,]  12   24
[2,]  11   23


, , 3

     [,1] [,2]
[1,]  12   24
[2,]  11   23
```

This extracts the elements at rows 3 and 2 (in that order), column 4, for all layers and for all array blocks. Consider the following final example:

```
R> BR[2,,1,]
     [,1] [,2] [,3]
[1,]   2    2    2
[2,]   5    5    5
[3,]   8    8    8
[4,]  11   11   11
```

Here you've asked R to return the entire second rows of the first layers of all the arrays stored in BR.

Deleting and overwriting elements in high-dimensional arrays follows the same rules as for stand-alone vectors and matrices. You specify the dimension positions the same way, using negative indexes (for deletion) or using the assignment operator for overwriting.

You can use the array function to create one-dimensional arrays (vectors) and two-dimensional arrays (matrices) should you want to (by setting the dim argument to be of length 1 or 2, respectively). Note, though, that vectors in particular may be treated differently by some functions if created with array instead of c (see the help file ?array for technical details). For this reason, and to make large sections of code more readable, it's more conventional in R programming to use the specific vector- and matrix-creation functions c and matrix.

## Exercise 3.3

a. Create and store a three-dimensional array with six layers of a $4 \times 2$ matrix, filled with a decreasing sequence of values between 4.8 and 0.1 of the appropriate length.

b. Extract and store as a new object the fourth- and first-row elements, in that order, of the second column only of all layers of (a).

c. Use a fourfold repetition of the second row of the matrix formed in (b) to fill a new array of dimensions $2 \times 2 \times 2 \times 3$.

d. Create a new array comprised of the results of deleting the sixth layer of (a).

e. Overwrite the second and fourth row elements of the second column of layers 1, 3, and 5 of (d) with $-99$.

## Important Code in This Chapter

| Function/operator | Brief description | First occurrence |
|---|---|---|
| matrix | Create a matrix | Section 3.1, p. 40 |
| rbind | Create a matrix (bind rows) | Section 3.1.2, p. 41 |
| cbind | Create a matrix (bind columns) | Section 3.1.2, p. 42 |
| dim | Get matrix dimensions | Section 3.1.3, p. 42 |
| nrow | Get number of rows | Section 3.1.3, p. 42 |
| ncol | Get number of columns | Section 3.1.3, p. 42 |
| [ , ] | Matrix/array subsetting | Section 3.2, p. 43 |
| diag | Diagonal elements/identity matrix | Section 3.2.1, p. 44 |
| t | Matrix transpose | Section 3.3.1, p. 47 |
| * | Scalar matrix multiple | Section 3.3.3, p. 49 |
| +, - | Matrix addition/subtraction | Section 3.3.4, p. 49 |
| %*% | Matrix multiplication | Section 3.3.5, p. 50 |
| solve | Matrix inversion | Section 3.3.6, p. 51 |
| array | Create an array | Section 3.4.1, p. 53 |

# 4

## NON-NUMERIC VALUES

So far, you've been working almost exclusively with numeric values. But statistical programming also requires non-numeric values. In this chapter, we'll consider three important non-numeric data types: logicals, characters, and factors. These data types play an important role in effective use of R, especially as we get into more complex R programming in Part II.

## 4.1 Logical Values

Logical values (also simply called *logicals*) are based on a simple premise: a logical-valued object can only be either TRUE or FALSE. These can be interpreted as yes/no, one/zero, satisfied/not satisfied, and so on. This is a concept that appears across all programming languages, and logical values have many important uses. Often, they signal whether a condition has been satisfied or whether a parameter should be switched on or off.

You encountered logical values briefly when you used the sort function in Section 2.3.2 and the matrix function in Section 3.1. When using sort, setting decreasing=TRUE returns a vector ordered from largest to smallest, and decreasing=FALSE sorts the vector the other way around. Similarly, when constructing a matrix, byrow=TRUE fills the matrix entries row-wise; otherwise, the matrix is filled column-wise. Now, you'll take a more detailed look at ways to use logicals.

### 4.1.1   TRUE or FALSE?

Logical values in R are written fully as TRUE and FALSE, but they are frequently abbreviated as T or F. The abbreviated version has no effect on the execution of the code, so, for example, using decreasing=T is equivalent to decreasing=TRUE in the sort function. (But do not create objects named T or F if you want to make use of this convenience—see Section 9.1.3.)

Assigning logical values to an object is the same as assigning numeric values.

```
R> foo <- TRUE
R> foo
[1] TRUE
R> bar <- F
R> bar
[1] FALSE
```

This gives you one object with the value TRUE and one with the value FALSE. Similarly, vectors can be filled with logical values.

```
R> baz <- c(T,F,F,F,T,F,T,T,T,F,T,F)
R> baz
 [1]  TRUE FALSE FALSE FALSE  TRUE FALSE  TRUE  TRUE  TRUE FALSE  TRUE FALSE
R> length(x=baz)
[1] 12
```

Matrices (and other higher-dimensional arrays) can be created with these values too. Using foo and baz from earlier, you could construct something like this:

```
R> qux <- matrix(data=baz,nrow=3,ncol=4,byrow=foo)
R> qux
     [,1]  [,2]  [,3]  [,4]
[1,] TRUE FALSE FALSE FALSE
[2,] TRUE FALSE  TRUE  TRUE
[3,] TRUE FALSE  TRUE FALSE
```

### 4.1.2   A Logical Outcome: Relational Operators

Logicals are commonly used to check relationships between values. For example, you might want to know whether some number $a$ is greater than a predefined threshold $b$. For this, you use the standard *relational operators* shown in Table 4-1, which produce logical values as results.

**Table 4-1:** Relational Operators

| Operator | Interpretation |
|----------|----------------|
| == | Equal to |
| != | Not equal to |
| > | Greater than |
| < | Less than |
| >= | Greater than or equal to |
| <= | Less than or equal to |

Typically, these operators are used on numeric values (though you'll look at some other possibilities in Section 4.2.1). Here's an example:

```
R> 1==2
[1] FALSE
R> 1>2
[1] FALSE
R> (2-1)<=2
[1] TRUE
R> 1!=(2+3)
[1] TRUE
```

The results should be unsurprising: 1 being equal to 2 is FALSE and 1 being greater than 2 is also FALSE, while the result of 2-1 being less than or equal to 2 is TRUE and it is also TRUE that 1 is not equal to 5 (2+3). These kinds of operations are much more useful when used on numbers that are variable in some way, as you'll see shortly.

You're already familiar with R's element-wise behavior when working with vectors. The same rules apply when using relational operators. To illustrate this, let's first create two vectors and double-check that they're of equal length.

```
R> foo <- c(3,2,1,4,1,2,1,-1,0,3)
R> bar <- c(4,1,2,1,1,0,0,3,0,4)
R> length(x=foo)==length(x=bar)
[1] TRUE
```

Now consider the following four evaluations:

```
R> foo==bar
 [1] FALSE FALSE FALSE FALSE  TRUE FALSE FALSE FALSE  TRUE FALSE
R> foo<bar
 [1]  TRUE FALSE  TRUE FALSE FALSE FALSE FALSE  TRUE FALSE  TRUE
R> foo<=bar
 [1]  TRUE FALSE  TRUE FALSE  TRUE FALSE FALSE  TRUE  TRUE  TRUE
R> foo<=(bar+10)
 [1] TRUE TRUE TRUE TRUE TRUE TRUE TRUE TRUE TRUE TRUE
```

The first line checks whether the entries in foo are equal to the corresponding entries in bar, which is true only for the fifth and ninth entries. The returned vector will contain a logical result for each pair of elements, so it will be the same length as the vectors being compared. The second line compares foo and bar in the same way, this time checking whether the entries in foo are less than the entries in bar. Contrast this result with the third comparison, which asks whether entries are less than *or equal to* one another. Finally, the fourth line checks whether foo's members are less than or equal to bar, when the elements of bar are increased by 10. Naturally, the results are all TRUE.

Vector recycling also applies to logicals. Let's use foo from earlier, along with a shorter vector, baz.

```
R> baz <- foo[c(10,3)]
R> baz
[1] 3 1
```

Here you create baz as a vector of length 2 comprised of the 10th and 3rd elements of foo. Now consider the following:

```
R> foo>baz
[1] FALSE  TRUE FALSE  TRUE FALSE  TRUE FALSE FALSE FALSE  TRUE
```

Here, the two elements of baz are recycled and checked against the 10 elements of foo. Elements 1 and 2 of foo are checked against 1 and 2 of baz, elements 3 and 4 of foo are checked against 1 and 2 of baz, and so on. You can also check all the values of a vector against a single value. Here's an example:

```
R> foo<3
[1] FALSE  TRUE  TRUE FALSE  TRUE  TRUE  TRUE  TRUE  TRUE FALSE
```

This is a typical operation when handling data sets in R.

Now let's rewrite the contents of foo and bar as 5 × 2 column-filled matrices.

```
R> foo.mat <- matrix(foo,nrow=5,ncol=2)
R> foo.mat
     [,1] [,2]
[1,]    3    2
[2,]    2    1
[3,]    1   -1
[4,]    4    0
[5,]    1    3
R> bar.mat <- matrix(bar,nrow=5,ncol=2)
R> bar.mat
     [,1] [,2]
[1,]    4    0
```

```
[2,]    1    0
[3,]    2    3
[4,]    1    0
[5,]    1    4
```

The same element-wise behavior applies here; if you compare the matrices, you get a matrix of the same size filled with logicals.

```
R> foo.mat<=bar.mat
       [,1]  [,2]
[1,]   TRUE FALSE
[2,]  FALSE FALSE
[3,]   TRUE  TRUE
[4,]  FALSE  TRUE
[5,]   TRUE  TRUE
R> foo.mat<3
       [,1]  [,2]
[1,]  FALSE  TRUE
[2,]   TRUE  TRUE
[3,]   TRUE  TRUE
[4,]  FALSE  TRUE
[5,]   TRUE FALSE
```

This kind of evaluation also applies to arrays of more than two dimensions.

There are two useful functions you can use to quickly inspect a collection of logical values: any and all. When examining a vector, any returns TRUE if any of the logicals in the vector are TRUE and returns FALSE otherwise. The function all returns a TRUE only if *all* of the logicals are TRUE, and returns FALSE otherwise. As a quick example, let's work with two of the logical vectors formed by the comparisons of foo and bar from the beginning of this section.

```
R> qux <- foo==bar
R> qux
 [1] FALSE FALSE FALSE FALSE  TRUE FALSE FALSE FALSE  TRUE FALSE
R> any(qux)
[1] TRUE
R> all(qux)
[1] FALSE
```

Here, the qux contains two TRUEs, and the rest are FALSE—so the result of any is of course TRUE, but the result of all is FALSE. Following the same rules, you get this:

```
R> quux <- foo<=(bar+10)
R> quux
 [1] TRUE TRUE TRUE TRUE TRUE TRUE TRUE TRUE TRUE TRUE
R> any(quux)
```

```
[1] TRUE
R> all(quux)
[1] TRUE
```

---

The any and all functions do the same thing for matrices and arrays of logical values.

<br>

> ## Exercise 4.1
>
> a. Store the following vector of 15 values as an object in your workspace: c(6,9,7,3,6,7,9,6,3,6,6,7,1,9,1). Identify the following elements:
>    i. Those equal to 6
>    ii. Those greater than or equal to 6
>    iii. Those less than 6 + 2
>    iv. Those not equal to 6
>
> b. Create a new vector from the one used in (a) by deleting its first three elements. With this new vector, fill a $2 \times 2 \times 3$ array. Examine the array for the following entries:
>    i. Those less than or equal to 6 divided by 2, plus 4
>    ii. Those less than or equal to 6 divided by 2, plus 4, *after* increasing every element in the array by 2
>
> c. Confirm the specific locations of elements equal to 0 in the $10 \times 10$ identity matrix $I_{10}$ (see Section 3.3).
>
> d. Check whether *any* of the values of the logical arrays created in (b) are TRUE. If they are, check whether they are *all* TRUE.
>
> e. By extracting the diagonal elements of the logical matrix created in (c), use any to confirm there are no TRUE entries.

## 4.1.3    Multiple Comparisons: Logical Operators

Logicals are especially useful when you want to examine whether multiple conditions are satisfied. Often you'll want to perform certain operations only if a number of different conditions have been met.

The previous section looked at relational operators, used to compare the literal values (that is, numeric or otherwise) of stored R objects. Now you'll look at *logical operators*, which are used to compare two TRUE or FALSE objects. These operators are based on the statements AND and OR. Table 4-2 summarizes the R syntax and the behavior of logical operators. The AND and OR operators each have a "single" and "element-wise" version—you'll see how they're different in a moment.

**Table 4-2:** Logical Operators Comparing Two Logical Values

| Operator | Interpretation | Results |
|---|---|---|
| & | AND (element-wise) | TRUE & TRUE is TRUE<br>TRUE & FALSE is FALSE<br>FALSE & TRUE is FALSE<br>FALSE & FALSE is FALSE |
| && | AND (single comparison) | Same as & above |
| \| | OR (element-wise) | TRUE\|TRUE is TRUE<br>TRUE\|FALSE is TRUE<br>FALSE\|TRUE is TRUE<br>FALSE\|FALSE is FALSE |
| \|\| | OR (single comparison) | Same as \| above |
| ! | NOT | !TRUE is FALSE<br>!FALSE is TRUE |

The result of using any logical operator is a logical value. An AND comparison is true only if *both* logicals are TRUE. An OR comparison is true if at least one of the logicals is TRUE. The NOT operator (!) simply returns the opposite of the logical value it's used on. You can combine these operators to examine multiple conditions at once.

```
R> FALSE||((T&&TRUE)||FALSE)
[1] TRUE
R> !TRUE&&TRUE
[1] FALSE
R> (T&&(TRUE||F))&&FALSE
[1] FALSE
R> (6<4)||(3!=1)
[1] TRUE
```

As with numeric arithmetic, there is an order of importance for logical operations in R. An AND statement has a higher precedence than an OR statement. It's helpful to place each comparative pair in parentheses to preserve the correct order of evaluation and make the code more readable. You can see this in the first line of this code, where the innermost comparison is the first to be carried out: T&&TRUE results in TRUE; this is then provided as one of the logical values for the next bracketed comparison where TRUE||FALSE results in TRUE. The final comparison is then FALSE||TRUE, and the result, TRUE, is printed to the console. The second line reads as NOT TRUE AND TRUE, which of course returns FALSE. In the third line, once again the innermost pair is evaluated first: TRUE||F is TRUE; T&&TRUE is TRUE; and finally TRUE&&FALSE

is FALSE. The fourth and final example evaluates two distinct conditions in parentheses, which are then compared using a logical operator. Since 6<4 is FALSE and 3!=1 is TRUE, that gives you a logical comparison of FALSE||TRUE and a final result of TRUE.

In Table 4-2, there is a short (&, |) and long (&&, ||) version of the AND and OR operators. The short versions are meant for element-wise comparisons, where you have two logical vectors and you want multiple logicals as a result. The long versions, which you've been using so far, are meant for comparing two individual values and will return a single logical value. This is important when programming conditional checks in R in an if statement, which you'll look at in Chapter 10. It's possible to compare a single pair of logicals using the short version—though it's considered better practice to use the longer versions when a single TRUE/FALSE result is needed.

Let's look at some examples of element-wise comparisons. Suppose you have two vectors of equal length, foo and bar:

```
R> foo <- c(T,F,F,F,T,F,T,T,T,F,T,F)
R> foo
 [1]  TRUE FALSE FALSE FALSE  TRUE FALSE  TRUE  TRUE  TRUE FALSE  TRUE FALSE
```

and

```
R> bar <- c(F,T,F,T,F,F,F,F,T,T,T,T)
R> bar
 [1] FALSE  TRUE FALSE  TRUE FALSE FALSE FALSE FALSE  TRUE  TRUE  TRUE  TRUE
```

The short versions of the logical operators match each pair of elements by position and return the result of the comparison.

```
R> foo&bar
 [1] FALSE FALSE FALSE FALSE FALSE FALSE FALSE FALSE  TRUE FALSE  TRUE FALSE
R> foo|bar
 [1]  TRUE  TRUE FALSE  TRUE  TRUE FALSE  TRUE  TRUE  TRUE  TRUE  TRUE  TRUE
```

Using the long version of the operators, on the other hand, means R carries out the comparison only on the first pair of logicals in the two vectors.

```
R> foo&&bar
[1] FALSE
R> foo||bar
[1] TRUE
```

Notice that the last two results match the first entries of the vectors you got using the short versions of the logical operators.

## Exercise 4.2

a. Store the vector c(7,1,7,10,5,9,10,3,10,8) as foo. Identify the elements greater than 5 OR equal to 2.

b. Store the vector c(8,8,4,4,5,1,5,6,6,8) as bar. Identify the elements less than or equal to 6 AND not equal to 4.

c. Identify the elements that satisfy (a) in foo AND satisfy (b) in bar.

d. Store a third vector called baz that is equal to the element-wise sum of foo and bar. Determine the following:

   i. The elements of baz greater than or equal to 14 but not equal to 15

   ii. The elements of the vector obtained via an element-wise division of baz by foo that are greater than 4 OR less than or equal to 2

e. Confirm that using the long version in all of the preceding exercises performs only the first comparison (that is, the results each match the first entries of the previously obtained vectors).

### 4.1.4   Logicals Are Numbers!

Because of the binary nature of logical values, they're often represented with TRUE as 1 and FALSE as 0. In fact, in R, if you perform elementary numeric operations on logical values, TRUE is treated like 1, and FALSE is treated like 0.

```
R> TRUE+TRUE
[1] 2
R> FALSE-TRUE
[1] -1
R> T+T+F+T+F+F+T
[1] 4
```

These operations turn out the same as if you had used the digits 1 and 0. In some situations when you'd use logicals, you can substitute the numeric values.

```
R> 1&&1
[1] TRUE
R> 1||0
[1] TRUE
R> 0&&1
[1] FALSE
```

Being able to interpret logicals as zeros and ones means you can use a variety of functions to summarize a logical vector, and you'll explore this further in Part III.

### 4.1.5   Logical Subsetting and Extraction

Logicals can also be used to extract and subset elements in vectors and other objects, in the same way as you've done so far with index vectors. Rather than entering explicit indexes in the square brackets, you can supply logical *flag* vectors, where an element is extracted if the corresponding entry in the flag vector is TRUE. As such, logical flag vectors should be the same length as the vector that's being accessed (though recycling does occur for shorter flag vectors, as a later example shows).

At the beginning of Section 2.3.3 you defined a vector of length 10 as follows:

```
R> myvec <- c(5,-2.3,4,4,4,6,8,10,40221,-8)
```

If you wanted to extract the two negative elements, you could either enter myvec[c(2,10)], or you could do the following using logical flags:

```
R> myvec[c(F,T,F,F,F,F,F,F,F,T)]
[1] -2.3 -8.0
```

This particular example may seem far too cumbersome for practical use. It becomes useful, however, when you want to extract elements based on whether they satisfy a certain condition (or several conditions). For example, you can easily use logicals to find negative elements in myvec by applying the condition <0.

```
R> myvec<0
 [1] FALSE  TRUE FALSE FALSE FALSE FALSE FALSE FALSE FALSE  TRUE
```

This a perfectly valid flag vector that you can use to subset myvec to get the same result as earlier.

```
R> myvec[myvec<0]
[1] -2.3 -8.0
```

As mentioned, R recycles the flag vector if it's too short. To extract every second element from myvec, starting with the first, you could enter the following:

```
R> myvec[c(T,F)]
[1]     5     4     4     8 40221
```

You can do more complicated extractions using relational and logical operators, such as:

```
R> myvec[(myvec>0)&(myvec<1000)]
[1]  5  4  4  4  6  8 10
```

This returns the positive elements that are less than 1,000. You can also overwrite specific elements using a logical flag vector, just as with index vectors.

```
R> myvec[myvec<0] <- -200
R> myvec
 [1]     5  -200     4     4     4     6     8    10 40221  -200
```

This replaces all existing negative entries with −200. Note, though, that you cannot directly use negative logical flag vectors to delete specific elements; this can be done only with numeric index vectors.

As you can see, logicals are therefore very useful for element extraction. You don't need to know beforehand which specific index positions to return, since the conditional check can find them for you. This is particularly valuable when you're dealing with large data sets and you want to inspect records or recode entries that match certain criteria.

In some cases, you might want to convert a logical flag vector into a numeric index vector. This is helpful when you need the explicit indexes of elements that were flagged TRUE. The R function which takes in a logical vector as the argument x and returns the indexes corresponding to the positions of any and all TRUE entries.

```
R> which(x=c(T,F,F,T,T))
[1] 1 4 5
```

You can use this to identify the index positions of myvec that meet a certain condition; for example, those containing negative numbers:

```
R> which(x=myvec<0)
[1]  2 10
```

The same can be done for the other myvec selections you experimented with. Note that a line of code such as myvec[which(x=myvec<0)] is redundant because that extraction can be made using the condition by itself, that is, via myvec[myvec<0], without using which. On the other hand, using which lets you delete elements based on logical flag vectors. You can simply use which to identify the numeric indexes you want to delete and render them negative. To omit the negative entries of myvec, you could execute the following:

```
R> myvec[-which(x=myvec<0)]
 [1]     5     4     4     4     6     8    10 40221
```

The same can be done with matrices and other arrays. In Section 3.2, you stored a $3 \times 3$ matrix as follows:

```
R> A <- matrix(c(0.3,4.5,55.3,91,0.1,105.5,-4.2,8.2,27.9),nrow=3,ncol=3)
R> A
     [,1]  [,2] [,3]
[1,]  0.3  91.0 -4.2
[2,]  4.5   0.1  8.2
[3,] 55.3 105.5 27.9
```

To extract the second and third column elements of the first row of A using numeric indexes, you could execute A[1,2:3]. To do this with logical flags, you could enter the following:

```
R> A[c(T,F,F),c(F,T,T)]
[1] 91.0 -4.2
```

Again, though, you usually wouldn't explicitly specify the logical vectors. Suppose for example you want to replace all elements in A that are less than 1 with −7. Performing this using numeric indexes is rather fiddly. It's much easier to use the logical flag matrix created with the following:

```
R> A<1
       [,1]  [,2]  [,3]
[1,]  TRUE FALSE  TRUE
[2,] FALSE  TRUE FALSE
[3,] FALSE FALSE FALSE
```

You can supply this logical matrix to the square bracket operators, and the replacement is done as follows:

```
R> A[A<1] <- -7
R> A
     [,1]  [,2] [,3]
[1,] -7.0  91.0 -7.0
[2,]  4.5  -7.0  8.2
[3,] 55.3 105.5 27.9
```

This is the first time you've subsetted a matrix without having to list row or column positions inside the square brackets, using commas to separate out dimensions (see Section 3.2). This is because the flag matrix has the same number of rows and columns as the target matrix, thereby providing all the relevant structural information.

If you use which to identify numeric indexes based on a logical flag structure, you have to be a little more careful when dealing with two-dimensional objects or higher. Suppose you want the index positions of the elements that are greater than 25. The appropriate logical matrix is as follows.

```
R> A>25
      [,1]  [,2]  [,3]
[1,] FALSE  TRUE FALSE
[2,] FALSE FALSE FALSE
[3,]  TRUE  TRUE  TRUE
```

Now, say you ask R the following:

```
R> which(x=A>25)
[1] 3 4 6 9
```

This returns the four indexes of the elements that satisfied the relational check, but they are provided as scalar values. How do these correspond to the row/column positioning of the matrix?

The answer lies in R's default behavior for the which function, which essentially treats the multidimensional object as a single vector (laid out column after column) and then returns the vector of corresponding indexes. Say the matrix A was arranged as a vector by stacking the columns first through third, using c(A[,1],A[,2],A[,3]). Then the indexes returned make more sense.

```
R> which(x=c(A[,1],A[,2],A[,3])>25)
[1] 3 4 6 9
```

With the columns laid out end to end, the elements that return TRUE are the third, fourth, sixth, and ninth elements in the list. This can be difficult to interpret, though, especially when dealing with higher-dimensional arrays. In this kind of situation, you can make which return dimension-specific indexes using the optional argument arr.ind (array indexes). By default, this argument is set to FALSE, resulting in the vector converted indexes. Setting arr.ind to TRUE, on the other hand, treats the object as a matrix or array rather than a vector, providing you with the row and column positions of the elements you requested.

```
R> which(x=A>25,arr.ind=T)
     row col
[1,]   3   1
[2,]   1   2
[3,]   3   2
[4,]   3   3
```

The returned object is now a matrix, where each row represents an element that satisfied the logical comparison and each column provides the position of the element. Comparing the output here with A, you can see these positions do indeed correspond to elements where A>25.

Both versions of the output (with arr.ind=T or arr.ind=F) can be useful—the correct choice depends on the application.

## Exercise 4.3

a. Store this vector of 10 values: foo <- c(7,5,6,1,2,10,8,3,8,2).
   Then, do the following:
   i. Extract the elements greater than or equal to 5, storing the
      result as bar.
   ii. Display the vector containing those elements from foo that
      remain after omitting all elements that are greater than or
      equal to 5.

b. Use bar from (a)(i) to construct a $2 \times 3$ matrix called baz, filled in
   a row-wise fashion. Then, do the following:
   i. Replace any elements that are equal to 8 with the *squared*
      value of the element in row 1, column 2 of baz itself.
   ii. Confirm that *all* values in baz are now less than or equal to 25
      AND greater than 4.

c. Create a $3 \times 2 \times 3$ array called qux using the following vector of
   18 values: c(10,5,1,4,7,4,3,3,1,3,4,3,1,7,8,3,7,3). Then, do the
   following:
   i. Identify the dimension-specific index positions of elements
      that are either 3 OR 4.
   ii. Replace all elements in qux that are less than 3 OR greater
      than or equal to 7 with the value 100.

d. Return to foo from (a). Use the vector c(F,T) to extract every
   second value from foo. In Section 4.1.4, you saw that in some
   situations, you can substitute 0 and 1 for TRUE and FALSE. Can you
   perform the same extraction from foo using the vector c(0,1)?
   Why or why not? What does R return in this case?

## 4.2 Characters

Character strings are another common data type, and are used to represent text. In R, strings are often used to specify folder locations or software options (as shown briefly in Section 1.2); to supply an argument to a function; and to annotate stored objects, provide textual output, or help clarify plots and graphics. In a simple way, they can also be used to define different groups making up a categorical variable, though as you'll see in see Section 4.3, *factors* are better suited for that.

**NOTE** *There are three different string formats in the R environment. The default string format is called an* extended regular expression; *the other variants are named* Perl *and* literal regular expressions. *The intricacies of these variants are beyond the scope of this book, so any mention of character strings from here on refers to an extended regular expression. For more technical details about other string formats, enter* ?regex *at the prompt.*

### 4.2.1 Creating a String

Character strings are indicated by double quotation marks, ". To create a string, just enter text between a pair of quotes.

```
R> foo <- "This is a character string!"
R> foo
[1] "This is a character string!"
R> length(x=foo)
[1] 1
```

R treats the string as a single entity. In other words, foo is a vector of length 1 because R counts only the total number of distinct strings rather than individual words or characters. To count the number of individual characters, you can use the nchar function. Here's an example using foo:

```
R> nchar(x=foo)
[1] 27
```

Almost any combination of characters, including numbers, can be a valid character string.

```
R> bar <- "23.3"
R> bar
[1] "23.3"
```

Note that in this form, the string has no numeric meaning, and it won't be treated like the number 23.3. Attempting to multiply it by 2, for example, results in an error.

```
R> bar*2
Error in bar * 2 : non-numeric argument to binary operator
```

This error occurs because * is expecting to operate on two numeric values (not one number and one string, which makes no sense).

Strings can be compared in several ways, the most common comparison being a check for equality.

```
R> "alpha"=="alpha"
[1] TRUE
R> "alpha"!="beta"
[1] TRUE
R> c("alpha","beta","gamma")=="beta"
[1] FALSE  TRUE FALSE
```

Other relational operators work as you might expect. For example, R considers letters that come later in the alphabet to be greater than earlier letters, meaning it can determine whether one string of letters is greater than another with respect to alphabetical order.

```
R> "alpha"<="beta"
[1] TRUE
R> "gamma">"Alpha"
[1] TRUE
```

Furthermore, uppercase letters are considered greater than lowercase letters.

```
R> "Alpha">"alpha"
[1] TRUE
R> "beta">="bEtA"
[1] FALSE
```

Most symbols can also be used in a string. The following string is valid, for example:

```
R> baz <- "&4 _ 3 **%.? $ymbolic non$en$e ,; "
R> baz
[1] "&4 _ 3 **%.? $ymbolic non$en$e ,; "
```

One important exception is the backslash \, also called an *escape*. When a backslash is used within the quotation marks of a string, it initiates some simple control over the printing or display of the string itself. You'll see how this works in a moment in Section 4.2.3. First let's look at two useful functions for combining strings.

### 4.2.2   Concatenation

There are two main functions used to *concatenate* (or glue together) one or more strings: cat and paste. The difference between the two lies in how their contents are returned. The first function, cat, sends its output directly to the console screen and doesn't formally *return* anything. The paste function concatenates its contents and then returns the final character string as a usable R object. This is useful when the result of a string concatenation needs to be passed to another function or used in some secondary way, as opposed to just being displayed. Consider the following vector of character strings:

```
R> qux <- c("awesome","R","is")
R> length(x=qux)
[1] 3
R> qux
[1] "awesome" "R"       "is"
```

As with numbers and logicals, you can also store any number of strings in a matrix or array structure if you want.

When calling cat or paste, you pass arguments to the function in the order you want them combined. The following lines show identical usage yet different types of output from the two functions:

```
R> cat(qux[2],qux[3],"totally",qux[1],"!")
R is totally awesome !
R> paste(qux[2],qux[3],"totally",qux[1],"!")
[1] "R is totally awesome !"
```

Here, you've used the three elements of qux as well as two additional strings, "totally" and "!", to produce the final concatenated string. In the output, note that cat has simply concatenated and printed the text to the screen. This means you cannot directly assign the result to a new variable and treat it as a character string. For paste, however, the [1] to the left of the output and the presence of the " quotes indicate the returned item is a vector containing a character string, and this can be assigned to an object and used in other functions.

**NOTE**    *There's a slight difference between OS X and Windows in the default handling of string concatenation when using the R GUI. After calling* cat *in Windows, the new R prompt awaiting your next command appears on the same line as the printed string, in which case you can just hit* ENTER *to move to the next line, or use an* escape sequence, *which you'll look at in Section 4.2.3. In OS X, the new prompt appears on the next line as usual.*

These two functions have an optional argument, sep, that's used as a separator between strings as they're concatenated. You pass sep a character string, and it will place this string between all other strings you've provided to paste or cat. For example:

```
R> paste(qux[2],qux[3],"totally",qux[1],"!",sep="---")
[1] "R---is---totally---awesome---!"
R> paste(qux[2],qux[3],"totally",qux[1],"!",sep="")
[1] "Ristotallyawesome!"
```

The same behavior would occur for cat. Note that if you don't want any separation, you set sep="", an empty string, as shown in the second example. The empty string separator can be used to achieve correct sentence spacing; note the gap between awesome and the exclamation mark in the previous code when you first used paste and cat. If the sep argument isn't included, R will insert a space between strings by default.

For example, using manual insertion of spaces where necessary, you can write the following:

```
R> cat("Do you think ",qux[2]," ",qux[3]," ",qux[1],"?",sep="")
Do you think R is awesome?
```

Concatenation can be useful when you want to neatly summarize the results from a certain function or set of calculations. Many kinds of R objects can be passed directly to paste or cat; the software will attempt to automatically *coerce* these items into character strings. This means R will convert the input into a string so the values can be included in the final concatenated string. This works particularly well with numeric objects, as the following examples demonstrate:

```
R> a <- 3
R> b <- 4.4
R> cat("The value stored as 'a' is ",a,".",sep="")
The value stored as 'a' is 3.
R> paste("The value stored as 'b' is ",b,".",sep="")
[1] "The value stored as 'b' is 4.4."
R> cat("The result of a+b is ",a,"+",b,"=",a+b,".",sep="")
The result of a+b is 3+4.4=7.4.
R> paste("Is ",a+b," less than 10? That's totally ",a+b<10,".",sep="")
[1] "Is 7.4 less than 10? That's totally TRUE."
```

Here, the values of the non-string objects are placed where you want them in the final string output. The results of calculations can also appear as fields, as shown with the arithmetic a+b and the logical comparison a+b<10. You'll see more details about coercion from one kind of value to another in Section 6.2.4.

### 4.2.3   Escape Sequences

In Section 4.2.1, I noted that a stand-alone backslash doesn't act like a normal character within a string. The \ is used to invoke an *escape sequence*. An escape sequence lets you enter characters that control the format and spacing of the string, rather than being interpreted as normal text. Table 4-3 describes some of the most common escape sequences, and you can find a full list by entering ?Quotes at the prompt.

**Table 4-3:** Common Escape Sequences for Use in Character Strings

| Escape sequence | Result |
| --- | --- |
| \n | Starts a newline |
| \t | Horizontal tab |
| \b | Invokes a backspace |
| \\ | Used as a single backslash |
| \" | Includes a double quote |

Escape sequences add flexibility to the display of character strings, which can be useful for summaries of results and plot annotations. You enter the sequence precisely where you want it to take effect. Let's look at an example.

```
R> cat("here is a string\nsplit\tto neww\b\n\n\tlines")
here is a string
split    to new

   lines
```

Since the signal for an escape is \ and the signal to begin and end a string is ", if you want either of these characters to be included in a string, you must also use an escape to have them be interpreted as a normal character.

```
R> cat("I really want a backslash: \\\nand a double quote: \"")
I really want a backslash: \
and a double quote: "
```

These escape sequences mean that you can't use a stand-alone backslash in file path strings in R. As noted in Section 1.2.3 (where you used getwd to print the current working directory and setwd to change it), folder separation must use a forward slash / and not a backslash.

```
R> setwd("/folder1/folder2/folder3/")
```

File path specification crops up when reading and writing files, which you'll explore in Chapter 8.

### 4.2.4   Substrings and Matching

*Pattern matching* lets you inspect a given string to identify smaller strings within it.

The function substr takes a string x and extracts the part of the string between two character positions (inclusive), indicated with numbers passed as start and stop arguments. Let's try it on the object foo from Section 4.2.1.

```
R> foo <- "This is a character string!"
R> substr(x=foo,start=21,stop=27)
[1] "string!"
```

Here, you've extracted the characters between positions 21 and 27, inclusive, to get "string!". The function substr can also be used with the assignment operator to directly substitute in a new set of characters. In this case, the replacement string should contain the same number of characters as the selected area.

```
R> substr(x=foo,start=1,stop=4) <- "Here"
R> foo
[1] "Here is a character string!"
```

If the replacement string is longer than the number of characters indicated by start and stop, then replacement still takes place, beginning at start and ending at stop. It cuts off any characters that overrun the number of characters you're replacing. If the string is shorter than the number of characters you're replacing, then replacement ends when the string is fully inserted, leaving the original characters up to stop untouched.

Substitution is more flexible using the functions sub and gsub. The sub function searches a given string x for a smaller string pattern contained within. It then replaces the first instance with a new string, given as the argument replacement. The gsub function does the same thing, but it replaces *every* instance of pattern. Here's an example:

```
R> bar <- "How much wood could a woodchuck chuck"
R> sub(pattern="chuck",replacement="hurl",x=bar)
[1] "How much wood could a woodhurl chuck"
R> gsub(pattern="chuck",replacement="hurl",x=bar)
[1] "How much wood could a woodhurl hurl"
```

With sub and gsub, the replacement value need not have the same number of characters as the pattern being replaced. These functions also have search options like case-sensitivity. The help files ?substr and ?sub have more details, as well as noting a handful of other pattern-matching functions and techniques. You might also want to check out the grep command and its variants; see the relevant help file ?grep.

## Exercise 4.4

a. Re-create exactly the following output:

```
"The quick brown fox
    jumped over
        the lazy dogs"
```

b. Suppose you've stored the values num1 <- 4 and num2 <- 0.75. Write a line of R code that returns the following string:

```
[1] "The result of multiplying 4 by 0.75 is 3"
```

Make sure your code produces a string with the correct multiplication result for *any* two numbers stored as num1 and num2.

c. On my local machine, the directory for my work on this book is specified in R as "/Users/tdavies/Documents/RBook/". Imagine it is your machine—write a line of code that replaces *tdavies* in this string with your first initial and surname.

d. In Section 4.2.4, you stored the following string:

```
R> bar <- "How much wood could a woodchuck chuck"
```

   i. Store a new string by gluing onto bar the words
      "if a woodchuck could chuck wood".
   ii. In the result of (i), replace all instances of wood with metal.

e. Store the string "Two 6-packs for $12.99". Then do the following:
   i. Use a check for equality to confirm that the substring
      beginning with character 5 and ending with character 10
      is "6-pack".
   ii. Make it a better deal by changing the price to $10.99.

## 4.3 Factors

In this section, you'll look at some simple functions related to creating,
handling, and inspecting *factors*. Factors are R's most natural way of representing data points that fit in only one of a finite number of distinct categories, rather than belonging to a continuum. Categorical data like this can
play an important role in data science, and you'll look at factors again in
more detail from a statistical perspective in Chapter 13.

### 4.3.1 Identifying Categories

To see how factors work, let's start with a simple data set. Suppose you
find eight people and record their first name, sex, and month of birth in
Table 4-4.

**Table 4-4:** An Example Data Set of
Eight Individuals

| Person | Sex | Month of birth |
|--------|--------|----------------|
| Liz | Female | April |
| Jolene | Female | January |
| Susan | Female | December |
| Boris | Male | September |
| Rochelle | Female | November |
| Tim | Male | July |
| Simon | Male | July |
| Amy | Female | June |

There's really only one sensible way to represent the name of each person in R—as a vector of character strings.

```
R> firstname <- c("Liz","Jolene","Susan","Boris","Rochelle","Tim","Simon",
        "Amy")
```

You have more flexibility when it comes to recording sex, however. Coding females as 0 and males as 1, a numeric option would be as follows:

```
R> sex.num <- c(0,0,0,1,0,1,1,0)
```

Of course, character strings are also possible, and many prefer this because you don't need to remember the numeric code for each group.

```
R> sex.char <- c("female","female","female","male","female","male","male",
        "female")
```

There is, however, a fundamental difference between an individual's name and their sex when stored as data. Where a person's name is a unique identifier that can take any one of an infinite number of possibilities, there are generally only two options for recording a person's sex. These kinds of data, where all possible values fall into a finite number of categories, are best represented in R using factors.

Factors are typically created from a numeric or a character vector (note that you cannot fill matrices or multidimensional arrays using factor values; factors can only take the form of vectors). To create a factor vector, use the function factor, as in this example working with sex.num and sex.char:

```
R> sex.num.fac <- factor(x=sex.num)
R> sex.num.fac
[1] 0 0 0 1 0 1 1 0
Levels: 0 1
R> sex.char.fac <- factor(x=sex.char)
R> sex.char.fac
[1] female female female male   female male   male   female
Levels: female male
```

Here, you obtain factor versions of the two vectors storing gender values.

At first glance, these objects don't look much different from the character and numeric vectors from which they were created. Indeed, factor objects work in much the same way as vectors, but with a little extra information attached (R's internal representation of factor objects is a little different as well). Functions like length and which work the same way on factor objects as with vectors, for example.

The most important extra piece of information (or *attribute*; see Section 6.2.1) that a factor object contains is its *levels*, which store the possible values in the factor. These levels are printed at the bottom of each factor vector. You can extract the levels as a vector of character strings using the levels function.

```
R> levels(x=sex.num.fac)
[1] "0" "1"
R> levels(x=sex.char.fac)
[1] "female" "male"
```

You can also relabel a factor using levels. Here's an example:

```
R> levels(x=sex.num.fac) <- c("1","2")
R> sex.num.fac
[1] 1 1 1 2 1 2 2 1
Levels: 1 2
```

This relabels the females 1 and the males 2.

Factor-valued vectors are subsetted in the same way as any other vector.

```
R> sex.char.fac[2:5]
[1] female female male   female
Levels: female male
R> sex.char.fac[c(1:3,5,8)]
[1] female female female female female
Levels: female male
```

Note that after subsetting a factor object, the object continues to store *all* defined levels even if some of the levels are no longer represented in the subsetted object.

If you want to subset from a factor using a logical flag vector, keep in mind that the levels of a factor are stored as character strings, even if the original data vector was numeric, so you need to use a string when requesting or testing for a particular level. To, for example, identify all the men using the newly relabeled sex.num.fac, use this:

```
R> sex.num.fac=="2"
[1] FALSE FALSE FALSE  TRUE FALSE  TRUE  TRUE FALSE
```

Since the elements in firstname and sex have corresponding positions in their factor vectors, you can then use this logical vector to obtain the names of all the men (this time using the "male"/"female" factor vector).

```
R> firstname[sex.char.fac=="male"]
[1] "Boris" "Tim"   "Simon"
```

Of course, this simple subsetting could have been achieved in much the same way with the raw numeric vector sex.num or the raw character vector sex.char. In the next section, you'll explore some more distinctive advantages to having categorical data represented as a factor in R.

## 4.3.2 Defining and Ordering Levels

The sex factor from the previous section represents the simplest kind of factor variable—there are only two possible levels with no ordering, in that one level is not intuitively considered "higher than" or "following" the other. Here you'll look at factors with levels that can be logically ordered; for example, month of birth (MOB), where there are 12 levels that have a natural order. Let's store the observed MOB data from earlier as a character vector.

```
R> mob <- c("Apr","Jan","Dec","Sep","Nov","Jul","Jul","Jun")
```

There are two problems with the data in this vector. First, not all possible categories are represented since mob contains only seven unique months. Second, this vector doesn't reflect the natural order of the months. If you compare January and December to see which is greater, you get:

```
R> mob[2]
[1] "Jan"
R> mob[3]
[1] "Dec"
R> mob[2]<mob[3]
[1] FALSE
```

Alphabetically, this result is of course correct—*J* doesn't occur before *D*. But in terms of the order of the calendar months, which is what we're interested in, the FALSE result is incorrect.

If you create a factor object from these values, you can deal with both of these problems by supplying additional arguments to the factor function. You can define additional levels by supplying a character vector of all possible values to the levels argument and then instruct R to order the values precisely as they appear in levels by setting the argument ordered to TRUE.

```
R> ms <- c("Jan","Feb","Mar","Apr","May","Jun","Jul","Aug","Sep","Oct","Nov",
           "Dec")
R> mob.fac <- factor(x=mob,levels=ms,ordered=TRUE)
R> mob.fac
[1] Apr Jan Dec Sep Nov Jul Jul Jun
Levels: Jan < Feb < Mar < Apr < May < Jun < Jul < Aug < Sep < Oct < Nov < Dec
```

Here, the mob.fac vector contains the same individual entries at the same index positions as the mob vector from earlier. But notice that this variable has 12 levels, even though you have not made any observations for the levels "Feb", "Mar", "May", "Aug", or "Oct". (Note that if your R console window is too narrow to print all the levels to the screen, you may see a ..., indicating there's more output that's been hidden. Just widen your window and reprint the object to see the hidden levels.) Also, the strict order of these levels is shown by the < symbol in the object output. Using this new factor object, you can perform the relational comparison from earlier and get the result you might expect.

```
R> mob.fac[2]
[1] Jan
Levels: Jan < Feb < Mar < Apr < May < Jun < Jul < Aug < Sep < Oct < Nov < Dec
R> mob.fac[3]
[1] Dec
Levels: Jan < Feb < Mar < Apr < May < Jun < Jul < Aug < Sep < Oct < Nov < Dec
R> mob.fac[2]<mob.fac[3]
[1] TRUE
```

These improvements are far from just cosmetic. There's a big difference, for example, between a data set with zero observations in some of the categories and the same data set defined with fewer categories to begin with. The choice of whether to instruct R to formally order a factor vector can also have important consequences in the implementation of various statistical methods, such as regression and other types of modeling.

### 4.3.3 Combining and Cutting

As you've seen, it's usually simple to combine multiple vectors of the same kind (whether numeric, logical, or character) using the c function. Here's an example:

```
R> foo <- c(5.1,3.3,3.1,4)
R> bar <- c(4.5,1.2)
R> c(foo,bar)
[1] 5.1 3.3 3.1 4.0 4.5 1.2
```

This combines the two numeric vectors into one.

However, the c function doesn't work the same way with factor-valued vectors. Let's see what happens when you use it on the data in Table 4-4 and the MOB factor vector mob.fac, from Section 4.3.2. Suppose you now observe three more individuals with MOB values "Oct", "Feb", and "Feb", which are stored as a factor object, as follows.

```
R> new.values <- factor(x=c("Oct","Feb","Feb"),levels=levels(mob.fac),
                        ordered=TRUE)
R> new.values
[1] Oct Feb Feb
Levels: Jan < Feb < Mar < Apr < May < Jun < Jul < Aug < Sep < Oct < Nov < Dec
```

Now you have mob.fac with the original eight observations and new.values with an additional three. Both are factor objects, defined with identical, ordered levels. You might expect that you can just use c to combine the two as follows.

```
R> c(mob.fac,new.values)
 [1]  4  1 12  9 11  7  7  6 10  2  2
```

Clearly, this has not done what you want it to do. Combining the two factor objects resulted in a numeric vector. This is because the c function interprets factors as integers. Comparing this with the defined levels, you can see that the numbers refer to the index of each month within the ordered levels.

```
R> levels(mob.fac)
 [1] "Jan" "Feb" "Mar" "Apr" "May" "Jun" "Jul" "Aug" "Sep" "Oct" "Nov" "Dec"
```

This means you can use these integers with levels(mob.fac) to retrieve a character vector of the complete observed data—the original eight observations plus the additional three.

```
R> levels(mob.fac)[c(mob.fac,new.values)]
 [1] "Apr" "Jan" "Dec" "Sep" "Nov" "Jul" "Jul" "Jun" "Oct" "Feb" "Feb"
```

Now you have all the observations stored in a vector, but they are currently stored as strings, not factor values. The final step is to turn this vector into a factor object.

```
R> mob.new <- levels(mob.fac)[c(mob.fac,new.values)]
R> mob.new.fac <- factor(x=mob.new,levels=levels(mob.fac),ordered=TRUE)
R> mob.new.fac
 [1] Apr Jan Dec Sep Nov Jul Jul Jun Oct Feb Feb
Levels: Jan < Feb < Mar < Apr < May < Jun < Jul < Aug < Sep < Oct < Nov < Dec
```

As this example shows, combining factors requires you to essentially deconstruct the two objects, obtaining the numeric index positions of each entry with respect to the factor levels, and then rebuild them together. This helps ensure that the levels are consistent and the observations are valid in the final product.

Factors are also often created from data that was originally measured on a continuum, for example the weight of a set of adults or the amount of a drug given to a patient. Sometimes you'll need to group (or *bin*) these types of observations into categories, like Small/Medium/Large or Low/High. In R, you can mold this kind of data into discrete factor categories using the cut function. Consider the following numeric vector of length 10:

```
R> Y <- c(0.53,5.4,1.5,3.33,0.45,0.01,2,4.2,1.99,1.01)
```

Suppose you want to bin the data as follows: *Small* refers to observations in the interval $[0, 2)$, *Medium* refers to $[2, 4)$, and *Large* refers to [4,6]. A square bracket refers to *inclusion* of its nearest value, and a parenthesis indicates *exclusion*, so an observation $y$ will fall in the Small interval if $0 \leq y < 2$, in Medium if $2 \leq y < 4$, or in Large if $4 \leq y \leq 6$. For this you'd use cut and supply your desired break intervals to the breaks argument:

```
R> br <- c(0,2,4,6)
R> cut(x=Y,breaks=br)
 [1] (0,2] (4,6] (0,2] (2,4] (0,2] (0,2] (0,2] (4,6] (0,2] (0,2]
Levels: (0,2] (2,4] (4,6]
```

This gives you a factor, with each observation now assigned an interval. However, notice that your boundary intervals are back-to-front—you want the boundary levels on the left like [0,2), rather than the right as they appear by default, (0,2]. You can fix this by setting the logical argument right to FALSE.

```
R> cut(x=Y,breaks=br,right=F)
 [1] [0,2) [4,6) [0,2) [2,4) [0,2) [0,2) [2,4) [4,6) [0,2) [0,2)
Levels: [0,2) [2,4) [4,6)
```

Now you've swapped which boundaries are inclusive and exclusive. This is important because it changes which categories the values fall into. Notice that the seventh observation has changed categories. But there's still a problem: the final interval currently *excludes* 6, and you want this maximum value to be *included* in the highest level. You can fix this with another logical argument: include.lowest. Even though it's called "include.lowest," this argument can also be used to include the *highest* value if right is FALSE, as indicated in the help file ?cut.

```
R> cut(x=Y,breaks=br,right=F,include.lowest=T)
 [1] [0,2) [4,6] [0,2) [2,4) [0,2) [0,2) [2,4) [4,6] [0,2) [0,2)
Levels: [0,2) [2,4) [4,6]
```

The intervals are now defined how you want. Finally, you want to add better labels to the categories, rather than using the interval levels that R applies by default, by passing a character string vector to the labels argument. The order of labels must match the order of the levels in the factor object.

```
R> lab <- c("Small","Medium","Large")
R> cut(x=Y,breaks=br,right=F,include.lowest=T,labels=lab)
 [1] Small  Large  Small  Medium Small  Small  Medium Large  Small  Small
Levels: Small Medium Large
```

## Exercise 4.5

The New Zealand government consists of the political parties National, Labour, Greens, and Māori, with several smaller parties labeled as Other. Suppose you asked 20 New Zealanders which of these they identified most with and obtained the following data:

- There were 12 males and 8 females; the individuals numbered 1, 5–7, 12, and 14–16 were females.

- The individuals numbered 1, 4, 12, 15, 16, and 19 identified with Labour; no one identified with Māori; the individuals numbered 6, 9, and 11 identified with Greens; 10 and 20 identified with Other; and the rest identified with National.

a. Use your knowledge of vectors (for example, subsetting and overwriting) to create two character vectors: sex with entries "M" (male) and "F" (female) and party with entries "National", "Labour", "Greens", "Maori", and "Other". Make sure the entries are placed in the correct positions as outlined earlier.

b. Create two different factor vectors based on sex and party. Does it make any sense to use ordered=TRUE in either case? How has R appeared to arrange the levels?

c. Use factor subsetting to do the following:
   i.  Return the factor vector of chosen parties for only the male participants.
   ii. Return the factor vector of genders for those who chose National.

d. Another six people joined the survey, with the results c("National","Maori","Maori","Labour","Greens","Labour") for the preferred party and c("M","M","F","F","F","M") as their gender. Combine these results with the original factors from (b).

Suppose you also asked all individuals to state how confident they were that Labour will win more seats in Parliament than National in the next election and to attach a subjective percentage to that confidence. The following 26 results were obtained: 93, 55, 29, 100, 52, 84, 56, 0, 33, 52, 35, 53, 55, 46, 40, 40, 56, 45, 64, 31, 10, 29, 40, 95, 18, 61.

e. Create a factor with levels of confidence as follows: Low for percentages [0,30]; Moderate for percentages (30,70]; and High for percentages (70,100].

f. From (e), extract the levels corresponding to those individuals who originally said they identified with Labour. Do this also for National. What do you notice?

## Important Code in This Chapter

| Function/operator | Brief description | First occurrence |
|---|---|---|
| TRUE, FALSE | Reserved logical values | Section 4.1.1, p. 60 |
| T, F | Unreserved versions of above | Section 4.1.1, p. 60 |
| ==, !=, >, <, >=, <= | relational operators | Section 4.1.2, p. 61 |
| any | Checks whether any entries are TRUE | Section 4.1.2, p. 63 |
| all | Checks whether all entries are TRUE | Section 4.1.2, p. 63 |
| &&, &, \|\|, \|, ! | logical operators | Section 4.1.3, p. 65 |
| which | Determines indexes of TRUEs | Section 4.1.5, p. 69 |
| " " | Creates a character string | Section 4.2.1, p. 73 |
| nchar | Gets number of characters in a string | Section 4.2.1, p. 73 |
| cat | Concatenates strings (no return) | Section 4.2.2, p. 74 |
| paste | Pastes strings (returns a string) | Section 4.2.2, p. 74 |
| \ | String escape | Section 4.2.3, p. 76 |
| substr | Subsets a string | Section 4.2.4, p. 77 |
| sub, gsub | String matching and replacement | Section 4.2.4, p. 78 |
| factor | Creates a factor vector | Section 4.3.1, p. 80 |
| levels | Gets levels of a factor | Section 4.3.1, p. 81 |
| cut | Creates factor from continuum | Section 4.3.3, p. 85 |

# 5

## LISTS AND DATA FRAMES

Vectors, matrices, and arrays are efficient and convenient data storage structures in R, but they have one distinct limitation: they can store only one type of data. In this chapter, you'll explore two more data structures, lists and data frames, which can store multiple types of values at once.

## 5.1   Lists of Objects

The *list* is an incredibly useful data structure. It can be used to group together any mix of R structures and objects. A single list could contain a numeric matrix, a logical array, a single character string, and a factor object. You can even have a list as a component of another list. In this section, you'll see how to create, modify, and access components of these flexible structures.

### 5.1.1   Definition and Component Access

Creating a list is much like creating a vector. You supply the elements that you want to include to the list function, separated by commas.

```
R> foo <- list(matrix(data=1:4,nrow=2,ncol=2),c(T,F,T,T),"hello")
R> foo
[[1]]
     [,1] [,2]
[1,]    1    3
[2,]    2    4

[[2]]
[1]  TRUE FALSE  TRUE  TRUE

[[3]]
[1] "hello"
```

In the list foo, you've stored a $2 \times 2$ numeric matrix, a logical vector, and a character string. These are printed in the order they were supplied to list. Just as with vectors, you can use the length function to check the number of components in a list.

```
R> length(x=foo)
[1] 3
```

You can retrieve components from a list using indexes, which are entered in double square brackets.

```
R> foo[[1]]
     [,1] [,2]
[1,]    1    3
[2,]    2    4
R> foo[[3]]
[1] "hello"
```

This action is known as a *member reference*. When you've retrieved a component this way, you can treat it just like a stand-alone object in the workspace; there's nothing special that needs to be done.

```
R> foo[[1]] + 5.5
     [,1] [,2]
[1,]  6.5  8.5
[2,]  7.5  9.5
R> foo[[1]][1,2]
[1] 3
R> foo[[1]][2,]
[1] 2 4
R> cat(foo[[3]],"you!")
hello you!
```

To overwrite a member of foo, you use the assignment operator.

```
R> foo[[3]]
[1] "hello"
R> foo[[3]] <- paste(foo[[3]],"you!")
R> foo
[[1]]
     [,1] [,2]
[1,]    1    3
[2,]    2    4

[[2]]
[1]  TRUE FALSE  TRUE  TRUE

[[3]]
[1] "hello you!"
```

Suppose now you want to access the second and third components of foo and store them as one object. Your first instinct might be to try something like this:

```
R> foo[[c(2,3)]]
[1] TRUE
```

But R hasn't done what you wanted. Instead, it returned the third element of the second component. This is because using double square brackets on a list is always interpreted with respect to a single member. Fortunately, member referencing with the double square brackets is not the only way to access components of a list. You can also use single square bracket notation. This is referred to as *list slicing*, and it lets you select multiple list items at once.

```
R> bar <- foo[c(2,3)]
R> bar
[[1]]
[1]  TRUE FALSE  TRUE  TRUE

[[2]]
[1] "hello you!"
```

Note that the result bar is itself a list with the two components stored in the order in which they were requested.

### 5.1.2 Naming

You can *name* list components to make the elements more recognizable and easy to work with. Just like the information stored about factor levels (as you saw in Section 4.3.1), a name is an R *attribute*.

Let's start by adding names to the list foo from earlier.

```
R> names(foo) <- c("mymatrix","mylogicals","mystring")
R> foo
$mymatrix
     [,1] [,2]
[1,]   1    3
[2,]   2    4

$mylogicals
[1]  TRUE FALSE  TRUE  TRUE

$mystring
[1] "hello you!"
```

This has changed how the object is printed to the console. Where earlier it printed [[1]], [[2]], and [[3]] before each component, now it prints the names you specified: $mymatrix, $mylogicals, and $mystring. You can now perform member referencing using these names and the dollar operator, rather than the double square brackets.

```
R> foo$mymatrix
     [,1] [,2]
[1,]   1    3
[2,]   2    4
```

This is the same as calling foo[[1]]. In fact, even when an object is named, you can still use the numeric index to obtain a member.

```
R> foo[[1]]
     [,1] [,2]
[1,]   1    3
[2,]   2    4
```

Subsetting named members also works the same way.

```
R> all(foo$mymatrix[,2]==foo[[1]][,2])
[1] TRUE
```

This confirms (using the all function you saw in Section 4.1.2) that these two ways of extracting the second column of the matrix in foo provide an identical result.

To name the components of a list as it's being created, assign a label to each component in the list command. Using some components of foo, create a new, named list.

```
R> baz <- list(tom=c(foo[[2]],T,T,T,F),dick="g'day mate",harry=foo$mymatrix*2)
R> baz
$tom
```

```
[1]   TRUE FALSE   TRUE   TRUE   TRUE   TRUE   TRUE FALSE

$dick
[1] "g'day mate"

$harry
      [,1] [,2]
[1,]    2    6
[2,]    4    8
```

The object baz now contains the three named components tom, dick, and harry.

```
R> names(baz)
[1] "tom"    "dick"   "harry"
```

If you want to rename these members, you can simply assign a character vector of length 3 to names(baz), the same way you did for foo earlier.

**NOTE** *When using the names function, the component names are always provided and returned as character strings in double quotes. However, if you're specifying names when a list is created (inside the list function) or using names to extract members with the dollar operator, the names are entered without quotes (in other words, they are* not *given as strings).*

### 5.1.3  Nesting

As noted earlier, a member of a list can itself be a list. When nesting lists like this, it's important to keep track of the depth of any member for subsetting or extraction later.

Note that you can add components to any existing list by using the dollar operator and a *new* name. Here's an example using foo and baz from earlier:

```
R> baz$bobby <- foo
R> baz
$tom
[1]   TRUE FALSE   TRUE   TRUE   TRUE   TRUE   TRUE FALSE

$dick
[1] "g'day mate"

$harry
      [,1] [,2]
[1,]    2    6
[2,]    4    8

$bobby
```

```
$bobby$mymatrix
     [,1] [,2]
[1,]    1    3
[2,]    2    4

$bobby$mylogicals
[1]  TRUE FALSE  TRUE  TRUE

$bobby$mystring
[1] "hello you!"
```

Here you've defined a fourth component to the list baz called bobby. The member bobby is assigned the entire list foo. As you can see by printing the new baz, there are now three components in bobby. Naming and indexes are now both layered, and you can use either (or combine them) to retrieve members of the inner list.

```
R> baz$bobby$mylogicals[1:3]
[1]  TRUE FALSE  TRUE
R> baz[[4]][[2]][1:3]
[1]  TRUE FALSE  TRUE
R> baz[[4]]$mylogicals[1:3]
[1]  TRUE FALSE  TRUE
```

These all instruct R to return the first three elements of the logical vector stored as the second component ([[2]], also named mylogicals) of the list bobby, which in turn is the fourth component of the list baz. As long as you're aware of what is returned at each layer of a subset, you can continue to subset as needed using names and numeric indexes. Consider the third line in this example. The first layer of the subset is baz[[4]], which is a list with three components. The second layer of subsetting extracts the component mylogicals from that list by calling baz[[4]]$mylogicals. This component represents a vector of length 4, so the third layer of subsetting retrieves the first three elements of that vector with the line baz[[4]]$mylogicals[1:3].

Lists are often used to return output from various R functions. But they can quickly become rather large objects in terms of system resources to store. It's generally recommended that when you have only one type of data, you should stick to using basic vector, matrix, or array structures to record and store the observations.

## Exercise 5.1

a. Create a list that contains, in this order, a sequence of 20 evenly spaced numbers between −4 and 4; a 3 × 3 matrix of the logical vector c(F,T,T,T,F,T,T,F,F) filled column-wise; a character vector with the two strings `"don"` and `"quixote"`; and a factor vector containing the observations `c("LOW","MED","LOW","MED","MED","HIGH")`. Then, do the following:

i. Extract row elements 2 and 1 of columns 2 and 3, in that order, of the logical matrix.

ii. Use `sub` to overwrite `"quixote"` with `"Quixote"` and `"don"` with `"Don"` inside the list. Then, using the newly overwritten list member, concatenate to the console screen the following statement exactly:

---

```
"Windmills! ATTACK!"
    -\Don Quixote/-
```

---

iii. Obtain all values from the sequence between −4 and 4 that are greater than 1.

iv. Using `which`, determine which indexes in the factor vector are assigned the `"MED"` level.

b. Create a new list with the factor vector from (a) as a component named `"facs"`; the numeric vector `c(3,2.1,3.3,4,1.5,4.9)` as a component named `"nums"`; and a nested list comprised of the first three members of the list from (a) (use list slicing to obtain this), named `"oldlist"`. Then, do the following:

i. Extract the elements of `"facs"` that correspond to elements of `"nums"` that are greater than or equal to 3.

ii. Add a new member to the list named `"flags"`. This member should be a logical vector of length 6, obtained as a twofold repetition of the third column of the logical matrix in the `"oldlist"` component.

iii. Use `"flags"` and the logical negation operator `!` to extract the entries of `"num"` corresponding to `FALSE`.

iv. Overwrite the character string vector component of `"oldlist"` with the single character string `"Don Quixote"`.

## 5.2 Data Frames

A *data frame* is R's most natural way of presenting a data set with a collection of recorded observations for one or more variables. Like lists, data frames have no restriction on the data types of the variables; you can store numeric data, factor data, and so on. The R data frame can be thought of as a list with some extra rules attached. The most important distinction is that in a data frame (unlike a list), the members must all be vectors of equal length.

The data frame is one of the most important and frequently used tools in R for statistical data analysis. In this section, you'll look at how to create data frames and learn about their general characteristics.

## 5.2.1 Construction

To create a data frame from scratch, use the data.frame function. You supply your data, grouped by variable, as vectors of the same length—the same way you would construct a named list. Consider the following example data set:

```
R> mydata <- data.frame(person=c("Peter","Lois","Meg","Chris","Stewie"),
                        age=c(42,40,17,14,1),
                        sex=factor(c("M","F","F","M","M")))
R> mydata
  person age sex
1  Peter  42   M
2   Lois  40   F
3    Meg  17   F
4  Chris  14   M
5 Stewie   1   M
```

Here, you've constructed a data frame with the first name, age in years, and sex of five individuals. The returned object should make it clear why vectors passed to data.frame must be of equal length: vectors of differing lengths wouldn't make sense in this context. If you pass vectors of unequal length to data.frame, then R will attempt to recycle any shorter vectors to match the longest, throwing your data off and potentially allocating observations to the wrong variable. Notice that data frames are printed to the console in rows and columns—they look more like a matrix than a named list. This natural spreadsheet style makes it easy to read and manipulate data sets. Each row in a data frame is called a *record*, and each column is a *variable*.

You can extract portions of the data by specifying row and column index positions (much as with a matrix). Here's an example:

```
R> mydata[2,2]
[1] 40
```

This gives you the element at row 2, column 2—the age of Lois. Now extract the third, fourth, and fifth elements of the third column:

```
R> mydata[3:5,3]
[1] F M M
Levels: F M
```

This returns a factor vector with the sex of Meg, Chris, and Stewie. The following extracts the entire third and first columns (in that order):

```
R> mydata[,c(3,1)]
  sex person
1   M  Peter
2   F   Lois
3   F    Meg
```

```
4   M Chris
5   M Stewie
```

This results in another data frame giving the sex and then the name of each person.

You can also use the names of the vectors that were passed to data.frame to access variables even if you don't know their column index positions, which can be useful for large data sets. You use the same dollar operator you used for member-referencing named lists.

```
R> mydata$age
[1] 42 40 17 14  1
```

You can subset this returned vector, too:

```
R> mydata$age[2]
[1] 40
```

This returns the same thing as the earlier call of mydata[2,2].

You can report the size of a data frame—the number of records and variables—just as you've seen for the dimensions of a matrix (first shown in Section 3.1.3).

```
R> nrow(mydata)
[1] 5
R> ncol(mydata)
[1] 3
R> dim(mydata)
[1] 5 3
```

The nrow function retrieves the number of rows (records), ncol retrieves the number of columns (variables), and dim retrieves both.

R's default behavior for character vectors passed to data.frame is to convert each variable into a factor object. Observe the following:

```
R> mydata$person
[1] Peter  Lois   Meg    Chris  Stewie
Levels: Chris Lois Meg Peter Stewie
```

Notice that this variable has levels, which shows it's being treated as a factor. But this isn't what you intended when you defined mydata earlier—you explicitly defined sex to be a factor but left person as a vector of character strings. To prevent this automatic conversion of character strings to factors when using data.frame, set the optional argument stringsAsFactors to FALSE (otherwise, it defaults to TRUE). Reconstructing mydata with this in place looks like this:

```
R> mydata <- data.frame(person=c("Peter","Lois","Meg","Chris","Stewie"),
                        age=c(42,40,17,14,1),
```

```
                        sex=factor(c("M","F","F","M","M")),
                        stringsAsFactors=FALSE)
R> mydata
  person age sex
1  Peter  42   M
2   Lois  40   F
3    Meg  17   F
4  Chris  14   M
5 Stewie   1   M
R> mydata$person
[1] "Peter"  "Lois"   "Meg"    "Chris"  "Stewie"
```

You now have person in the desired, nonfactor form.

## 5.2.2   Adding Data Columns and Combining Data Frames

Say you want to add data to an existing data frame. This could be a set of
observations for a new variable (adding to the number of columns), or it
could be more records (adding to the number of rows). Once again, you
can use some of the functions you've already seen applied to matrices.

Recall the rbind and cbind functions from Section 3.1.2, which let you
append rows and columns, respectively. These same functions can be used
to extend data frames intuitively. For example, suppose you had another
record to include in mydata: the age and sex of another individual, Brian.
The first step is to create a new data frame that contains Brian's information.

```
R> newrecord <- data.frame(person="Brian",age=7,
                        sex=factor("M",levels=levels(mydata$sex)))
R> newrecord
  person age sex
1  Brian   7   M
```

To avoid any confusion, it's important to make sure the variable names
and the data types match the data frame you're planning to add this to.
Note that for a factor, you can extract the levels of the existing factor variable using levels.

Now, you can simply call the following:

```
R> mydata <- rbind(mydata,newrecord)
R> mydata
  person age sex
1  Peter  42   M
2   Lois  40   F
3    Meg  17   F
4  Chris  14   M
5 Stewie   1   M
6  Brian   7   M
```

Using rbind, you combined mydata with the new record and overwrote mydata with the result.

Adding a variable to a data frame is also quite straightforward. Let's say you're now given data on the classification of how funny these six individuals are, defined as a "degree of funniness." The degree of funniness can take three possible values: Low, Med (medium), and High. Suppose Peter, Lois, and Stewie have a high degree of funniness, Chris and Brian have a medium degree of funniness, and Meg has a low degree of funniness. In R, you'd have a factor vector like this:

```
R> funny <- c("High","High","Low","Med","High","Med")
R> funny <- factor(x=funny,levels=c("Low","Med","High"))
R> funny
[1] High High Low  Med  High Med
Levels: Low Med High
```

The first line creates the basic character vector as funny, and the second line overwrites funny by turning it into a factor. The order of these elements must correspond to the records in your data frame. Now, you can simply use cbind to append this factor vector as a column to the existing mydata.

```
R> mydata <- cbind(mydata,funny)
R> mydata
  person age sex funny
1  Peter  42   M  High
2   Lois  40   F  High
3    Meg  17   F   Low
4  Chris  14   M   Med
5 Stewie   1   M  High
6  Brian   7   M   Med
```

The rbind and cbind functions aren't the only ways to extend a data frame. One useful alternative for adding a variable is to use the dollar operator, much like adding a new member to a named list, as in Section 5.1.3. Suppose now you want to add another variable to mydata by including a column with the age of the individuals in months, not years, calling this new variable age.mon.

```
R> mydata$age.mon <- mydata$age*12
R> mydata
  person age sex funny age.mon
1  Peter  42   M  High     504
2   Lois  40   F  High     480
3    Meg  17   F   Low     204
4  Chris  14   M   Med     168
5 Stewie   1   M  High      12
6  Brian   7   M   Med      84
```

This creates a new age.mon column with the dollar operator and at the same time assigns it the vector of ages in years (already stored as age) multiplied by 12.

### 5.2.3   Logical Record Subsets

In Section 4.1.5, you saw how to use logical flag vectors to subset data structures. This is a particularly useful technique with data frames, where you'll often want to examine a subset of entries that meet certain criteria. For example, when working with data from a clinical drug trial, a researcher might want to examine the results for just male participants and compare them to the results for females. Or the researcher might want to look at the characteristics of individuals who responded most positively to the drug.

Let's continue to work with mydata. Say you want to examine all records corresponding to males. From Section 4.3.1, you know that the following line will identify the relevant positions in the sex factor vector:

```
R> mydata$sex=="M"
[1]  TRUE FALSE FALSE  TRUE  TRUE  TRUE
```

This flags the male records. You can use this with the matrix-like syntax you saw in Section 5.2.1 to get the male-only subset.

```
R> mydata[mydata$sex=="M",]
  person age sex funny age.mon
1  Peter  42   M  High     504
4  Chris  14   M   Med     168
5 Stewie   1   M  High      12
6  Brian   7   M   Med      84
```

This returns data for all variables for only the male participants. You can use the same behavior to pick and choose which variables to return in the subset. For example, since you know you are selecting the males only, you could omit sex from the result using a negative numeric index in the column dimension.

```
R> mydata[mydata$sex=="M",-3]
  person age funny age.mon
1  Peter  42  High     504
4  Chris  14   Med     168
5 Stewie   1  High      12
6  Brian   7   Med      84
```

If you don't have the column number or if you want to have more control over the returned columns, you can use a character vector of variable names instead.

```
R> mydata[mydata$sex=="M",c("person","age","funny","age.mon")]
  person age funny age.mon
1  Peter  42  High     504
4  Chris  14   Med     168
5 Stewie   1  High      12
6  Brian   7   Med      84
```

The logical conditions you use to subset a data frame can be as simple or as complicated as you need them to be. The logical flag vector you place in the square brackets just has to match the number of records in the data frame. Let's extract from mydata the full records for individuals who are more than 10 years old OR have a high degree of funniness.

```
R> mydata[mydata$age>10|mydata$funny=="High",]
  person age sex funny age.mon
1  Peter  42   M  High     504
2   Lois  40   F  High     480
3    Meg  17   F   Low     204
4  Chris  14   M   Med     168
5 Stewie   1   M  High      12
```

Sometimes, asking for a subset will yield no records. In this case, R returns a data frame with zero rows, which looks like this:

```
R> mydata[mydata$age>45,]
[1] person   age      sex      funny    age.mon
<0 rows> (or 0-length row.names)
```

In this example, no records are returned from mydata because there are no individuals older than 45. To check whether a subset will contain any records, you can use nrow on the result—if this is equal to zero, then no records have satisfied the specified condition(s).

## Exercise 5.2

a.  Create and store this data frame as dframe in your R workspace:

| person   | sex | funny |
|----------|-----|-------|
| Stan     | M   | High  |
| Francine | F   | Med   |
| Steve    | M   | Low   |
| Roger    | M   | High  |
| Hayley   | F   | Med   |
| Klaus    | M   | Med   |

The variables person, sex, and funny should be identical in nature to the variables in the mydata object studied throughout Section 5.2. That is, person should be a character vector, sex should be a factor with levels F and M, and funny should be a factor with levels Low, Med, and High.

b. Stan and Francine are 41 years old, Steve is 15, Hayley is 21, and Klaus is 60. Roger is extremely old—1,600 years. Append these data as a new numeric column variable in dframe called age.

c. Use your knowledge of reordering the column variables based on column index positions to overwrite dframe, bringing it in line with mydata. That is, the first column should be person, the second column age, the third column sex, and the fourth column funny.

d. Turn your attention to mydata as it was left after you included the age.mon variable in Section 5.2.2. Create a new version of mydata called mydata2 by deleting the age.mon column.

e. Now, combine mydata2 with dframe, naming the resulting object mydataframe.

f. Write a single line of code that will extract from mydataframe just the names and ages of any records where the individual is female and has a level of funniness equal to Med OR High.

g. Use your knowledge of handling character strings in R to extract all records from mydataframe that correspond to people whose names start with S. Hint: Recall substr from Section 4.2.4 (note that substr can be applied to a vector of multiple character strings).

## Important Code in This Chapter

| Function/operator | Brief description | First occurrence |
|---|---|---|
| list | Create a list | Section 5.1.1, p. 89 |
| [[ ]] | Unnamed member reference | Section 5.1.1, p. 90 |
| [ ] | List slicing (multiple members) | Section 5.1.1, p. 91 |
| $ | Get named member/variable | Section 5.1.2, p. 92 |
| data.frame | Create a data frame | Section 5.2.1, p. 96 |
| [ , ] | Extract data frame row/columns | Section 5.2.1, p. 96 |

# 6

# SPECIAL VALUES, CLASSES, AND COERCION

You've now learned about numeric values, logicals, character strings, and factors, as well as their unique properties and applications. Now you'll look at some special values in R that aren't as well-defined. You'll see how they might come about and how to handle and test for them. Then you'll look at different data types in R and some general object class concepts.

## 6.1   Some Special Values

Many situations in R call for special values. For example, when a data set has missing observations or when a practically infinite number is calculated, the software has some unique terms that it reserves for these situations. These special values can be used to mark abnormal or missing values in vectors, arrays, or other data structures.

### 6.1.1  Infinity

In Section 2.1, I mentioned that R imposes limits on how extreme a number can be before the software cannot reliably represent it. When a number is too large for R to represent, the value is deemed to be *infinite*. Of course, the mathematical concept of infinity ($\infty$) does not correspond to a specific number—R simply has to define an extreme cutoff point. The precise cutoff value varies from system to system and is governed in part by the amount of memory R has access to. This value is represented by the special object Inf, which is case sensitive. Because it represents a numeric value, Inf can be associated only with numeric vectors. Let's create some objects to test it out.

```
R> foo <- Inf
R> foo
[1] Inf
R> bar <- c(3401,Inf,3.1,-555,Inf,43)
R> bar
[1] 3401.0    Inf    3.1 -555.0    Inf   43.0
R> baz <- 90000^100
R> baz
[1] Inf
```

Here, you've defined an object foo that is a single instance of an infinite value. You've also defined a numeric vector, bar, with two infinite elements, and then raised 90,000 to a power of 100 in baz to produce a result R deems infinite.

R can also represent negative infinity, with -Inf.

```
R> qux <- c(-42,565,-Inf,-Inf,Inf,-45632.3)
R> qux
[1]     -42.0    565.0     -Inf     -Inf      Inf -45632.3
```

This creates a vector with two negative-infinite values and one positive-infinite value.

Though infinity does not represent any specific value, to a certain extent you can still perform mathematical operations on infinite values in R. For example, multiplying Inf by any negative value will result in -Inf.

```
R> Inf*-9
[1] -Inf
```

If you add to or multiply infinity, you also get infinity as a result.

```
R> Inf+1
[1] Inf
R> 4*-Inf
[1] -Inf
R> -45.2-Inf
[1] -Inf
```

```
R> Inf-45.2
[1] Inf
R> Inf+Inf
[1] Inf
R> Inf/23
[1] Inf
```

Zero and infinity go hand in hand when it comes to division. Any (finite) numeric value divided by infinity, positive or negative, will result in zero.

```
R> -59/Inf
[1] 0
R> -59/-Inf
[1] 0
```

Though it isn't mathematically defined, note that in R, any nonzero value divided by zero will result in infinity (positive or negative depending on the sign of the numerator).

```
R> -59/0
[1] -Inf
R> 59/0
[1] Inf
R> Inf/0
[1] Inf
```

Often, you'll simply want to detect infinite values in a data structure. The functions is.infinite and is.finite take in a collection of values, typically a vector, and return for each element a logical value answering the question posed. Here's an example using qux from earlier:

```
R> qux
[1]    -42.0    565.0     -Inf     -Inf      Inf -45632.3
R> is.infinite(x=qux)
[1] FALSE FALSE  TRUE  TRUE  TRUE FALSE
R> is.finite(x=qux)
[1]  TRUE  TRUE FALSE FALSE FALSE  TRUE
```

Note that these functions do not distinguish between positive or negative infinity, and the result of is.finite will always be the opposite (the negation) of the result of is.infinite.

Finally, relational operators function as you might expect.

```
R> -Inf<Inf
[1] TRUE
R> Inf>Inf
[1] FALSE
```

```
R> qux==Inf
[1] FALSE FALSE FALSE FALSE  TRUE FALSE
R> qux==-Inf
[1] FALSE FALSE  TRUE  TRUE FALSE FALSE
```

Here, the first line confirms that -Inf is indeed treated as less than Inf, and the second line shows that Inf is not greater than Inf. The third and fourth lines, again using qux, test for equality, which is a useful way to distinguish between positive and negative infinity if you need to.

### 6.1.2   NaN

In some situations, it's impossible to express the result of a calculation using a number, Inf, or -Inf. These difficult-to-quantify special values are labeled NaN in R, which stands for *Not a Number*.

As with infinite values, NaN values are associated only with numeric observations. It's possible to define or include a NaN value directly, but this is rarely the way they're encountered.

```
R> foo <- NaN
R> foo
[1] NaN
R> bar <- c(NaN,54.3,-2,NaN,90094.123,-Inf,55)
R> bar
[1]       NaN    54.30    -2.00       NaN 90094.12     -Inf    55.00
```

Typically, NaN is the unintended result of attempting a calculation that's impossible to perform with the specified values.

In Section 6.1.1, you saw that adding or subtracting from Inf or -Inf will simply result again in Inf or -Inf. However, if you attempt to cancel representations of infinity in any way, the result will be NaN.

```
R> -Inf+Inf
[1] NaN
R> Inf/Inf
[1] NaN
```

Here, the first line won't result in zero because positive and negative infinity can't be interpreted in that numeric sense, so you get NaN as a result. The same thing happens if you attempt to divide Inf by itself. In addition, although you saw earlier that a nonzero value divided by zero will result in positive or negative infinity, NaN results when *zero* is divided by zero.

```
R> 0/0
[1] NaN
```

Note that any mathematical operation involving NaN will simply result in NaN.

```
R> NaN+1
[1] NaN
R> 2+6*(4-4)/0
[1] NaN
R> 3.5^(-Inf/Inf)
[1] NaN
```

In the first line, adding 1 to "not a number" is still NaN. In the second line, you obtain NaN from the (4-4)/0, which is clearly 0/0, so the result is also NaN. In the third line, NaN results from -Inf/Inf, so the result of the remaining calculation is again NaN. This begins to give you an idea of how NaN or infinite values might unintentionally crop up. If you have a function where various values are passed to a fixed calculation and you don't take care to prevent, for example, 0/0 from occurring, then the code will return NaN.

Like with Inf, a special function (is.nan) is used to detect the presence of NaN values. Unlike infinite values, however, relational operators cannot be used with NaN. Here's an example using bar, which was defined earlier:

```
R> bar
[1]      NaN    54.30    -2.00      NaN 90094.12    -Inf    55.00
R> is.nan(x=bar)
[1]  TRUE FALSE FALSE  TRUE FALSE FALSE FALSE
R> !is.nan(x=bar)
[1] FALSE  TRUE  TRUE FALSE  TRUE  TRUE  TRUE
R> is.nan(x=bar)|is.infinite(x=bar)
[1]  TRUE FALSE FALSE  TRUE FALSE  TRUE FALSE
R> bar[-(which(is.nan(x=bar)|is.infinite(x=bar)))]
[1]    54.30    -2.00 90094.12    55.00
```

Using the is.nan function on bar flags the two NaN positions as TRUE. In the second example, you use the negation operator ! to flag the positions where the elements are NOT NaN. Using the element-wise OR, | (see Section 4.1.3), you then identify elements that are either NaN OR infinite. Finally, the last line uses which to convert these logical values into numeric index positions so that you can remove them with negative indexes in square brackets (see Section 4.1.5 for a refresher on using which).

You can find more details on the functionality and behavior of NaN and Inf in the R help file by entering ?Inf at the prompt.

## Exercise 6.1

a.  Store the following vector:

```
foo <- c(13563,-14156,-14319,16981,12921,11979,9568,8833,-12968,
         8133)
```

Then, do the following:

   i.  Output all elements of foo that, when raised to a power of 75, are NOT infinite.

   ii.  Return the elements of foo, excluding those that result in negative infinity when raised to a power of 75.

b.  Store the following $3 \times 4$ matrix as the object bar:

$$\begin{bmatrix} 77875.40 & 27551.45 & 23764.30 & -36478.88 \\ -35466.25 & -73333.85 & 36599.69 & -70585.69 \\ -39803.81 & 55976.34 & 76694.82 & 47032.00 \end{bmatrix}$$

Now, do the following:

   i.  Identify the coordinate-specific indexes of the entries of bar that are NaN when you raise bar to a power of 65 and divide by infinity.

   ii.  Return the values in bar that are NOT NaN when bar is raised to a power of 67 and infinity is added to the result. Confirm this is identical to identifying those values in bar that, when raised to a power of 67, are not equal to negative infinity.

   iii.  Identify those values in bar that are either negative infinity OR finite when you raise bar to a power of 67.

### 6.1.3  NA

In statistical analyses, data sets often contain missing values. For example, someone filling out a questionnaire may not respond to a particular item, or a researcher may record some observations from an experiment incorrectly. Identifying and handling missing values is important so that you can still use the rest of the data. R provides a standard special term to represent missing values, NA, which reads as *Not Available*.

NA entries are not the same as NaN entries. Whereas NaN is used only with respect to numeric operations, missing values can occur for any type of observation. As such, NAs can exist in both numeric and non-numeric settings. Here's an example:

```
R> foo <- c("character","a",NA,"with","string",NA)
R> foo
[1] "character" "a"         NA          "with"      "string"    NA
R> bar <- factor(c("blue",NA,NA,"blue","green","blue",NA,"red","red",NA,
            "green"))
R> bar
 [1] blue  <NA>  <NA>  blue  green blue  <NA>  red   red   <NA>  green
Levels: blue green red
R> baz <- matrix(c(1:3,NA,5,6,NA,8,NA),nrow=3,ncol=3)
```

```
R> baz
     [,1] [,2] [,3]
[1,]    1   NA   NA
[2,]    2    5    8
[3,]    3    6   NA
```

The object foo is a character vector with entries 3 and 6 missing; bar is a factor vector of length 11 with elements 2, 3, 7, and 10 missing; and baz is a numeric matrix with row 1, columns 2 and 3, and row 3, column 3, elements missing. In the factor vector, note that the NAs are printed as <NA>. This is to differentiate between bona fide levels of the factor and the missing observations, to prevent NA from being mistakenly interpreted as one of the levels.

Like the other special values so far, you can identify NA elements using the function is.na. This is often useful for removing or replacing NA values. Consider the following numeric vector:

```
R> qux <- c(NA,5.89,Inf,NA,9.43,-2.35,NaN,2.10,-8.53,-7.58,NA,-4.58,2.01,NaN)
R> qux
 [1]    NA  5.89   Inf    NA  9.43 -2.35   NaN  2.10 -8.53 -7.58    NA -4.58
[13]  2.01   NaN
```

This vector has a total of 14 entries, including NA, NaN, and Inf.

```
R> is.na(x=qux)
 [1]  TRUE FALSE FALSE  TRUE FALSE FALSE  TRUE FALSE FALSE FALSE  TRUE FALSE
[13] FALSE  TRUE
```

As you can see, is.na flags the corresponding NA entries in qux as TRUE. But this is not all—note that it also flags elements 7 and 14, which are NaN, not NA. Strictly speaking, NA and NaN are different entities, but numericly they are practically the same since there is almost nothing you can do with either value. Using is.na labels both as TRUE, allowing the user to remove or recode both at the same time.

If you want to identify NA and NaN entries separately, you can use is.nan in conjunction with logical operators. Here's an example:

```
R> which(x=is.nan(x=qux))
[1]  7 14
```

This identifies the index positions whose elements are specifically NaN. If you want to identify NA entries only, try the following:

```
R> which(x=(is.na(x=qux)&!is.nan(x=qux)))
[1]  1  4 11
```

This identifies the element indexes for only the NA entries (by checking for entries where is.na is TRUE AND where is.nan is NOT TRUE).

After locating the offending elements, you could use negative indexes in square brackets to remove them, though R offers a more direct option. The function na.omit will take a structure and delete all NAs from it; na.omit will also apply to NaNs if the elements are numeric.

```
R> quux <- na.omit(object=qux)
R> quux
[1]  5.89   Inf  9.43 -2.35  2.10 -8.53 -7.58 -4.58  2.01
attr(,"na.action")
[1]  1  4  7 11 14
attr(,"class")
[1] "omit"
```

Note that the structure passed to na.omit is given as the argument object and that some additional output is displayed in printing the returned object. These extra details are provided to inform the user that there were elements in the original vector that were removed (in this case, the element positions provided in the attribute na.action). Attributes will be discussed more in Section 6.2.1.

Similar to NaN, arithmetic calculations with NA result in NA. Using relational operators with either NaN or NA will also result in NA.

```
R> 3+2.1*NA-4
[1] NA
R> 3*c(1,2,NA,NA,NaN,6)
[1]  3  6 NA NA NaN  18
R> NA>76
[1] NA
R> 76>NaN
[1] NA
```

You can find more details on the usage and finer technicalities of NA values by entering ?NA.

## 6.1.4   NULL

Finally, you'll look at the *null* value, written as NULL. This value is often used to explicitly define an "empty" entity, which is quite different from a "missing" entity specified with NA. An instance of NA clearly denotes an existing position that can be accessed and/or overwritten if necessary—not so for NULL. You can see an indication of this if you compare the assignment of NA with the assignment of a NULL.

```
R> foo <- NULL
R> foo
NULL
R> bar <- NA
```

```
R> bar
[1] NA
```

Note that bar, the NA object, is printed with an index position [1]. This suggests you have a vector with a single element. In contrast, you explicitly instructed foo to be empty with NULL. Printing this object doesn't provide a position index because there is no position to access.

This interpretation of NULL also applies to vectors that have other well-defined items. Consider the following two lines of code:

```
R> c(2,4,NA,8)
[1]  2  4 NA  8
R> c(2,4,NULL,8)
[1] 2 4 8
```

The first line creates a vector of length 4, with the third position coded as NA. The second line creates a similar vector but using NULL instead of NA. The result is a vector with a length of only 3. That's because NULL cannot take up a position in the vector. As such, it makes no sense to assign NULL to multiple positions in a vector (or any other structure). Again, here's an example:

```
R> c(NA,NA,NA)
[1] NA NA NA
R> c(NULL,NULL,NULL)
NULL
```

The first line can be interpreted as "three possible slots with unrecorded observations." The second line simply provides "emptiness three times," which is interpreted as one single, unsubsettable, empty object.

At this point, you might wonder why there is even a need for NULL. If something is empty and doesn't exist, why define it in the first place? The answer lies in the need to be able to explicitly state or check whether a certain object has been defined. This occurs often when calling functions in R. For example, when a function contains optional arguments, internally the function has to check which of those arguments have been supplied and which are missing or empty. The NULL value is a useful and flexible tool that the author of a function can use to facilitate such checks. You'll see examples of this later on in Chapter 11.

The is.null function is used to check whether something is explicitly NULL. Suppose you have a function with an optional argument named opt.arg and that, if supplied, opt.arg should be a character vector of length 3. Let's say a user calls this function with the following.

```
R> opt.arg <- c("string1","string2","string3")
```

Now if you check whether the argument was supplied using NA, you might call this:

```
R> is.na(x=opt.arg)
[1] FALSE FALSE FALSE
```

The position-specific nature of NA means that this check is element-wise and returns an answer for each value in opt.arg. This is problematic because you want only a single answer—is opt.arg empty or is it supplied? This is when NULL comes to the party.

```
R> is.null(x=opt.arg)
[1] FALSE
```

Quite clearly opt.arg is not empty, and the function can proceed as necessary. If the argument is empty, using NULL over NA for the check is again better for these purposes.

```
R> opt.arg <- c(NA,NA,NA)
R> is.na(x=opt.arg)
[1] TRUE TRUE TRUE

R> opt.arg <- c(NULL,NULL,NULL)
R> is.null(x=opt.arg)
[1] TRUE
```

As noted earlier, filling a vector with NULL isn't usual practice; it's done here just for illustration. But usage of NULL is far from specific to this particular example. It's commonly used throughout both ready-to-use and user-contributed functionality in R.

The empty NULL has an interesting effect if it's included in arithmetic or relational comparisons.

```
R> NULL+53
numeric(0)
R> 53<=NULL
logical(0)
```

Rather than NULL as you might expect, the result is an "empty" vector of a type determined by the nature of the operation attempted. NULL typically dominates any arithmetic, even if it includes other special values.

```
R> NaN-NULL+NA/Inf
numeric(0)
```

NULL also occurs naturally when examining lists and data frames. For example, define a new list foo.

```
R> foo <- list(member1=c(33,1,5.2,7),member2="NA or NULL?")
R> foo
$member1
[1] 33.0  1.0  5.2  7.0

$member2
[1] "NA or NULL?"
```

This list obviously doesn't include a member called member3. Look at what happens when you try to access a member in foo by that name:

```
R> foo$member3
NULL
```

The result of NULL signals that a member called member3 in foo doesn't exist, or in R terms, is empty. Therefore, it can be filled with whatever you want.

```
R> foo$member3 <- NA
R> foo
$member1
[1] 33.0  1.0  5.2  7.0

$member2
[1] "NA or NULL?"

$member3
[1] NA
```

The same principle applies when querying a data frame for a nonexistent column or variable using the dollar operator (as in Section 5.2.2).

For more technical details on how NULL and is.null are handled by R, see the help file accessed by ?NULL.

---

## Exercise 6.2

a. Consider the following line of code:

```
foo <- c(4.3,2.2,NULL,2.4,NaN,3.3,3.1,NULL,3.4,NA)
```

Decide yourself which of the following statements are true and which are false and then use R to confirm:

i. The length of foo is 8.

ii. Calling which(x=is.na(x=foo)) will not result in 4 and 8.

   iii. Checking is.null(x=foo) will provide you with the locations of the two NULL values that are present.

   iv. Executing is.na(x=foo[8])+4/NULL will not result in NA.

 b. Create and store a list containing a single member: the vector c(7,7,NA,3,NA,1,1,5,NA). Then, do the following:

   i. Name the member "alpha".

   ii. Confirm that the list doesn't have a member with the name "beta" using the appropriate logical valued function.

   iii. Add a new member called beta, which is the vector obtained by identifying the index positions of alpha that are NA.

## 6.2 Understanding Types, Classes, and Coercion

By now, you've studied many of the fundamental features in the R language for representing, storing, and handling data. In this section, you'll examine how to formally distinguish between different kinds of values and structures and look at some simple examples of conversion from one type to another.

### 6.2.1 Attributes

Each R object you create has additional information about the nature of the object itself. This additional information is referred to as the object's *attributes.* You've see a few attributes already. In Section 3.1.3, you identified the dimensions attribute of a matrix using dim. In Section 4.3.1, you used levels to get the levels attribute of a factor. It was also noted that names can get the member names of a list in Section 5.1.2, and in Section 6.1.3, that an attribute annotates the result of applying na.omit.

  In general, you can think of attributes as either *explicit* or *implicit.* Explicit attributes are immediately visible to the user, while R determines implicit attributes internally. You can print explicit attributes for a given object with the attributes function, which takes any object and returns a named list. Consider, for example, the following $3 \times 3$ matrix:

```
R> foo <- matrix(data=1:9,nrow=3,ncol=3)
R> foo
     [,1] [,2] [,3]
[1,]    1    4    7
[2,]    2    5    8
[3,]    3    6    9

R> attributes(foo)
$dim
[1] 3 3
```

Here, calling attributes returns a list with one member: dim. Of course, you can retrieve the contents of dim with attributes(foo)$dim, but if you know the name of an attribute, you can also use attr:

```
R> attr(x=foo,which="dim")
[1] 3 3
```

This function takes the object in as x and the name of the attribute as which. Recall that names are specified as character strings in R. To make things even more convenient, the most common attributes have their own functions (usually named after the attribute) to access the corresponding value. For the dimensions of a matrix, you've already seen the function dim.

```
R> dim(foo)
[1] 3 3
```

These attribute-specific functions are useful because they also allow access to implicit attributes, which, while still controllable by the user, are set automatically by the software as a matter of necessity. The names and levels functions mentioned earlier are also both attribute-specific functions.

Explicit attributes are often optional; if they aren't specified, they are NULL. For example, when building a matrix with the matrix function, you can use the optional argument dimnames to annotate the rows and columns with names. You pass dimnames a list made up of two members, both character vectors of the appropriate lengths—the first giving row names and the second giving column names. Let's define the matrix bar as follows:

```
R> bar <- matrix(data=1:9,nrow=3,ncol=3,dimnames=list(c("A","B","C"),
                 c("D","E","F")))
R> bar
  D E F
A 1 4 7
B 2 5 8
C 3 6 9
```

Because the dimension names are attributes, the dimnames appear when you call attributes(bar).

```
R> attributes(bar)
$dim
[1] 3 3

$dimnames
$dimnames[[1]]
[1] "A" "B" "C"

$dimnames[[2]]
[1] "D" "E" "F"
```

Note that `dimnames` is itself a list, nested inside the larger attributes list. Again, to extract the values of this attribute, you can use list member referencing, you can use `attr` as shown earlier, or you can use the attribute-specific function.

```
R> dimnames(bar)
[[1]]
[1] "A" "B" "C"

[[2]]
[1] "D" "E" "F"
```

Some attributes can be modified after an object has been created (as you saw already in Section 5.1.2, where you renamed members of a list). Here, to make `foo` match `bar` exactly, you can give `foo` some `dimnames` by assigning them to the attribute-specific function:

```
R> dimnames(foo) <- list(c("A","B","C"),c("D","E","F"))
R> foo
  D E F
A 1 4 7
B 2 5 8
C 3 6 9
```

I've used matrices in the discussion here, but optional attributes for other objects in R are treated the same way. Attributes are not restricted to built-in R objects, either. Objects you build yourself can be defined with their own attributes and attribute-specific functions. Just remember that the role of an attribute is typically to provide descriptive data about an object, or you could end up overcomplicating your object structures unnecessarily.

### 6.2.2   Object Class

An object's *class* is one of the most useful attributes for describing an entity in R. Every object you create is identified, either implicitly or explicitly, with at least one class. R is an *object-oriented* programming language, meaning entities are stored as objects and have methods that act upon them. In such a language, class identification is formally referred to as *inheritance*.

**NOTE**    *This section will focus on the most common classing structure used in R, called S3. There is another structure, S4, which is essentially a more formal set of rules for the identification and treatment of different objects. For most practical intents and certainly for beginners, understanding and using S3 will be sufficient. You can find further details in R's online documentation.*

The class of an object is explicit in situations where you have user-defined object structures or an object such as a factor vector or data frame where other attributes play an important part in the handling of the object itself—for example, level labels of a factor vector, or variable names in a data frame, are modifiable attributes that play a primary role in accessing the observations of each object. Elementary R objects such as vectors, matrices, and arrays, on the other hand, are implicitly classed, which means the class is not identified with the attributes function. Whether implicit or explicit, the class of a given object can always be retrieved using the attribute-specific function class.

### Stand-Alone Vectors

Let's create some simple vectors to use as examples.

```
R> num.vec1 <- 1:4
R> num.vec1
[1] 1 2 3 4
R> num.vec2 <- seq(from=1,to=4,length=6)
R> num.vec2
[1] 1.0 1.6 2.2 2.8 3.4 4.0
R> char.vec <- c("a","few","strings","here")
R> char.vec
[1] "a"       "few"     "strings" "here"
R> logic.vec <- c(T,F,F,F,T,F,T,T)
R> logic.vec
[1]  TRUE FALSE FALSE FALSE  TRUE FALSE  TRUE  TRUE
R> fac.vec <- factor(c("Blue","Blue","Green","Red","Green","Yellow"))
R> fac.vec
[1] Blue   Blue   Green  Red    Green  Yellow
Levels: Blue Green Red Yellow
```

You can pass any object to the class function, and it returns a character vector as output. Here are examples using the vectors just created:

```
R> class(num.vec1)
[1] "integer"
R> class(num.vec2)
[1] "numeric"
R> class(char.vec)
[1] "character"
R> class(logic.vec)
[1] "logical"
R> class(fac.vec)
[1] "factor"
```

The output from using class on the character vector, the logical vector, and the factor vector simply match the kind of data that has been stored. The output from the number vectors is a little more intricate, however. So far, I've referred to any object with an arithmetically valid set of numbers as "numeric." If all the numbers stored in a vector are whole, then R identifies the vector as "integer". Numbers with decimal places (called *floating-point* numbers), on the other hand, are identified as "numeric". This distinction is necessary because some tasks strictly require integers, not floating-point numbers. Colloquially, I'll continue to refer to both types as "numeric" and in fact, the is.numeric function will return TRUE for both integer and floating-point structures, as you'll see in Section 6.2.3.

## Other Data Structures

As mentioned earlier, R's classes are essentially designed to facilitate object-oriented programming. As such, class usually reports on the nature of the data *structure*, rather than the type of data that's stored—it returns the data type only when used on stand-alone vectors. Let's try it on some matrices.

```
R> num.mat1 <- matrix(data=num.vec1,nrow=2,ncol=2)
R> num.mat1
     [,1] [,2]
[1,]    1    3
[2,]    2    4
R> num.mat2 <- matrix(data=num.vec2,nrow=2,ncol=3)
R> num.mat2
     [,1] [,2] [,3]
[1,]  1.0  2.2  3.4
[2,]  1.6  2.8  4.0
R> char.mat <- matrix(data=char.vec,nrow=2,ncol=2)
R> char.mat
     [,1]  [,2]
[1,] "a"   "strings"
[2,] "few" "here"
R> logic.mat <- matrix(data=logic.vec,nrow=4,ncol=2)
R> logic.mat
      [,1]  [,2]
[1,]  TRUE  TRUE
[2,] FALSE FALSE
[3,] FALSE  TRUE
[4,] FALSE  TRUE
```

Note from Section 4.3.1 that factors are used only in vector form, so fac.vec is not included here. Now check these matrices with class.

```
R> class(num.mat1)
[1] "matrix"
R> class(num.mat2)
[1] "matrix"
R> class(char.mat)
[1] "matrix"
R> class(logic.mat)
[1] "matrix"
```

You see that regardless of the data type, class reports the structure of the object itself—all matrices. The same is true for other object structures, like arrays, lists, and data frames.

## Multiple Classes

Certain objects will have multiple classes. A variant on a standard form of an object, such as an ordered factor vector, will inherit the usual factor class and also contain the additional ordered class. Both are returned if you use the class function.

```
R> ordfac.vec <- factor(x=c("Small","Large","Large","Regular","Small"),
                        levels=c("Small","Regular","Large"),
                        ordered=TRUE)
R> ordfac.vec
[1] Small    Large    Large    Regular Small
Levels: Small < Regular < Large
R> class(ordfac.vec)
[1] "ordered" "factor"
```

Earlier, fac.vec was identified as "factor" only, but the class of ordfac.vec has two components. It's still identified as "factor", but it also includes "ordered", which identifies the variant of the "factor" class also present in the object. Here, you can think of "ordered" as a *subclass* of "factor". In other words, it is a special case that inherits from, and therefore behaves like, a "factor". For further technical details on R subclasses, I recommend Chapter 9 of *The Art of R Programming* by Matloff (2011).

**NOTE**    *I have focused on the class function here because it's directly relevant to the object-oriented programming style exercised in this text, especially in Part II. There are other functions that show some of the complexities of R's classing rules. For example, the function typeof reports the type of data contained within an object, not just for vectors but also for matrices and arrays. Note, however, that the terminology in the output of typeof doesn't always match the output of class. See the help file ?typeof for details on the values it returns.*

To summarize, an object's class is first and foremost a descriptor of the data structure, though for simple vectors, the class function reports the type of data stored. If the vector entries are exclusively whole numbers, then R classes the vector as "integer", whereas "numeric" is used to label a vector with floating-point numbers.

## *6.2.3   Is-Dot Object-Checking Functions*

Identifying the class of an object is essential for functions that operate on stored objects, especially those that behave differently depending on the class of the object. To check whether the object is a specific class or data type, you can use the *is-dot* functions on the object and it will return a TRUE or FALSE logical value.

Is-dot functions exist for almost any sensible check you can think of. For example, consider once more the num.vec1 vector from Section 6.2.2 and the following six checks:

```
R> num.vec1 <- 1:4
R> num.vec1
[1] 1 2 3 4
R> is.integer(num.vec1)
[1] TRUE
R> is.numeric(num.vec1)
[1] TRUE
R> is.matrix(num.vec1)
[1] FALSE
R> is.data.frame(num.vec1)
[1] FALSE
R> is.vector(num.vec1)
[1] TRUE
R> is.logical(num.vec1)
[1] FALSE
```

The first, second, and sixth is-dot functions check the kind of data stored in the object, while the others check the structure of the object itself. The results are to be expected: num.vec1 *is* an "integer" (and *is* "numeric"), and it *is* a "vector." It's not a matrix or a data frame, nor is it logical.

Briefly, it's worth noting that these checks use more general categories than the formal classes identified with class. Recall that num.vec1 was identified solely as "integer" in Section 6.2.2, but using is.numeric here still returns TRUE. In this example, the num.vec1 with integer data is generalized to be "numeric". Similarly, for a data frame, an object of class "data.frame" will return TRUE for is.data.frame *and* is.list because a data frame is intuitively generalized to a list.

There's a difference between the object is-dot functions detailed here and functions such as is.na discussed in Section 6.1. The functions to check for the special values like NA should be thought of as a check for equality; they exist because it is not legal syntax to write something like foo==NA. Those functions from Section 6.1 thus operate in R's element-wise fashion, whereas the object is-dot functions inspect the object *itself*, returning only a single logical value.

## 6.2.4   *As-Dot Coercion Functions*

You've seen different ways to modify an object after it's been created—by accessing and overwriting elements, for example. But what about the structure of the object itself and the type of data contained within?

Converting from one object or data type to another is referred to as *coercion*. Like other features of R you've met so far, coercion is performed either implicitly or explicitly. Implicit coercion occurs automatically when elements need to be converted to another type in order for an operation to complete. In fact, you've come across this behavior already, in Section 4.1.4, for example, when you used numeric values for logical values. Remember that logical values can be thought of as integers—one for TRUE and zero for FALSE. Implicit coercion of logical values to their numeric counterparts occurs in lines of code like this:

```
R> 1:4+c(T,F,F,T)
[1] 2 2 3 5
```

In this operation, R recognizes that you're attempting an arithmetic calculation with +, so it expects numeric quantities. Since the logical vector is not in this form, the software internally coerces it to ones and zeros before completing the task.

Another frequent example of implicit coercion is when paste and cat are used to glue together character strings, as explored in Section 4.2.2. Noncharacter entries are automatically coerced to strings before the concatenation takes place. Here's an example:

```
R> foo <- 34
R> bar <- T
R> paste("Definitely foo: ",foo,"; definitely bar: ",bar,".",sep="")
[1] "Definitely foo: 34; definitely bar: TRUE."
```

Here, the integer 34 and the logical T are implicitly coerced to characters since R knows the output of paste must be a string.

In other situations, coercion won't happen automatically and must be carried out by the user. This explicit coercion can be achieved with the *as-dot* functions. Like is-dot functions, as-dot functions exist for most typical R data types and object classes. The previous two examples can be coerced explicitly, as follows.

```
R> as.numeric(c(T,F,F,T))
[1] 1 0 0 1
R> 1:4+as.numeric(c(T,F,F,T))
[1] 2 2 3 5
R> foo <- 34
R> foo.ch <- as.character(foo)
R> foo.ch
[1] "34"
R> bar <- T
R> bar.ch <- as.character(bar)
R> bar.ch
[1] "TRUE"
R> paste("Definitely foo: ",foo.ch,"; definitely bar: ",bar.ch,".",sep="")
[1] "Definitely foo: 34; definitely bar: TRUE."
```

Coercions are possible in most cases that "make sense." For example, it's easy to see why R is able to read something like this:

```
R> as.numeric("32.4")
[1] 32.4
```

However, the following conversion makes no sense:

```
R> as.numeric("g'day mate")
[1] NA
Warning message:
NAs introduced by coercion
```

Since there is no logical way to translate "g'day mate" into numbers, the entry is returned as NA (in this case, R has also issued a warning message). This means that in certain cases, multiple coercions are needed to attain the final result. Suppose, for example, you have the character vector c("1","0","1","0","0") and you want to coerce it to a logical-valued vector. Direct character to logical coercion is not possible, because even if all the character strings contained numbers, there is no guarantee in general that they would all be ones and zeros.

```
R> as.logical(c("1","0","1","0","0"))
[1] NA NA NA NA NA
```

However, you know that character string numbers can be converted to a numeric data type, and you know that ones and zeros are easily coerced to logicals. So, you can perform the coercion in those two steps, as follows:

```
R> as.logical(as.numeric(c("1","0","1","0","0")))
[1]  TRUE FALSE  TRUE FALSE FALSE
```

Not all data-type coercion is entirely straightforward. Factors, for example, are trickier because R treats the levels as integers. In other words, regardless of how the levels of a given factor are actually labeled, the software will refer to them internally as level 1, level 2, and so on. This is clear if you try to coerce a factor to a numeric data type.

```
R> baz <- factor(x=c("male","male","female","male"))
R> baz
[1] male   male   female male
Levels: female male
R> as.numeric(baz)
[1] 2 2 1 2
```

Here, you see that R has assigned the numeric representation of the factor in the stored order of the factor labels (alphabetic by default). Level 1 refers to female, and level 2 refers to male. This example is simple enough, though it's important to be aware of the behavior since coercion from factors with numeric levels can cause confusion.

```
R> qux <- factor(x=c(2,2,3,5))
R> qux
[1] 2 2 3 5
Levels: 2 3 5
R> as.numeric(qux)
[1] 1 1 2 3
```

The numeric representation of the factor qux is c(1,1,2,3). This highlights again that the levels of qux are simply treated as level 1 (even though it has a label of 2), level 2 (which has a label of 3), and level 3 (which has a label of 5).

Coercion between object classes and structures can also be useful. For example, you might need to store the contents of a matrix as a single vector.

```
R> foo <- matrix(data=1:4,nrow=2,ncol=2)
R> foo
     [,1] [,2]
[1,]    1    3
[2,]    2    4
R> as.vector(foo)
[1] 1 2 3 4
```

Note that as.vector has coerced the matrix by "stacking" the columns into a single vector. The same column-wise deconstruction occurs for higher-dimensional arrays, in order of layer or block.

```
R> bar <- array(data=c(8,1,9,5,5,1,3,4,3,9,8,8),dim=c(2,3,2))
R> bar
, , 1

     [,1] [,2] [,3]
[1,]    8    9    5
[2,]    1    5    1

, , 2

     [,1] [,2] [,3]
[1,]    3    3    8
[2,]    4    9    8

R> as.matrix(bar)
     [,1]
[1,]    8
[2,]    1
[3,]    9
```

```
[4,]    5
[5,]    5
[6,]    1
[7,]    3
[8,]    4
[9,]    3
[10,]   9
[11,]   8
[12,]   8

R> as.vector(bar)
[1] 8 1 9 5 5 1 3 4 3 9 8 8
```

You can see that as.matrix stores the array as a $12 \times 1$ matrix, and as.vector stores it as a single vector. Similar commonsense rules for data types apply to coercion when working with object structures. For example, coercing the following list baz to a data frame produces an error:

```
R> baz <- list(var1=foo,var2=c(T,F,T),var3=factor(x=c(2,3,4,4,2)))
R> baz
$var1
     [,1] [,2]
[1,]    1    3
[2,]    2    4


$var2
[1]  TRUE FALSE  TRUE

$var3
[1] 2 3 4 4 2
Levels: 2 3 4

R> as.data.frame(baz)
Error in data.frame(var1 = 1:4, var2 = c(TRUE, FALSE, TRUE), var3 = c(1L,  :
  arguments imply differing number of rows: 2, 3, 5
```

The error occurs because the variables do not have matching lengths. But there is no problem with coercing the list qux, shown here, which has equal-length members:

```
R> qux <- list(var1=c(3,4,5,1),var2=c(T,F,T,T),var3=factor(x=c(4,4,2,1)))
R> qux
$var1
[1] 3 4 5 1

$var2
[1]  TRUE FALSE  TRUE  TRUE
```

```
$var3
[1] 4 4 2 1
Levels: 1 2 4

R> as.data.frame(qux)
  var1  var2 var3
1    3  TRUE    4
2    4 FALSE    4
3    5  TRUE    2
4    1  TRUE    1
```

This stores the variables as a data set in a column-wise fashion, in the order that your list supplies them as members.

This discussion on object classes, data types, and coercion is not exhaustive, but it serves as a useful introduction to how R deals with issues surrounding the formal identification, description, and handling of the objects you create—issues that are present for most high-level languages. Once you're more familiar with R, the help files (such as the one accessed by entering ?as at the prompt) provide further details about object handling in the software.

---

### Exercise 6.3

a. Identify the class of the following objects. For each object, also state whether the class is explicitly or implicitly defined.

   i.   `foo <- array(data=1:36,dim=c(3,3,4))`

   ii.  `bar <- as.vector(foo)`

   iii. `baz <- as.character(bar)`

   iv. `qux <- as.factor(baz)`

   v.  `quux <- bar+c(-0.1,0.1)`

b. For each object defined in (a), find the sum of the result of calling is.numeric and is.integer on it separately. For example, is.numeric(foo)+is.integer(foo) would compute the sum for (i). Turn the collection of five results into a factor with levels 0, 1, and 2, identified by the results themselves. Compare this factor vector with the result of coercing it to a numeric vector.

c. Turn the following:

```
    [,1] [,2] [,3] [,4]
[1,]   2    5    8   11
[2,]   3    6    9   12
[3,]   4    7   10   13
```

into the following:

```
[1] "2"  "5"  "8"  "11" "3"  "6"  "9"  "12" "4"  "7"  "10" "13"
```

d. Store the following matrix:

$$\begin{bmatrix} 34 & 0 & 1 \\ 23 & 1 & 2 \\ 33 & 1 & 1 \\ 42 & 0 & 1 \\ 41 & 0 & 2 \end{bmatrix}$$

Then, do the following:

i. Coerce the matrix to a data frame.
ii. As a data frame, coerce the second column to be logical-valued.
iii. As a data frame, coerce the third column to be factor-valued.

## Important Code in This Chapter

| Function/operator | Brief description | First occurrence |
|---|---|---|
| Inf, -Inf | Value for ±infinity | Section 6.1.1, p. 104 |
| is.infinite | Element-wise check for Inf | Section 6.1.1, p. 105 |
| is.finite | Element-wise check for finiteness | Section 6.1.1, p. 105 |
| NaN | Value for invalid numerics | Section 6.1.2, p. 106 |
| is.nan | Element-wise check for NaN | Section 6.1.2, p. 107 |
| NA | Value for missing observation | Section 6.1.3, p. 108 |
| is.na | Element-wise check for NA OR NaN | Section 6.1.3, p. 109 |
| na.omit | Delete all NAs and NaNs | Section 6.1.3, p. 110 |
| NULL | Value for "empty" | Section 6.1.4, p. 110 |
| is.null | Check for NULL | Section 6.1.4, p. 111 |
| attributes | List explicit attributes | Section 6.2.1, p. 114 |
| attr | Obtain specific attribute | Section 6.2.1, p. 115 |
| dimnames | Get array dimension names | Section 6.2.1, p. 116 |
| class | Get object class (S3) | Section 6.2.2, p. 117 |
| is._ | Object-checking functions | Section 6.2.3, p. 120 |
| as._ | Object-coercion functions | Section 6.2.4, p. 121 |

# 7

## BASIC PLOTTING

One particularly popular feature of R is its incredibly flexible plotting tools for data and model visualization. This is what draws many to R in the first place. Mastering R's graphical functionality does require practice, but the fundamental concepts are straightforward. In this chapter, I'll provide an overview of the plot function and some useful options for controlling the appearance of the final graph. Then I'll cover the basics of using ggplot2, a powerful library for visualizing data in R. This chapter will cover just the basics of plotting, and then you'll learn more about creating different types of statistical plots in Chapter 14, and about advanced plotting techniques in Part V.

## 7.1   Using plot with Coordinate Vectors

The easiest way to think about generating plots in R is to treat your screen as a blank, two-dimensional canvas. You can plot points and lines using $x$- and $y$-coordinates. On paper, these coordinates are usually represented with points written as a pair: ($x$ value, $y$ value). The R function plot, on the other hand, takes in two vectors—one vector of $x$ locations and one vector of $y$ locations—and opens a *graphics device* where it displays the result. If a graphics device is already open, R's default behavior is to refresh the device, overwriting the current contents with the new plot.

For example, let's say you wanted to plot the points $(1.1, 2)$, $(2, 2.2)$, $(3.5, -1.3)$, $(3.9, 0)$, and $(4.2, 0.2)$. In plot, you must provide the vector of $x$ locations first, and the $y$ locations second. Let's define these as foo and bar, respectively:

```
R> foo <- c(1.1,2,3.5,3.9,4.2)
R> bar <- c(2,2.2,-1.3,0,0.2)
R> plot(foo,bar)
```

Figure 7-1 shows the resulting graphics device with the plot (I'll use this simple data set as a working example throughout this section).

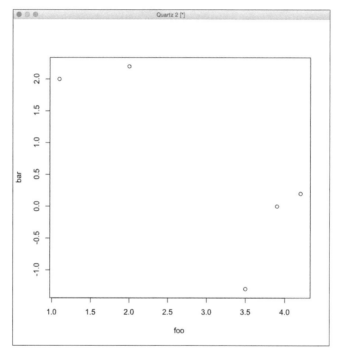

Figure 7-1: The five plotted points using R's default behavior

The $x$ and $y$ locations don't necessarily need to be specified as separate vectors. You can also supply coordinates in the form of a matrix, with the $x$ values in the first column and the $y$ values in the second column, or as a list. For example, setting up a matrix of the five points, the following code exactly reproduces Figure 7-1 (note the window pane will look slightly different depending on your operating system):

```
R> baz <- cbind(foo,bar)
R> baz
     foo  bar
```

```
[1,] 1.1  2.0
[2,] 2.0  2.2
[3,] 3.5 -1.3
[4,] 3.9  0.0
[5,] 4.2  0.2
R> plot(baz)
```

The `plot` function is one of R's versatile *generic* functions. It works differently for different objects and allows users to define their own methods for handling objects (including user-defined object classes). Technically, the version of the `plot` command that you've just used is internally identified as `plot.default`. The help file `?plot.default` provides additional details on this *scatterplot* style of data visualization.

## 7.2 Graphical Parameters

There are a wide range of *graphical parameters* that can be supplied as arguments to the `plot` function (or other plotting functions, such as those in Section 7.3). These parameters invoke simple visual enhancements, like coloring the points and adding axis labels, and can also control technical aspects of the graphics device (Chapter 23 covers the latter in more detail). Some of the most commonly used graphical parameters are listed here; I'll briefly discuss each of these in turn in the following sections:

**type**   Tells R how to plot the supplied coordinates (for example, as stand-alone points or joined by lines or both dots and lines).

**main, xlab, ylab**   Options to include plot title, the horizontal axis label, and the vertical axis label, respectively.

**col**   Color (or colors) to use for plotting points and lines.

**pch**   Stands for *point character*. This selects which character to use for plotting individual points.

**cex**   Stands for *character expansion*. This controls the size of plotted point characters.

**lty**   Stands for *line type*. This specifies the type of line to use to connect the points (for example, solid, dotted, or dashed).

**lwd**   Stands for *line width*. This controls the thickness of plotted lines.

**xlim, ylim**   This provides limits for the horizontal range and vertical range (respectively) of the plotting region.

### 7.2.1  Automatic Plot Types

By default, the `plot` function will plot individual points, as shown in Figure 7-1. This is the default plot type, but other plot types will have a different appearance. To control the plot type, you can specify a single character-valued option for the argument `type`.

For example, in many cases it makes sense to show lines connecting each coordinate, such as when plotting time series data. For this, you would specify plot type "l". Using foo and bar from Section 7.1, the following produces the plot in the left panel of Figure 7-2:

```
R> plot(foo,bar,type="l")
```

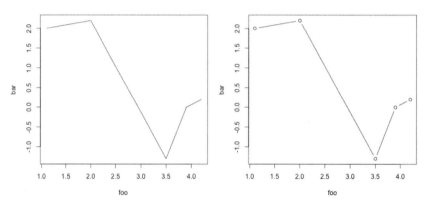

*Figure 7-2: A line plot produced using five adjoined coordinates, setting* type="l" *(left) or* type="b" *(right)*

The default value for type is "p", which can be interpreted as "points only." Since you didn't specify anything different, this is what was used for the graph in Figure 7-1. In this last example, on the other hand, you've set type="l" (meaning "lines only"). Other options include "b" for both points *and* lines (shown in the right panel of Figure 7-2) and "o" for overplotting the points with lines (this eliminates the gaps between points and lines visible for type="b"). The option type="n" results in no points or lines plotted, creating an empty plot, which can be useful for complicated plots that must be constructed in steps.

### 7.2.2   Title and Axis Labels

By default, a basic plot won't have a main title, and its axes will be labeled with the names of the vectors being plotted. But a main title and more descriptive axis labels often make the plotted data easier to interpret. You can add these by supplying text as character strings to main for a title, xlab for the *x*-axis label, and ylab for the *y*-axis label. Note that these strings may include escape sequences (discussed in Section 4.2.3). The following code produces the plots in Figure 7-3:

```
R> plot(foo,bar,type="b",main="My lovely plot",xlab="x axis label",
        ylab="location y")
R> plot(foo,bar,type="b",main="My lovely plot\ntitle on two lines",xlab="",
        ylab="")
```

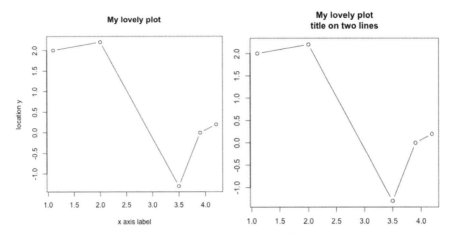

Figure 7-3: Two examples of plots with axis labels and titles

In the second plot, note how the new line escape sequence splits the title into two lines. In that plot, xlab and ylab are also set to the empty string "" to prevent R from labeling the axes with the names of the *x* and *y* vectors.

### 7.2.3   Color

Adding color to a graph is far from just an aesthetic consideration. Color can make data much clearer—for example by distinguishing factor levels or emphasizing important numeric limits. You can set colors with the col parameter in a number of ways. The simplest options are to use an integer selector or a character string. There are a number of color string values recognized by R, which you can see by entering colors() at the prompt. The default color is integer 1 or the character string "black". The top row of Figure 7-4 shows two examples of colored graphs, created by the following code:

```
R> plot(foo,bar,type="b",main="My lovely plot",xlab="",ylab="",col=2)
R> plot(foo,bar,type="b",main="My lovely plot",xlab="",ylab="",col="seagreen4")
```

There are eight possible integer values (shown in the leftmost plot of Figure 7-5) and around 650 character strings to specify color. But you aren't limited to these options since you can also specify colors using RGB (red, green, and blue) levels and by creating your own palettes. I'll talk more about the last two options in Chapter 25.

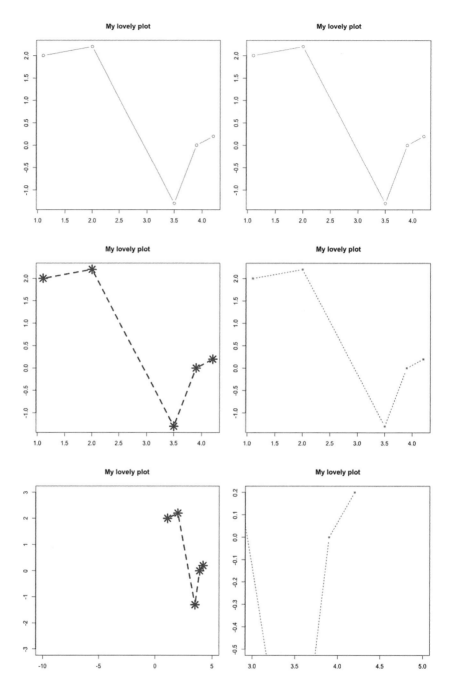

Figure 7-4: Experimenting with basic R plotting. Top row: Two examples of colored plots with col=2 (left) and col="seagreen4" (right). Middle row: Two further examples making use of pch, lty, cex, and lwd. Bottom row: Setting plotting region limits xlim=c(-10,5), ylim=c(-3,3) (left), and xlim=c(3,5), ylim=c(-0.5,0.2) (right).

## 7.2.4 Line and Point Appearances

To alter the appearance of the plotted points you would use pch, and to alter the lines you would use lty. The pch parameter controls the character used to plot individual data points. You can specify a single character to use for each point, or you can specify a value between 1 and 25 (inclusive). The symbols corresponding to each integer are shown in the middle plot of Figure 7-5. The lty parameter, which affects the type of line drawn, can take the values 1 through 6. These options are shown in the rightmost plot of Figure 7-5.

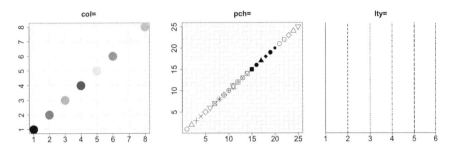

Figure 7-5: Some reference plots giving the results of possible integer options of col (left), pch (middle), and lty (right)

You can also control the size of plotted points using cex and the thickness of lines using lwd. The default size and thickness for both of these is 1. To request half-size points, for example, you'd specify cex=0.5; to specify double-thick lines, use lwd=2.

The following two lines produce the two plots in the middle row of Figure 7-4, showing off pch, lty, cex, and lwd:

```
R> plot(foo,bar,type="b",main="My lovely plot",xlab="",ylab="",
        col=4,pch=8,lty=2,cex=2.3,lwd=3.3)
R> plot(foo,bar,type="b",main="My lovely plot",xlab="",ylab="",
        col=6,pch=15,lty=3,cex=0.7,lwd=2)
```

## 7.2.5 Plotting Region Limits

As you can see in the plots of foo and bar, by default R sets the range of each axis by using the range of the supplied *x* and *y* values (plus a small constant to pad a little area around the outermost points). But you might need more space than this to, for example, annotate individual points, add a legend, or plot additional points that fall outside the original ranges (as you'll see in Section 7.3). You can set custom plotting area limits using xlim and ylim. Both parameters require a numeric vector of length 2, provided as c(*lower,upper*).

Consider the plots in the bottom row of Figure 7-4, created with the following two commands:

```
R> plot(foo,bar,type="b",main="My lovely plot",xlab="",ylab="",
        col=4,pch=8,lty=2,cex=2.3,lwd=3.3,xlim=c(-10,5),ylim=c(-3,3))
R> plot(foo,bar,type="b",main="My lovely plot",xlab="",ylab="",
        col=6,pch=15,lty=3,cex=0.7,lwd=2,xlim=c(3,5),ylim=c(-0.5,0.2))
```

These plots are exactly the same as the two in the middle row, except for one important difference. In the bottom-left plot of Figure 7-4, the *x*- and *y*-axes are set to be much wider than the observed data, and the plot on the right restricts the plotting window so that only a portion of the data is displayed.

## 7.3   Adding Points, Lines, and Text to an Existing Plot

Generally speaking, each call to plot will refresh the active graphics device for a new plotting region. But this is not always desired—to build more complicated plots, it's easiest to start with an empty plotting region and progressively add any required points, lines, text, and legends to this canvas. Here are some useful, ready-to-use functions in R that will add to a plot without refreshing or clearing the window:

**points**   Adds points

**lines, abline, segments**   Adds lines

**text**   Writes text

**arrows**   Adds arrows

**legend**   Adds a legend

The syntax for calling and setting parameters for these functions is the same as plot. The best way to see how these work is through an extended example, which I'll base on some hypothetical data made up of 20 $(x, y)$ locations.

```
R> x <- 1:20
R> y <- c(-1.49,3.37,2.59,-2.78,-3.94,-0.92,6.43,8.51,3.41,-8.23,
        -12.01,-6.58,2.87,14.12,9.63,-4.58,-14.78,-11.67,1.17,15.62)
```

Using these data, you'll build up the plot shown in Figure 7-6 (note that you may need to manually enlarge your graphics device and replot to ensure the legend doesn't overlap other features of the image). It's worth remembering a generally accepted rule in plotting: "keep it clear and simple." Figure 7-6 is an exception for the sake of demonstrating the R commands used.

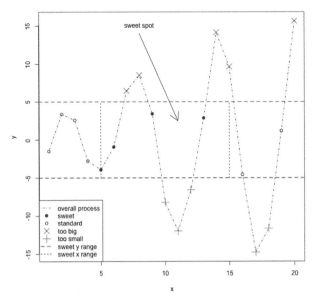

*Figure 7-6: An elaborate final plot of some hypothetical data*

In Figure 7-6, the data points will be plotted differently according to their *x* and *y* locations, depending on their relation to the "sweet spot" pointed out in the figure. Points with a *y* value greater than 5 are marked with a purple ×; points with a *y* value less than −5 are marked with a green +. Points between these two *y* values but still outside of the sweet spot are marked with a ○. Finally, points in the sweet spot (with *x* between 5 and 15 *and* with *y* between −5 and 5) are marked as a blue ●. Red horizontal and vertical lines delineate the sweet spot, which is labeled with an arrow, and there's also a legend.

Ten lines of code were used to build this plot in its entirety (plus one additional line to add the legend). The plot, as it looks at each step, is given in Figure 7-7. The lines of code are detailed next.

1.  The first step is to create the empty plotting region where you can add points and draw lines. This first line tells R to plot the data in x and y, though the option type is set to "n". As mentioned in Section 7.2, this opens or refreshes the graphics device and sets the axes to the appropriate lengths (with labels and axes), but it doesn't plot any points or lines.

    ```
    R> plot(x,y,type="n",main="")
    ```

2.  The abline function is a simple way to add straight lines spanning a plot. The line (or lines) can be specified with *slope* and *intercept* values (see the later discussions on regression in Chapter 20). You can also simply add horizontal or vertical lines. This line of code adds two separate horizontal lines, one at *y* = 5 and the other at *y* = 5, using h=c(-5,5). The three parameters (covered in Section 7.2) make these

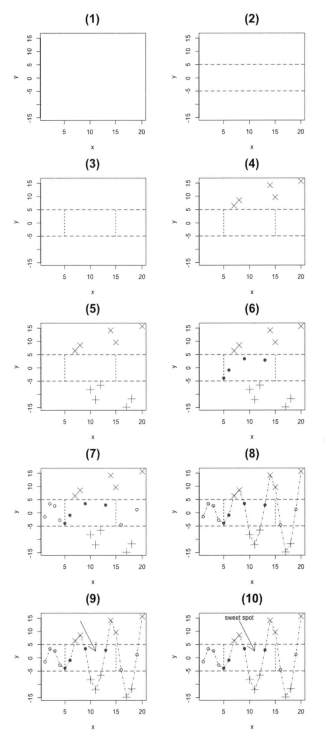

Figure 7-7: Building the final plot given in Figure 7-6. The plots (1) through (10) correspond to the itemized lines of code in the text.

two lines red, dashed, and double-thickness. For vertical lines, you could have written v=c(-5,5), which would have drawn them at $x = -5$ and $x = 5$.

```
R> abline(h=c(-5,5),col="red",lty=2,lwd=2)
```

3. The third line of code adds shorter vertical lines between the horizontal ones drawn in step 2 to form a box. For this you use segments, not abline, since you don't want these lines to span the entire plotting region. The segments command takes a "from" coordinate (given as x0 and y0) and a "to" coordinate (as x1 and y1) and draws the corresponding line. The vector-oriented behavior of R matches up the two sets of "from" and "to" coordinates. Both lines are red and dotted and have double-thickness. (You could also supply vectors of length 2 to these parameters, in which case the first segment would use the first parameter value and the second segment would use the second value.)

```
R> segments(x0=c(5,15),y0=c(-5,-5),x1=c(5,15),y1=c(5,5),col="red",lty=3,
       lwd=2)
```

4. As step 4, you use points to begin adding specific coordinates from x and y to the plot. Just like plot, points takes two vectors of equal lengths with $x$ and $y$ values. In this case, you want points plotted differently according to their location, so you use logical vector subsetting (see Section 4.1.5) to identify and extract elements of x and y where the $y$ value is greater than or equal to 5. These (and only these) points are added as purple × symbols and are enlarged by a factor of 2 with cex.

```
R> points(x[y>=5],y[y>=5],pch=4,col="darkmagenta",cex=2)
```

5. The fifth line of code is much like the fourth; this time it extracts the coordinates where $y$ values are less than or equal to −5. A + point character is used, and you set the color to dark green.

```
R> points(x[y<=-5],y[y<=-5],pch=3,col="darkgreen",cex=2)
```

6. The sixth step adds the blue "sweet spot" points, which are identified with (x>=5&x<=15)&(y>-5&y<5). This slightly more complicated set of conditions extracts the points whose $x$ location lies between 5 and 15 (inclusive) AND whose $y$ location lies between −5 and 5 (exclusive). Note that this line uses the "short" form of the logical operator & throughout since you want element-wise comparisons here (see Section 4.1.3).

```
R> points(x[(x>=5&x<=15)&(y>-5&y<5)],y[(x>=5&x<=15)&(y>-5&y<5)],pch=19,
       col="blue")
```

7. This next command identifies the remaining points in the data set (with an $x$ value that is either less than 5 OR greater than 15 AND a $y$ value between −5 and 5). No graphical parameters are specified, so these points are plotted with the default black ∘.

```
R> points(x[(x<5|x>15)&(y>-5&y<5)],y[(x<5|x>15)&(y>-5&y<5)])
```

8. To draw lines connecting the coordinates in x and y, you use lines. Here you've also set lty to 4, which draws a dash-dot-dash style line.

```
R> lines(x,y,lty=4)
```

9. The ninth line of code adds the arrow pointing to the sweet spot. The function arrows is used just like segments, where you provide a "from" coordinate (x0, y0) and a "to" coordinate (x1, y1). By default, the head of the arrow is located at the "to" coordinate, though this (and other options such as the angle and length of the head) can be altered using optional arguments described in the help file ?arrows.

```
R> arrows(x0=8,y0=14,x1=11,y1=2.5)
```

10. The tenth line prints a label on the plot at the top of the arrow. As per the default behavior of text, the string supplied as labels is *centered* on the coordinates provided with the arguments x and y.

```
R> text(x=8,y=15,labels="sweet spot")
```

As a finishing touch, you can add the legend with the legend function, which gives you the final product shown in Figure 7-6.

```
legend("bottomleft",
       legend=c("overall process","sweet","standard",
                "too big","too small","sweet y range","sweet x range"),
       pch=c(NA,19,1,4,3,NA,NA),lty=c(4,NA,NA,NA,NA,2,3),
       col=c("black","blue","black","darkmagenta","darkgreen","red","red"),
       lwd=c(1,NA,NA,NA,NA,2,2),pt.cex=c(NA,1,1,2,2,NA,NA))
```

The first argument sets where the legend should be placed. There are various ways to do this (including setting exact *x*- and *y*-coordinates), but it often suffices to pick a corner using one of the four following character strings: "topleft", "topright", "bottomleft", or "bottomright". Next you supply the labels as a vector of character strings to the legend argument. Then you need to supply the remaining argument values in vectors of the same length so that the right elements match up with each label.

For example, for the first label ("overall process"), you want a line of type 4 with default thickness and color. So, in the first positions of the remaining argument vectors, you set pch=NA, lty=4, col="black", lwd=1, and pt.cex=NA (all of these are default values, except for lty). Here, pt.cex simply refers to the cex parameter when calling points (using just cex in legend would expand the text used, not the points).

Note that you have to fill in some elements in these vectors with NA when you don't want to set the corresponding graphical parameter. This is just to preserve the equal lengths of the vectors supplied so R can track which parameter values correspond to each particular reference. As you work through this book, you'll see plenty more examples using legend.

## Exercise 7.1

a. As closely as you can, re-create the following plot:

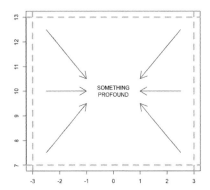

b. With the following data, create a plot of weight on the *x*-axis and height on the *y*-axis. Use different point characters or colors to distinguish between males and females and provide a matching legend. Label the axes and give the plot a title.

| Weight (kg) | Height (cm) | Sex |
|---|---|---|
| 55 | 161 | female |
| 85 | 185 | male |
| 75 | 174 | male |
| 42 | 154 | female |
| 93 | 188 | male |
| 63 | 178 | male |
| 58 | 170 | female |
| 75 | 167 | male |
| 89 | 181 | male |
| 67 | 178 | female |

## 7.4 The ggplot2 Package

This chapter so far has shown off R's built-in graphical tools (often called *base R graphics* or *traditional R graphics*). Now, let's look at another important suite of graphical tools: ggplot2, a prominent contributed package by Hadley Wickham (2009). Available on CRAN like any other contributed package, ggplot2 offers particularly powerful alternatives to the standard plotting procedures in R. The *gg* stands for *grammar of graphics*—a particular approach to graphical production described by Wilkinson (2005). In following this approach, ggplot2 standardizes the production of different plot and graph types, streamlines some of the more fiddly aspects of adding to existing plots (such as including a legend), and lets you build plots by defining and manipulating *layers*. For the moment, let's see the elementary behavior of ggplot2 using the same simple examples in Sections 7.1–7.3. You'll get familiar with the basic plotting function qplot and how it differs from the generic plot function used earlier. I'll return to the topic of ggplot2 when I cover statistical plots in Chapter 14, and you'll investigate even more advanced abilities in Chapter 24.

### 7.4.1  A Quick Plot with qplot

First, you must install the ggplot2 package by downloading it manually or simply entering install.packages("ggplot2") at the prompt (see Section A.2.3). Then, load the package with the following:

```
R> library("ggplot2")
```

Now, let's go back to the five data points originally stored in Section 7.1 as foo and bar.

```
R> foo <- c(1.1,2,3.5,3.9,4.2)
R> bar <- c(2,2.2,-1.3,0,0.2)
```

You can produce ggplot2's version of Figure 7-1 using its "quick plot" function qplot.

```
R> qplot(foo,bar)
```

The result is shown in the left image of Figure 7-8. There are some obvious differences between this image and the one produced using plot, but the basic syntax of qplot is the same as earlier. The first two arguments passed to qplot are vectors of equal length, with the *x*-coordinates in foo supplied first, followed by the *y*-coordinates in bar.

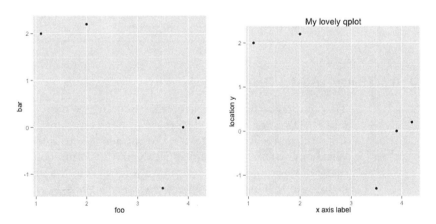

Figure 7-8: Five plotted points using ggplot2's default behavior for the qplot function (left) and with title and axis labels added (right)

Adding a title and axis labels also uses the same arguments you already saw with plot in Section 7.2.

```
R> qplot(foo,bar,main="My lovely qplot",xlab="x axis label",ylab="location y")
```

This produces the right panel of Figure 7-8.

Underneath this basic similarity in syntax, though, there is a fundamental difference between how ggplot2 and base R graphics create plots. Constructing plots using the built-in graphics tools is essentially a live, step-by-step process. This was particularly noticeable in Section 7.3, where you treated the graphics device as an active canvas where you added points, lines, and other features one by one. By contrast, ggplot2 plots are stored as objects, which means they have an underlying, static representation until you *change* the object—what you essentially visualize with qplot is the printed object at any given time. To highlight this, enter the following code:

```
R> baz <- plot(foo,bar)
R> baz
NULL
R> qux <- qplot(foo,bar)
R> qux
```

The first assignment uses the built-in plot function. When you run that line of code, the plot in Figure 7-1 pops up. Since nothing is actually stored in the workspace, printing the supposed object baz yields the empty NULL value. On the other hand, it makes sense to store the qplot content (stored as the object qux here). This time, when you perform the assignment, no plot is displayed. The graphic, which matches Figure 7-8, is displayed only when you enter qux at the prompt, which invokes the print method for that object. This may seem like a minor point, but the fact that you can save a plot this way before displaying it opens up new ways to modify or enhance plots before displaying them (as you will see in a moment), and it can be a distinct advantage over base R graphics.

## 7.4.2 Setting Appearance Constants with Geoms

To add and customize points and lines in a ggplot2 graphic, you alter the object itself, rather than using a long list of arguments or secondary functions executed separately (such as points or lines). You can modify the object using ggplot2's convenient suite of *geometric modifiers*, known as *geoms*. Let's say you want to connect the five points in foo and bar with a line, just as you did in Section 7.1. You can first create a blank plot object and then use geometric modifiers on it like this:

```
R> qplot(foo,bar,geom="blank") + geom_point() + geom_line()
```

The resulting plot is shown on the left of Figure 7-9. In the first call to qplot, you create an empty plot object by setting the initial geometric modifier as geom="blank" (if you displayed this plot, you would just see the gray background and the axes). Then you layer on the two other geoms as geom_point() and geom_line(). As indicated by the parentheses, these geoms are functions that result in their own specialized objects. You can add geoms to the qplot object using the + operator. Here, you haven't supplied any arguments to either geom, which means they'll operate on the same data originally supplied to qplot (foo and bar) and they'll stick to the default settings for any other features, such as color or point/line type. You can control those features by specifying optional arguments, as shown here:

```
R> qplot(foo,bar,geom="blank") + geom_point(size=3,shape=6,color="blue") +
        geom_line(color="red",linetype=2)
```

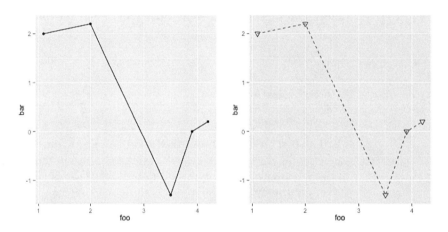

Figure 7-9: Two simple plots that use geometric modifiers to alter the appearance of a qplot object. Left: Adding points and lines using default settings. Right: Using the geoms to affect point character, size, and color, and line type and color.

Note that some of ggplot2's argument names used here for things such as point characters and size (shape and size) are different from the base R graphics arguments (pch and cex). But ggplot2 is actually compatible with many of the common graphical parameters used in R's standard plot function, so you can use those arguments here too if you prefer. For instance, setting cex=3 and pch=6 in geom_point in this example would result in the same image.

The object-oriented nature of ggplot2 graphics means tweaking a plot or experimenting with different visual features no longer requires you to rerun every plotting command each time you change something. This is facilitated by geoms. Say you like the line type used on the right side of Figure 7-9 but want a different point character. To experiment, you could first store the qplot object you created earlier and then use geom_point with that object to try different point styles.

```
R> myqplot <- qplot(foo,bar,geom="blank") + geom_line(color="red",linetype=2)
R> myqplot + geom_point(size=3,shape=3,color="blue")
R> myqplot + geom_point(size=3,shape=7,color="blue")
```

The first line stores the original plot in myqplot, and the next lines call myqplot with different point shapes. The second and third lines produce the graphics on the left and right of Figure 7-10, respectively.

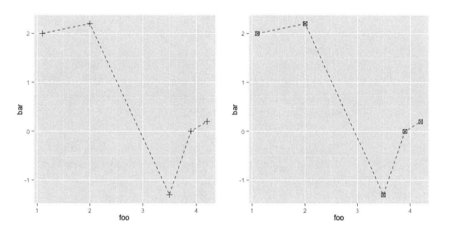

Figure 7-10: Using the object-oriented nature of ggplot2 graphics to experiment with different point characters

There are a number of geometric modifiers that can be called using a function name beginning with geom_ in ggplot2. To obtain a list, simply ensure the package is loaded and enter ??"geom_" as a help search at the prompt.

### 7.4.3   Aesthetic Mapping with Geoms

Geoms and ggplot2 also provide efficient, automated ways to apply different styles to different subsets of a plot. If you split a data set into categories using a factor object, ggplot2 can automatically apply particular styles to different categories. In ggplot2's documentation, the factor that holds these categories is called a *variable*, which ggplot2 can *map* to *aesthetic* values. This gets rid of much of the effort that goes into isolating subsets of data and plotting them separately using base R graphics (as you did in Section 7.3).

All this is best illustrated with an example. Let's return to the 20 observations you manually plotted, step-by-step, to produce the elaborate plot in Figure 7-6.

```
R> x <- 1:20
R> y <- c(-1.49,3.37,2.59,-2.78,-3.94,-0.92,6.43,8.51,3.41,-8.23,
        -12.01,-6.58,2.87,14.12,9.63,-4.58,-14.78,-11.67,1.17,15.62)
```

In Section 7.3, you defined several categories that classified each observation as either "standard," "sweet," "too big," or "too small" based on their x and y values. Using those same classification rules, let's explicitly define a factor to correspond to x and y.

```
R> ptype <- rep(NA,length(x=x))
R> ptype[y>=5] <- "too_big"
R> ptype[y<=-5] <- "too_small"
R> ptype[(x>=5&x<=15)&(y>-5&y<5)] <- "sweet"
R> ptype[(x<5|x>15)&(y>-5&y<5)] <- "standard"
R> ptype <- factor(x=ptype)
R> ptype
 [1] standard  standard  standard  standard  sweet      sweet      too_big
 [8] too_big   sweet      too_small too_small too_small sweet      too_big
[15] too_big   standard  too_small too_small standard  too_big
Levels: standard sweet too_big too_small
```

Now you have a factor with 20 values sorted into four levels. You'll use this factor to tell qplot how to map your aesthetics. Here's a simple way to do that:

```
R> qplot(x,y,color=ptype,shape=ptype)
```

This single line of code produces the left plot in Figure 7-11, which separates the four categories by color and point character and even provides a legend. This was all done by the aesthetic mapping in the call to qplot, where you set color and shape to be mapped to the ptype variable.

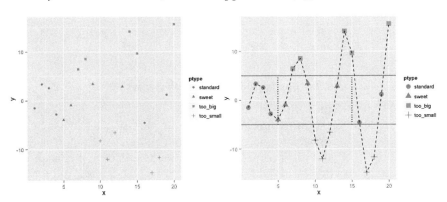

Figure 7-11: Demonstration of aesthetic mapping using qplot and geoms in ggplot2. Left: The initial call to qplot, which maps point character and color using ptype. Right: Augmenting the left plot using various geoms to override the default mappings.

Now, let's replot these data using the same `qplot` object along with a suite of geom modifications in order to get something more like Figure 7-6. Executing the following produces the plot on the right of Figure 7-11:

```
R> qplot(x,y,color=ptype,shape=ptype) + geom_point(size=4) +
      geom_line(mapping=aes(group=1),color="black",lty=2) +
      geom_hline(mapping=aes(yintercept=c(-5,5)),color="red") +
      geom_segment(mapping=aes(x=5,y=-5,xend=5,yend=5),color="red",lty=3) +
      geom_segment(mapping=aes(x=15,y=-5,xend=15,yend=5),color="red",lty=3)
```

In the first line, you add `geom_point(size=4)` to increase the size of all the points on the graph. In the lines that follow, you add a line connecting all the points, plus horizontal and vertical lines to mark out the sweet spot. For those last four lines, you have to use aes to set alternate aesthetic mappings for the point categories. Let's look a little closer at what's going on there.

Since you used `ptype` for aesthetic mapping in the initial call to `qplot`, by default all other geoms will be mapped to each category in the same way, *unless* you override that default mapping with aes. For example, when you call `geom_line` to connect all the points, if you were to stick with the default mapping to `ptype` instead of including `mapping=aes(group=1)`, this geom would draw lines connecting points within each category. You would see four separate dashed lines—one connecting all "standard" points, another connecting all "sweet" points, and so on. But that's not what you want here; you want a line that connects all of the points, from left to right. So, you tell `geom_line` to treat all the observations as one group by entering `aes(group=1)`.

After that, you use the `geom_hline` function to draw horizontal lines at $y = -5$ and $y = 5$ using its `yintercept` argument, again passed to aes to redefine that geom's `mapping`. In this case, you need to redefine the mapping to operate on the vector `c(-5,5)`, rather than using the observed data in x and y. Similarly, you end by using `geom_segment` to draw the two vertical dotted line segments. `geom_segment` operates much like `segments`—you redefine the mapping based on a "from" coordinate (arguments x and y) and a "to" coordinate (xend and yend here). Since the first geom, `geom_point(size=4)`, sets a constant enlarged size for every plotted point, it doesn't matter how the geom is mapped because it simply makes a uniform change to each point.

Plotting in R, from base graphics to contributed packages like `ggplot2`, stays true to the nature of the language. The element-wise matching allows you to create intricate plots with a handful of straightforward and intuitive functions. Once you display a plot, you can save it to the hard drive by selecting the graphics device and choosing File → Save. However, you can also write plots to a file directly, as you'll see momentarily in Section 8.3.

The graphical capabilities explored in this section are merely the tip of the iceberg, and you'll continue to use data visualizations from this point onward.

## Exercise 7.2

In Exercise 7.1 (b), you used base R graphics to plot some weight and height data, distinguishing males and females using different points or colors. Repeat this task using ggplot2.

## Important Code in This Chapter

| Function/operator | Brief description | First occurrence |
|---|---|---|
| plot | Create/display base R plot | Section 7.1, p. 128 |
| type | Set plot type | Section 7.2.1, p. 130 |
| main, xlab, ylab | Set axis labels | Section 7.2.2, p. 130 |
| col | Set point/line color | Section 7.2.3, p. 131 |
| pch, cex | Set point type/size | Section 7.2.4, p. 133 |
| lty, lwd | Set line type/width | Section 7.2.4, p. 133 |
| xlim, ylim | Set plot region limits | Section 7.2.5, p. 134 |
| abline | Add vertical/horizontal line | Section 7.3, p. 137 |
| segments | Add specific line segments | Section 7.3, p. 137 |
| points | Add points | Section 7.3, p. 137 |
| lines | Add lines following coords | Section 7.3, p. 138 |
| arrows | Add arrows | Section 7.3, p. 138 |
| text | Add text | Section 7.3, p. 138 |
| legend | Add/control legend | Section 7.3, p. 138 |
| qplot | Create ggplot2 "quick plot" | Section 7.4.1, p. 140 |
| geom_point | Add points geom | Section 7.4.2, p. 141 |
| geom_line | Add lines geom | Section 7.4.2, p. 141 |
| size, shape, color | Set geom constants | Section 7.4.2, p. 142 |
| linetype | Set geom line type | Section 7.4.2, p. 142 |
| mapping, aes | Geom aesthetic mapping | Section 7.4.3, p. 145 |
| geom_hline | Add horizontal lines geom | Section 7.4.3, p. 145 |
| geom_segment | Add line segments geom | Section 7.4.3, p. 145 |

# 8

## READING AND WRITING FILES

Now I'll cover one more fundamental aspect of working with R: loading and saving data in an active workspace by reading and writing files. Typically, to work with a large data set, you'll need to read in the data from an external file, whether it's stored as plain text, in a spreadsheet file, or on a website. R provides command line functions you can use to import these data sets, usually as a data frame object. You can also export data frames from R by writing a new file on your computer, plus you can save any plots you create as image files. In this chapter, I'll go over some useful command-based read and write operations for importing and exporting data.

## 8.1   R-Ready Data Sets

First, let's take a brief look at some of the data sets that are built into the software or are part of user-contributed packages. These data sets are useful samples to practice with and to experiment with functionality.

Enter data() at the prompt to bring up a window listing these ready-to-use data sets along with a one-line description. These data sets are organized in alphabetical order by name and grouped by package (the exact list that appears will depend on what contributed packages have been installed from CRAN; see Section A.2).

## 8.1.1   Built-in Data Sets

There are a number of data sets contained within the built-in, automatically loaded package datasets. To see a summary of the data sets contained in the package, you can use the library function as follows:

```
R> library(help="datasets")
```

R-ready data sets have a corresponding help file where you can find important details about the data and how it's organized. For example, one of the built-in data sets is named ChickWeight. If you enter ?ChickWeight at the prompt, you'll see the window in Figure 8-1.

Figure 8-1: The help file for the ChickWeight data set

As you can see, this file explains the variables and their values; it notes that the data are stored in a data frame with 578 rows and 4 columns. Since the objects in datasets are built in, all you have to do to access ChickWeight is enter its name at the prompt. Let's look at the first 15 records.

```
R> ChickWeight[1:15,]
   weight Time Chick Diet
1      42    0     1    1
2      51    2     1    1
3      59    4     1    1
4      64    6     1    1
5      76    8     1    1
```

| | | | | |
|---|---|---|---|---|
| 6 | 93 | 10 | 1 | 1 |
| 7 | 106 | 12 | 1 | 1 |
| 8 | 125 | 14 | 1 | 1 |
| 9 | 149 | 16 | 1 | 1 |
| 10 | 171 | 18 | 1 | 1 |
| 11 | 199 | 20 | 1 | 1 |
| 12 | 205 | 21 | 1 | 1 |
| 13 | 40 | 0 | 2 | 1 |
| 14 | 49 | 2 | 2 | 1 |
| 15 | 58 | 4 | 2 | 1 |

You can treat this data set like any other data frame you've created in R—note the use of [1:15,] to access the desired rows from such an object, as detailed in Section 5.2.1.

## 8.1.2   Contributed Data Sets

There are many more R-ready data sets that come as part of contributed packages. To access them, first install and load the relevant package. Consider the data set ice.river, which is in the contributed package tseries by Trapletti and Hornik (2013). First, you have to install the package, which you can do by running the line install.packages("tseries") at the prompt. Then, to access the components of the package, load it using library:

```
R> library("tseries")

    'tseries' version: 0.10-32

    'tseries' is a package for time series analysis and computational finance.

    See 'library(help="tseries")' for details.
```

Now you can enter library(help="tseries") to see the list of data sets in this package, and you can enter ?ice.river to find more details about the data set you want to work with here. The help file describes ice.river as a "time series object" comprised of river flow, precipitation, and temperature measurements—data initially reported in Tong (1990). To access this object itself, you must explicitly load it using the data function. Then you can work with ice.river in your workspace as usual. Here are the first five records:

```
R> data(ice.river)
R> ice.river[1:5,]
      flow.vat flow.jok prec temp
 [1,]    16.10     30.2  8.1  0.9
 [2,]    19.20     29.0  4.4  1.6
 [3,]    14.50     28.4  7.0  0.1
 [4,]    11.00     27.8  0.0  0.6
 [5,]    13.60     27.8  0.0  2.0
```

The availability and convenience of these R-ready data sets make it easy to test code, and I'll use them in subsequent chapters for demonstrations. To analyze your own data, however, you'll often have to import them from some external file. Let's see how to do that.

## 8.2   Reading in External Data Files

R has a variety of functions for reading characters from stored files and making sense of them. You'll look at how to read *table-format* files, which are among the easiest for R to read and import.

### 8.2.1   The Table Format

Table-format files are best thought of as plain-text files with three key features that fully define how R should read the data.

**Header**   If a *header* is present, it's always the first line of the file. This optional feature is used to provide names for each column of data. When importing a file into R, you need to tell the software whether a header is present so that it knows whether to treat the first line as variable names or, alternatively, observed data values.

**Delimiter**   The all-important *delimiter* is a character used to separate the entries in each line. The delimiter character cannot be used for anything else in the file. This tells R when a specific entry begins and ends (in other words, its exact position in the table).

**Missing value**   This is another unique character string used exclusively to denote a missing value. When reading the file, R will turn these entries into the form it recognizes: NA.

Typically, these files have a *.txt* extension (highlighting the plain-text style) or *.csv* (for *comma-separated values*).

Let's try an example, using a variation on the data frame mydata as defined at the end of Section 5.2.2. Figure 8-2 shows an appropriate table-format file called *mydatafile.txt*, which has the data from that data frame with a few values now marked as missing. This data file can be found on the book's website at *https://www.nostarch.com/bookofr/*, or you can create it yourself from Figure 8-2 using a text editor.

Figure 8-2: A plain-text table-format file

Note that the first line is the header, the values are delimited with a single space, and missing values are denoted with an asterisk (∗). Also, note that each new record is required to start on a new line. Suppose you're handed this plain-text file for data analysis in R. The ready-to-use command read.table imports table-format files, producing a data frame object, as follows:

```
R> mydatafile <- read.table(file="/Users/tdavies/mydatafile.txt",
                            header=TRUE,sep=" ",na.strings="*",
                            stringsAsFactors=FALSE)
R> mydatafile
  person age sex funny age.mon
1  Peter  NA   M  High     504
2   Lois  40   F  <NA>     480
3    Meg  17   F   Low     204
4  Chris  14   M   Med     168
5 Stewie   1   M  High      NA
6  Brian  NA   M   Med      NA
```

In a call to read.table, file takes a character string with the filename and folder location (using forward slashes), header is a logical value telling R whether file has a header (TRUE in this case), sep takes a character string providing the delimiter (a single space, " ", in this case), and na.strings requests the characters used to denote missing values ("∗" in this case).

If you're reading in multiple files and don't want to type the entire folder location each time, it's possible to first set your working directory via setwd (Section 1.2.3) and then simply use the filename and its extension as the character string supplied to the file argument. However, both approaches require you to know exactly where your file is located when you're working at the R prompt. Fortunately, R possesses some useful additional tools should you forget your file's precise location. You can view textual output of the contents of any folder by using list.files. The following example betrays the messiness of my local user directory.

```
R> list.files("/Users/tdavies")
 [1] "bands-SCHIST1L200.txt" "Brass"               "Desktop"
 [4] "Documents"             "DOS Games"           "Downloads"
 [7] "Dropbox"               "Exercise2-20Data.txt" "Google Drive"
[10] "iCloud"                "Library"             "log.txt"
[13] "Movies"                "Music"               "mydatafile.txt"
[16] "OneDrive"              "peritonitis.sav"     "peritonitis.txt"
[19] "Personal9414"          "Pictures"            "Public"
[22] "Research"              "Rintro.tex"          "Rprofile.txt"
[25] "Rstartup.R"            "spreadsheetfile.csv" "spreadsheetfile.xlsx"
[28] "TakeHome_template.tex" "WISE-P2L"            "WISE-P2S.txt"
[31] "WISE-SCHIST1L200.txt"
```

One important feature to note here, though, is that it can be difficult to distinguish between files and folders. Files will typically have an extension, and folders won't; however, WISE-P2L is a file that happens to have no extension and looks no different from any of the listed folders.

You can also find files interactively from R. The file.choose command opens your filesystem viewer directly from the R prompt—just as any other program does when you want to open something. Then, you can navigate to the folder of interest, and after you select your file (see Figure 8-3), only a character string is returned.

```
R> file.choose()
[1] "/Users/tdavies/mydatafile.txt"
```

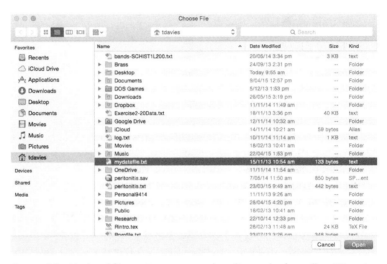

Figure 8-3: My local file navigator opened as the result of a call to file.choose. When the file of interest is opened, the R command returns the full file path to that file as a character string.

This command is particularly useful, as it returns the character string of the directory in precisely the format that's required for a command such as read.table. So, calling the following line and selecting *mydatafile.txt*, as in Figure 8-3, will produce an identical result to the explicit use of the file path in file, shown earlier:

```
R> mydatafile <- read.table(file=file.choose(),header=TRUE,sep=" ",
                            na.strings="*",stringsAsFactors=FALSE)
```

If your file has been successfully loaded, you should be returned to the R prompt without receiving any error messages. You can check this with a call to mydatafile, which should return the data frame. When importing data into data frames, keep in mind the difference between character string observations and factor observations. No factor attribute information is stored in the plain-text file, but `read.table` will convert non-numeric values into factors by default. Here, you want to keep some of your data saved as strings, so set `stringsAsFactors=FALSE`, which prevents R from treating all non-numeric elements as factors. This way, `person`, `sex`, and `funny` are all stored as character strings.

You can then overwrite `sex` and `funny` with factor versions of themselves if you want them as that data type.

```
R> mydatafile$sex <- as.factor(mydatafile$sex)
R> mydatafile$funny <- factor(x=mydatafile$funny,levels=c("Low","Med","High"))
```

### 8.2.2 Spreadsheet Workbooks

Next, let's examine some ubiquitous spreadsheet software file formats. The standard file format for Microsoft Office Excel is *.xls* or *.xlsx*. In general, these files are not directly compatible with R. There are some contributed package functions that attempt to bridge this gap—see, for example, gdata by Warnes et al. (2014) or XLConnect by Mirai Solutions GmbH (2014)—but it's generally preferable to first export the spreadsheet file to a table format, such as CSV. Consider the hypothetical data from Exercise 7.1 (b), which has been stored in an Excel file called *spreadsheetfile.xlsx*, shown in Figure 8-4.

Figure 8-4: A spreadsheet file of the data
from Exercise 7.1 (b)

To read this spreadsheet with R, you should first convert it to a table format. In Excel, File → Save As... provides a wealth of options. Save the spreadsheet as a comma-separated file, called *spreadsheet.csv*. R has a shortcut version of `read.table`, `read.csv`, for these files.

```
R> spread <- read.csv(file="/Users/tdavies/spreadsheetfile.csv",
                      header=FALSE,stringsAsFactors=TRUE)
R> spread
   V1  V2     V3
1  55 161 female
```

```
2  85 185   male
3  75 174   male
4  42 154 female
5  93 188   male
6  63 178   male
7  58 170 female
8  75 167   male
9  89 181   male
10 67 178 female
```

Here, the `file` argument again specifies the desired file, which has no header, so `header=FALSE`. You set `stringsAsFactors=TRUE` because you do want to treat the `sex` variable (the only non-numeric variable) as a factor. There are no missing values, so you don't need to specify `na.strings` (though if there were, this argument is simply used in the same way as earlier), and by definition, *.csv* files are comma-delimited, which `read.csv` correctly implements by default, so you don't need the `sep` argument. The resulting data frame, `spread`, can then be printed in your R console.

As you can see, reading tabular data into R is fairly straightforward—you just need to be aware of how the data file is headed and delimited and how missing entries are identified. The simple table format is a natural and common way for data sets to be stored, but if you need to read in a file with a more complicated structure, R and its contributed packages make available some more sophisticated functions. See, for example, the documentation for the `scan` and `readLines` functions, which provide advanced control over how to parse a file. You can also find documentation on `read.table` and `read.csv` by accessing `?read.table` from the prompt.

### 8.2.3  Web-Based Files

With an Internet connection, R can read in files from a website with the same `read.table` command. All the same rules concerning headers, delimiters, and missing values remain in place; you just have to specify the URL address of the file instead of a local folder location.

As an example, you'll use the online repository of data sets made available by the *Journal of Statistics Education (JSE)* through the American Statistical Association at *http://www.amstat.org/publications/jse/jse_data_archive.htm*.

One of the first files linked to at the top of this page is the table-format data set *4cdata.txt* (*http://www.amstat.org/publications/jse/v9n2/4cdata.txt*), which contains data on the characteristics of 308 diamonds from an analysis by Chu (2001) based on an advertisement in a Singaporean newspaper. Figure 8-5 shows the data.

You can look at the documentation file (*4c.txt*) and the accompanying article linked from the JSE site for details on what is recorded in this table. Note that of the five columns, the first and fifth are numeric, and the others would be well represented by factors. The delimiter is blank whitespace, there's no header, and there are no missing values (so you don't have to specify a value used to represent them).

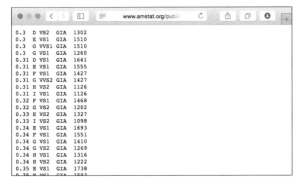

*Figure 8-5: A table-format data file found online*

With this in mind, you can create a data frame directly from the R prompt simply with the following lines:

```
R> dia.url <- "http://www.amstat.org/publications/jse/v9n2/4cdata.txt"
R> diamonds <- read.table(dia.url)
```

Note that you haven't supplied any extra values in this call to `read.table` because the defaults all work just fine. Because there's no header in the table, you can leave the default `header` value `FALSE`. The default value for `sep` is `""`, meaning whitespace (not to be confused with `" "`, meaning an explicit space character), which is exactly what this table uses. The default value for `stringsAsFactors` is `TRUE`, which is what you want for your character string columns. Following the import, you can supply names (based on the information in the documentation) to each column as follows:

```
R> names(diamonds) <- c("Carat","Color","Clarity","Cert","Price")
R> diamonds[1:5,]
  Carat Color Clarity Cert Price
1  0.30     D     VS2  GIA  1302
2  0.30     E     VS1  GIA  1510
3  0.30     G    VVS1  GIA  1510
4  0.30     G     VS1  GIA  1260
5  0.31     D     VS1  GIA  1641
```

Viewing the first five records shows that the data frame is displayed as you intended.

### 8.2.4   Other File Formats

There are other file formats besides *.txt* or *.csv* files that can be read into R, such as the data file format *.dat*. These files can also be imported using `read.table`, though they may contain extra information at the top that must be skipped using the optional `skip` argument. The `skip` argument asks for the number of lines at the top of the file that should be ignored before R begins the import.

As mentioned in Section 8.2.2, there are also contributed packages that can cope with other statistical software files; however, if there are multiple worksheets within a file it can complicate things. The R package foreign (R Core Team, 2015), available from CRAN, provides support for reading data files used by statistical programs such as Stata, SAS, Minitab, and SPSS.

Other contributed packages on CRAN can help R handle files from various database management systems (DBMSs). For example, the RODBC package (Ripley and Lapsley, 2013) lets you query Microsoft Access databases and return the results as a data frame object. Other interfaces include the packages RMySQL (James and DebRoy, 2012) and RJDBC (Urbanek, 2013).

## 8.3   Writing Out Data Files and Plots

Writing out new files from data frame objects with R is just as easy as reading in files. R's vector-oriented behavior is a fast and convenient way to recode data sets, so it's perfect for reading in data, restructuring it, and writing it back out to a file.

### 8.3.1   Data Sets

The function for writing table-format files to your computer is write.table. You supply a data frame object as x, and this function writes its contents to a new file with a specified name, delimiter, and missing value string. For example, the following line takes the mydatafile object from Section 8.2 and writes it to a file:

```
R> write.table(x=mydatafile,file="/Users/tdavies/somenewfile.txt",
               sep="@",na="??",quote=FALSE,row.names=FALSE)
```

You provide file with the folder location, ending in the filename you want for your new data file. This command creates a new table-format file called *somenewfile.txt* in the specified folder location, delimited by @ and with missing values denoted with ?? (because you're actually creating a new file, the file.choose command doesn't tend to be used here). Since mydatafile has variable names, these are automatically written to the file as a header. The optional logical argument quote determines whether to encapsulate each non-numeric entry in double quotes (if you explicitly need them in your file for, say, formatting requirements of other software); request no quotes by setting the argument to FALSE. Another optional logical argument, row.names, asks whether to include the row names of mydatafile (in this example, this would just be the numbers 1 to 6), which you also omit with FALSE. The resulting file, shown in Figure 8-6, can be opened in a text editor.

Like read.csv, write.csv is a shortcut version of the write.table function designed specifically for *.csv* files.

*Figure 8-6: The contents of* somenewfile.txt

## 8.3.2 Plots and Graphics Files

Plots can also be written directly to a file. In Chapter 7, you created and displayed plots in an active graphics device. This graphics device needn't be a screen window; it can be a specified file. Instead of displaying the plot immediately on the screen, you can have R follow these steps: open a "file" graphics device, run any plotting commands to create the final plot, and close the device. R supports direct writing to *.jpeg*, *.bmp*, *.png*, and *.tiff* files using functions of the same names. For example, the following code uses these three steps to create a *.jpeg* file:

```
R> jpeg(filename="/Users/tdavies/myjpegplot.jpeg",width=600,height=600)
R> plot(1:5,6:10,ylab="a nice ylab",xlab="here's an xlab",
        main="a saved .jpeg plot")
R> points(1:5,10:6,cex=2,pch=4,col=2)
R> dev.off()
null device
          1
```

The file graphics device is opened by a call to jpeg, where you provide the intended name of the file and its folder location as filename. By default, the dimensions of the device are set to 480 × 480 pixels, but here you change them to 600 × 600. You could also set these dimensions by supplying other units (inches, centimeters, or millimeters) to width and height and by specifying the unit with an optional units argument. Once the file is opened, you execute any R plotting commands you need in order to create the image—this example plots some points and then includes some additional points with a second command. The final graphical result is silently written to the specified file just as it would have been displayed on the screen. When you've finished plotting, you must explicitly close the file device with a call to dev.off(), which prints information on the remaining active device (here, "null device" can be loosely interpreted as "nothing is left open"). If dev.off() isn't called, then R will continue to output any subsequent plotting commands to the file, and possibly overwrite what you have there. The left plot in Figure 8-7 shows the resulting file created in this example.

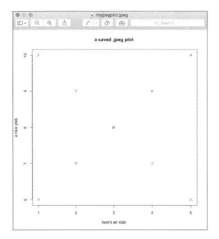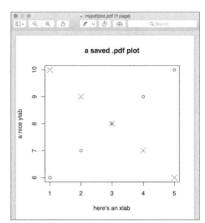

Figure 8-7: R plots that have been written directly to disk: a .jpeg version (left)
and a .pdf version (right) of the same plotting commands

You can also store R plots as other file types, such as PDFs (using the
pdf function) and EPS files (using the postscript function). Though some
argument names and default values are different for these functions, they
follow the same basic premise. You specify a folder location, a filename, and
width and height dimensions; enter your plotting commands; and then close
the device with dev.off(). The right panel of Figure 8-7 shows the *.pdf* file
created with the following code:

```
R> pdf(file="/Users/tdavies/mypdfplot.pdf",width=5,height=5)
R> plot(1:5,6:10,ylab="a nice ylab",xlab="here's an xlab",
        main="a saved .pdf plot")
R> points(1:5,10:6,cex=2,pch=4,col=2)
R> dev.off()
null device
          1
```

Here, you use the same plotting commands as before, and there are just
a few minor differences in the code. The argument for the file is file (as
opposed to filename), and the units for width and height default to inches
in pdf. The difference of appearance between the two images in Figure 8-7
results primarily from these differences in width and height.

This same process also works for ggplot2 images. True to style, however,
ggplot2 provides a convenient alternative. The ggsave function can be used
to write the most recently plotted ggplot2 graphic to file and performs the
device open/close action in one line.

For example, the following code creates and displays a ggplot2 object
from a simple data set.

```
R> foo <- c(1.1,2,3.5,3.9,4.2)
R> bar <- c(2,2.2,-1.3,0,0.2)
```

```
R> qplot(foo,bar,geom="blank")
    + geom_point(size=3,shape=8,color="darkgreen")
    + geom_line(color="orange",linetype=4)
```

Now, to save this plot to a file, all you need is the following line:

```
R> ggsave(filename="/Users/tdavies/mypngqplot.png")
Saving 7 x 7 in image
```

This writes the image to a *.png* file in the specified `filename` directory. (Note that dimensions are reported if you don't explicitly set them using `width` and `height`; these will vary depending on the size of your graphics device.) The result is shown in Figure 8-8.

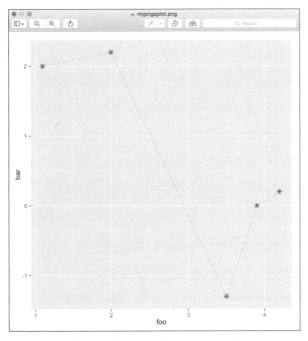

*Figure 8-8: The .png file created using* ggplot2's ggsave *command*

Beyond just being concise, ggsave is convenient in a few other ways. For one, you can use the same command to create a variety of image file types—the type is simply determined by the extension you supply in the `filename` argument. Also, ggsave has a range of optional arguments if you want to control the size of the image and the quality or scaling of the graphic.

For more details on saving images from base R graphics, see the `?jpeg`, `?pdf`, and `?postscript` help files. You can consult `?ggsave` for more on saving images with ggplot2.

## 8.4 Ad Hoc Object Read/Write Operations

For the typical R user, the most common input/output operations will probably revolve around data sets and plot images. But if you need to read or write other kinds of R objects, such as lists or arrays, you'll need the dput and dget commands, which can handle objects in a more ad hoc style.

Suppose, for example, you create this list in the current session:

```
R> somelist <- list(foo=c(5,2,45),
                    bar=matrix(data=c(T,T,F,F,F,F,T,F,T),nrow=3,ncol=3),
                    baz=factor(c(1,2,2,3,1,1,3),levels=1:3,ordered=T))
R> somelist
$foo
[1]  5  2 45

$bar
      [,1]  [,2]  [,3]
[1,]  TRUE FALSE  TRUE
[2,]  TRUE FALSE FALSE
[3,] FALSE FALSE  TRUE

$baz
[1] 1 2 2 3 1 1 3
Levels: 1 < 2 < 3
```

This object can itself be written to a file, which is useful if you want to pass it to a colleague or open it in a new R session elsewhere. Using dput, the following line stores the object as a plain-text file that is interpretable by R:

```
R> dput(x=somelist,file="/Users/tdavies/myRobject.txt")
```

In technical terms, this command creates an American Standard Code for Information Interchange (ASCII) representation of the object. As you call dput, the object you want to write is specified as x, and the folder location and name of the new plain-text file are passed to file. Figure 8-9 shows the contents of the resulting file.

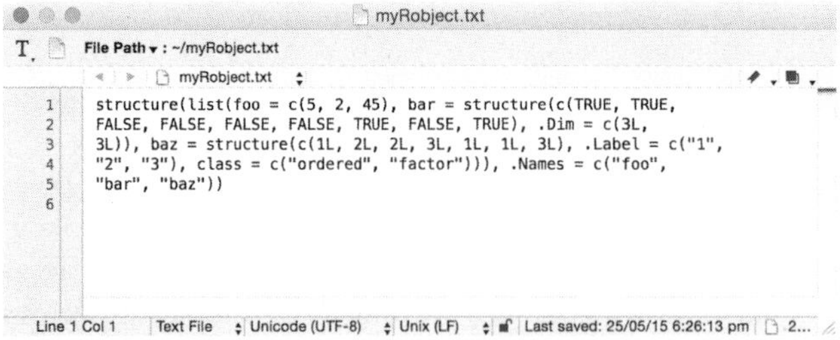

*Figure 8-9:* myRobject.txt *created by using* dput *on* somelist

Notice that `dput` stores all of the members of the object plus any other relevant information, such as attributes. The third element of `somelist`, for example, is an ordered factor, so it isn't enough to simply represent it in the text file as a stand-alone vector.

Now, let's say you want to import this list into an R workspace. If a file has been created with `dput`, then it can be read into any other workspace using `dget`.

```
R> newobject <- dget(file="/Users/tdavies/myRobject.txt")
R> newobject
$foo
[1]  5  2 45

$bar
      [,1]  [,2]  [,3]
[1,]  TRUE FALSE  TRUE
[2,]  TRUE FALSE FALSE
[3,] FALSE FALSE  TRUE

$baz
[1] 1 2 2 3 1 1 3
Levels: 1 < 2 < 3
```

You read the object from the *myRobject.txt* file using `dget` and assign it to `newobject`. This object is the same as the original R object `somelist`, with all structures and attributes present.

There are some drawbacks to using these commands. For starters, `dput` is not as reliable a command as `write.table` because it's sometimes quite difficult for R to create the necessary plain-text representation for an object (fundamental object classes typically cause no problems, but complex user-defined classes can). Also, because they need to store structural information, files created using `dput` are relatively inefficient both in terms of required space and in terms of how long it takes to execute read and write operations. This becomes more noticeable for objects that contain a lot of data. Nevertheless, `dput` and `dget` are useful ways to store or transfer specific objects without having to save an entire workspace.

## Exercise 8.1

a.  In R's built-in datasets library is the data frame `quakes`. Make sure you can access this object and view the corresponding help file to get an idea of what this data represents. Then, do the following:

   i.  Select only those records that correspond to a magnitude (`mag`) of greater than or equal to 5 and write them to a table-format file called *q5.txt* in an existing folder on your machine. Use a delimiting character of ! and do not include any row names.

ii.    Read the file back into your R workspace, naming the object `q5.dframe`.

b.    In the contributed package car, there's a data frame called `Duncan`, which provides historical data on perceived job prestige in 1950. Install the car package and access the `Duncan` data set and its help file. Then, do the following:

i.    Write R code that will plot `education` on the *x*-axis and `income` on the *y*-axis, with both *x*- and *y*-axis limits fixed to be $[0, 100]$. Provide appropriate axis labels. For jobs with a prestige value of less than or equal to 80, use a black ∘ as the point character. For jobs with prestige greater than 80, use a blue •.

ii.    Add a legend explaining the difference between the two types of points and then save a $500 \times 500$ pixel *.png* file of the image.

c.    Create a list called `exer` that contains the three data sets `quakes`, `q5.dframe`, and `Duncan`. Then, do the following:

i.    Write the list object directly to disk, calling it *Exercise8-1.txt*. Briefly inspect the contents of the file in a text editor.

ii.    Read *Exercise8-1.txt* back into your workspace; call the resulting object `list.of.dataframes`. Check that `list.of.dataframes` does indeed contain the three data frame objects.

d.    In Section 7.4.3, you created a `ggplot2` graphic of 20 observations displayed as the bottom image of Figure 7-11 on page 144. Use ggsave to save a copy of this plot as a *.tiff* file.

## Important Code in This Chapter

| Function/operator | Brief description | First occurrence |
| --- | --- | --- |
| data | Load contributed data set | Section 8.1.2, p. 149 |
| read.table | Import table-format data file | Section 8.2.1, p. 151 |
| list.files | Print specific folder contents | Section 8.2.1, p. 151 |
| file.choose | Interactive file selection | Section 8.2.1, p. 152 |
| read.csv | Import comma-delimited file | Section 8.2.2, p. 153 |
| write.table | Write table-format file to disk | Section 8.3.1, p. 156 |
| jpeg, bmp, png, tiff | Write image/plot file to disk | Section 8.3.2, p. 157 |
| dev.off | Close file graphics device | Section 8.3.2, p. 157 |
| pdf, postscript | Write image/plot file to disk | Section 8.3.2, p. 158 |
| ggsave | Write ggplot2 plot file to disk | Section 8.3.2, p. 159 |
| dput | Write R object to file (ASCII) | Section 8.4, p. 160 |
| dget | Import ASCII object file | Section 8.4, p. 161 |

# PART II

## PROGRAMMING

# 9

## CALLING FUNCTIONS

Before you start writing your own functions in R, it's useful to understand how functions are called and interpreted in an R session. First, you'll look at how variable names are compartmentalized in R. You'll see R's rules for naming arguments and objects, and how R searches for arguments and other variables when a function is called. Then you'll look at some alternative ways to specify arguments when calling a function.

## 9.1 Scoping

To begin with, it's important to understand R's *scoping rules*, which determine how the language compartmentalizes objects and retrieves them in a given session. This framework also defines the situations in which duplicate object names can exist at once. For example, you've used the argument data when calling matrix (Section 3.1), but data is also the name of a ready-to-use function that loads data sets from contributed packages (Section 8.1.2). In this section, you'll gain an introductory understanding of how R behaves internally in these circumstances, which will help you later on when it comes to programming and executing your own functions and those of other packages.

## 9.1.1  Environments

R enforces scoping rules with virtual *environments*. You can think of environments as separate compartments where data structures and functions are stored. They allow R to distinguish between identical names that are associated with different scopes and therefore stored in different environments. Environments are dynamic entities—new environments can be created, and existing environments can be manipulated or removed.

**NOTE**  *Technically speaking, environments don't actually contain items. Rather, they have pointers to the location of those items in the computer's memory. But using the "compartment" metaphor and thinking of objects "belonging to" these compartments is useful when you're first getting a general sense of how environments work.*

There are three important kinds of environments: global environments, package environments and namespaces, and local or lexical environments.

### Global Environment

The *global environment* is the compartment set aside for user-defined objects. Every object you've created or overwritten so far has resided in the global environment of your current R session. In Section 1.3.1, I mentioned that a call to ls() lists all the objects, variables, and user-defined functions in the active workspace—more precisely, ls() prints the names of everything in the current global environment.

Starting with a new R workspace, the following code creates two objects and confirms their existence in the global environment:

```
R> foo <- 4+5
R> bar <- "stringtastic"
R> ls()
[1] "bar" "foo"
```

But what about all the ready-to-use objects and functions? Why aren't those printed alongside foo and bar as members of this environment? In fact, those objects and functions belong to package-specific environments, described next.

### Package Environments and Namespaces

For simplicity, I'll use the term *package environment* rather loosely to refer to the items made available by each package in R. In fact, the structure of R packages in terms of scoping is a bit more complicated. Each package environment actually represents several environments that control different aspects of a search for a given object. A package *namespace*, for example, essentially defines the visibility of its functions. (A package can have visible functions that a user is able to use and invisible functions that provide internal support to the visible functions.) Another part of the package environment handles *imports* designations, dealing with any functions or objects from other libraries that the package needs to import for its own functionality.

To clarify this, you can think of all the ready-to-use functions and objects you're working with in this book as belonging to specific package environments. The same is true for the functions and objects of any contributed packages you've explicitly loaded with a call to `library`. You can use `ls` to list the items in a package environment as follows:

```
R> ls("package:graphics")
 [1] "abline"           "arrows"           "assocplot"        "axis"
 [5] "Axis"             "axis.Date"        "axis.POSIXct"     "axTicks"
 [9] "barplot"          "barplot.default"  "box"              "boxplot"
[13] "boxplot.default"  "boxplot.matrix"   "bxp"              "cdplot"
[17] "clip"             "close.screen"     "co.intervals"     "contour"
[21] "contour.default"  "coplot"           "curve"            "dotchart"
[25] "erase.screen"     "filled.contour"   "fourfoldplot"     "frame"
[29] "grconvertX"       "grconvertY"       "grid"             "hist"
[33] "hist.default"     "identify"         "image"            "image.default"
[37] "layout"           "layout.show"      "lcm"              "legend"
[41] "lines"            "lines.default"    "locator"          "matlines"
[45] "matplot"          "matpoints"        "mosaicplot"       "mtext"
[49] "pairs"            "pairs.default"    "panel.smooth"     "par"
[53] "persp"            "pie"              "plot"             "plot.default"
[57] "plot.design"      "plot.function"    "plot.new"         "plot.window"
[61] "plot.xy"          "points"           "points.default"   "polygon"
[65] "polypath"         "rasterImage"      "rect"             "rug"
[69] "screen"           "segments"         "smoothScatter"    "spineplot"
[73] "split.screen"     "stars"            "stem"             "strheight"
[77] "stripchart"       "strwidth"         "sunflowerplot"    "symbols"
[81] "text"             "text.default"     "title"            "xinch"
[85] "xspline"          "xyinch"           "yinch"
```

The `ls` command lists all of the visible objects contained in the graphics package environment. Note that this list includes some of the functions you used in Chapter 7, such as `arrows`, `plot`, and `segments`.

### Local Environments

Each time a function is called in R, a new environment is created called the *local environment*, sometimes referred to as the *lexical environment*. This local environment contains all the objects and variables created in and visible to the function, including any arguments you've supplied to the function upon execution. It's this feature that allows the presence of argument names that are identical to other object names accessible in a given workspace.

For example, say you call `matrix` and pass in the argument `data`, as follows:

```
R> youthspeak <- matrix(data=c("OMG","LOL","WTF","YOLO"),nrow=2,ncol=2)
R> youthspeak
     [,1]  [,2]
[1,] "OMG" "WTF"
[2,] "LOL" "YOLO"
```

Calling this function creates a local environment containing the `data` vector. When you execute the function, it begins by looking for `data` in this local environment. That means R isn't confused by other objects or functions named `data` in other environments (such as the `data` function automatically loaded from the `utils` package environment). If a required item isn't found in the local environment, only then does R begin to widen its search for that item (I'll discuss this feature a little more in Section 9.1.2). Once the function has completed, this local environment is automatically removed. The same comments apply to the `nrow` and `ncol` arguments.

## 9.1.2   Search Path

To access data structures and functions from environments other than the immediate global environment, R follows a *search path*. The search path lays out all the environments that a given R session has available to it.

The search path is basically a list of the environments that R will search when an object is requested. If the object isn't found in one environment, R proceeds to the next one. You can view R's search path at any time using `search()`.

```
R> search()
 [1] ".GlobalEnv"        "tools:RGUI"         "package:stats"
 [4] "package:graphics"  "package:grDevices"  "package:utils"
 [7] "package:datasets"  "package:methods"    "Autoloads"
[10] "package:base"
```

From the command prompt, this path will always begin at the global user environment (`.GlobalEnv`) and end after the base package environment (`package:base`). You can think of these as belonging to a hierarchy, with an arrow pointing from left to right between each pair of environments. For my current session, if I request a certain object at the R prompt, the program will inspect `.GlobalEnv` → `tools:RGUI` → `package:stats` → ... → `package:base` in turn, stopping the search when the desired object is found and retrieved. Note that, depending on your operating system and whether you're using the built-in GUI, `tools:RGUI` might not be included in your search path.

If R doesn't find what it's looking for by following the environments in the search path, the *empty environment* is reached. The empty environment is not explicitly listed in the output from `search()`, but it's always the final destination after package:base. This environment is special because it marks the end of the search path.

For example, if you call the following, a number of things happen internally:

```
R> baz <- seq(from=0,to=3,length.out=5)
R> baz
[1] 0.00 0.75 1.50 2.25 3.00
```

R first searches the global environment for a function called seq, and when this isn't found, it goes on to search in the enclosing environment, which is the next level up in the search path (according to the left-to-right arrows mentioned earlier). It doesn't find it there, so R keeps going through the path to the next environment, searching the packages that have been loaded (automatically or otherwise) until it finds what it's looking for. In this example, R locates seq in the built-in base package environment. Then it executes the seq function (creating a temporary local environment) and assigns the results to a new object, baz, which resides in the global environment. In the subsequent call to print baz, R begins by searching the global environment and immediately finds the requested object.

You can look up the enclosing environment of any function using environment, as follows:

```
R> environment(seq)
<environment: namespace:base>
R> environment(arrows)
<environment: namespace:graphics>
```

Here, I've identified the package namespace of base as the owner of the seq function and the graphics package as the owner of the arrows function.

Each environment has a *parent*, to direct the order of the search path. Examining the earlier output from the call search(), you can see that the parent of package:stats, for example, is package:graphics. The specific parent-child structure is dynamic in the sense that the search path changes when additional libraries are loaded or data frames are attached. When you load a contributed package with a call to library, this essentially just inserts the desired package in the search path. For example, in Exercise 8.1 on page 161, you installed the contributed package car. After loading this package, your search path will include its contents.

```
R> library("car")
R> search()
 [1] ".GlobalEnv"       "package:car"        "tools:RGUI"
 [4] "package:stats"    "package:graphics"   "package:grDevices"
 [7] "package:utils"    "package:datasets"   "package:methods"
[10] "Autoloads"        "package:base"
```

Note the position of the car package environment in the path—inserted directly after the global environment. This is where each subsequently loaded package will be placed (followed by any additional packages it depends upon for its own functionality).

As noted earlier, R will stop searching once it has exhausted the entire search path and reached the empty environment. If you request a function or object that you haven't defined, that doesn't exist, or that is perhaps in a contributed package that you've forgotten to load (this is quite a common little mistake), then an error is thrown. These "cannot find" errors are recognizable for both functions and other objects.

```
R> neither.here()
Error: could not find function "neither.here"
R> nor.there
Error: object 'nor.there' not found
```

Environments help compartmentalize the huge amount of functionality in R. This becomes particularly important when there are functions with the same name in different packages in the search path. At that point, *masking*, discussed in Section 12.3, comes into play.

As you get more comfortable with R and want more precise control over how it operates, it's worth investigating in full how R handles environments. For more technical details on this, Gupta (2012) provides a particularly well-written online article.

### 9.1.3   Reserved and Protected Names

A few key terms are strictly forbidden from being used as object names in R. These *reserved* names are necessary in order to protect fundamental operations and data types frequently used in the language.

The following identifiers are reserved:

- if and else

- for, while, and in

- function

- repeat, break, and next

- TRUE and FALSE

- Inf and -Inf

- NA, NaN, and NULL

I haven't yet covered some of the terms on this list. These items represent the core tools for programming in the R language, and you'll begin to explore them in the following chapter. The last three bullet points include the familiar logical values (Section 4.1) and special terms used to represent things like infinity and missing entries (Section 6.1).

If you try to assign a new value to any of these reserved terms, an error occurs.

```
R> NaN <- 5
Error in NaN <- 5 : invalid (do_set) left-hand side to assignment
```

Because R is case sensitive, it's possible to assign values to any case-variant of the reserved names, but this can be confusing and is generally not advisable.

```
R> False <- "confusing"
R> nan <- "this is"
R> cat(nan,False)
this is confusing
```

Also be wary of assigning values to T and F, the abbreviations of TRUE and FALSE. The full identifiers TRUE and FALSE are reserved, but the abbreviated versions are not.

```
R> T <- 42
R> F <- TRUE
R> F&&TRUE
[1] TRUE
```

Assigning values to T and F this way will affect any subsequent code that intends to use T and F to refer to TRUE and FALSE. The second assignment (F <- TRUE) is perfectly legal in R's eyes, but it's extremely confusing given the normal usage of F as an abbreviation: the line F&&TRUE now represents a TRUE&&TRUE comparison! It's best to simply avoid these types of assignments.

If you've been following along with the examples in your R console, it's prudent at this point to clear the global environment (thereby deleting the objects False, nan, T, and F from your workspace). To do this, use the rm function as shown next. Using ls(), supply a character vector of all objects in the global environment as the argument list.

```
R> ls()
[1] "bar"        "baz"        "F"        "False"      "foo"      "nan"
[7] "T"          "youthspeak"
R> rm(list=ls())
R> ls()
character(0)
```

Now the global environment is empty, and calling ls() returns an empty character vector (character(0)).

a.  Identify the first 20 items contained in the built-in and auto-
    matically loaded methods package. How many items are there in
    total?

b.  Determine the environment that owns each of the following
    functions:
    i.   read.table
    ii.  data
    iii. matrix
    iv.  jpeg

c.  Use ls and a test for character string equality to confirm the
    function smoothScatter is part of the graphics package.

## 9.2   Argument Matching

Another set of rules that determine how R interprets function calls has to
do with *argument matching*. Argument matching conditions allow you to provide arguments to functions either with abbreviated names or without names
at all.

### 9.2.1   Exact

So far, you've mostly been using *exact* matching of arguments, where each
argument tag is written out in full. This is the most exhaustive way to call a
function. It's helpful to write out full argument names this way when first
getting to know R or a new function.

Other benefits of exact matching include the following:

- Exact matching is less prone to mis-specification of arguments than
  other matching styles.

- The order in which arguments are supplied doesn't matter.

- Exact matching is useful when a function has many possible arguments
  but you want to specify only a few.

The main drawbacks of exact matching are clear:

- It can be cumbersome for relatively simple operations.

- Exact matching requires the user to remember or look up the full, case-
  sensitive tags.

As an example, in Section 6.2.1, you used exact matching to execute the following:

```
R> bar <- matrix(data=1:9,nrow=3,ncol=3,dimnames=list(c("A","B","C"),
                                             c("D","E","F")))
R> bar
  D E F
A 1 4 7
B 2 5 8
C 3 6 9
```

This creates a $3\times3$ matrix object bar with a dimnames attribute for the rows and columns. Since the argument tags are fully specified, the order of the arguments doesn't matter. You could switch around the arguments, and the function still has all the information it requires.

```
R> bar <- matrix(nrow=3,dimnames=list(c("A","B","C"),c("D","E","F")),ncol=3,
           data=1:9)
R> bar
  D E F
A 1 4 7
B 2 5 8
C 3 6 9
```

This behaves the same way as the previous function call. For the sake of consistency, you usually won't switch around arguments each time you call a function, but this example shows a benefit of exact matching: you don't have to worry about the order of any optional arguments or about skipping them.

### 9.2.2  Partial

*Partial* matching lets you identify arguments with an abbreviated tag. This can shorten your code, and it still lets you provide arguments in any order.

Here is another way to call matrix that takes advantage of partial matching:

```
R> bar <- matrix(nr=3,di=list(c("A","B","C"),c("D","E","F")),nc=3,dat=1:9)
R> bar
  D E F
A 1 4 7
B 2 5 8
C 3 6 9
```

Notice I've shortened the `nrow`, `dimnames`, and `ncol` argument tags to the first two letters and shortened the `data` argument to the first three. For partial matching, there's no set number of letters you have to provide, as long as each argument is still uniquely identifiable for the function being called. Partial matching has the following benefits:

- It requires less code than exact matching.
- Argument tags are still visible (which limits the possibility of mis-specification).
- The order of supplied arguments still doesn't matter.

But partial matching also has some limitations. For one, it gets trickier if there are multiple arguments whose tags start with the same letters. Here's an example:

```
R> bar <- matrix(nr=3,di=list(c("A","B","C"),c("D","E","F")),nc=3,d=1:9)
Error in matrix(nr = 3, di = list(c("A", "B", "C"), c("D", "E", "F")),  :
  argument 4 matches multiple formal arguments
```

An error has occurred. The fourth argument tag is designated simply as `d`, which is meant to stand for `data`. This is illegal because another argument, namely `dimnames`, also starts with `d`. Even though `dimnames` is specified separately as `di` earlier in the same line, the call isn't valid.

Drawbacks of partial matching include the following:

- The user must be aware of other potential arguments that can be matched by the shortened tag (even if they aren't specified in the call or have a default value assigned).
- Each tag must have a unique identification, which can be difficult to remember.

### 9.2.3 Positional

The most compact mode of function calling in R is *positional matching*. This is when you supply arguments without tags, and R interprets them based solely on their order.

Positional matching is usually used for relatively simple functions with only a few arguments, or functions that are very familiar to the user. For this type of matching, you *must* be aware of the precise positions of each argument. You can find that information in the "Usage" section of the function's help file, or it can be printed to the console with the `args` function. Here's an example:

```
R> args(matrix)
function (data = NA, nrow = 1, ncol = 1, byrow = FALSE, dimnames = NULL)
NULL
```

This shows the defined order of arguments of the matrix function, as well as the default value for each argument. To construct the matrix bar with positional matching, execute the following:

```
R> bar <- matrix(1:9,3,3,F,list(c("A","B","C"),c("D","E","F")))
R> bar
  D E F
A 1 4 7
B 2 5 8
C 3 6 9
```

The benefits of positional matching are as follows:

- Shorter, cleaner code, particularly for routine tasks
- No need to remember specific argument tags

Notice that when using exact and partial matching, you didn't need to supply anything for the byrow argument, which, by default, is set to FALSE. With positional matching, you must provide a value (given here as F) for byrow as the fourth argument because R relies on position alone to interpret the function call. If you leave out the argument, you get an error, as follows:

```
R> bar <- matrix(1:9,3,3,list(c("A","B","C"),c("D","E","F")))
Error in matrix(1:9, 3, 3, list(c("A", "B", "C"), c("D", "E", "F"))) :
  invalid 'byrow' argument
```

Here R has tried to assign the fourth argument (the list you intended for dimnames) as the value for the logical byrow argument. This brings us to the drawbacks of positional matching:

- You must look up and exactly match the defined order of arguments.
- Reading code written by someone else can be more difficult, especially when it includes unfamiliar functions.

### 9.2.4 Mixed

Since each matching style has pros and cons, it's quite common, and perfectly legal, to mix these three styles in a single function call.

For instance, you can avoid the type of error shown in the previous example like so:

```
R> bar <- matrix(1:9,3,3,dim=list(c("A","B","C"),c("D","E","F")))
R> bar
  D E F
A 1 4 7
B 2 5 8
C 3 6 9
```

Here I've used positional matching for the first three arguments, which are by now familiar to you. At the same time, I've used partial matching to explicitly tell R that the list is meant as a `dimnames` value, not for `byrow`.

## 9.2.5 Dot-Dot-Dot: Use of Ellipses

Many functions exhibit *variadic* behavior. That is, they can accept any number of arguments, and it's up to the user to decide how many arguments to provide. The functions `c`, `data.frame`, and `list` are all like this. When you call a function like `data.frame`, you can specify any number of members as arguments.

This flexibility is achieved in R through the special *dot-dot-dot* designation (...), also called the *ellipsis*. This construct allows the user to supply any number of data vectors (these become the columns in the final data frame). You can see whether an ellipsis is used in a function on the function's help page or with `args`. Looking at `data.frame`, notice the first argument slot is an ellipsis:

```
R> args(data.frame)
function (..., row.names = NULL, check.rows = FALSE, check.names = TRUE,
    stringsAsFactors = default.stringsAsFactors())
NULL
```

When you call a function and supply an argument that can't be matched with one of the function's defined argument tags, normally this would produce an error. But if the function is defined with an ellipsis, any arguments that aren't matched to other argument tags are matched to the ellipsis.

Functions that employ ellipses generally fall into two groups. The first group includes functions such as `c`, `data.frame`, and `list`, where the ellipsis always represents the "main ingredients" in the function call. That is, the objective of the function is to use contents of the ellipsis in the resulting object or output. The second group consists of functions where the ellipsis is meant as a *supplementary* or *potential* repository of optional arguments. This is common when the function of interest calls other *subfunctions* that themselves require additional arguments depending upon the originally supplied items. Rather than explicitly copy all the arguments desired by the subfunction into the argument list of the "parent" function, the parent function can instead be defined including an ellipsis that is subsequently provided to the subfunction.

Here's an example of the ellipsis used for supplementary arguments with the generic `plot` function:

```
R> args(plot)
function (x, y, ...)
NULL
```

From examining the arguments, it's clear that optional arguments such as point size (argument tag `cex`) or line type (argument tag `lty`), if supplied, are matched to the ellipsis. These optional arguments are then passed in to the function to be used by various methods that tweak graphical parameters.

Ellipses are a convenient programming tool for writing variadic functions or functions where an unknown number of arguments may be supplied. This will become clearer when you start writing your own functions in Chapter 11. However, when writing functions like this, it's important to properly document the intended use of ... so the potential users of the function know exactly which arguments can be passed to it and what those arguments are subsequently used for in execution.

---

### Exercise 9.2

a. Use positional matching with seq to create a sequence of values between −4 and 4 that progresses in steps of 0.2.

b. In each of the following lines of code, identify which style of argument matching is being used: exact, partial, positional, or mixed. If mixed, identify which arguments are specified in each style.
   i.   `array(8:1,dim=c(2,2,2))`
   ii.  `rep(1:2,3)`
   iii. `seq(from=10,to=8,length=5)`
   iv.  `sort(decreasing=T,x=c(2,1,1,2,0.3,3,1.3))`
   v.   `which(matrix(c(T,F,T,T),2,2))`
   vi.  `which(matrix(c(T,F,T,T),2,2),a=T)`

c. Suppose you explicitly ran the plotting function plot.default and supplied values to arguments tagged type, pch, xlab, ylab, lwd, lty, and col. Use the function documentation to determine which of these arguments fall under the umbrella of the ellipsis.

---

## Important Code in This Chapter

| Function/operator | Brief description | First occurrence |
|---|---|---|
| ls | Inspect environment objects | Section 9.1.1, p. 167 |
| search | Current search path | Section 9.1.2, p. 168 |
| environment | Function environment properties | Section 9.1.2, p. 169 |
| rm | Delete objects in workspace | Section 9.1.3, p. 171 |
| args | Show function arguments | Section 9.2.3, p. 174 |

# 10

## CONDITIONS AND LOOPS

To write more sophisticated programs with R, you'll need to control the flow and order of execution in your code. One fundamental way to do this is to make the execution of certain sections of code dependent on a *condition*. Another basic control mechanism is the *loop*, which repeats a block of code a certain number of times. In this chapter, we'll explore these core programming techniques using if-else statements, for and while loops, and other control structures.

## 10.1   if Statements

The if statement is the key to controlling exactly which operations are carried out in a given chunk of code. An if statement runs a block of code only if a certain condition is true. These constructs allow a program to respond differently depending on whether a condition is TRUE or FALSE.

### 10.1.1 Stand-Alone Statement

Let's start with the stand-alone `if` statement, which looks something like this:

```
if(condition){
    do any code here
}
```

The `condition` is placed in parentheses after the `if` keyword. This condition must be an expression that yields a single logical value (TRUE or FALSE). If it's TRUE, the code in the braces, {}, will be executed. If the condition isn't satisfied, the code in the braces is skipped, and R does nothing (or continues on to execute any code after the closing brace).

Here's a simple example. In the console, store the following:

```
R> a <- 3
R> mynumber <- 4
```

Now, in the R editor, write the following code chunk:

```
if(a<=mynumber){
    a <- a^2
}
```

When this chunk is executed, what will the value of a be? It depends on the condition defining the `if` statement, as well as what's actually specified in the braced area. In this case, when the condition `a<=mynumber` is evaluated, the result is TRUE since 3 is indeed less than or equal to 4. That means the code inside the braces is executed, which sets a to `a^2`, or 9.

Now highlight the entire chunk of code in the editor and send it to the console for evaluation. Remember, you can do this in several ways:

- Copy and paste the selected text from the editor directly into the console.

- From the menu, select **Edit → Run line or selection** in Windows or select **Edit → Execute** in OS X.

- Use the keystroke shortcut such as CTRL-R in Windows or ⌘-RETURN on a Mac.

Once you execute the code in the console, you'll see something like this:

```
R> if(a<=mynumber){
+     a <- a^2
+ }
```

Then, look at the object a, shown here:

```
R> a
[1] 9
```

Next, suppose you execute the same if statement again right away. Will a be squared once more, giving 81? Nope! Since a is now 9 and mynumber is still 4, the condition a<=mynumber will be FALSE, and the code in the braces will not be executed; a will remain at 9.

Note that after you send the if statement to the console, each line after the first is prefaced by a +. These + signs do not represent any kind of arithmetic addition; rather, they indicate that R is expecting more input before it begins execution. For example, when a left brace is opened, R will not begin any kind of execution until that section is closed with a right brace. To avoid redundancy, in future examples I won't show this repetition of code sent from the editor to the console.

**NOTE** *You can change the + symbol by assigning a different character string to the* continue *component of R's* options *command, in the way you reset the prompt in Section 1.2.1.*

The if statement offers a huge amount of flexibility—you can place any kind of code in the braced area, including more if statements (see the upcoming discussion of nesting in Section 10.1.4), enabling your program to make a sequence of decisions.

To illustrate a more complicated if statement, consider the following two new objects:

```
R> myvec <- c(2.73,5.40,2.15,5.29,1.36,2.16,1.41,6.97,7.99,9.52)
R> myvec
 [1] 2.73 5.40 2.15 5.29 1.36 2.16 1.41 6.97 7.99 9.52
R> mymat <- matrix(c(2,0,1,2,3,0,3,0,1,1),5,2)
R> mymat
     [,1] [,2]
[1,]    2    0
[2,]    0    3
[3,]    1    0
[4,]    2    1
[5,]    3    1
```

Use these two objects in the code chunk given here:

```
if(any((myvec-1)>9)||matrix(myvec,2,5)[2,1]<=6){
    cat("Condition satisfied --\n")
    new.myvec <- myvec
    new.myvec[seq(1,9,2)] <- NA
    mylist <- list(aa=new.myvec,bb=mymat+0.5)
    cat("-- a list with",length(mylist),"members now exists.")
}
```

Send this to the console, and it produces the following output:

```
Condition satisfied --
-- a list with 2 members now exists.
```

Indeed, an object mylist has been created that you can examine.

```
R> mylist
$aa
 [1]    NA 5.40    NA 5.29    NA 2.16    NA 6.97    NA 9.52

$bb
     [,1] [,2]
[1,]  2.5  0.5
[2,]  0.5  3.5
[3,]  1.5  0.5
[4,]  2.5  1.5
[5,]  3.5  1.5
```

In this example, the condition consists of two parts separated by an OR statement using ||, which produces a single logical result. Let's walk through it.

- The first part of the condition looks at myvec, takes 1 away from each element, and checks whether any of the results are greater than 9. If you run this part on its own, it yields FALSE.

```
R> myvec-1
[1] 1.73 4.40 1.15 4.29 0.36 1.16 0.41 5.97 6.99 8.52
R> (myvec-1)>9
[1] FALSE FALSE FALSE FALSE FALSE FALSE FALSE FALSE FALSE FALSE
R> any((myvec-1)>9)
[1] FALSE
```

- The second part of the condition uses positional matching in a call to matrix to construct a two-row, five-column, column-filled matrix using entries of the original myvec. Then, the number in the second row of the first column of that result is checked to see whether it's less than or equal to 6, which it is.

```
R> matrix(myvec,2,5)
     [,1] [,2] [,3] [,4] [,5]
[1,] 2.73 2.15 1.36 1.41 7.99
[2,] 5.40 5.29 2.16 6.97 9.52
R> matrix(myvec,2,5)[2,1]
[1] 5.4
R> matrix(myvec,2,5)[2,1]<=6
[1] TRUE
```

This means the overall condition being checked by the if statement will be FALSE||TRUE, which evaluates as TRUE.

```
R> any((myvec-1)>9)||matrix(myvec,2,5)[2,1]<=6
[1] TRUE
```

As a result, the code inside the braces is accessed and executed. First, it prints the "Condition satisfied" string and copies myvec to new.myvec. Using seq, it then accesses the odd-numbered indexes of new.myvec and overwrites them with NA. Next, it creates mylist. In this list, new.myvec is stored in a member named aa, and then it takes the original mymat, increases all its elements by 0.5, and stores the result in bb. Lastly, it prints the length of the resulting list.

Note that if statements don't have to match the exact style I'm using here. Some programmers, for example, prefer to open the left brace on a new line after the condition, or some may prefer a different amount of indentation.

### 10.1.2   else Statements

The if statement executes a chunk of code if and only if a defined condition is TRUE. If you want something different to happen when the condition is FALSE, you can add an else declaration. Here's an example in pseudocode:

```
if(condition){
    do any code in here if condition is TRUE
} else {
    do any code in here if condition is FALSE
}
```

You set the condition, then in the first set of braces you place the code to run if the condition is TRUE. After this, you declare else followed by a new set of braces where you can place code to run if the condition is FALSE.

Let's return to the first example in Section 10.1.1, once more storing these values at the console prompt.

```
R> a <- 3
R> mynumber <- 4
```

In the editor, create a new version of the earlier if statement.

```
if(a<=mynumber){
    cat("Condition was",a<=mynumber)
    a <- a^2
} else {
    cat("Condition was",a<=mynumber)
    a <- a-3.5
```

```
}
a
```

Here, you again square a if the condition a<=mynumber is TRUE, but if FALSE, a is overwritten by the result of itself minus 3.5. You also print text to the console stating whether the condition was met. After resetting a and mynumber to their original values, the first run of the if loop computes a as 9, just as earlier, outputting the following:

```
Condition was TRUE
R> a
[1] 9
```

Now, immediately highlight and execute the entire statement again. This time around, a<=mynumber will evaluate to FALSE and execute the code after else.

```
Condition was FALSE
R> a
[1] 5.5
```

### 10.1.3   Using ifelse for Element-wise Checks

An if statement can check the condition of only a single logical value. If you pass in, for example, a vector of logicals for the condition, the if statement will only check (and operate based on) the very first element. It will issue a warning saying as much, as the following dummy example shows:

```
R> if(c(FALSE,TRUE,FALSE,TRUE,TRUE)){}
Warning message:
In if (c(FALSE, TRUE, FALSE, TRUE, TRUE)) { :
  the condition has length > 1 and only the first element will be used
```

There is, however, a shortcut function available, ifelse, which can perform this kind of vector-oriented check in relatively simple cases. To demonstrate how it works, consider the objects x and y defined as follows:

```
R> x <- 5
R> y <- -5:5
R> y
 [1] -5 -4 -3 -2 -1  0  1  2  3  4  5
```

Now, suppose you want to produce the result of x/y but with any instance of Inf (that is, any instance where x is divided by zero) replaced with NA. In other words, for each element in y, you want to check whether y is zero. If so, you want the code to output NA, and if not, it should output the result of x/y.

As you've just seen, a simple `if` statement won't work here. Since it accepts only a single logical value, it can't run through the entire logical vector produced by `y==0`.

```
R> y==0
 [1] FALSE FALSE FALSE FALSE FALSE  TRUE FALSE FALSE FALSE FALSE FALSE
```

Instead, you can use the element-wise `ifelse` function for this kind of scenario.

```
R> result <- ifelse(test=y==0,yes=NA,no=x/y)
R> result
 [1] -1.000000 -1.250000 -1.666667 -2.500000 -5.000000   NA  5.000000  2.500000
 [9]  1.666667  1.250000  1.000000
```

Using exact matching, this command creates the desired `result` vector in one line. Three arguments must be specified: `test` takes a logical-valued data structure, `yes` provides the element to return if the condition is satisfied, and `no` gives the element to return if the condition is `FALSE`. As noted in the function documentation (which you can access with `?ifelse`), the returned structure will be of "the same length and attributes as test."

## Exercise 10.1

a. Create the following two vectors:

```
vec1 <- c(2,1,1,3,2,1,0)
vec2 <- c(3,8,2,2,0,0,0)
```

Without executing them, determine which of the following `if` statements would result in the string being printed to the console. Then confirm your answers in R.

i. `if((vec1[1]+vec2[2])==10){ cat("Print me!") }`

ii. `if(vec1[1]>=2&&vec2[1]>=2){ cat("Print me!") }`

iii. `if(all((vec2-vec1)[c(2,6)]<7)){ cat("Print me!") }`

iv. `if(!is.na(vec2[3])){ cat("Print me!") }`

b. Using `vec1` and `vec2` from (a), write and execute a line of code that multiplies the corresponding elements of the two vectors together *if* their sum is greater than 3. Otherwise, the code should simply sum the two elements.

c. In the editor, write R code that takes a square character matrix and checks *if* any of the character strings on the diagonal (top left to bottom right) begin with the letter *g*, lowercase or uppercase. If satisfied, these specific entries should be overwritten with the string "HERE". Otherwise, the entire matrix should be replaced with an identity matrix of the same dimensions. Then, try your code on the following matrices, checking the result each time:

i.
```
mymat <- matrix(as.character(1:16),4,4)
```

ii.
```
mymat <- matrix(c("DANDELION","Hyacinthus","Gerbera",
                  "MARIGOLD","geranium","ligularia",
                  "Pachysandra","SNAPDRAGON","GLADIOLUS"),3,3)
```

iii.
```
mymat <- matrix(c("GREAT","exercises","right","here"),2,2,
                byrow=T)
```

Hint: This requires some thought—you will find the functions diag from Section 3.2.1 and substr from Section 4.2.4 useful.

### 10.1.4  Nesting and Stacking Statements

An if statement can itself be placed within the outcome of another if statement. By *nesting* or *stacking* several statements, you can weave intricate paths of decision-making by checking a number of conditions at various stages during execution.

In the editor, modify the mynumber example once more as follows:

```
if(a<=mynumber){
    cat("First condition was TRUE\n")
    a <- a^2
    if(mynumber>3){
        cat("Second condition was TRUE")
        b <- seq(1,a,length=mynumber)
    } else {
        cat("Second condition was FALSE")
        b <- a*mynumber
    }
} else {
    cat("First condition was FALSE\n")
    a <- a-3.5
    if(mynumber>=4){
        cat("Second condition was TRUE")
        b <- a^(3-mynumber)
    } else {
        cat("Second condition was FALSE")
        b <- rep(a+mynumber,times=3)
```

```
    }
}
a
b
```

Here you see the same initial decision being made as earlier. The value a is squared if it's less than or equal to mynumber; if not, it has 3.5 subtracted from it. But now there's another if statement within each braced area. If the first condition is satisfied and a is squared, you then check whether mynumber is greater than 3. If TRUE, b is assigned seq(1,a,length=mynumber). If FALSE, b is assigned a*mynumber.

If the first condition fails and you subtract 3.5 from a, then you check a second condition to see whether mynumber is greater than or equal to 4. If it is, then b becomes a^(3-mynumber). If it's not, b becomes rep(a+mynumber,times=3). Note that I've indented the code within each subsequent braced area to make it easier to see which lines are relevant to each possible decision.

Now, reset a <- 3 and mynumber <- 4 either directly in the console or from the editor. When you run the mynumber example code, you'll get the following output:

```
First condition was TRUE
Second condition was TRUE
R> a
[1] 9
R> b
[1] 1.000000 3.666667 6.333333 9.000000
```

The result indicates exactly which code was invoked—the first condition and second condition were both TRUE. Trying another run of the same code, after first setting

```
R> a <- 6
R> mynumber <- 4
```

you see this output:

```
First condition was FALSE
Second condition was TRUE
R> a
[1] 2.5
R> b
[1] 0.4
```

This time the first condition fails, but the second condition checked inside the else statement is TRUE.

Alternatively, you could accomplish the same thing by sequentially *stacking* if statements and using a combination of logical expressions in each condition. In the following example, you check for the same four situations, but this time you stack `if` statements by placing a new `if` declaration immediately following an `else` declaration:

```
if(a<=mynumber && mynumber>3){
    cat("Same as 'first condition TRUE and second TRUE'")
    a <- a^2
    b <- seq(1,a,length=mynumber)
} else if(a<=mynumber && mynumber<=3){
    cat("Same as 'first condition TRUE and second FALSE'")
    a <- a^2
    b <- a*mynumber
} else if(mynumber>=4){
    cat("Same as 'first condition FALSE and second TRUE'")
    a <- a-3.5
    b <- a^(3-mynumber)
} else {
    cat("Same as 'first condition FALSE and second FALSE'")
    a <- a-3.5
    b <- rep(a+mynumber,times=3)
}
a
b
```

Just as before, only one of the four braced areas will end up being executed. Comparing this to the nested version, the first two braced areas correspond to what was originally the first condition (`a<=mynumber`) being satisfied, but this time you use `&&` to check two expressions at once. If neither of those two situations is met, this means the first condition is false, so in the third statement, you just have to check whether `mynumber>4`. For the final `else` statement, you don't need to check any conditions because that statement will be executed only if all the previous conditions were not met.

If you again reset a and `mynumber` to 3 and 4, respectively, and execute the stacked statements shown earlier, you get the following result:

```
Same as 'first condition TRUE and second TRUE'
R> a
[1] 9
R> b
[1] 1.000000 3.666667 6.333333 9.000000
```

This produces the same values for a and b as earlier. If you execute the code again using the second set of initial values (a as 6 and `mynumber` as 4), you get the following:

```
Same as 'first condition FALSE and second TRUE'
R> a
[1] 2.5
```

```
R> b
[1] 0.4
```

This again matches the results of using the nested version of the code.

### 10.1.5   The switch Function

Let's say you need to choose which code to run based on the value of a single object (a common scenario). One option is to use a series of if statements, where you compare the object with various possible values to produce a logical value for each condition. Here's an example:

```
if(mystring=="Homer"){
    foo <- 12
} else if(mystring=="Marge"){
    foo <- 34
} else if(mystring=="Bart"){
    foo <- 56
} else if(mystring=="Lisa"){
    foo <- 78
} else if(mystring=="Maggie"){
    foo <- 90
} else {
    foo <- NA
}
```

The goal of this code is simply to assign a numeric value to an object foo, where the exact number depends on the value of mystring. The mystring object can take one of the five possibilities shown, or if mystring doesn't match any of these, foo is assigned NA.

This code works just fine as it is. For example, setting

```
R> mystring <- "Lisa"
```

and executing the chunk, you'll see this:

```
R> foo
[1] 78
```

Setting the following

```
R> mystring <- "Peter"
```

and executing the chunk again, you'll see this:

```
R> foo
[1] NA
```

This setup using if-else statements is quite cumbersome for such a basic operation, though. R can handle this type of multiple-choice decision in a far more compact form via the switch function. For example, you could rewrite the stacked if statements as a much shorter switch statement as follows:

```
R> mystring <- "Lisa"
R> foo <- switch(EXPR=mystring,Homer=12,Marge=34,Bart=56,Lisa=78,Maggie=90,NA)
R> foo
[1] 78
```

and

```
R> mystring <- "Peter"
R> foo <- switch(EXPR=mystring,Homer=12,Marge=34,Bart=56,Lisa=78,Maggie=90,NA)
R> foo
[1] NA
```

The first argument, EXPR, is the object of interest and can be either a numeric or a character string. The remaining arguments provide the values or operations to carry out based on the value of EXPR. If EXPR is a string, these argument tags must *exactly* match the possible results of EXPR. Here, the switch statement evaluates to 12 if mystring is "Homer", 34 if mystring is "Marge", and so on. The final, untagged value, NA, indicates the result if mystring doesn't match any of the preceding items.

The integer version of switch works in a slightly different way. Instead of using tags, the outcome is determined purely with positional matching. Consider the following example:

```
R> mynum <- 3
R> foo <- switch(mynum,12,34,56,78,NA)
R> foo
[1] 56
```

Here, you provide an integer mynum as the first argument, and it's positionally matched to EXPR. The example code then shows five untagged arguments: 12 to NA. The switch function simply returns the value in the specific position requested by mynum. Since mynum is 3, the statement assigns 56 to foo. Had mynum been 1, 2, 4, or 5, foo would've been assigned 12, 34, 78, or NA, respectively. Any other value of mynum (less than 1 or greater than 5) will return NULL.

```
R> mynum <- 0
R> foo <- switch(mynum,12,34,56,78,NA)
R> foo
NULL
```

In these types of situations, the `switch` function behaves the same way as a set of stacked `if` statements, so it can serve as a convenient shortcut. However, if you need to examine multiple conditions at once or you need to execute a more complicated set of operations based on this decision, you'll need to use the explicit `if` and `else` control structures.

## Exercise 10.2

a.  Write an explicit stacked set of `if` statements that does the same thing as the integer version of the `switch` function illustrated earlier. Test it with `mynum <- 3` and `mynum <- 0`, as in the text.

b.  Suppose you are tasked with computing the precise dosage amounts of a certain drug in a collection of hypothetical scientific experiments. These amounts depend upon some predetermined set of "dosage thresholds" (`lowdose`, `meddose`, and `highdose`), as well as a predetermined dose level factor vector named `doselevel`. Look at the following items (i–iv) to see the intended form of these objects. Then write a set of nested `if` statements that produce a new numeric vector called `dosage`, according to the following rules:

–   First, *if* there are any instances of `"High"` in `doselevel`, perform the following operations:

*   Check *if* `lowdose` is greater than or equal to 10. If so, overwrite `lowdose` with 10; *otherwise*, overwrite `lowdose` by itself divided by 2.

*   Check *if* `meddose` is greater than or equal to 26. If so, overwrite `meddose` by 26.

*   Check *if* `highdose` is less than 60. If so, overwrite `highdose` with 60; *otherwise*, overwrite `highdose` by itself multiplied by 1.5.

*   Create a vector named `dosage` with the value of `lowdose` repeated (`rep`) to match the `length` of `doselevel`.

*   Overwrite the elements in `dosage` corresponding to the index positions of instances of `"Med"` in `doselevel` by `meddose`.

*   Overwrite the elements in `dosage` corresponding to the index positions of instances of `"High"` in `doselevel` by `highdose`.

–   *Otherwise* (in other words, if there are no instances of `"High"` in `doselevel`), perform the following operations:

*   Create a new version of `doselevel`, a factor vector with levels `"Low"` and `"Med"` only, and label these with `"Small"` and `"Large"`, respectively (refer to Section 4.3 for details or see `?factor`).

* Check to see *if* lowdose is less than 15 AND meddose is less than 35. If so, overwrite lowdose by itself multiplied by 2 and overwrite meddose by itself plus highdose.
* Create a vector named dosage, which is the value of lowdose repeated (rep) to match the length of doselevel.
* Overwrite the elements in dosage corresponding to the index positions of instances of "Large" in doselevel by meddose.

Now, confirm the following:

i. Given

```
lowdose <- 12.5
meddose <- 25.3
highdose <- 58.1
doselevel <- factor(c("Low","High","High","High","Low","Med",
                      "Med"),levels=c("Low","Med","High"))
```

the result of dosage after running the nested if statements is as follows:

```
R> dosage
[1] 10.0 60.0 60.0 60.0 10.0 25.3 25.3
```

ii. Using the same lowdose, meddose, and highdose thresholds as in (i), given

```
doselevel <- factor(c("Low","Low","Low","Med","Low","Med",
                      "Med"),levels=c("Low","Med","High"))
```

the result of dosage after running the nested if statements is as follows:

```
R> dosage
[1] 25.0 25.0 25.0 83.4 25.0 83.4 83.4
```

Also, doselevel has been overwritten as follows:

```
R> doselevel
[1] Small Small Small Large Small Large Large
Levels: Small Large
```

iii. Given

```
lowdose <- 9
meddose <- 49
highdose <- 61
doselevel <- factor(c("Low","Med","Med"),
                    levels=c("Low","Med","High"))
```

the result of dosage after running the nested if statements is as follows:

```
R> dosage
[1]  9 49 49
```

Also, doselevel has been overwritten as follows:

```
R> doselevel
[1] Small Large Large
Levels: Small Large
```

iv. Using the same lowdose, meddose, and highdose thresholds as (iii), as well as the same doselevel as (i), the result of dosage after running the nested if statements is as follows:

```
R> dosage
[1]  4.5 91.5 91.5 91.5  4.5 26.0 26.0
```

c. Assume the object mynum will only ever be a single integer between 0 and 9. Use ifelse and switch to produce a command that takes in mynum and returns a matching character string for all possible values 0, 1, ..., 9. Supplied with 3, for example, it should return "three"; supplied with 0, it should return "zero".

## 10.2 Coding Loops

Another core programming mechanism is the *loop*, which repeats a specified section of code, often while incrementing an index or counter. There are two styles of looping: the for loop repeats code as it works its way through a vector, element by element; the while loop simply repeats code until a specific condition evaluates to FALSE. Looplike behavior can also be achieved with R's suite of apply functions, which are discussed in Section 10.2.3.

### 10.2.1 *for Loops*

The R for loop always takes the following general form:

```
for(loopindex in loopvector){
    do any code in here
}
```

Here, the *loopindex* is a placeholder that represents an element in the *loopvector*—it starts off as the first element in the vector and moves to the next element with each loop repetition. When the for loop begins, it runs the code in the braced area, replacing any occurrence of the *loopindex* with the first element of the *loopvector*. When the loop reaches the closing brace, the *loopindex* is incremented, taking on the second element of the *loopvector*, and the braced area is repeated. This continues until the loop reaches the final element of the *loopvector*, at which point the braced code is executed for the final time, and the loop exits.

Here's a simple example written in the editor:

```
for(myitem in 5:7){
    cat("--BRACED AREA BEGINS--\n")
    cat("the current item is",myitem,"\n")
    cat("--BRACED AREA ENDS--\n\n")
}
```

This loop prints the current value of the *loopindex* (which I've named myitem here) as it increments from 5 to 7. Here's the output after sending to the console:

```
--BRACED AREA BEGINS--
the current item is 5
--BRACED AREA ENDS--

--BRACED AREA BEGINS--
the current item is 6
--BRACED AREA ENDS--

--BRACED AREA BEGINS--
the current item is 7
--BRACED AREA ENDS--
```

You can use loops to manipulate objects that exist outside the loop. Consider the following example:

```
R> counter <- 0
R> for(myitem in 5:7){
+     counter <- counter+1
+     cat("The item in run",counter,"is",myitem,"\n")
+ }
The item in run 1 is 5
The item in run 2 is 6
The item in run 3 is 7
```

Here, I've initially defined an object, counter, and set it to zero in the workspace. Then, inside the loop, counter is overwritten by itself plus 1. Each time the loop repeats, counter increases, and the current value is printed to the console.

## Looping via Index or Value

Note the difference between using the *loopindex* to directly represent elements in the *loopvector* and using it to represent *indexes* of a vector. The

following two loops use these two different approaches to print double each
number in myvec:

```
R> myvec <- c(0.4,1.1,0.34,0.55)
R> for(i in myvec){
+    print(2*i)
+ }
[1] 0.8
[1] 2.2
[1] 0.68
[1] 1.1
R> for(i in 1:length(myvec)){
+    print(2*myvec[i])
+ }
[1] 0.8
[1] 2.2
[1] 0.68
[1] 1.1
```

The first loop uses the *loopindex* i to directly represent the elements in
myvec, printing the value of each element times 2. In the second loop, on the
other hand, you use i to represent integers in the sequence 1:length(myvec).
These integers form all the possible index positions of myvec, and you use
these indexes to extract myvec's elements (once again multiplying each element by 2 and printing the result). Though it takes a slightly longer form,
using vector index positions provides more flexibility in terms of how you
can use the *loopindex*. This will become clearer when your needs demand
more complicated for loops, such as in the next example.

Suppose you want to write some code that will inspect any list object and
gather information about any matrix objects stored as members in the list.
Consider the following list:

```
R> foo <- list(aa=c(3.4,1),bb=matrix(1:4,2,2),cc=matrix(c(T,T,F,T,F,F),3,2),
              dd="string here",ee=matrix(c("red","green","blue","yellow")))
R> foo
$aa
[1] 3.4 1.0

$bb
     [,1] [,2]
[1,]    1    3
[2,]    2    4

$cc
       [,1]  [,2]
[1,]   TRUE  TRUE
[2,]   TRUE FALSE
[3,]  FALSE FALSE
```

```
$dd
[1] "string here"

$ee
     [,1]
[1,] "red"
[2,] "green"
[3,] "blue"
[4,] "yellow"
```

Here you've created foo, which contains three matrices of varying dimensions and data types. You'll write a for loop that goes through each member of a list like this one and checks whether the member is a matrix. If it is, the loop will retrieve the number of rows and columns and the data type of the matrix.

Before you write the for loop, you should create some vectors that will store information about the list members: name for the list member names, is.mat to indicate whether each member is a matrix (with "Yes" or "No"), nc and nr to store the number of rows and columns for each matrix, and data.type to store the data type of each matrix.

```
R> name <- names(foo)
R> name
[1] "aa" "bb" "cc" "dd" "ee"
R> is.mat <- rep(NA,length(foo))
R> is.mat
[1] NA NA NA NA NA
R> nr <- is.mat
R> nc <- is.mat
R> data.type <- is.mat
```

Here, you store the names of the members of foo as name. You also set up is.mat, nr, nc, and data.type, which are all assigned vectors of length length(foo) filled with NAs. These values will be updated as appropriate by your for loop, which you're now ready to write. Enter the following code in the editor:

```
for(i in 1:length(foo)){
    member <- foo[[i]]
    if(is.matrix(member)){
        is.mat[i] <- "Yes"
        nr[i] <- nrow(member)
        nc[i] <- ncol(member)
        data.type[i] <- class(as.vector(member))
    } else {
        is.mat[i] <- "No"
```

```
        }
}
bar <- data.frame(name,is.mat,nr,nc,data.type,stringsAsFactors=FALSE)
```

Initially, set up the *loopindex* i so that it will increment through the index positions of foo (the sequence 1:length(foo)). In the braced code, the first command is to write the member of foo at position i to an object member. Next, you can check whether that member is a matrix using is.matrix (refer to Section 6.2.3). If TRUE, you do the following: the ith position of is.mat vector is set as "Yes"; the ith element of nr and nc is set as the number of rows and number of columns of member, respectively; and the ith element of data.type is set as the result of class(as.vector(member)). This final command first coerces the matrix into a vector with as.vector and then uses the class function (covered in Section 6.2.2) to find the data type of the elements.

If member isn't a matrix and the if condition fails, the corresponding entry in is.mat is set to "No", and the entries in the other vectors aren't changed (so they will remain NA).

After the loop is run, a data frame bar is created from the vectors (note the use of stringsAsFactors=FALSE in order to prevent the character string vectors in bar being automatically converted to factors; see Section 5.2.1). After executing the code, bar looks like this:

```
R> bar
  name is.mat nr nc data.type
1   aa     No NA NA      <NA>
2   bb    Yes  2  2   integer
3   cc    Yes  3  2   logical
4   dd     No NA NA      <NA>
5   ee    Yes  4  1 character
```

As you can see, this matches the nature of the matrices present in the list foo.

### Nesting for Loops

You can also nest for loops, just like if statements. When a for loop is nested in another for loop, the inner loop is executed in full before the outer loop *loopindex* is incremented, at which point the inner loop is executed all over again. Create the following objects in your R console:

```
R> loopvec1 <- 5:7
R> loopvec1
[1] 5 6 7
R> loopvec2 <- 9:6
R> loopvec2
[1] 9 8 7 6
R> foo <- matrix(NA,length(loopvec1),length(loopvec2))
```

```
R> foo
      [,1] [,2] [,3] [,4]
[1,]    NA   NA   NA   NA
[2,]    NA   NA   NA   NA
[3,]    NA   NA   NA   NA
```

The following nested loop fills foo with the result of multiplying each integer in loopvec1 by each integer in loopvec2:

```
R> for(i in 1:length(loopvec1)){
+   for(j in 1:length(loopvec2)){
+      foo[i,j] <- loopvec1[i]*loopvec2[j]
+   }
+ }
R> foo
      [,1] [,2] [,3] [,4]
[1,]    45   40   35   30
[2,]    54   48   42   36
[3,]    63   56   49   42
```

Note that nested loops require a unique *loopindex* for each use of for. In this case, the *loopindex* is i for the outer loop and j for the inner loop. When the code is executed, i is first assigned 1, the inner loop begins, and then j is also assigned 1. The only command in the inner loop is to take the product of the ith element of loopvec1 and the jth element of loopvec2 and assign it to row i, column j of foo. The inner loop repeats until j reaches length(loopvec2) and fills the first row of foo; then i increments, and the inner loop is started up again. The entire procedure is complete after i reaches length(loopvec1) and the matrix is filled.

Inner *loopvectors* can even be defined to match the current value of the *loopindex* of the outer loop. Using loopvec1 and loopvec2 from earlier, here's an example:

```
R> foo <- matrix(NA,length(loopvec1),length(loopvec2))
R> foo
      [,1] [,2] [,3] [,4]
[1,]    NA   NA   NA   NA
[2,]    NA   NA   NA   NA
[3,]    NA   NA   NA   NA
R> for(i in 1:length(loopvec1)){
+   for(j in 1:i){
+      foo[i,j] <- loopvec1[i]+loopvec2[j]
+   }
+ }
R> foo
      [,1] [,2] [,3] [,4]
[1,]    14   NA   NA   NA
```

| | | | | |
|---|---|---|---|---|
| [2,] | 15 | 14 | NA | NA |
| [3,] | 16 | 15 | 14 | NA |

Here, the ith row, jth column element of foo is filled with the sum of loopvec1[i] and loopvec2[j]. However, the inner loop values for j are now decided based on the value of i. For example, when i is 1, the inner *loopvector* is 1:1, so the inner loop executes only once before returning to the outer loop. With i as 2, the inner *loopvector* is then 1:2, and so on. This makes it so each row of foo is only partially filled. Extra care must be taken when programming loops this way. Here, for example, the values for j depend on the length of loopvec1, so an error will occur if length(loopvec1) is greater than length(loopvec2).

Any number of for loops can be nested, but the computational expense can become a problem if nested loops are used unwisely. Loops in general add some computational cost, so to produce more efficient code in R, you should always ask "Can I do this in a vector-oriented fashion?" Only when the individual operations are not possible or straightforward to achieve en masse should you explore an iterative, looped approach. You can find some relevant and valuable comments on R loops and associated best-practice coding in the "R Help Desk" article by Ligges and Fox (2008).

## Exercise 10.3

a. In the interests of efficient coding, rewrite the nested loop example from this section, where the matrix foo was filled with the multiples of the elements of loopvec1 and loopvec2, using only a single for loop.

b. In Section 10.1.5, you used the command

```
switch(EXPR=mystring,Homer=12,Marge=34,Bart=56,Lisa=78,Maggie=90,
    NA)
```

to return a number based on the supplied value of a single character string. This line won't work if mystring is a character vector. Write some code that will take a character vector and return a vector of the appropriate numeric values. Test it on the following vector:

```
c("Peter","Homer","Lois","Stewie","Maggie","Bart")
```

c. Suppose you have a list named mylist that can contain other lists as members, but assume those "member lists" cannot themselves contain lists. Write nested loops that can search any possible mylist defined in this way and count how many matrices are present. Hint: Simply set up a counter before commencing the loops that is incremented each time a matrix is found, regardless of whether it is a straightforward member of mylist or it is a member of a member list of mylist.

Then confirm the following:

i. That the answer is 4 if you have the following:

```
mylist <- list(aa=c(3.4,1),bb=matrix(1:4,2,2),
               cc=matrix(c(T,T,F,T,F,F),3,2),dd="string here",
               ee=list(c("hello","you"),matrix(c("hello",
                                                  "there"))),
               ff=matrix(c("red","green","blue","yellow"))))
```

ii. That the answer is 0 if you have the following:

```
mylist <- list("tricked you",as.vector(matrix(1:6,3,2)))
```

iii. That the answer is 2 if you have the following:

```
mylist <- list(list(1,2,3),list(c(3,2),2),
               list(c(1,2),matrix(c(1,2)))),
               rbind(1:10,100:91))
```

## 10.2.2   *while Loops*

To use for loops, you must know, or be able to easily calculate, the number of times the loop should repeat. In situations where you don't know how many times the desired operations need to be run, you can turn to the while loop. A while loop runs and repeats while a specified condition returns TRUE, and takes the following general form:

```
while(loopcondition){
    do any code in here
}
```

A while loop uses a single logical-valued *loopcondition* to control how many times it repeats. Upon execution, the *loopcondition* is evaluated. If the condition is found to be TRUE, the braced area code is executed line by line as usual until complete, at which point the *loopcondition* is checked again. The loop terminates only when the condition evaluates to FALSE, and it does so immediately—the braced code is *not* run one last time.

This means the operations carried out in the braced area must somehow cause the loop to exit, either by affecting the *loopcondition* somehow or by declaring break, which you'll see a little later. If not, the loop will keep repeating forever, creating an *infinite loop*, which will freeze the console (and, depending on the operations specified inside the braced area, R can crash because of memory constraints). If that occurs, you can terminate the loop in the R user interface by clicking the Stop button in the top menu or by pressing ESC.

As a simple example of a while loop, consider the following code:

```
myval <- 5
while(myval<10){
    myval <- myval+1
    cat("\n'myval' is now",myval,"\n")
    cat("'mycondition' is now",myval<10,"\n")
}
```

Here, you set a new object `myval` to 5. Then you start a while loop with the condition `myval<10`. Since this is `TRUE` to begin with, you enter the braced area. Inside the loop you increment `myval` by 1, print its current value, and print the logical value of the condition `myval<5`. The loop continues until the condition `myval<10` is `FALSE` at the next evaluation. Execute the code chunk, and you see the following:

```
'myval' is now 6
'mycondition' is now TRUE

'myval' is now 7
'mycondition' is now TRUE

'myval' is now 8
'mycondition' is now TRUE

'myval' is now 9
'mycondition' is now TRUE

'myval' is now 10
'mycondition' is now FALSE
```

As expected, the loop repeats until `myval` is set to `10`, at which point `myval<10` returns `FALSE`, causing the loop to exit because the initial condition is no longer `TRUE`.

In more complicated settings, it's often useful to set the *loopcondition* to be a separate object so that you can modify it as necessary within the braced area. For the next example, you'll use a while loop to iterate through an integer vector and create an identity matrix (see Section 3.3.2) with the dimension matching the current integer. This loop should stop when it reaches a number in the vector that's greater than 5 or when it reaches the end of the integer vector.

In the editor, define some initial objects, followed by the loop itself.

```
mylist <- list()
counter <- 1
mynumbers <- c(4,5,1,2,6,2,4,6,6,2)
mycondition <- mynumbers[counter]<=5
while(mycondition){
```

```
        mylist[[counter]] <- diag(mynumbers[counter])
        counter <- counter+1
        if(counter<=length(mynumbers)){
            mycondition <- mynumbers[counter]<=5
        } else {
            mycondition <- FALSE
        }
    }
```

The first object, mylist, will store all the matrices that the loop creates. You'll use the vector mynumbers to provide the matrix sizes, and you'll use counter and mycondition to control the loop.

The *loopcondition*, mycondition, is initially set to TRUE since the first element of mynumbers is less than or equal to 5. Inside the loop beginning at while, the first line uses double square brackets and the value of counter to dynamically create a new entry at that position in mylist (you did this earlier with named lists in Section 5.1.3). This entry is assigned an identity matrix whose size matches the corresponding element of mynumbers. Next, the counter is incremented, and now you have to update mycondition. Here you want to check whether mynumbers[counter]<=5, but you also need to check whether you've reached the end of the integer vector (otherwise, you can end up with an error by trying to retrieve an index position outside the range of mynumbers). So, you can use an if statement to first check the condition counter<=length(mynumbers). If TRUE, then set mycondition to the outcome of mynumbers[counter]<=5. If not, this means you've reached the end of mynumbers, so you make sure the loop exits by setting mycondition <- FALSE.

Execute the loop with those predefined objects, and it will produce the mylist object shown here:

```
R> mylist
[[1]]
     [,1] [,2] [,3] [,4]
[1,]    1    0    0    0
[2,]    0    1    0    0
[3,]    0    0    1    0
[4,]    0    0    0    1

[[2]]
     [,1] [,2] [,3] [,4] [,5]
[1,]    1    0    0    0    0
[2,]    0    1    0    0    0
[3,]    0    0    1    0    0
[4,]    0    0    0    1    0
[5,]    0    0    0    0    1

[[3]]
     [,1]
[1,]    1
```

```
[[4]]
    [,1] [,2]
[1,]   1    0
[2,]   0    1
```

As expected, you have a list with four members—identity matrices of size $4 \times 4$, $5 \times 5$, $1 \times 1$, and $2 \times 2$—matching the first four elements of mynumbers. The loop stopped executing when it reached the fifth element of mynumbers (6) since that's greater than 5.

---

### Exercise 10.4

a.  Based on the most recent example of storing identity matrices in a list, determine what the resulting mylist would look like for each of the following possible mynumbers vectors, without executing anything:

   i.   mynumbers <- c(2,2,2,2,5,2)
   ii.  mynumbers <- 2:20
   iii. mynumbers <- c(10,1,10,1,2)

    Then, confirm your answers in R (note you'll also have to reset the initial values of mylist, counter, and mycondition each time, just as in the text).

b.  For this problem, I'll introduce the *factorial* operator. The factorial of a non-negative integer $x$, expressed as $x!$, refers to $x$ multiplied by the product of all integers less than $x$, down to 1. Formally, it is written like this:

$$\text{``}x \text{ factorial''} = x! = x \times (x - 1) \times (x - 2) \times \ldots \times 1$$

    Note that there is a special case of *zero factorial*, which is always 1. That is:

$$0! = 1$$

    For example, to work out 3 factorial, you compute the following:

$$3 \times 2 \times 1 = 6$$

    To work out 7 factorial, you compute the following:

$$7 \times 6 \times 5 \times 4 \times 3 \times 2 \times 1 = 5040$$

    Write a while loop that computes and stores as a new object the factorial of any non-negative integer mynum by decrementing mynum by 1 at each repetition of the braced code.

Using your loop, confirm the following:

i. That the result of using mynum <- 5 is 120
ii. That using mynum <- 12 yields 479001600
iii. That having mynum <- 0 correctly returns 1

c. Consider the following code, where the operations in the braced area of the while loop have been omitted:

```
mystring <- "R fever"
index <- 1
ecount <- 0
result <- mystring
while(ecount<2 && index<=nchar(mystring)){
    # several omitted operations #
}
result
```

Your task is to complete the code in the braced area so it inspects mystring character by character until it reaches the second instance of the letter *e* or the end of the string, whichever comes first. The result object should be the entire character string if there is no second *e* or the character string made up of all the characters up to, but not including, the second *e* if there is one. For example, mystring <- "R fever" should provide result as "R fev". This must be achieved by following these operations in the braces:

1. Use substr (Section 4.2.4) to extract the single character of mystring at position index.
2. Use a check for equality to determine whether this single-character string is either "e" OR "E". If so, increase ecount by 1.
3. Next, perform a separate check to see whether ecount is equal to 2. If so, use substr to set result equal to the characters between 1 and index-1 inclusive.
4. Increment index by 1.

Test your code—ensure the previous result for mystring <- "R fever". Furthermore, confirm the following:

- Using mystring <- "beautiful" provides result as "beautiful"
- Using mystring <- "ECCENTRIC" provides result as "ECC"
- Using mystring <- "ElAbOrAte" provides result as "ElAbOrAt"
- Using mystring <- "eeeeek!" provides result as "e"

### 10.2.3   Implicit Looping with apply

In some situations, especially for relatively routine for loops (such as executing some function on each member of a list), you can avoid some of the details associated with explicit looping by using the apply function. The apply function is the most basic form of implicit looping—it takes a function and applies it to each *margin* of an array.

For a simple illustrative example, let's say you have the following matrix:

```
R> foo <- matrix(1:12,4,3)
R> foo
     [,1] [,2] [,3]
[1,]    1    5    9
[2,]    2    6   10
[3,]    3    7   11
[4,]    4    8   12
```

Say you want to find the sum of each row. If you call the following, you just get the grand total of all elements, which is not what you want.

```
R> sum(foo)
[1] 78
```

Instead, you could use a for loop like this one:

```
R> row.totals <- rep(NA,times=nrow(foo))
R> for(i in 1:nrow(foo)){
+    row.totals[i] <- sum(foo[i,])
+ }
R> row.totals
[1] 15 18 21 24
```

This cycles through each row and stores the sum in row.totals. But you can use apply to get the same result in a more compact form. To call apply, you have to specify at least three arguments. The first argument, X, is the object you want to cycle through. The next argument, MARGIN, takes an integer that flags which margin of X to operate on (rows, columns, etc.). Finally, FUN provides the function you want to perform on each margin. With the following call, you get the same result as the earlier for loop.

```
R> row.totals2 <- apply(X=foo,MARGIN=1,FUN=sum)
R> row.totals2
[1] 15 18 21 24
```

The MARGIN index follows the positional order of the dimension for matrices and arrays, as discussed in Chapter 3—1 always refers to rows, 2 to columns, 3 to layers, 4 to blocks, and so on. To instruct R to sum each column of foo instead, simply change the MARGIN argument to 2.

```
R> apply(X=foo,MARGIN=2,FUN=sum)
[1] 10 26 42
```

The operations supplied to FUN should be appropriate for the MARGIN selected. So, if you select rows or columns with MARGIN=1 or MARGIN=2, make sure the FUN function is appropriate for vectors. Or if you have a three-dimensional array and use apply with MARGIN=3, be sure to set FUN to a function appropriate for matrices. Here's an example for you to enter:

```
R> bar <- array(1:18,dim=c(3,3,2))
R> bar
, , 1

     [,1] [,2] [,3]
[1,]   1    4    7
[2,]   2    5    8
[3,]   3    6    9

, , 2

     [,1] [,2] [,3]
[1,]   10   13   16
[2,]   11   14   17
[3,]   12   15   18
```

Then, make the following call:

```
R> apply(bar,3,FUN=diag)
     [,1] [,2]
[1,]   1    10
[2,]   5    14
[3,]   9    18
```

This extracts the diagonal elements of each of the matrix layers of bar. Each call to diag on a matrix returns a vector, and these vectors are returned as columns of a new matrix. The FUN argument can also be any appropriate user-defined function, and you'll look at some examples of using apply with your own functions in Chapter 11.

## Other apply Functions

There are different flavors of the basic apply function. The tapply function, for example, performs operations on subsets of the object of interest, where those subsets are defined in terms of one or more factor vectors. As an example, let's return to the code from Section 8.2.3, which reads in a web-based data file on diamond pricing, sets appropriate variable names of the data frame, and displays the first five records.

```
R> dia.url <- "http://www.amstat.org/publications/jse/v9n2/4cdata.txt"
R> diamonds <- read.table(dia.url)
R> names(diamonds) <- c("Carat","Color","Clarity","Cert","Price")
R> diamonds[1:5,]
```

```
  Carat Color Clarity Cert Price
1  0.30    D     VS2   GIA  1302
2  0.30    E     VS1   GIA  1510
3  0.30    G    VVS1   GIA  1510
4  0.30    G     VS1   GIA  1260
5  0.31    D     VS1   GIA  1641
```

To add up the total value of the diamonds present for the full data set but separated according to Color, you can use tapply like this:

```
R> tapply(diamonds$Price,INDEX=diamonds$Color,FUN=sum)
     D      E      F      G      H      I
113598 242349 392485 287702 302866 207001
```

This sums the relevant elements of the target vector diamonds$Price. The corresponding factor vector diamonds$Color is passed to INDEX, and the function of interest is specified with FUN=sum exactly as earlier.

Another particularly useful alternative is lapply, which can operate member by member on a list. In Section 10.2.1, recall you wrote a for loop to inspect matrices in the following list:

```
R> baz <- list(aa=c(3.4,1),bb=matrix(1:4,2,2),cc=matrix(c(T,T,F,T,F,F),3,2),
              dd="string here",ee=matrix(c("red","green","blue","yellow")))
```

Using lapply, you can check for matrices in the list with a single short line of code.

```
R> lapply(baz,FUN=is.matrix)
$aa
[1] FALSE

$bb
[1] TRUE

$cc
[1] TRUE

$dd
[1] FALSE

$ee
[1] TRUE
```

Note that no margin or index information is required for lapply; R knows to apply FUN to each member of the specified list. The returned value is itself a list. Another variant, sapply, returns the same results as lapply but in an array form.

```
R> sapply(baz,FUN=is.matrix)
   aa    bb    cc    dd    ee
FALSE  TRUE  TRUE FALSE  TRUE
```

Here, the result is provided as a vector. In this example, baz has a names attribute that is copied to the corresponding entries of the returned object.

Other variants of apply include vapply, which is similar to sapply albeit with some relatively subtle differences, and mapply, which can operate on multiple vectors or lists at once. To learn more about mapply, see the ?mapply help file; vapply and sapply are both covered in the ?lapply help file.

All of R's apply functions allow for additional arguments to be passed to FUN; most of them do this via an ellipsis. For example, take another look at the matrix foo:

```
R> apply(foo,1,sort,decreasing=TRUE)
     [,1] [,2] [,3] [,4]
[1,]    9   10   11   12
[2,]    5    6    7    8
[3,]    1    2    3    4
```

Here you've applied sort to each row of the matrix and supplied the additional argument decreasing=TRUE to sort the rows from largest to smallest.

Some programmers prefer using the suite of apply functions wherever possible to improve the compactness and neatness of their code. However, note that these functions generally do not offer any substantial improvement in terms of computational speed or efficiency over an explicit loop (this is particularly the case with more recent versions of R). Plus, when you're first learning the R language, explicit loops can be easier to read and follow since the operations are laid out clearly line by line.

## Exercise 10.5

a.  Continuing on from the most recent example in the text, write an implicit loop that calculates the product of all the column elements of the matrix returned by the call to apply(foo,1,sort,decreasing=TRUE).

b.  Convert the following for loop to an implicit loop that does exactly the same thing:

```
matlist <- list(matrix(c(T,F,T,T),2,2),
                matrix(c("a","c","b","z","p","q"),3,2),
                matrix(1:8,2,4))
matlist
for(i in 1:length(matlist)){
    matlist[[i]] <- t(matlist[[i]])
```

```
}
matlist
```

c.  In R, store the following $4 \times 4 \times 2 \times 3$ array as the object qux:

```
R> qux <- array(96:1,dim=c(4,4,2,3))
```

      That is, it is a four-dimensional array comprised of three blocks, with each block being an array made up of two layers of $4 \times 4$ matrices. Then, do the following:

i.  Write an implicit loop that obtains the diagonal elements of all second-layer matrices only to produce the following matrix:

```
      [,1] [,2] [,3]
[1,]   80   48   16
[2,]   75   43   11
[3,]   70   38    6
[4,]   65   33    1
```

ii.  Write an implicit loop that will return the dimensions of each of the three matrices formed by accessing the fourth column of every matrix in qux, regardless of layer or block, wrapped by another implicit loop that finds the row sums of that returned structure, resulting simply in the following vector:

```
[1] 12  6
```

## 10.3   Other Control Flow Mechanisms

To round off this chapter, you'll look at three more control flow mechanisms: break, next, and repeat. These mechanisms are often used in conjunction with the loops and if statements you've seen already.

### 10.3.1   Declaring break or next

Normally a for loop will exit only when the *loopindex* exhausts the *loopvector*, and a while loop will exit only when the *loopcondition* evaluates to FALSE. But you can also preemptively terminate a loop by declaring break.

      For example, say you have a number, foo, that you want to divide by each element in a numeric vector bar.

```
R> foo <- 5
R> bar <- c(2,3,1.1,4,0,4.1,3)
```

      Furthermore, let's say you want to divide foo by bar element by element but want to halt execution if one of the results evaluates to Inf (which will result if dividing by zero). To do this, you can check each iteration with the is.finite function (Section 6.1.1), and you can issue a break command to terminate the loop if it returns FALSE.

```
R> loop1.result <- rep(NA,length(bar))
R> loop1.result
[1] NA NA NA NA NA NA NA
R> for(i in 1:length(bar)){
+   temp <- foo/bar[i]
+   if(is.finite(temp)){
+       loop1.result[i] <- temp
+   } else {
+       break
+   }
+ }
R> loop1.result
[1] 2.500000 1.666667 4.545455 1.250000       NA       NA       NA
```

Here, the loop divides the numbers normally until it reaches the fifth element of bar and divides by zero, resulting in Inf. Upon the resulting conditional check, the loop ends immediately, leaving the remaining entries of loop1.result as they were originally set—NAs.

Invoking break is a fairly drastic move. Often, a programmer will include it only as a safety catch that's meant to highlight or prevent unintended calculations. For more routine operations, it's best to use another method. For instance, the example loop could easily be replicated as a while loop or the vector-oriented ifelse function, rather than relying on a break.

Instead of breaking and completely ending a loop, you can use next to simply advance to the next iteration and continue execution. Consider the following, where using next avoids division by zero:

```
R> loop2.result <- rep(NA,length(bar))
R> loop2.result
[1] NA NA NA NA NA NA NA
R> for(i in 1:length(bar)){
+   if(bar[i]==0){
+       next
+   }
+   loop2.result[i] <- foo/bar[i]
+ }
R> loop2.result
[1] 2.500000 1.666667 4.545455 1.250000       NA 1.219512 1.666667
```

First, the loop checks to see whether the ith element of bar is zero. If it is, next is declared, and as a result, R ignores any subsequent lines of code in the braced area of the loop and returns to the top, automatically advancing to the next value of the *loopindex*. In the current example, the loop skips the fifth entry of bar (leaving the original NA value for that place) and continues through the rest of bar.

Note that if you use either break or next in a nested loop, the command will apply only to the innermost loop. Only that inner loop will exit or advance to the next iteration, and any outer loops will continue as normal. For example, let's return to the nested for loops from Section 10.2.1 that you used to fill a matrix with multiples of two vectors. This time you'll use next in the inner loop to skip certain values.

```
R> loopvec1 <- 5:7
R> loopvec1
[1] 5 6 7
R> loopvec2 <- 9:6
R> loopvec2
[1] 9 8 7 6
R> baz <- matrix(NA,length(loopvec1),length(loopvec2))
R> baz
     [,1] [,2] [,3] [,4]
[1,]  NA   NA   NA   NA
[2,]  NA   NA   NA   NA
[3,]  NA   NA   NA   NA
R> for(i in 1:length(loopvec1)){
+    for(j in 1:length(loopvec2)){
+       temp <- loopvec1[i]*loopvec2[j]
+       if(temp>=54){
+          next
+       }
+       baz[i,j] <- temp
+    }
+ }
R> baz
     [,1] [,2] [,3] [,4]
[1,]  45   40   35   30
[2,]  NA   48   42   36
[3,]  NA   NA   49   42
```

The inner loop skips to the next iteration if the product of the current elements is greater than or equal to 54. Note the effect applies *only to that innermost loop*—that is, only the j *loopindex* is preemptively incremented, while i is left untouched, and the outer loop continues normally.

I've been using for loops to illustrate next and break, but they behave the same way inside while loops.

### 10.3.2   *The repeat Statement*

Another option for repeating a set of operations is the repeat statement. The general definition is simple.

```
repeat{
    do any code in here
}
```

Notice that a repeat statement doesn't include any kind of *loopindex* or *loopcondition*. To stop repeating the code inside the braces, you must use a break declaration inside the braced area (usually within an if statement); without it, the braced code will repeat without end, creating an infinite loop. To avoid this, you must make sure the operations will at some point cause the loop to reach a break.

To see repeat in action, you'll use it to calculate the famous mathematical series the *Fibonacci sequence*. The Fibonacci sequence is an infinite series of integers beginning with $1, 1, 2, 3, 5, 8, 13, \ldots$ where each term in the series is determined by the sum of the two previous terms. Formally, if $F_n$ represents the $n$th Fibonacci number, then you have:

$$F_{n+1} = F_n + F_{n-1}; \qquad n = 2, 3, 4, 5, \ldots,$$

where

$$F_1 = F_2 = 1.$$

The following repeat statement computes and prints the Fibonacci sequence, ending when it reaches a term greater than 150:

```
R> fib.a <- 1
R> fib.b <- 1
R> repeat{
+    temp <- fib.a+fib.b
+    fib.a <- fib.b
+    fib.b <- temp
+    cat(fib.b,", ",sep="")
+    if(fib.b>150){
+        cat("BREAK NOW...\n")
+        break
+    }
+ }
2, 3, 5, 8, 13, 21, 34, 55, 89, 144, 233, BREAK NOW...
```

First, the sequence is initialized by storing the first two terms, both 1, as fib.a and fib.b. Then, the repeat statement is entered, and it uses fib.a and fib.b to compute the next term in the sequence, stored as temp. Next, fib.a is overwritten to be fib.b, and fib.b is overwritten to be temp so that the two variables move forward through the series. That is, fib.b becomes the newly calculated Fibonacci number, and fib.a becomes the second-to-last number in the series so far. Use of cat then prints the new value of fib.b to the console. Finally, a check is made to see whether the latest term is greater than 150, and if it is, break is declared.

When you run the code, the braced area is repeated over and over until `fib.b` reaches the first number that is greater than 150, namely, 89 + 144 = 233. Once that happens, the `if` statement condition evaluates as `TRUE`, and R runs into `break`, terminating the loop.

The `repeat` statement is not as commonly used as the standard `while` or `for` loops, but it's useful if you don't want to be bound by formally specifying the *loopindex* and *loopvector* of a `for` loop or the *loopcondition* of a `while` loop. However, with `repeat`, you have to take a bit more caution to prevent infinite loops.

---

### Exercise 10.6

a. Using the same objects from Section 10.3.1,

```
foo <- 5
bar <- c(2,3,1.1,4,0,4.1,3)
```

do the following:

i. Write a `while` loop—*without* using `break` or `next`—that will reach exactly the same result as the `break` example in Section 10.3.1. That is, produce the same vector as `loop2.result` in the text.

ii. Obtain the same result as `loop3.result`, the example concerning `next`, using an `ifelse` function instead of a loop.

b. To demonstrate `while` loops in Section 10.2.2, you used the vector

```
mynumbers <- c(4,5,1,2,6,2,4,6,6,2)
```

to progressively fill `mylist` with identity matrices whose dimensions matched the values in `mynumbers`. The loop was instructed to stop when it reached the end of the numeric vector or a number that was greater than 5.

i. Write a `for` loop using a `break` declaration that does the same thing.

ii. Write a `repeat` statement that does the same thing.

c. Suppose you have two lists, `matlist1` and `matlist2`, both filled with numeric matrices as their members. Assume that all members have finite, nonmissing values, but *do not* assume that the dimensions of the matrices are the same throughout. Write a nested pair of `for` loops that aim to create a result list, `reslist`, of all possible *matrix products* (refer to Section 3.3.5) of the members of the two lists according to the following guidelines:

– The `matlist1` object should be indexed/searched in the outer loop, and the `matlist2` object should be indexed/searched in the inner loop.

- You're interested only in the possible matrix products of the members of matlist1 with the members of matlist2 in that order.
- If a particular multiple isn't possible (that is, if the ncol of a member of matlist1 doesn't match the nrow of a member of matlist2), then you should skip that multiplication, store the string "not possible" at the relevant position in reslist, and proceed directly to the next matrix multiplication.
- You can define a counter that is incremented at each comparison (inside the inner loop) to keep track of the current position of reslist.

Note, therefore, that the length of reslist will be equal to length(matlist1)*length(matlist2). Now, confirm the following results:

i. If you have

```
matlist1 <- list(matrix(1:4,2,2),matrix(1:4),matrix(1:8,4,2))
matlist2 <- matlist1
```

then all members of reslist should be "not possible" apart from members [[1]] and [[7]].

ii. If you have

```
matlist1 <- list(matrix(1:4,2,2),matrix(2:5,2,2),
                 matrix(1:16,4,2))
matlist2 <- list(matrix(1:8,2,4),matrix(10:7,2,2),
                 matrix(9:2,4,2))
```

then only the "not possible" members of reslist should be [[3]], [[6]], and [[9]].

## Important Code in This Chapter

| Function/operator | Brief description | First occurrence |
|---|---|---|
| if( ){ } | Conditional check | Section 10.1.1, p. 180 |
| if( ){ } else { } | Check and alternative | Section 10.1.2, p. 183 |
| ifelse | Element-wise if-else check | Section 10.1.3, p. 185 |
| switch | Multiple if choices | Section 10.1.5, p. 190 |
| for( ){ } | Iterative loop | Section 10.2.1, p. 194 |
| while( ){ } | Conditional loop | Section 10.2.2, p. 200 |
| apply | Implicit loop by margin | Section 10.2.3, p. 205 |
| tapply | Implicit loop by factor | Section 10.2.3, p. 207 |
| lapply | Implicit loop by member | Section 10.2.3, p. 207 |
| sapply | As lapply, array returned | Section 10.2.3, p. 207 |
| break | Exit explicit loop | Section 10.3.1, p. 210 |
| next | Skip to next loop iteration | Section 10.3.1, p. 210 |
| repeat{ } | Repeat code until break | Section 10.3.2, p. 212 |

# 11

# WRITING FUNCTIONS

Defining a function allows you to reuse a chunk of code without endlessly copying and pasting. It also allows other users to use your functions to carry out the same computations on their own data or objects. In this chapter, you'll learn about writing your own R functions. You'll learn how to define and use arguments, how to return output from a function, and how to specialize your functions in other ways.

## 11.1 The function Command

To define a function, use the `function` command and assign the results to an object name. Once you've done this, you can call the function using that object name just like any other built-in or contributed function in the workspace. This section will walk you through the basics of function creation and discuss some associated issues, such as returning objects and specifying arguments.

### 11.1.1   Function Creation

A function definition always follows this standard format:

```
functionname <- function(arg1,arg2,arg3,...){
    do any code in here when called
    return(returnobject)
}
```

The *functionname* placeholder can be any valid R object name, which is
what you'll ultimately use to call the function. Assign to this *functionname*
a call to function, followed by parentheses with any arguments you want
the function to have. The pseudocode includes three argument place-
holders plus an ellipsis. Of course, the number of arguments, their tags,
and whether to include an ellipsis all depend on the particular function
you're defining. If the function does not require any arguments, simply
include empty parentheses: (). If you do include arguments in this definition, note that they are not objects in the workspace and they do not have
any type or class attributes associated with them—they are merely a declaration of argument names that will be required by *functionname*.

When the function is called, it runs the code in the braced area (also
called the *function body* or *body code*). It can include if statements, loops,
and even other function calls. When encountering an internal function call
during execution, R follows the search rules discussed in Chapter 9. In the
braced area, you can use *arg1*, *arg2*, and *arg3*, and they are treated as objects
in the function's lexical environment.

Depending on how those declared arguments are used in the body
code, each argument may require a certain data type and object structure.
If you're writing functions that you intend for others to use, it's important to
have sound documentation to say what the function expects.

Often, the function body will include one or more calls to the return
command. When R encounters a return statement during execution, the
function exits, returning control to the user at the command prompt. This
mechanism is what allows you to pass results from operations in the function
back to the user. This output is denoted in the pseudocode by *returnobject*,
which is typically assigned an object created or calculated earlier in the function body. If there is no return statement, the function will simply return the
object created by the last executed expression (I'll discuss this feature more
in Section 11.1.2).

It's time for an example. Let's take the Fibonacci sequence generator
from Section 10.3.2 and turn it into a function in the editor.

```
myfib <- function(){
    fib.a <- 1
    fib.b <- 1
    cat(fib.a,", ",fib.b,", ",sep="")
    repeat{
        temp <- fib.a+fib.b
```

```
    fib.a <- fib.b
    fib.b <- temp
    cat(fib.b,", ",sep="")
    if(fib.b>150){
        cat("BREAK NOW...")
        break
    }
  }
}
```

I've named the function myfib, and it doesn't use or require any arguments. The body code is identical to the example in Section 10.3.2, except I've added the third line, cat(fib.a,", ",fib.b,", ",sep=""), to ensure the first two terms, 1 and 1, are also printed to the screen.

Before you can call myfib from the console, you have to send the function definition there. Highlight the code in the editor and press CTRL-R or ⌘-RETURN.

```
R> myfib <- function(){
+    fib.a <- 1
+    fib.b <- 1
+    cat(fib.a,", ",fib.b,", ",sep="")
+    repeat{
+        temp <- fib.a+fib.b
+        fib.a <- fib.b
+        fib.b <- temp
+        cat(fib.b,", ",sep="")
+        if(fib.b>150){
+            cat("BREAK NOW...")
+            break
+        }
+    }
+ }
```

This imports the function into the workspace (if you enter ls() at the command prompt, "myfib" will now appear in the list of present objects). This step is required anytime you create or modify a function and want to use it from the command prompt.

Now you can call the function from the console.

```
R> myfib()
1, 1, 2, 3, 5, 8, 13, 21, 34, 55, 89, 144, 233, BREAK NOW...
```

It computes and prints the Fibonacci sequence up to 250, just as instructed.

## Adding Arguments

Rather than printing a fixed set of terms, let's add an argument to control how many Fibonacci numbers are printed. Consider the following new function, `myfib2`, with this modification:

```
myfib2 <- function(thresh){
    fib.a <- 1
    fib.b <- 1
    cat(fib.a,", ",fib.b,", ",sep="")
    repeat{
        temp <- fib.a+fib.b
        fib.a <- fib.b
        fib.b <- temp
        cat(fib.b,", ",sep="")
        if(fib.b>thresh){
            cat("BREAK NOW...")
            break
        }
    }
}
```

This version now takes a single argument, `thresh`. In the body code, `thresh` acts as a threshold determining when to end the repeat procedure, halt printing, and complete the function—once a value of `fib.b` that is greater than `thresh` is calculated, the repeat statement will exit after encountering the call to break. Therefore, the output printed to the console will be the Fibonacci sequence up to and including the first `fib.b` value bigger than `thresh`. This means that `thresh` must be supplied as a single numeric value—supplying a character string, for example, would make no sense.

After importing the definition of `myfib2` into the console, note the same results as given by the original `myfib` when you set `thresh=150`.

```
R> myfib2(thresh=150)
1, 1, 2, 3, 5, 8, 13, 21, 34, 55, 89, 144, 233, BREAK NOW...
```

But now you can print the sequence to any limit you want (this time using positional matching to specify the argument):

```
R> myfib2(1000000)
1, 1, 2, 3, 5, 8, 13, 21, 34, 55, 89, 144, 233, 377, 610, 987, 1597, 2584,
4181, 6765, 10946, 17711, 28657, 46368, 75025, 121393, 196418, 317811,
514229, 832040, 1346269, BREAK NOW...
```

## Returning Results

If you want to use the results of a function in future operations (rather than just printing output to the console), you need to return content to the user.

Continuing with the current example, here's a Fibonacci function that stores the sequence in a vector and returns it:

```
myfib3 <- function(thresh){
    fibseq <- c(1,1)
    counter <- 2
    repeat{
        fibseq <- c(fibseq,fibseq[counter-1]+fibseq[counter])
        counter <- counter+1
        if(fibseq[counter]>thresh){
            break
        }
    }
    return(fibseq)
}
```

First you create the vector `fibseq` and assign it the first two terms of the sequence. This vector will ultimately become the *returnobject*. You also create a `counter` initialized to 2 to keep track of the current position in `fibseq`. Then the function enters a `repeat` statement, which overwrites `fibseq` with `c(fibseq,fibseq[counter-1]+fibseq[counter])`. That expression constructs a new `fibseq` by appending the sum of the most recent two terms to the contents of what is already stored in `fibseq`. For example, with `counter` starting at 2, the first run of this line will sum `fibseq[1]` and `fibseq[2]`, appending the result as a third entry onto the original `fibseq`.

Next, `counter` is incremented, and the condition is checked. If the most recent value of `fibseq[counter]` is not greater than `thresh`, the loop repeats. If it is greater, the loop breaks, and you reach the final line of `myfib3`. Calling `return` ends the function and passes out the specified *returnobject* (in this case, the final contents of `fibseq`).

After importing `myfib3`, consider the following code:

```
R> myfib3(150)
 [1]   1   1   2   3   5   8  13  21  34  55  89 144 233
R> foo <- myfib3(10000)
R> foo
 [1]     1     1     2     3     5     8    13    21    34    55    89   144
[13]   233   377   610   987  1597  2584  4181  6765 10946
R> bar <- foo[1:5]
R> bar
[1] 1 1 2 3 5
```

Here, the first line calls `myfib3` with `thresh` assigned 150. The output is still printed to the screen, but this isn't the result of the `cat` command as it was earlier; it is the *returnobject*. You can assign this *returnobject* to a variable, such as `foo`, and `foo` is now just another R object in the global environment that you can manipulate. For example, you use it to create `bar` with a simple vector subset. This would not have been possible with either `myfib` or `myfib2`.

### 11.1.2 Using return

If there's no return statement inside a function, the function will end when the last line in the body code has been run, at which point it will return the most recently assigned or created object in the function. If nothing is created, such as in myfib and myfib2 from earlier, the function returns NULL. To demonstrate this point, enter the following two dummy functions in the editor:

```
dummy1 <- function(){
    aa <- 2.5
    bb <- "string me along"
    cc <- "string 'em up"
    dd <- 4:8
}

dummy2 <- function(){
    aa <- 2.5
    bb <- "string me along"
    cc <- "string 'em up"
    dd <- 4:8
    return(dd)
}
```

The first function, dummy1, simply assigns four different objects in its lexical environment (not the global environment) and doesn't explicitly return anything. On the other hand, dummy2 creates the same four objects and explicitly returns the last one, dd. If you import and run the two functions, both provide the same return object.

```
R> foo <- dummy1()
R> foo
[1] 4 5 6 7 8
R> bar <- dummy2()
R> bar
[1] 4 5 6 7 8
```

A function will end as soon as it evaluates a return command, without executing any remaining code in the function body. To emphasize this, consider one more version of the dummy function:

```
dummy3 <- function(){
    aa <- 2.5
    bb <- "string me along"
    return(aa)
    cc <- "string 'em up"
    dd <- 4:8
```

```
    return(bb)
}
```

Here, `dummy3` has two calls to return: one in the middle and one at the end. But when you import and execute the function, it returns only one value.

```
R> baz <- dummy3()
R> baz
[1] 2.5
```

Executing `dummy3` returns only the object `aa` because only the first instance of return is executed and the function exits immediately at that point. In the current definition of `dummy3`, the last three lines (the assignment of `cc` and `dd` and the return of `bb`) will never be executed.

Using return adds another function call to your code, so technically, it introduces a little extra computational expense. Because of this, some argue that return statements should be avoided unless absolutely necessary. But the additional computational cost of the call to return is small enough to be negligible for most purposes. Plus, return statements can make code more readable, making it easier to see where the author of a function intends it to complete and precisely what is intended to be supplied as output. I'll use return throughout the remainder of this work.

## Exercise 11.1

a. Write another Fibonacci sequence function, naming it `myfib4`. This function should provide an option to perform either the operations available in `myfib2`, where the sequence is simply printed to the console, or the operations in `myfib3`, where a vector of the sequence is formally returned. Your function should take two arguments: the first, `thresh`, should define the limit of the sequence (just as in `myfib2` or `myfib3`); the second, `printme`, should be a logical value. If `TRUE`, then `myfib4` should just print; if `FALSE`, then `myfib4` should return a vector. Confirm the correct results arise from the following calls:
   - `myfib4(thresh=150,printme=TRUE)`
   - `myfib4(1000000,T)`
   - `myfib4(150,FALSE)`
   - `myfib4(1000000,printme=F)`

b. In Exercise 10.4 on page 203, you were tasked with writing a while loop to perform integer factorial calculations.
   i. Using your factorial while loop (or writing one if you didn't do so earlier), write your own R function, `myfac`, to compute the factorial of an integer argument `int` (you may assume `int`

will always be supplied as a non-negative integer). Perform a quick test of the function by computing 5 factorial, which is 120; 12 factorial, which is 479,001,600; and 0 factorial, which is 1.

ii. Write another version of your factorial function, naming it myfac2. This time, you may still assume int will be supplied as an integer but not that it will be non-negative. If negative, the function should return NaN. Test myfac2 on the same three values as previously, but also try using int=-6.

## 11.2 Arguments

Arguments are an essential part of most R functions. In this section, you'll consider how R evaluates arguments. You'll also see how to write functions that have default argument values, how to make functions handle missing argument values, and how to pass extra arguments into an internal function call with ellipses.

### 11.2.1 Lazy Evaluation

An important concept related to handling arguments in many high-level programming languages is *lazy evaluation.* Generally, this refers to the fact that expressions are evaluated only when they are needed. This applies to arguments in the sense that they are accessed and used only at the point they appear in the function body.

Let's see exactly how R functions recognize and use arguments during execution. As a working example to be used throughout this section, you'll write a function to search through a specified list for matrix objects and attempt to post-multiply each with another matrix specified as a second argument (refer back to Section 3.3.5 for details on matrix multiplication). The function will store and return the result in a new list. If no matrices are in the supplied list or if no appropriate matrices (given the dimensions of the multiplying matrix) are present, the function should return a character string informing the user of these facts. You can assume that if there are matrices in the specified list, they will be numeric. Consider the following function, which I'll call multiples1:

```
multiples1 <- function(x,mat,str1,str2){
    matrix.flags <- sapply(x,FUN=is.matrix)

    if(!any(matrix.flags)){
        return(str1)
    }

    indexes <- which(matrix.flags)
    counter <- 0
```

```
result <- list()
for(i in indexes){
    temp <- x[[i]]
    if(ncol(temp)==nrow(mat)){
        counter <- counter+1
        result[[counter]] <- temp%*%mat
    }
}

if(counter==0){
    return(str2)
} else {
    return(result)
}
}
```

This function takes four arguments, with no default values assigned. The target list to search is intended to be supplied to x; the post-multiplying matrix is supplied to mat; and two other arguments, str1 and str2, take character strings to return if x has no suitable members.

Inside the body code, a vector called matrix.flags is created with the sapply implicit looping function. This applies the function is.matrix to the list argument x. The result is a logical vector of equal length as x, with TRUE elements where the corresponding member of x is in fact a matrix. If there are no matrices in x, the function hits a return statement, which exits the function and outputs the argument str1.

If the function did not exit at that point, this means there are indeed matrices in x. The next step is to retrieve the matrix member indexes by applying which to matrix.flags. A counter is initialized to 0 to keep track of how many successful matrix multiplications are carried out, and an empty list (result) is created to store any results.

Next, you enter a for loop. For each member of indexes, the loop stores the matrix member at that position as temp and checks to see whether it's possible to perform post-multiplication of temp by the argument mat (to perform the operation, ncol(temp) must equal nrow(mat)). If the matrices are compatible, counter is incremented, and this position of result is filled with the relevant calculation. If FALSE, nothing is done. The indexer, i, then takes on the next value of indexes and repeats until completion.

The final procedure in multiples1 checks whether the for loop actually found any compatible matrix products. If no compatibility existed, the braced if statement code inside the for loop would never have been executed, and the counter would remain set to zero. So, if counter is still equal to zero upon completion of the loop, the function simply returns the str2 argument. Otherwise, if compatible matrices were found, appropriate results will have been computed, and multiples1 returns the result list, which would have at least one member.

It's time to import and then test the function. You'll use the following three list objects:

```
R> foo <- list(matrix(1:4,2,2),"not a matrix",
                "definitely not a matrix",matrix(1:8,2,4),matrix(1:8,4,2))
R> bar <- list(1:4,"not a matrix",c(F,T,T,T),"??")
R> baz <- list(1:4,"not a matrix",c(F,T,T,T),"??",matrix(1:8,2,4))
```

You'll set the argument mat to the $2 \times 2$ identity matrix (post-multiplying any appropriate matrix by this will simply return the original matrix), and you'll pass in appropriate string messages for str1 and str2. Here's how the function works on foo:

```
R> multiples1(x=foo,mat=diag(2),str1="no matrices in 'x'",
             str2="matrices in 'x' but none of appropriate dimensions given
             'mat'")
[[1]]
     [,1] [,2]
[1,]    1    3
[2,]    2    4

[[2]]
     [,1] [,2]
[1,]    1    5
[2,]    2    6
[3,]    3    7
[4,]    4    8
```

The function has returned result with the two compatible matrices of foo (members [[1]] and [[5]]). Now let's try it on bar using the same arguments.

```
R> multiples1(x=bar,mat=diag(2),str1="no matrices in 'x'",
             str2="matrices in 'x' but none of appropriate dimensions given
             'mat'")
[1] "no matrices in 'x'"
```

This time, the value of str1 has been returned. The initial check identified that there are no matrices in the list supplied to x, so the function has exited before the for loop. Finally, let's try baz.

```
R> multiples1(x=baz,mat=diag(2),str1="no matrices in 'x'",
             str2="matrices in 'x' but none of appropriate dimensions given
             'mat'")
[1] "matrices in 'x' but none of appropriate dimensions given 'mat'"
```

Here the value of str2 was returned. Though there is a matrix in baz and the for loop in the body code of multiples1 has been executed, the matrix is not compatible for post-multiplication by mat.

Notice that the string arguments str1 and str2 are used only when the argument x does not contain a matrix with the appropriate dimensions. When you applied multiples1 to x=foo, for example, there was no need to use str1 or str2. R evaluates the defined expressions lazily, dictating that argument values are sought only at the moment they are required during execution. In this function, str1 and str2 are required only when the input list doesn't have suitable matrices, so you could lazily ignore providing values for these arguments when x=foo.

```
R> multiples1(x=foo,mat=diag(2))
[[1]]
     [,1] [,2]
[1,]   1    3
[2,]   2    4

[[2]]
     [,1] [,2]
[1,]   1    5
[2,]   2    6
[3,]   3    7
[4,]   4    8
```

This returns the same results as before with no problem whatsoever. Attempting this with bar, on the other hand, doesn't work.

```
R> multiples1(x=bar,mat=diag(2))
Error in multiples1(x = bar, mat = diag(2)) :
  argument "str1" is missing, with no default
```

Here we are quite rightly chastised by R because it requires the value for str1. It informs us that the value is missing and there is no default.

## 11.2.2   Setting Defaults

The previous example shows one case where it's useful to set default values for certain arguments. Default argument values are also sensible in many other situations, such as when the function has a large number of arguments or when arguments have natural values that are used more often than not. Let's write a new version of the multiples1 function from Section 11.2.1, multiples2, which now includes default values for str1 and str2.

```
multiples2 <- function(x,mat,str1="no valid matrices",str2=str1){
    matrix.flags <- sapply(x,FUN=is.matrix)

    if(!any(matrix.flags)){
```

```
        return(str1)
    }

    indexes <- which(matrix.flags)
    counter <- 0
    result <- list()
    for(i in indexes){
        temp <- x[[i]]
        if(ncol(temp)==nrow(mat)){
            counter <- counter+1
            result[[counter]] <- temp%*%mat
        }
    }

    if(counter==0){
        return(str2)
    } else {
        return(result)
    }
}
```

Here, you have given str1 a default value of "no valid matrices" by assigning the string value in the formal definition of the arguments. You've also set a default for str2 by assigning str1 to it. If you import and execute this function again on the three lists, you no longer need to explicitly provide values for those arguments.

```
R> multiples2(foo,mat=diag(2))
[[1]]
     [,1] [,2]
[1,]    1    3
[2,]    2    4

[[2]]
     [,1] [,2]
[1,]    1    5
[2,]    2    6
[3,]    3    7
[4,]    4    8

R> multiples2(bar,mat=diag(2))
[1] "no valid matrices"
R> multiples2(baz,mat=diag(2))
[1] "no valid matrices"
```

You can now call the function, whatever the outcome, without being required to specify every argument in full. If you don't want to use the default arguments in a specific call, you can still specify different values for those arguments when calling the function, and those values will overwrite the defaults.

### 11.2.3   Checking for Missing Arguments

The missing function checks the arguments of a function to see if all required arguments have been supplied. It takes an argument tag and returns a single logical value of TRUE if the specified argument isn't found. You can use missing to avoid the error you saw in an earlier call to multiples1, when str1 was required but not supplied.

In some situations, the missing function can be particularly useful in the body code. Consider another modification to the example function:

```
multiples3 <- function(x,mat,str1,str2){
    matrix.flags <- sapply(x,FUN=is.matrix)

    if(!any(matrix.flags)){
        if(missing(str1)){
            return("'str1' was missing, so this is the message")
        } else {
            return(str1)
        }
    }

    indexes <- which(matrix.flags)
    counter <- 0
    result <- list()
    for(i in indexes){
        temp <- x[[i]]
        if(ncol(temp)==nrow(mat)){
            counter <- counter+1
            result[[counter]] <- temp%*%mat
        }
    }

    if(counter==0){
        if(missing(str2)){
            return("'str2' was missing, so this is the message")
        } else {
            return(str2)
        }
    } else {
        return(result)
    }
}
```

The only differences between this version and multiples1 are in the first and last if statements. The first if statement checks whether there are no matrices in x, in which case it returns a string message. In multiples1, that message was always str1, but now you use another if statement with missing(str1) to see whether the str1 argument actually has a value first. If not, the function returns another character string saying that str1 was missing. A similar alternative is defined for str2. Here it is once more importing the function and using foo, bar, and baz:

```
R> multiples3(foo,diag(2))
[[1]]
     [,1] [,2]
[1,]   1    3
[2,]   2    4

[[2]]
     [,1] [,2]
[1,]   1    5
[2,]   2    6
[3,]   3    7
[4,]   4    8

R> multiples3(bar,diag(2))
[1] "'str1' was missing, so this is the message"
R> multiples3(baz,diag(2))
[1] "'str2' was missing, so this is the message"
```

Using missing this way permits arguments to be left unsupplied in a given function call. It is primarily used when it's difficult to choose a default value for a certain argument, yet the function still needs to handle cases when that argument isn't provided. In the current example, it makes more sense to define defaults for str1 and str2, as you did for multiples2, and avoid the extra code required to implement missing.

### 11.2.4   Dealing with Ellipses

In Section 9.2.5, I introduced the ellipsis, also called dot-dot-dot notation. The ellipsis allows you to pass in extra arguments without having to first define them in the argument list, and these arguments can then be passed to another function call within the code body. When included in a function definition, the ellipsis is often (but not always) placed in the last position because it represents a variable number of arguments.

Building on the myfib3 function from Section 11.1.1, let's use the ellipsis to write a function that can plot the specified Fibonacci numbers.

```
myfibplot <- function(thresh,plotit=TRUE,...){
    fibseq <- c(1,1)
    counter <- 2
```

```
    repeat{
        fibseq <- c(fibseq,fibseq[counter-1]+fibseq[counter])
        counter <- counter+1
        if(fibseq[counter]>thresh){
            break
        }
    }

    if(plotit){
        plot(1:length(fibseq),fibseq,...)
    } else {
        return(fibseq)
    }
}
```

In this function, an if statement checks to see whether the plotit argument is TRUE (which is the default value). If so, then you call plot, passing in 1:length(fibseq) for the *x*-axis coordinates and the Fibonacci numbers themselves for the *y*-axis. After these coordinates, you also pass the ellipsis directly into plot. In this case, the ellipsis represents any additional arguments a user might pass in to control the execution of plot.

Importing myfibplot and executing the following line, the plot in Figure 11-1 pops up in a graphics device.

```
R> myfibplot(150)
```

Here you used positional matching to assign 150 to thresh, leaving the default value for the plotit argument. The ellipsis is empty in this call.

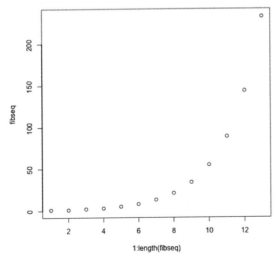

*Figure 11-1: The default plot produced by a call to* myfibplot, *with* thresh=150

Since you didn't specify otherwise, R has simply followed the default behavior of plot. You can spruce things up by specifying more plotting options. The following line produces the plot in Figure 11-2:

```
R> myfibplot(150,type="b",pch=4,lty=2,main="Terms of the Fibonacci sequence",
            ylab="Fibonacci number",xlab="Term (n)")
```

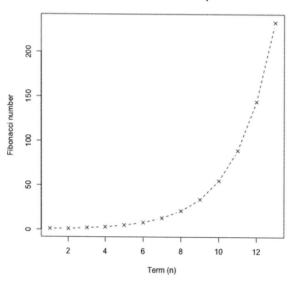

Figure 11-2: A plot produced by a call to myfibplot with graphical parameters passed in using the ellipsis

Here the ellipsis allows you to pass arguments to plot through the call to myfibplot, even though the particular graphical parameters are not explicitly defined arguments of myfibplot.

Ellipses can be convenient, but they require care. The ambiguous ... can represent any number of mysterious arguments. Good function documentation is key to indicate the appropriate usage.

If you want to unpack the arguments passed in through an ellipsis, you can use the list function to convert those arguments into a list. Here's an example:

```
unpackme <- function(...){
    x <- list(...)
    cat("Here is ... in its entirety as a list:\n")
    print(x)
    cat("\nThe names of ... are:",names(x),"\n\n")
    cat("\nThe classes of ... are:",sapply(x,class))
}
```

This dummy function simply takes an ellipsis and converts it to a list with x <- list(...). This subsequently allows the object x to be treated the same way as any other list. In this case, you can summarize the object by providing its names and class attributes. Here's a sample run:

```
R> unpackme(aa=matrix(1:4,2,2),bb=TRUE,cc=c("two","strings"),
          dd=factor(c(1,1,2,1)))
Here is ... in its entirety as a list:
$aa
     [,1] [,2]
[1,]   1    3
[2,]   2    4

$bb
[1] TRUE

$cc
[1] "two"      "strings"

$dd
[1] 1 1 2 1
Levels: 1 2


The names of ... are: aa bb cc dd


The classes of ... are: matrix logical character factor
```

Four tagged arguments, aa, bb, cc, and dd, are provided as the contents of the ellipsis, and they are explicitly identified within unpackme by using the simple list(...) operation. This construction can be useful for identifying or extracting specific arguments supplied through ... in a given call.

## Exercise 11.2

a. Accruing annual compound interest is a common financial benefit for investors. Given a principal investment amount $P$, an interest rate per annum $i$ (expressed as a percentage), and a frequency of interest paid per year $t$, the final amount $F$ after $y$ years is given as follows:

$$F = P\left(1 + \frac{i}{100t}\right)^{ty}$$

Write a function that can compute $F$ as per the following notes:

- Arguments must be present for $P, i, t,$ and $y$. The argument for $t$ should have a default value of 12.
- Another argument giving a logical value that determines whether to plot the amount $F$ at each integer time should be included. For example, if plotit=TRUE (the default) and $y$ is 5 years, the plot should show the amount $F$ at $y = 1,2,3,4,5$.
- If this function is plotted, the plot should always be a step-plot, so plot should always be called with type="s".
- If plotit=FALSE, the final amount $F$ should be returned as a numeric vector corresponding to the same integer times, as shown earlier.
- An ellipsis should also be included to control other details of plotting, if it takes place.

Now, using your function, do the following:

i. Work out the final amount after a 10-year investment of a principal of $5000, at an interest rate of 4.4 percent per annum compounded monthly.

ii. Re-create the following step-plot, which shows the result of $100 invested at 22.9 percent per annum, compounded monthly, for 20 years:

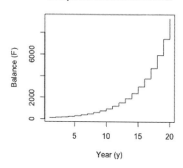

**Compound interest calculator**

iii. Perform another calculation based on the same parameters as in (ii), but this time, assume the interest is compounded annually. Return and store the results as a numeric vector. Then, use lines to add a second step-line, corresponding to this annually accrued amount, to the plot created previously. Use a different color or line type and make use of the legend function so the two lines can be differentiated.

b. A *quadratic equation* in the variable $x$ is often expressed in the following form:

$$k_1 x^2 + k_2 x + k_3 = 0$$

Here, $k_1$, $k_2$, and $k_3$ are constants. Given values for these constants, you can attempt to find up to two *real roots*—values of $x$ that satisfy the equation. Write a function that takes $k_1$, $k_2$, and $k_3$ as arguments and finds and returns any solutions (as a numeric vector) in such a situation. This is achieved as follows:

- Evaluate $k_2^2 - 4k_1k_3$. If this is negative, there are no solutions, and an appropriate message should be printed to the console.
- If $k_2^2 - 4k_1k_3$ is zero, then there is one solution, computed by $-k_2/2k_1$.
- If $k_2^2 - 4k_1k_3$ is positive, then there are two solutions, given by $(-k_2 - (k_2^2 - 4k_1k_3)^{0.5})/2k_1$ and $(-k_2 + (k_2^2 - 4k_1k_3)^{0.5})/2k_1$.
- No default values are needed for the three arguments, but the function should check to see whether any are missing. If so, an appropriate character string message should be returned to the user, informing the user that the calculations are not possible.

Now, test your function.

i. Confirm the following:
   * $2x^2 - x - 5$ has roots 1.850781 and −1.350781.
   * $x^2 + x + 1$ has no real roots.

ii. Attempt to find solutions to the following quadratic equations:
   * $1.3x^2 - 8x - 3.13$
   * $2.25x^2 - 3x + 1$
   * $1.4x^2 - 2.2x - 5.1$
   * $-5x^2 + 10.11x - 9.9$

iii. Test your programmed response in the function if one of the arguments is missing.

## 11.3   Specialized Functions

In this section, you'll look at three kinds of specialized user-defined R functions. First, you'll look at helper functions, which are designed to be called multiple times by another function (and they can even be defined inside the body of a parent function). Next, you'll look at disposable functions, which can be directly defined as an argument to another function call. Finally, you'll look at recursive functions, which call themselves.

### 11.3.1   Helper Functions

It is common for R functions to call other functions from within their body code. A *helper function* is a general term used to describe functions written and used specifically to facilitate the computations carried out by another function. They're a good way to improve the readability of complicated functions.

A helper function can be either defined internally (within another function definition) or externally (within the global environment). In this section, you'll see an example of each.

### Externally Defined

Building on the multiples2 function from Section 11.2.2, here's a new version that splits the functionality over two separate functions, one of which is an externally defined helper function:

```
multiples_helper_ext <- function(x,matrix.flags,mat){
    indexes <- which(matrix.flags)
    counter <- 0
    result <- list()
    for(i in indexes){
        temp <- x[[i]]
        if(ncol(temp)==nrow(mat)){
            counter <- counter+1
            result[[counter]] <- temp%*%mat
        }
    }
    return(list(result,counter))
}

multiples4 <- function(x,mat,str1="no valid matrices",str2=str1){
    matrix.flags <- sapply(x,FUN=is.matrix)

    if(!any(matrix.flags)){
        return(str1)
    }

    helper.call <- multiples_helper_ext(x,matrix.flags,mat)
    result <- helper.call[[1]]
    counter <- helper.call[[2]]

    if(counter==0){
        return(str2)
    } else {
        return(result)
    }
}
```

If you import and execute this code on the sample lists from earlier, it behaves the same way as the previous version. All you've done here is moved the matrix-checking loop to an external function. The multiples4 function now calls a helper function named multiples_helper_ext. Once the code in `multiples4` makes sure that there are in fact matrices in the list `x` to be checked, it calls `multiples_helper_ext` to execute the required loop. This helper function is defined externally, meaning that it exists in the global environment for any other function to call, making it easier to reuse.

### Internally Defined

If the helper function is intended to be used for only one particular function, it makes more sense to define the helper function internally, within the lexical environment of the function that calls it. The fifth version of the matrix multiplication function does just that, shifting the definition to within the body code.

```
multiples5 <- function(x,mat,str1="no valid matrices",str2=str1){
    matrix.flags <- sapply(x,FUN=is.matrix)

    if(!any(matrix.flags)){
        return(str1)
    }

    multiples_helper_int <- function(x,matrix.flags,mat){
        indexes <- which(matrix.flags)
        counter <- 0
        result <- list()
        for(i in indexes){
            temp <- x[[i]]
            if(ncol(temp)==nrow(mat)){
                counter <- counter+1
                result[[counter]] <- temp%*%mat
            }
        }
        return(list(result,counter))
    }

    helper.call <- multiples_helper_int(x,matrix.flags,mat)
    result <- helper.call[[1]]
    counter <- helper.call[[2]]

    if(counter==0){
        return(str2)
    } else {
        return(result)
    }
}
```

Now the helper function `multiples_helper_int` is defined within `multiples5`. That means it's visible only within the lexical environment as opposed to residing in the global environment like `multiples_helper_ext`. It makes sense to internally define a helper function when (a) it's used only by a single parent function, and (b) it's called multiple times within the parent function. (Of course, multiples5 satisfies only (a), and it's provided here just for the sake of illustration.)

## 11.3.2 Disposable Functions

Often, you may need a function that performs a simple, one-line task. For example, when you use apply, you'll typically just want to pass in a short, simple function as an argument. That's where *disposable* (or *anonymous*) functions come in—they allow you to define a function intended for use in a single instance without explicitly creating a new object in your global environment.

Say you have a numeric matrix whose columns you want to repeat twice and then sort.

```
R> foo <- matrix(c(2,3,3,4,2,4,7,3,3,6,7,2),3,4)
R> foo
     [,1] [,2] [,3] [,4]
[1,]   2    4    7    6
[2,]   3    2    3    7
[3,]   3    4    3    2
```

This is a perfect task for apply, which can apply a function to each column of the matrix. This function simply has to take in a vector, repeat it, and sort the result. Rather than define that short function separately, you can define a disposable function right in the argument of apply using the function command.

```
R> apply(foo,MARGIN=2,FUN=function(x){sort(rep(x,2))})
     [,1] [,2] [,3] [,4]
[1,]   2    2    3    2
[2,]   2    2    3    2
[3,]   3    4    3    6
[4,]   3    4    3    6
[5,]   3    4    7    7
[6,]   3    4    7    7
```

The function is defined in the standard format directly in the call to apply. This function is defined, called, and then immediately forgotten once apply is complete. It is disposable in the sense that it exists only for the one instance where it is actually used.

Using the function command this way is a shortcut more than anything else; plus, it avoids the unnecessary creation and storage of a function object in the global environment.

### 11.3.3 Recursive Functions

*Recursion* is when a function calls itself. This technique isn't commonly used in statistical analyses, but it pays to be aware of it. This section will briefly illustrate what it means for a function to call itself.

Suppose you want to write a function that takes a single positive integer argument n and returns the corresponding *n*th term of the Fibonacci sequence (where $n = 1$ and $n = 2$ correspond to the initial two terms 1 and 1, respectively). Earlier you built up the Fibonacci sequence in an *iterative* fashion by using a loop. In a recursive function, instead of using a loop to repeat an operation, the function calls itself multiple times. Consider the following:

```
myfibrec <- function(n){
    if(n==1||n==2){
        return(1)
    } else {
        return(myfibrec(n-1)+myfibrec(n-2))
    }
}
```

The recursive `myfibrec` checks a single `if` statement that defines a *stopping condition*. If either 1 or 2 is supplied to the function (requesting the first or second Fibonacci number), then `myfibrec` directly returns 1. Otherwise, the function returns the sum of `myfibrec(n-1)` and `myfibrec(n-2)`. That means if you call `myfibrec` with n greater than 2, the function generates two more calls to `myfibrec`, using n-1 and n-2. The recursion continues until it reaches a call for the 1st or 2nd term, triggering the stopping condition, `if(n==1||n==2)`, which simply returns 1. Here's a sample call that retrieves the fifth Fibonacci number:

```
R> myfibrec(5)
[1] 5
```

Figure 11-3 shows the structure of this recursive call.

Note that an accessible stopping rule is critical to any recursive function. Without one, recursion will continue indefinitely. For example, the current definition of `myfibrec` works as long as the user supplies a positive integer for the argument n. But if n is negative, the stopping rule condition will never be satisfied, and the function will recur indefinitely (though R has some automated safeguards to help prevent this and should just return an error message rather than getting stuck in an infinite loop).

Recursion is a powerful approach, especially when you don't know ahead of time how many times a function needs be called to complete a task. For many sort and search algorithms, recursion provides the speediest and most efficient solution. But in simpler cases, such as the Fibonacci example here, the recursive approach often requires more computational expense than an iterative looping approach. For beginners, I recommended sticking with explicit loops unless recursion is strictly required.

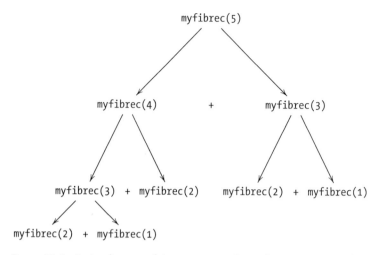

Figure 11-3: A visualization of the recursive calls made to myfibrec with n=5

## Exercise 11.3

a. Given a list whose members are character string vectors of varying lengths, use a disposable function with lapply to paste an exclamation mark onto the end of each element of each member, with an empty string as the separation character (note that the default behavior of paste when applied to character vectors is to perform the concatenation on each element). Execute your line of code on the list given by the following:

```
foo <- list("a",c("b","c","d","e"),"f",c("g","h","i"))
```

b. Write a recursive version of a function implementing the non-negative integer factorial operator (see Exercise 10.4 on page 203 for details of the factorial operator). The stopping rule should return the value 1 if the supplied integer is 0. Confirm that your function produces the same results as earlier.

   i.   5 factorial is 120.
   ii.  12 factorial is 479,001,600.
   iii. 0 factorial is 1.

c. For this problem, I'll introduce the *geometric mean*. The geometric mean is a particular measure of centrality, different from the more common arithmetic mean. Given $n$ observations denoted $x_1, x_2, \ldots, x_n$, their geometric mean $\bar{g}$ is computed as follows:

$$\bar{g} = (x_1 \times x_2 \times \ldots \times x_n)^{1/n} = \left( \prod_{i=1}^{n} x_i \right)^{1/n}$$

For example, to find the geometric mean of the data 4.3, 2.1, 2.2, 3.1, calculate the following:

$$\bar{g} = (4.3 \times 2.1 \times 2.2 \times 3.1)^{1/4} = 61.5846^{0.25} = 2.8$$

(This is rounded to 1 d.p.)

Write a function named geolist that can search through a specified list and compute the geometric means of each member per the following guidelines:

- Your function must define and use an internal helper function that returns the geometric mean of a vector argument.
- Assume the list can only have numeric vectors or numeric matrices as its members. Your function should contain an appropriate loop to inspect each member in turn.
- If the member is a vector, compute the geometric mean of that vector, overwriting the member with the result, which should be a single number.
- If the member is a matrix, use an implicit loop to compute the geometric mean *of each row* of the matrix, overwriting the member with the results.
- The final list should be returned to the user.

Now, as a quick test, check that your function matches the following two calls:

i.

```
R> foo <- list(1:3,matrix(c(3.3,3.2,2.8,2.1,4.6,4.5,3.1,9.4),4,2),
             matrix(c(3.3,3.2,2.8,2.1,4.6,4.5,3.1,9.4),2,4))
R> geolist(foo)
[[1]]
[1] 1.817121

[[2]]
[1] 3.896152 3.794733 2.946184 4.442972

[[3]]
[1] 3.388035 4.106080
```

ii.

```
R> bar <- list(1:9,matrix(1:9,1,9),matrix(1:9,9,1),matrix(1:9,3,3))
R> geolist(bar)
[[1]]
[1] 4.147166

[[2]]
[1] 4.147166
```

```
[[3]]
[1] 1 2 3 4 5 6 7 8 9

[[4]]
[1] 3.036589 4.308869 5.451362
```

## Important Code in This Chapter

| Function/operator | Brief description | First occurrence |
|---|---|---|
| function | Function creation | Section 11.1.1, p. 216 |
| return | Function return objects | Section 11.1.1, p. 219 |
| missing | Argument check | Section 11.2.3, p. 227 |
| ... | Ellipsis (as argument) | Section 11.2.4, p. 228 |

# 12

## EXCEPTIONS, TIMINGS, AND VISIBILITY

Now that you've seen how to write your own functions in R, let's examine some common function augmentations and behaviors. In this chapter, you'll learn how to make your functions throw an error or warning when they receive unexpected input. You'll also see some simple ways to measure completion time and check progress for computationally expensive functions. Finally, you'll see how R masks functions when two have the same name but reside in different packages.

## 12.1 Exception Handling

When there's an unexpected problem during execution of a function, R will notify you with either a *warning* or an *error*. In this section, I'll demonstrate how to build these constructs into your own functions where appropriate. I'll also show how to *try* a calculation to check whether it's possible without an error (that is, to see whether it'll even work).

### 12.1.1   Formal Notifications: Errors and Warnings

In Chapter 11, you made your functions print a string (for example,
"no valid matrices") when they couldn't perform certain operations. Warnings and errors are more formal mechanisms designed to convey these types
of messages and handle subsequent operations. An error forces the function
to immediately terminate at the point it occurs. A warning is less severe. It
indicates that the function is being run in an atypical way but tries to work
around the issue and continue executing. In R, you can issue warnings with
the warning command, and you can throw errors with the stop command.
The following two functions show an example of each:

```
warn_test <- function(x){
    if(x<=0){
        warning("'x' is less than or equal to 0 but setting it to 1 and
                continuing")
        x <- 1
    }
    return(5/x)
}

error_test <- function(x){
    if(x<=0){
        stop("'x' is less than or equal to 0... TERMINATE")
    }
    return(5/x)
}
```

Both warn_test and error_test divide 5 by the argument x. They also both
expect x to be positive. In warn_test, if x is nonpositive, the function issues a
warning, and x is overwritten to be 1. In error_test, on the other hand, if x is
nonpositive, the function throws an error and terminates immediately. The
two commands warning and stop are used with a character string argument,
which becomes the message printed to the console.

You can see these notifications by importing and calling the functions as
follows:

```
R> warn_test(0)
[1] 5
Warning message:
In warn_test(0) :
  'x' is less than or equal to 0 but setting it to 1 and continuing
R> error_test(0)
Error in error_test(0) : 'x' is less than or equal to 0... TERMINATE
```

Notice that warn_test has continued to execute and returned the
value 5—the result of 5/1 after setting x to 1. The call to error_test did not
return anything because R exited the function at the stop command.

Warnings are useful when there's a natural way for a function to try to save itself even when it doesn't get the input it expects. For example, in Section 10.1.3, R issued a warning when you supplied a logical vector of elements to the if statement. Remember that the if statement expects a single logical value, but rather than quit when a logical vector is provided instead, it continues execution using just the first entry in the supplied vector. That said, sometimes it's more appropriate to actually throw an error and stop execution altogether.

Let's go back to myfibrec from Section 11.3.3. This function expects a positive integer (the position of the Fibonacci number it should return). Suppose you assume that if the user supplies a negative integer, the user actually means the positive version of that term. You can add a warning to handle this situation. Meanwhile, if the user enters 0, which doesn't correspond to any position in the Fibonacci series, the code will throw an error. Consider these modifications:

```
myfibrec2 <- function(n){
    if(n<0){
        warning("Assuming you meant 'n' to be positive -- doing that instead")
        n <- n*-1
    } else if(n==0){
        stop("'n' is uninterpretable at 0")
    }

    if(n==1||n==2){
        return(1)
    } else {
        return(myfibrec2(n-1)+myfibrec2(n-2))
    }
}
```

In myfibrec2, you now check whether n is negative or zero. If it's negative, the function issues a warning and continues executing after swapping the argument's sign. If n is zero, an error halts execution with a corresponding message. Here you can see the responses for a few different arguments:

```
R> myfibrec2(6)
[1] 8
R> myfibrec2(-3)
[1] 2
Warning message:
In myfibrec2(-3) :
  Assuming you meant 'n' to be positive -- doing that instead
R> myfibrec2(0)
Error in myfibrec2(0) : 'n' is uninterpretable at 0
```

Note that the call to myfibrec2(-3) has returned the third Fibonacci number.

Broadly speaking, both errors and warnings signal that something has gone wrong. If you're using a certain function or running chunks of code and you encounter these kinds of messages, you should look carefully at what has been run and what may have occurred to spark them.

**NOTE** *Identifying and repairing erroneous code is referred to as* debugging, *for which there are various strategies. One of the most basic strategies involves including* print *or* cat *commands to inspect various quantities as they are calculated during live execution. R does have some more sophisticated debugging tools; if you're interested, check out the excellent discussion of them provided in Chapter 13 of* The Art of R Programming *by Matloff (2011). A more general discussion can be found in* The Art of Debugging *by Matloff and Salzman (2008). As you gain more experience in R, understanding error messages or locating potential problems in code before they arise becomes easier and easier, a benefit you get partly because of R's interpretative style.*

### 12.1.2   Catching Errors with try Statements

When a function terminates from an error, it also terminates any parent functions. For example, if function A calls function B and function B halts because of an error, this halts execution of A at the same point. To avoid this severe consequence, you can use a try statement to attempt a function call and check whether it produces an error. You can also use an if statement to specify alternative operations, rather than allowing all processes to cease.

For example, if you call the myfibrec2 function from earlier and pass it 0, the function throws an error and terminates. But watch what happens when you pass that function call as the first argument to try:

```
R> attempt1 <- try(myfibrec2(0),silent=TRUE)
```

Nothing seems to happen. What's happened to the error? In fact, the error has still occurred, but try has suppressed the printing of an error message to the console because you passed it the argument silent set to TRUE. The error information is now stored in the object attempt1, which is of class "try-error". To see the error, simply print attempt1 to the console:

```
R> attempt1
[1] "Error in myfibrec2(0) : 'n' is uninterpretable at 0\n"
attr(,"class")
[1] "try-error"
attr(,"condition")
<simpleError in myfibrec2(0): 'n' is uninterpretable at 0>
```

You would have seen this printed to the console if you'd left silent set to FALSE. Catching an error this way can be really handy, especially when a function produces the error in the body code of another function. Using try, you can handle the error without terminating that parent function.

Meanwhile, if you pass a function to try and it doesn't throw an error, then try has no effect, and you simply get the normal return value.

```
R> attempt2 <- try(myfibrec2(6),silent=TRUE)
R> attempt2
[1] 8
```

Here, you executed myfibrec2 with a valid argument, n=6. Since this call doesn't result in an error, the result passed to attempt2 is the normal return value from myfibrec2, in this case 8.

## Using try in the Body of a Function

Let's see a more complete example of how you could use try in a larger function. The following myfibvector function takes a vector of indexes as the argument nvec and provides the corresponding terms from the Fibonacci sequence:

```
myfibvector <- function(nvec){
    nterms <- length(nvec)
    result <- rep(0,nterms)
    for(i in 1:nterms){
        result[i] <- myfibrec2(nvec[i])
    }
    return(result)
}
```

This function uses a for loop to work through nvec element by element, computing the corresponding Fibonacci number with the earlier function, myfibrec2. As long as all the values in nvec are nonzero, myfibvector works just fine. For example, the following call obtains the first, the second, the tenth, and the eighth Fibonacci number:

```
R> foo <- myfibvector(nvec=c(1,2,10,8))
R> foo
[1]  1  1 55 21
```

Suppose, however, there's a mistake and one of the entries in nvec ends up being zero.

```
R> bar <- myfibvector(nvec=c(3,2,7,0,9,13))
Error in myfibrec2(nvec[i]) : 'n' is uninterpretable at 0
```

The internal call to myfibrec2 has thrown an error when it's called on n=0, and this has terminated execution of myfibvector. Nothing is returned, and the entire call has failed.

You can prevent this outright failure by using try within the for loop to check each call to myfibrec2 and have it catch any errors. The following function, myfibvectorTRY, does just that.

```
myfibvectorTRY <- function(nvec){
    nterms <- length(nvec)
    result <- rep(0,nterms)
    for(i in 1:nterms){
        attempt <- try(myfibrec2(nvec[i]),silent=T)
        if(class(attempt)=="try-error"){
            result[i] <- NA
        } else {
            result[i] <- attempt
        }
    }
    return(result)
}
```

Here, within the for loop, you use attempt to store the result of trying each call to myfibrec2. Then, you inspect attempt. If this object's class is "try-error", that means myfibrec2 produced an error, and you fill the corresponding slot in the result vector with NA. Otherwise, attempt will represent a valid return value from myfibrec2, so you place it in the corresponding slot of the result vector. Now if you import and call myfibvectorTRY on the same nvec, you see a complete set of results.

```
R> baz <- myfibvectorTRY(nvec=c(3,2,7,0,9,13))
R> baz
[1]   2   1  13  NA  34 233
```

The error that would have otherwise terminated everything was silently caught, and the alternative response in this situation, NA, was inserted into the result vector.

**NOTE**    *The* try *command is a simplification of R's more complex* tryCatch *function, which is beyond the scope of this book, but it provides more precise control over how you test and execute chunks of code. If you're interested in learning more, enter* ?tryCatch *in the console.*

## Suppressing Warning Messages

In all the try calls I've shown so far, I've set the silent argument to TRUE, which stops any error messages from being printed. If you leave silent set to FALSE (the default value), the error message will be printed, but the error will still be caught without terminating execution.

Note that setting silent=TRUE only suppresses error messages, not warnings. Observe the following:

```
R> attempt3 <- try(myfibrec2(-3),silent=TRUE)
Warning message:
In myfibrec2(-3) :
```

```
  Assuming you meant 'n' to be positive -- doing that instead
R> attempt3
[1] 2
```

Although silent was TRUE, the warning (for negative values of n in this example) is still issued and printed. Warnings are treated separately from errors in this type of situation, as they should be—they can highlight other unforeseen issues with your code during execution. If you are absolutely sure you don't want to see any warnings, you can use suppressWarnings.

```
R> attempt4 <- suppressWarnings(myfibrec2(-3))
R> attempt4
[1] 2
```

The suppressWarnings function should be used only if you are certain that every warning in a given call can be safely ignored and you want to keep the output tidy.

## Exercise 12.1

a.  In Exercise 11.3 (b) on page 238, your task was to write a recursive R function to compute integer factorials, given some supplied non-negative integer x. Now, modify your function so that it throws an error (with an appropriate message) if x is negative. Test your new function responses by using the following:

    i.   x as 5
    ii.  x as 8
    iii. x as -8

b.  The idea of *matrix inversion*, briefly discussed in Section 3.3.6, is possible only for certain square matrices (those with an equal number of columns as rows). These inversions can be computed using the solve function, for example:

```
R> solve(matrix(1:4,2,2))
     [,1] [,2]
[1,]   -2  1.5
[2,]    1 -0.5
```

    Note that solve throws an error if the supplied matrix cannot be inverted. With this in mind, write an R function that attempts to invert each matrix in a list, according to the following guidelines:

    –   The function should take four arguments.
        *   The list x whose members are to be tested for matrix inversion

- * A value noninv to fill in results where a given matrix member of x cannot be inverted, defaulting to NA
- * A character string nonmat to be the result if a given member of x is not a matrix, defaulting to "not a matrix"
- * A logical value silent, defaulting to TRUE, to be passed to try in the body code

- The function should first check whether x is in fact a list. If not, it should throw an error with an appropriate message.
- Then, the function should ensure that x has at least one member. If not, it should throw an error with an appropriate message.
- Next, the function should check whether nonmat is a character string. If not, it should try to coerce it to a character string using an appropriate "as-dot" function (see Section 6.2.4), and it should issue an appropriate warning.
- After these checks, a loop should search each member i of the list x.
  - * If member i is a matrix, attempt to invert it with try. If it's invertible without error, overwrite member i of x with the result. If an error is caught, then member i of x should be overwritten with the value of noninv.
  - * If member i is not a matrix, then member i of x should be overwritten with the value of nonmat.
- Finally, the modified list x should be returned.

Now, test your function using the following argument values to make sure it responds as expected:

i. x as

```
list(1:4,matrix(1:4,1,4),matrix(1:4,4,1),matrix(1:4,2,2))
```

and all other arguments at default.

ii. x as in (i), noninv as Inf, nonmat as 666, silent at default.
iii. Repeat (ii), but now with silent=FALSE.
iv. x as

```
list(diag(9),matrix(c(0.2,0.4,0.2,0.1,0.1,0.2),3,3),
     rbind(c(5,5,1,2),c(2,2,1,8),c(6,1,5,5),c(1,0,2,0)),
     matrix(1:6,2,3),cbind(c(3,5),c(6,5)),as.vector(diag(2)))
```

and noninv as "unsuitable matrix"; all other values at default.

Finally, test the error messages by attempting calls to your function with the following:

v. x as "hello"
vi. x as list()

## 12.2　Progress and Timing

R is often used for lengthy numeric exercises, such as simulation or random variate generation. For these complex, time-consuming operations, it's often useful to keep track of progress or see how long a certain task took to complete. For example, you may want to compare the speed of two different programming approaches to a given problem. In this section, you'll look at ways to time code execution and show its progress.

### 12.2.1　Textual Progress Bars: Are We There Yet?

A *progress bar* shows how far along R is as it executes a set of operations. To show how this works, you need to run code that takes a while to execute, which you'll do by making R *sleep*. The Sys.sleep command makes R pause for a specified amount of time, in seconds, before continuing.

```
R> Sys.sleep(3)
```

If you run this code, R will pause for three seconds before you can continue using the console. Sleeping will be used in this section as a surrogate for the delay caused by computationally expensive operations, which is where progress bars are most useful.

To use Sys.sleep in a more common fashion, consider the following:

```
sleep_test <- function(n){
    result <- 0
    for(i in 1:n){
        result <- result + 1
        Sys.sleep(0.5)
    }
    return(result)
}
```

The sleep_test function is basic—it takes a positive integer n and adds 1 to the result value for n iterations. At each iteration, you also tell the loop to sleep for a half second. Because of that sleep command, executing the following code takes about four seconds to return a result:

```
R> sleep_test(8)
[1] 8
```

Now, say you want to track the progress of this type of function as it executes. You can implement a textual progress bar with three steps: initialize the bar object with txtProgressBar, update the bar with setTxtProgressBar, and terminate the bar with close. The next function, prog_test, modifies sleep_test to include those three commands.

```
prog_test <- function(n){
    result <- 0
    progbar <- txtProgressBar(min=0,max=n,style=1,char="=")
    for(i in 1:n){
        result <- result + 1
        Sys.sleep(0.5)
        setTxtProgressBar(progbar,value=i)
    }
    close(progbar)
    return(result)
}
```

Before the for loop, you create an object named progbar by calling txtProgressBar with four arguments. The min and max arguments are
numeric values that define the limits of the bar. In this case, you set max=n,
which matches the number of iterations of the impending for loop. The
style argument (integer, either 1, 2, or 3) and the char argument (character
string, usually a single character) govern the appearance of the bar. Setting
style=1 means the bar will simply display a line of char; with char="=" it'll be a
series of equal signs.

Once this object is created, you have to instruct the bar to actually
progress during execution with a call to setTxtProgressBar. You pass in the
bar object to update (progbar) and the value it should update to (in this
case, i). Once complete (after exiting the loop), the progress bar must be
terminated with a call to close, passing in the bar object of interest. Import
and execute prog_test, and you'll see the line of "=" drawn in steps as the
loop completes.

```
R> prog_test(8)
==================================================================
[1] 8
```

The width of the bar is, by default, determined by the width of the R
console pane upon execution of the txtProgressBar command. You can customize the bar a bit by changing the style and char arguments. Choosing
style=3, for example, shows the bar as well as a "percent completed" counter.
Some packages offer more elaborate options too, such as pop-up widgets,
but the textual version is the simplest and most universally compatible version across different systems.

### 12.2.2  Measuring Completion Time: How Long Did It Take?

If you want to know how long a computation takes to complete, you can use
the Sys.time command. This command outputs an object that details current
date and time information based on your system.

```
R> Sys.time()
[1] "2016-03-06 16:39:27 NZDT"
```

You can store objects like these before and after some code and then compare them to see how much time has passed. Enter this in the editor:

```
t1 <- Sys.time()
Sys.sleep(3)
t2 <- Sys.time()
t2-t1
```

Now highlight all four lines and execute them in the console.

```
R> t1 <- Sys.time()
R> Sys.sleep(3)
R> t2 <- Sys.time()
R> t2-t1
Time difference of 3.012889 secs
```

By executing this entire code block together, you get an easy measure of the total completion time in a nicely formatted string printed to the console. Note that there's a small time cost for interpreting and invoking any commands, in addition to the three seconds you tell R to sleep. This time will vary between computers.

If you need more detailed timing reports, there are more sophisticated tools. For example, you can use proc.time() to receive not just the total elapsed "wall clock" time but also computer-related CPU timings (see the definitions in the help file ?proc.time). To time a single expression, you can also use the system.time function (which uses the same detail of output as proc.time). There are also *benchmarking* tools (formal or systematic comparisons of different approaches) for timing your code; see, for example, the rbenchmark package (Kusnierczyk, 2012). However, for everyday use, the time-object differencing approach used here is easy to interpret and provides a good indication of the computational expense.

---

### Exercise 12.2

a.  Modify prog_test from Section 12.2.1 to include an ellipsis in its argument list, intended to take values for the additional arguments in txtProgressBar; name the new function prog_test_fancy. Time how long it takes a call to prog_test_fancy to execute. Set 50 as n, instruct the progress bar (through the ellipsis) to use style=3, and set the bar character to be "r".

---

b. In Section 12.1.2, you defined a function named `myfibvectorTRY`
   (which itself calls `myfibrec2` from Section 12.1.1) to return multiple terms from the Fibonacci sequence based on a supplied
   "term vector" `nvec`. Write a new version of `myfibvectorTRY` that
   includes a progress bar of `style=3` and a character of your choosing that increments at each pass of the internal `for` loop. Then,
   do the following:

   i.   Use your new function to reproduce the results from the text
        where `nvec=c(3,2,7,0,9,13)`.
   ii.  Time how long it takes to use your new function to return
        the first 35 terms of the Fibonacci sequence. What do you
        notice, and what does this say about your recursive Fibonacci
        functions?

c. Remain with the Fibonacci sequence. Write a stand-alone `for`
   loop that can compute, and store in a vector, the same first 35
   terms as in (b)(ii). Time it. Which approach would you prefer?

# 12.3 Masking

With the plethora of built-in and contributed data and functionality available for R, it is virtually inevitable that at some point you will come across
objects, usually functions, that share the same name in distinctly different
loaded packages.

So, what happens in those instances? For example, say you define a function with the same name as a function in an R package that you have already
loaded. R responds by *masking* one of the objects—that is, one object or
function will take precedence over the other and assume the object or function name, while the masked function must be called with an additional
command. This protects objects from overwriting or blocking one another.
In this section, you'll look at the two most common masking situations in R.

## 12.3.1 *Function and Object Distinction*

When two functions or objects in different environments have the same
name, the object that comes earlier in the search path will mask the later
one. That is, when the object is sought, R will use the object or function it
finds first, and you'll need extra code to access the other, masked version.
Remember, you can see the current search path by executing search().

```
R> search()
 [1] ".GlobalEnv"        "tools:RGUI"         "package:stats"
 [4] "package:graphics"  "package:grDevices"  "package:utils"
 [7] "package:datasets"  "package:methods"    "Autoloads"
[10] "package:base"
```

When R searches, the function or object that falls closest to the start of the search path (the global environment) is reached first and masks the function or object of the same name that occurs somewhere later in the search path. To see a simple example of masking, you'll define a function with the same name as a function in the base package: sum. Here's how sum works normally, adding up all the elements in the vector foo:

```
R> foo <- c(4,1.5,3)
R> sum(foo)
[1] 8.5
```

Now, suppose you were to enter the following function:

```
sum <- function(x){
    result <- 0
    for(i in 1:length(x)){
        result <- result + x[i]^2
    }
    return(result)
}
```

This version of sum takes in a vector x and uses a for loop to square each element before summing them and returning the result. This can be imported into the R console without any problem, but clearly, it doesn't offer the same functionality as the (original) built-in version of sum. Now, after importing the function, if you make a call to sum, your version is used.

```
R> sum(foo)
[1] 27.25
```

This happens because the user-defined function is stored in the global environment (.GlobalEnv), which always comes first in the search path. R's built-in function is part of the base package, which comes at the end of the search path. In this case, the user-defined function is masking the original.

Now, if you want R to run the base version of sum, you have to include the name of its package in the call, with a double colon.

```
R> base::sum(foo)
[1] 8.5
```

This tells R to use the version in base, even though there's another version of the function in the global environment.

To avoid any confusion, let's remove the sum function from the global environment.

```
R> rm(sum)
```

## When Package Objects Clash

When you load a package, R will notify you if any objects in the package clash with other objects that are accessible in the present session. To illustrate this, I'll make use of two contributed packages: the car package (you saw this earlier in Exercise 8.1 (b) on page 162) and the spatstat package (you'll use this in Part V). After ensuring these two packages are installed, when I load them in the following order, I see this message:

```
R> library("spatstat")

spatstat 1.40-0        (nickname: 'Do The Maths')
For an introduction to spatstat, type 'beginner'
R> library("car")

Attaching package: 'car'

The following object is masked from 'package:spatstat':

    ellipse
```

This indicates that the two packages each have an object with the same name—ellipse. R has automatically notified you that this object is being masked. Note that the functionality of both car and spatstat remains completely available; it's just that the ellipse objects require some distinction should they be needed. Using ellipse at the prompt will access car's object since that package was loaded more recently. To use spatstat's version, you must type spatstat::ellipse. These rules also apply to accessing the respective help files.

A similar notification occurs when you load a package with an object that's masked by a global environment object (a global environment object will always take precedence over a package object). To see an example, you can load the MASS package (Venables and Ripley, 2002), which is included with R but isn't automatically loaded. Continuing in the current R session, create the following object:

```
R> cats <- "meow"
```

Now, suppose you need to load MASS.

```
R> library("MASS")

Attaching package: 'MASS'

The following object is masked _by_ '.GlobalEnv':

    cats
```

```
The following object is masked from 'package:spatstat':

    area
```

Upon loading the package, you're informed that the cats object you've just created is masking an object of the same name in MASS. (As you can see with ?MASS::cats, this object is a data frame with weight measurements of household felines.) Furthermore, it appears MASS also shares an object name with spatstat—area. The same kind of "package masking" message as shown earlier is also displayed for that particular item.

### Unmounting Packages

You can unmount loaded packages from the search path. With the packages loaded in this discussion, my current search path looks like this:

```
R> search()
 [1] ".GlobalEnv"        "package:MASS"      "package:car"
 [4] "package:spatstat"  "tools:RGUI"        "package:stats"
 [7] "package:graphics"  "package:grDevices" "package:utils"
[10] "package:datasets"  "package:methods"   "Autoloads"
[13] "package:base"
```

Now, suppose you don't need car anymore. You can remove it with the detach function as follows.

```
R> detach("package:car",unload=TRUE)
R> search()
 [1] ".GlobalEnv"        "package:MASS"      "package:spatstat"
 [4] "tools:RGUI"        "package:stats"     "package:graphics"
 [7] "package:grDevices" "package:utils"     "package:datasets"
[10] "package:methods"   "Autoloads"         "package:base"
```

This removes the elected package from the path, unloading its namespace. Now, the functionality of car is no longer immediately available, and spatstat's ellipsis function is no longer masked.

**NOTE**    *As contributed packages get updated by their maintainers, they may include new objects that spark new maskings or remove or rename objects that previously caused maskings (when compared with other contributed packages). The specific maskings illustrated here among car, spatstat, and MASS occur at the time of writing with the versions available and may change in the future.*

## 12.3.2   Data Frame Variable Distinction

There's one other common situation in which you'll be explicitly notified of masking: when you add a data frame to the search path. Let's see how this works. Continuing in the current workspace, define the following data frame:

```
R> foo <- data.frame(surname=c("a","b","c","d"),
                     sex=c(0,1,1,0),height=c(170,168,181,180),
                     stringsAsFactors=F)
R> foo
  surname sex height
1       a   0    170
2       b   1    168
3       c   1    181
4       d   0    180
```

The data frame foo has three column variables: person, sex, and height. To access one of these columns, normally you need to use the $ operator and enter something like foo$surname. However, you can *attach* a data frame directly to your search path, which makes it easier to access a variable.

```
R> attach(foo)
R> search()
 [1] ".GlobalEnv"        "foo"               "package:MASS"
 [4] "package:spatstat"  "tools:RGUI"        "package:stats"
 [7] "package:graphics"  "package:grDevices" "package:utils"
[10] "package:datasets"  "package:methods"   "Autoloads"
[13] "package:base"
```

Now the surname variable is directly accessible.

```
R> surname
[1] "a" "b" "c" "d"
```

This saves you from having to enter foo$ every time you want to access a variable, which can be a handy shortcut if your analysis deals exclusively with one static, unchanging data frame. However, if you forget about your attached objects, they can cause problems later, especially if you continue to mount more objects onto the search path in the same session. For example, say you enter another data frame.

```
R> bar <- data.frame(surname=c("e","f","g","h"),
                     sex=c(1,0,1,0),weight=c(55,70,87,79),
                     stringsAsFactors=F)
R> bar
  surname sex weight
1       e   1     55
2       f   0     70
3       g   1     87
4       h   0     79
```

Then add it to the search path too.

```
R> attach(bar)
The following objects are masked from foo:

    sex, surname
```

The notification tells you that the bar object now precedes foo in the search path.

```
R> search()
 [1] ".GlobalEnv"       "bar"                "foo"
 [4] "package:MASS"     "package:spatstat"   "tools:RGUI"
 [7] "package:stats"    "package:graphics"   "package:grDevices"
[10] "package:utils"    "package:datasets"   "package:methods"
[13] "Autoloads"        "package:base"
```

As a result, any direct use of either sex or surname will now access bar's contents, not foo's. Meanwhile, the unmasked variable height from foo is still directly accessible.

```
R> height
[1] 170 168 181 180
```

This is a pretty simple example, but it highlights the potential for confusion when data frames, lists, or other objects are added to the search path. Mounting objects this way can quickly become difficult to track, especially for large data sets with many different variables. For this reason, it's best to avoid attaching objects this way as a general guideline—unless, as stated earlier, you're working exclusively with one data frame.

Note that detach can be used to remove objects from the search path, in a similar way as you saw with packages. In this case, you can simply enter the object name itself.

```
R> detach(foo)
R> search()
 [1] ".GlobalEnv"         "bar"                "package:MASS"
 [4] "package:spatstat"   "tools:RGUI"         "package:stats"
 [7] "package:graphics"   "package:grDevices"  "package:utils"
[10] "package:datasets"   "package:methods"    "Autoloads"
[13] "package:base"
```

## Important Code in This Chapter

| Function/operator | Brief description | First occurrence |
| --- | --- | --- |
| warning | Issue warning | Section 12.1.1, p. 242 |
| stop | Throw error | Section 12.1.1, p. 242 |
| try | Attempt error catch | Section 12.1.2, p. 244 |
| Sys.sleep | Sleep (pause) execution | Section 12.2.1, p. 249 |
| txtProgressBar | Initialize progress bar | Section 12.2.1, p. 249 |
| setTxtProgressBar | Increment progress bar | Section 12.2.1, p. 249 |
| close | Close progress bar | Section 12.2.1, p. 249 |
| Sys.time | Get local system time | Section 12.2.2, p. 250 |
| detach | Remove library/object from path | Section 12.3.1, p. 255 |
| attach | Attach object to search path | Section 12.3.2, p. 256 |

# PART III

## STATISTICS AND PROBABILITY

# 13

## ELEMENTARY STATISTICS

Statistics is the practice of turning *data* into *information* to identify trends and understand features of populations. This chapter will cover some basic definitions and use R to demonstrate their application.

## 13.1 Describing Raw Data

Often, the first thing statistical analysts are faced with is raw data—in other words, the records or observations that make up a sample. Depending on the nature of the intended analysis, these data could be stored in a specialized R object, often a data frame (Chapter 5), possibly read in from an external file using techniques from Chapter 8. Before you can begin summarizing or modeling your data, however, it is important to clearly identify your available variables.

A *variable* is a characteristic of an individual in a population, the value of which can differ between entities within that population. For example, in Section 5.2, you experimented with an illustrative data frame `mydata`.

You recorded the age, sex, and humor level for a sample of people. These characteristics are your variables; the values measured will differ between the individuals.

Variables can take on a number of forms, determined by the nature of the values they may take. Before jumping into R, you'll look at some standard ways in which variables are described.

### 13.1.1 Numeric Variables

A *numeric* variable is one whose observations are naturally recorded as numbers. There are two types of numeric variables: continuous and discrete.

A *continuous* variable can be recorded as any value in some interval, up to any number of decimals (which technically gives an infinite number of possible values, even if the continuum is restricted in range). For example, if you were observing rainfall amount, a value of 15 mm would make sense, but so would a value of 15.42135 mm. Any degree of measurement precision gives a valid observation.

A *discrete* variable, on the other hand, may take on only distinct numeric values—and if the range is restricted, then the number of possible values is finite. For example, if you were observing the number of heads in 20 flips of a coin, only whole numbers would make sense. It would not make sense to observe 15.42135 heads; the possible outcomes are restricted to the integers from 0 to 20 (inclusive).

### 13.1.2 Categorical Variables

Though numeric observations are common for many variables, it's also important to consider *categorical* variables. Like some discrete variables, categorical variables may take only one of a finite number of possibilities. Unlike discrete variables, however, categorical observations are not always recorded as numeric values.

There are two types of categorical variables. Those that cannot be logically ranked are called *nominal*. A good example of a categorical-nominal variable is sex. In most data sets, it has two fixed possible values, male and female, and the order of these categories is irrelevant. Categorical variables that can be naturally ranked are called *ordinal*. An example of a categorical-ordinal variable would be the dose of a drug, with the possible values low, medium, and high. These values can be ordered in either increasing or decreasing amounts, and the ordering might be relevant to the research.

**NOTE** *Some statistical texts blur the definitions of discrete and categorical variables or even use them interchangeably. While this practice is not necessarily incorrect, I prefer to keep the definitions separate, for clarity. That is, I'll say "discrete" when referring to a naturally numeric variable that cannot be expressed on a continuous scale (such as a count), and I'll say "categorical" when the possible outcomes for a given individual are not necessarily numeric and the number of possible values is always finite.*

Once you know what to look for, identifying the types of variables in a given data set is straightforward. Take the data frame chickwts, which is available in the automatically loaded datasets package. At the prompt, directly entering the following gives you the first five records of this data set.

```
R> chickwts[1:5,]
  weight       feed
1    179 horsebean
2    160 horsebean
3    136 horsebean
4    227 horsebean
5    217 horsebean
```

R's help file (?chickwts) describes these data as comprising the weights of 71 chicks (in grams) after six weeks, based on the type of food provided to them. Now let's take a look at the two columns in their entirety as vectors:

```
R> chickwts$weight
 [1] 179 160 136 227 217 168 108 124 143 140 309 229 181 141 260 203 148 169
[19] 213 257 244 271 243 230 248 327 329 250 193 271 316 267 199 171 158 248
[37] 423 340 392 339 341 226 320 295 334 322 297 318 325 257 303 315 380 153
[55] 263 242 206 344 258 368 390 379 260 404 318 352 359 216 222 283 332
R> chickwts$feed
 [1] horsebean horsebean horsebean horsebean horsebean horsebean horsebean
 [8] horsebean horsebean horsebean linseed   linseed   linseed   linseed
[15] linseed   linseed   linseed   linseed   linseed   linseed   linseed
[22] linseed   soybean   soybean   soybean   soybean   soybean   soybean
[29] soybean   soybean   soybean   soybean   soybean   soybean   soybean
[36] soybean   sunflower sunflower sunflower sunflower sunflower sunflower
[43] sunflower sunflower sunflower sunflower sunflower sunflower meatmeal
[50] meatmeal  meatmeal  meatmeal  meatmeal  meatmeal  meatmeal  meatmeal
[57] meatmeal  meatmeal  meatmeal  casein    casein    casein    casein
[64] casein    casein    casein    casein    casein    casein    casein
[71] casein
Levels: casein horsebean linseed meatmeal soybean sunflower
```

weight is a numeric measurement that can fall anywhere on a continuum, so this is a numeric-continuous variable. The fact that the chick weights appear to have been rounded or recorded to the nearest gram does not affect this definition because in reality the weights can be any figure (within reason). feed is clearly a categorical variable because it has only six possible outcomes, which aren't numeric. The absence of any natural or easily identifiable ordering leads to the conclusion that feed is a categorical-nominal variable.

### 13.1.3 Univariate and Multivariate Data

When discussing or analyzing data related to only one dimension, you're dealing with *univariate* data. For example, the weight variable in the earlier example is univariate since each measurement can be expressed with one component—a single number.

When it's necessary to consider data with respect to variables that exist in more than one dimension (in other words, with more than one component or measurement associated with each observation), your data are considered *multivariate*. Multivariate measurements are arguably most relevant when the individual components aren't as useful when considered on their own (in other words, as univariate quantities) in any given statistical analysis.

An ideal example is that of spatial coordinates, which must be considered in terms of at least two components—a horizontal $x$-coordinate and a vertical $y$-coordinate. The univariate data alone—for example, the $x$-axis values only—aren't especially useful. Consider the quakes data set (like chickwts, this is automatically available through the datasets package), which contains observations on 1,000 seismic events recorded off the coast of Fiji. If you look at the first five records and read the descriptions in the help file ?quakes, you quickly get a good understanding of what's presented.

```
R> quakes[1:5,]
      lat    long depth mag stations
1 -20.42 181.62   562 4.8       41
2 -20.62 181.03   650 4.2       15
3 -26.00 184.10    42 5.4       43
4 -17.97 181.66   626 4.1       19
5 -20.42 181.96   649 4.0       11
```

The columns lat and long provide the latitude and longitude of the event, depth provides the depth of the event (in kilometers), mag provides the magnitude on the Richter scale, and stations provides the number of observation stations that detected the event. If you're interested in the spatial dispersion of these earthquakes, then examining only the latitude or the longitude is rather uninformative. The location of each event is described with two components: a latitude *and* a longitude value. You can easily plot these 1,000 events; Figure 13-1 shows the result of the following code:

```
R> plot(quakes$long,quakes$lat,xlab="Longitude",ylab="Latitude")
```

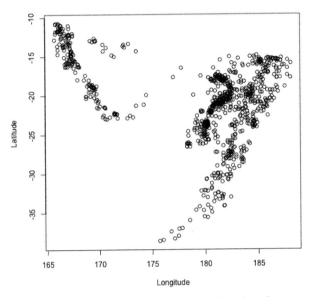

*Figure 13-1: Plotting the spatial locations of earthquakes using a bivariate (multivariate with two components) variable*

## 13.1.4 Parameter or Statistic?

As already noted, statistics as a discipline is concerned with understanding features of an overall *population*, defined as the entire collection of individuals or entities of interest. The characteristics of that population are referred to as *parameters*. Because researchers are rarely able to access relevant data on every single member of the population of interest, they typically collect a *sample* of entities to represent the population and record relevant data from these entities. They may then estimate the parameters of interest using the sample data—and those estimates are the *statistics*.

For example, if you were interested in the average age of women in the United States who own cats, the population of interest would be all women residing in the United States who own at least one cat. The parameter of interest is the true mean age of women in the United States who own at least one cat. Of course, obtaining the age of every single female American with a cat would be a difficult feat. A more feasible approach would be to randomly identify a smaller number of cat-owning American women and take data from them—this is your sample, and the mean age of the women in the sample is your statistic.

Thus, the key difference between a statistic and a parameter is whether the characteristic refers to the sample you drew your data from or the wider population. Figure 13-2 illustrates this, with the mean $\mu$ of a measure for individuals in a population as the parameter and with the mean $\bar{x}$ of a sample of individuals taken from that population as the statistic.

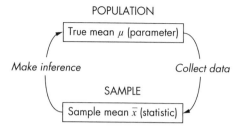

POPULATION

True mean $\mu$ (parameter)

Make inference

Collect data

SAMPLE

Sample mean $\bar{x}$ (statistic)

Figure 13-2: A conceptualization of statistical practice to illustrate the definitions of parameter and statistic, using the mean as an example

## Exercise 13.1

a. For each of the following, identify the type of variable described: numeric-continuous, numeric-discrete, categorical-nominal, or categorical-ordinal:

   i. The number of blemishes on the hood of a car coming off a production line

   ii. A survey question that asks the participant to select from Strongly agree, Agree, Neutral, Disagree, and Strongly disagree

   iii. The noise level (in decibels) at a concert

   iv. The noise level out of three possible choices: high, medium, low

   v. A choice of primary color

   vi. The distance between a cat and a mouse

b. For each of the following, identify whether the quantity discussed is a population parameter or a sample statistic. If the latter, also identify what the corresponding population parameter is.

   i. The percentage of 50 New Zealanders who own a gaming console

   ii. The average number of blemishes found on the hoods of three cars in the No Dodgy Carz yard

   iii. The proportion of domestic cats in the United States that wear a collar

   iv. The average number of times per day a vending machine is used in a year

   v. The average number of times per day a vending machine is used in a year, based on data collected on three distinct days in that year

## 13.2 Summary Statistics

Now that you've learned the basic terminology, you're ready to calculate some statistics with R. In this section, you'll look at the most common types of statistics used to summarize the different types of variables I've discussed.

### 13.2.1 Centrality: Mean, Median, Mode

*Measures of centrality* are commonly used to explain large collections of data by describing where numeric observations are centered. One of the most common measures of centrality is of course the arithmetic *mean*. It's considered to be the central "balance point" of a collection of observations.

For a set of $n$ numeric measurements labeled $x = \{x_1, x_2, \ldots, x_n\}$, you find the sample mean $\bar{x}$ as follows:

$$\bar{x} = \frac{(x_1 + x_2 + \ldots + x_n)}{n} = \frac{1}{n}\sum_{i=1}^{n} x_i \tag{13.1}$$

So, for example, if you observe the data $2, 4.4, 3, 3, 2, 2.2, 2, 4$, the mean is calculated like this:

$$\frac{2 + 4.4 + 3 + 3 + 2 + 2.2 + 2 + 4}{8} = 2.825$$

The *median* is the "middle magnitude" of your observations, so if you place your observations in order from smallest to largest, you can find the median by either taking the middle value (if there's an odd number of observations) or finding the mean of the two middle values (if there's an even number of observations). Using the notation for $n$ measurements labeled $x = \{x_1, x_2, \ldots, x_n\}$, you find the sample median $\bar{m}_x$ as follows:

- Sort the observations from smallest to largest to give the "order statistics" $x_i^{(1)}, x_j^{(2)}, \ldots, x_k^{(n)}$, where $x_i^{(t)}$ denotes the $t$th smallest observation, regardless of observation number $i, j, k, \ldots$.
- Then, do the following:

$$\bar{m}_x = \begin{cases} x_i^{\left(\frac{n+1}{2}\right)}, & \text{if } n \text{ is odd} \\ \left(x_i^{\left(\frac{n}{2}\right)} + x_j^{\left(\frac{n}{2}+1\right)}\right)/2, & \text{if } n \text{ is even} \end{cases} \tag{13.2}$$

For the same data, sorting them from smallest to largest yields $2, 2, 2, 2.2, 3, 3, 4, 4.4$. With $n = 8$ observations, you have $n/2 = 4$. The median is therefore as follows:

$$\left(x_i^{(4)} + x_j^{(5)}\right)/2 = (2.2 + 3)/2 = 2.6$$

The *mode* is simply the "most common" observation. This statistic is more often used with numeric-discrete data than with numeric-continuous, though it is used with reference to *intervals* of the latter (commonly when discussing probability density functions—see Chapters 15 and 16). It's possible for a collection of $n$ numeric measurements $x_1, x_2, \ldots, x_n$ to have no mode (where each observation is unique) or to have more than one mode (where more than one particular value occurs the largest number of times). To find the mode $\bar{d}_x$, simply tabulate the frequency of each measurement.

Again using the eight observations from the example, you can see the frequencies here:

| Observation | 2 | 2.2 | 3 | 4 | 4.4 |
|---|---|---|---|---|---|
| Frequency | 3 | 1 | 2 | 1 | 1 |

The value 2 occurs three times, which is more frequent than any other value, so the single mode for these data is the value 2.

In R, it's easy to compute the arithmetic mean and the median with built-in functions of the same names. First, store the eight observations as the numeric vector xdata.

```
R> xdata <- c(2,4.4,3,3,2,2.2,2,4)
```

Then compute the statistics.

```
R> x.bar <- mean(xdata)
R> x.bar
[1] 2.825
R> m.bar <- median(xdata)
R> m.bar
[1] 2.6
```

Finding a mode is perhaps most easily achieved by using R's table function, which gives you the frequencies you need.

```
R> xtab <- table(xdata)
R> xtab
xdata
  2 2.2   3   4 4.4
  3   1   2   1   1
```

Though this clearly shows the mode for a small data set, it's good practice to write code that can automatically identify the most frequent observations for any table. The min and max functions will report the smallest and largest values, with range returning both in a vector of length 2.

```
R> min(xdata)
[1] 2
R> max(xdata)
[1] 4.4
R> range(xdata)
[1] 2.0 4.4
```

When applied to a table, these commands operate on the reported frequencies.

```
R> max(xtab)
[1] 3
```

Finally, therefore, you can construct a logical flag vector to get the mode from table.

```
R> d.bar <- xtab[xtab==max(xtab)]
R> d.bar
2
3
```

Here, 2 is the value and 3 is the frequency of that value.

Let's return to the chickwts data set explored earlier in Section 13.1.2. The mean and median weights of the chicks are as follows:

```
R> mean(chickwts$weight)
[1] 261.3099
R> median(chickwts$weight)
[1] 258
```

You can also look at the quakes data set explored in Section 13.1.3. The most common magnitude of earthquake in the data set is identified with the following, which indicates that there were 107 occurrences of a 4.5 magnitude event:

```
R> Qtab <- table(quakes$mag)
R> Qtab[Qtab==max(Qtab)]
4.5
107
```

**NOTE** *Several methods are available to compute medians, though the impact on results is usually negligible for most practical purposes. Here I've simply used the default "sample" version used by R.*

Many of the functions R uses to compute statistics from a numeric structure will not run if the data set includes missing or undefined values (NAs or NaNs). Here's an example:

```
R> mean(c(1,4,NA))
[1] NA
R> mean(c(1,4,NaN))
[1] NaN
```

To prevent unintended NaNs or forgotten NAs being ignored without the user's knowledge, R does not by default ignore these special values when running functions such as mean—and therefore will not return the intended numeric results. You can, however, set an optional argument na.rm to TRUE, which will force the function to operate only on the numeric values that are present.

```
R> mean(c(1,4,NA),na.rm=TRUE)
[1] 2.5
R> mean(c(1,4,NaN),na.rm=TRUE)
[1] 2.5
```

You should use this argument only if you're aware there might be missing values and that the result will be computed based on only those values that *have* been observed. Functions that I've discussed already such as sum, prod, mean, median, max, min, and range—essentially anything that calculates a numeric statistic based on a numeric vector—all have the na.rm argument available to them.

Lastly, in calculating simple summary statistics, it's useful to remind yourself of the tapply function (see Section 10.2.3), used to compute statistics grouped by a specific categorical variable. Suppose, for example, you wanted to find the mean weight of the chicks grouped by feed type. One solution would be to use the mean function on each specific subset.

```
R> mean(chickwts$weight[chickwts$feed=="casein"])
[1] 323.5833
R> mean(chickwts$weight[chickwts$feed=="horsebean"])
[1] 160.2
R> mean(chickwts$weight[chickwts$feed=="linseed"])
[1] 218.75
R> mean(chickwts$weight[chickwts$feed=="meatmeal"])
[1] 276.9091
R> mean(chickwts$weight[chickwts$feed=="soybean"])
[1] 246.4286
R> mean(chickwts$weight[chickwts$feed=="sunflower"])
[1] 328.9167
```

This is cumbersome and lengthy. Using tapply, however, you can calculate the same values by category using just one line of code.

```
R> tapply(chickwts$weight,INDEX=chickwts$feed,FUN=mean)
   casein horsebean   linseed  meatmeal   soybean sunflower
 323.5833  160.2000  218.7500  276.9091  246.4286  328.9167
```

Here, the first argument is the numeric vector upon which to operate, the INDEX argument specifies the grouping variable, and the FUN argument gives the name of the function to be performed on the data in the first argument as per the subsets defined by INDEX. Like other functions you've seen that request the user to specify *another* function to govern operations, tapply includes an ellipsis (see Sections 9.2.5 and 11.2.4) to allow the user to supply further arguments directly to FUN if required.

## 13.2.2 Counts, Percentages, and Proportions

In this section, you'll look at the summary of data that aren't necessarily numeric. It makes little sense, for example, to ask R to compute the mean of a categorical variable, but it is sometimes useful to count the number of observations that fall within each category—these *counts* or *frequencies* represent the most elementary summary statistic of categorical data.

This uses the same count summary that was necessary for the mode calculation in Section 13.2.1, so again you can use the table command to obtain frequencies. Recall there are six feed types making up the diet of the chicks in the chickwts data frame. Getting these factor-level counts is as straightforward as this:

```
R> table(chickwts$feed)
```

```
  casein horsebean   linseed meatmeal   soybean sunflower
      12        10        12       11        14        12
```

You can gather more information from these counts by identifying the *proportion* of observations that fall into each category. This will give you comparable measures across multiple data sets. Proportions represent the fraction of observations in each category, usually expressed as a decimal (floating-point) number between 0 and 1 (inclusive). To calculate proportions, you only need to modify the previous count function by dividing the count (or frequency) by the overall sample size (obtained here by using nrow on the appropriate data frame object; see Section 5.2).

```
R> table(chickwts$feed)/nrow(chickwts)
```

```
   casein horsebean   linseed  meatmeal   soybean sunflower
0.1690141 0.1408451 0.1690141 0.1549296 0.1971831 0.1690141
```

Of course, you needn't do everything associated with counts via table. A simple sum of an appropriate logical flag vector can be just as useful—recall that TRUEs are automatically treated as 1 and FALSEs as 0 in any arithmetic treatment of logical structures in R (refer to Section 4.1.4). Such a sum will provide you with the desired frequency, but to get a proportion, you still need to divide by the total sample size. Furthermore, this is actually equivalent to finding the mean of a logical flag vector. For example, to find the proportion of chicks fed soybean, note that the following two calculations give identical results of around 0.197:

```
R> sum(chickwts$feed=="soybean")/nrow(chickwts)
[1] 0.1971831
```

```
R> mean(chickwts$feed=="soybean")
[1] 0.1971831
```

You can also use this approach to calculate the proportion of entities in combined groups, achieved easily through logical operators (see Section 4.1.3). The proportion of chicks fed either soybean *or* horsebean is as follows:

```
R> mean(chickwts$feed=="soybean"|chickwts$feed=="horsebean")
[1] 0.3380282
```

Yet again, the tapply function can prove useful. This time, to get the proportions of chicks on each diet, you'll define the FUN argument to be an anonymous function (refer to Section 11.3.2) that performs the required calculation.

```
R> tapply(chickwts$weight,INDEX=chickwts$feed,
        FUN=function(x) length(x)/nrow(chickwts))
  casein horsebean   linseed  meatmeal   soybean sunflower
0.1690141 0.1408451 0.1690141 0.1549296 0.1971831 0.1690141
```

The disposable function here is defined with a dummy argument x, which you're using to represent the vector of weights in each feed group to which FUN applies. Finding the desired proportion is therefore a case of dividing the number of observations in x by the total number of observations.

The last function to note is the round function, which rounds numeric data output to a certain number of decimal places. You need only supply to round your numeric vector (or matrix or any other appropriate data structure) and however many decimal places (as the argument digits) you want your figures rounded to.

```
R> round(table(chickwts$feed)/nrow(chickwts),digits=3)

  casein horsebean   linseed  meatmeal   soybean sunflower
   0.169     0.141     0.169     0.155     0.197     0.169
```

This provides output that's easier to read at a glance. If you set digits=0 (the default), output is rounded to the nearest integer.

Before the next exercise, it's worth briefly remarking on the relationship between a proportion and a percentage. The two represent the same thing. The only difference is the scale; the *percentage* is merely the proportion multiplied by 100. The percentage of chicks on a soybean diet is therefore approximately 19.7 percent.

```
R> round(mean(chickwts$feed=="soybean")*100,1)
[1] 19.7
```

Since proportions always lie in the interval $[0,1]$, percentages always lie within $[0,100]$.

Most statisticians use proportions over percentages because of the role proportions play in the direct representation of *probabilities* (discussed in Chapter 15). However, there are situations in which percentages are preferred, such as basic data summaries or in the definition of *percentiles*, which will be detailed in Section 13.2.3.

---

## Exercise 13.2

a. Obtain, rounded to two decimal places, the proportion of seismic events in the quakes data frame that occurred at a depth of 300 km or deeper.

b. Remaining with the quakes data set, calculate the mean and median magnitudes of the events that occurred at a depth of 300 km or deeper.

c. Using the chickwts data set, write a for loop that gives you the mean weight of chicks for each feed type—the same as the results given by the tapply function in Section 13.2.1. Display the results rounded to one decimal place and, when printing, ensure each mean is labeled with the appropriate feed type.

Another ready-to-use data set (in the automatically loaded datasets package) is InsectSprays. It contains data on the number of insects found on various agricultural units, as well as the type of insect spray that was used on each unit. Ensure you can access the data frame at the prompt; then study the help file ?InsectSprays to get an idea of R's representation of the two variables.

d. Identify the two variable types in InsectSprays (as per the definitions in Section 13.1.1 and Section 13.1.2).

e. Calculate the modes of the distribution of insect counts, regardless of spray type.

f. Use tapply to report the total insect counts by each spray type.

g. Using the same kind of for loop as in (c), compute the percentage of agricultural units in each spray type group that had at least five bugs on them. When printing to the screen, round the percentages to the nearest whole number.

h. Obtain the same numeric results as in (g), with rounding, but use tapply and a disposable function.

### 13.2.3  Quantiles, Percentiles, and the Five-Number Summary

Let's return, once more, to thinking about raw numeric observations. An understanding of how observations are *distributed* is an important statistical concept, and this will form a key feature of discussions in Chapter 15 onward.

You can gain more insight into the distribution of a set of observations by examining quantiles. A *quantile* is a value computed from a collection of numeric measurements that indicates an observation's rank when compared to all the other present observations. For example, the median (Section 13.2.1) is itself a quantile—it gives you a value below which half of the measurements lie—it's the 0.5th quantile. Alternatively, quantiles can be expressed as a *percentile*—this is identical but on a "percent scale" of 0 to 100. In other words, the $p$th quantile is equivalent to the $100 \times p$th percentile. The median, therefore, is the 50th percentile.

There are a number of different algorithms that can be used to compute quantiles and percentiles. They all work by sorting the observations from smallest to largest and using some form of weighted average to find the numeric value that corresponds to $p$, but results may vary slightly in other statistical software.

Obtaining quantiles and percentiles in R is done with the quantile function. Using the eight observations stored as the vector xdata, the 0.8th quantile (or 80th percentile) is confirmed as 3.6:

```
R> xdata <- c(2,4.4,3,3,2,2.2,2,4)
R> quantile(xdata,prob=0.8)
80%
3.6
```

As you can see, quantile takes the data vector of interest as its first argument, followed by a numeric value supplied to prob, giving the quantile of interest. In fact, prob can take a numeric vector of quantile values. This is convenient when multiple quantiles are desired.

```
R> quantile(xdata,prob=c(0,0.25,0.5,0.75,1))
  0%   25%   50%   75%  100%
2.00 2.00 2.60 3.25 4.40
```

Here, you've used quantile to obtain what's called the *five-number summary* of xdata, comprised of the 0th percentile (the minimum), the 25th percentile, the 50th percentile, the 75th percentile, and the 100th percentile (the maximum). The 0.25th quantile is referred to as the *first* or *lower quartile*, and the 0.75th quantile is referred to as the *third* or *upper quartile*. Also note that the 0.5th quantile of xdata is equivalent to the median (2.6, calculated in Section 13.2.1 using median). The median is the second quartile, with the maximum value being the fourth quartile.

There are ways to obtain the five-number summary other than using quantile; when applied to a numeric vector, the summary function also provides these statistics, along with the mean, automatically.

```
R> summary(xdata)
   Min. 1st Qu.  Median   Mean 3rd Qu.    Max.
  2.000   2.000   2.600  2.825   3.250   4.400
```

To look at some examples using real data, let's compute the lower and upper quartiles of the weights of the chicks in the chickwts.

```
R> quantile(chickwts$weight,prob=c(0.25,0.75))
   25%   75%
204.5 323.5
```

This indicates that 25 percent of the weights lie at or below 204.5 grams and that 75 percent of the weights lie at or below 323.5 grams.

Let's also compute the five-number summary (along with the mean) of the magnitude of the seismic events off the coast of Fiji that occurred at a depth of less than 400 km, using the quakes data frame.

```
R> summary(quakes$mag[quakes$depth<400])
   Min. 1st Qu.  Median   Mean 3rd Qu.    Max.
   4.00    4.40    4.60   4.67    4.90    6.40
```

This begins to highlight how useful quantiles are for interpreting the distribution of numeric measurements. From these results, you can see that most of the magnitudes of events at a depth of less than 400 km lie around 4.6, the median, and the first and third quartiles are just 4.4 and 4.9, respectively. But you can also see that the maximum value is much further away from the upper quartile than the minimum is from the lower quartile, suggesting a *skewed* distribution, one that stretches more positively (in other words, to the right) from its center than negatively (in other words, to the left). This notion is also supported by the fact that the mean is greater than the median—the mean is being "dragged upward" by the larger values.

You'll explore this further in Chapter 14 when you investigate data sets using basic statistical plots, and some of the associated terminology will be formalized in Chapter 15.

### 13.2.4   Spread: Variance, Standard Deviation, and the Interquartile Range

The measures of centrality explored in Section 13.2.1 offer a good indication of where your numeric measurements are massed, but the mean, median, and mode do nothing to describe how *dispersed* your data are. For this, measures of *spread* are needed.

In addition to your vector of eight hypothetical observations, given again here,

```
R> xdata <- c(2,4.4,3,3,2,2.2,2,4)
```

you'll also look at another eight observations stored as follows:

```
R> ydata <- c(1,4.4,1,3,2,2.2,2,7)
```

Although these are two different collections of numbers, note that they have an identical arithmetic mean.

```
R> mean(xdata)
[1] 2.825
R> mean(ydata)
[1] 2.825
```

Now let's plot these two data vectors side by side, each one on a horizontal line, by executing the following:

```
R> plot(xdata,type="n",xlab="",ylab="data vector",yaxt="n",bty="n")
R> abline(h=c(3,3.5),lty=2,col="gray")
R> abline(v=2.825,lwd=2,lty=3)
R> text(c(0.8,0.8),c(3,3.5),labels=c("x","y"))
R> points(jitter(c(xdata,ydata)),c(rep(3,length(xdata)),
                                   rep(3.5,length(ydata))))
```

You saw how to use these base R graphics functions in Chapter 7, though it should be explained that because some of the observations in xdata and in ydata occur more than once, you can randomly alter them slightly to prevent overplotting, which aids in the visual interpretation. This step is known as *jittering* and is achieved by passing the numeric vector of interest to the jitter function prior to plotting with points. Additionally, note that you can use yaxt="n" in any call to plot to suppress the y-axis; similarly, bty="n" removes the typical box that's placed around a plot (you'll focus more on this type of plot customization in Chapter 23).

The result, shown in Figure 13-3, provides you with valuable information. Though the mean is the same for both xdata and ydata, you can easily see that the observations in ydata are more "spread out" around the measure of centrality than the observations in xdata. To quantify spread, you use values such as the variance, the standard deviation, and the interquartile range.

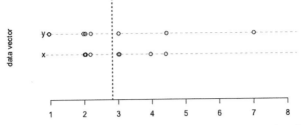

Figure 13-3: Comparing two hypothetical data vectors that share an identical arithmetic mean (marked by the vertical dotted line) but have different magnitudes of spread. Identical observations are jittered slightly.

The sample *variance* measures the degree of the spread of numeric observations around their arithmetic mean. The variance is a particular representation of the *average squared distance* of each observation when compared to the mean. For a set of $n$ numeric measurements labeled $x = \{x_1, x_2, \ldots, x_n\}$, the sample variance $s_x^2$ is given by the following, where $\bar{x}$ is the sample mean described in Equation (13.1):

$$s_x^2 = \frac{(x_1 - \bar{x})^2 + (x_2 - \bar{x})^2 + \ldots + (x_n - \bar{x})^2}{n-1} = \frac{1}{n-1} \sum_{i=1}^{n} (x_i - \bar{x})^2 \qquad (13.3)$$

For example, if you take the eight illustrative observations $2, 4.4, 3, 3, 2, 2.2, 2, 4$, their sample variance is as follows when rounded to three decimal places (some terms are hidden with ... for readability):

$$\frac{(2 - 2.825)^2 + (4.4 - 2.825)^2 + \ldots + (4 - 2.825)^2}{7}$$
$$= \frac{(-0.825)^2 + (1.575)^2 + \ldots + (1.175)^2}{7}$$
$$= \frac{6.355}{7} = 0.908$$

The *standard deviation* is simply the square root of the variance. Since the variance is a representation of the average squared distance, the standard deviation provides a value interpretable with respect to the scale of the original observations. With the same notation for a sample of $n$ observations, the sample standard deviation $s$ is found by taking the square root of Equation (13.3).

$$s_x = \sqrt{s^2} = \sqrt{\frac{1}{n-1} \sum_{i=1}^{n} (x_i - \bar{x})^2} \qquad (13.4)$$

For example, based on the sample variance calculated earlier, the standard deviation of the eight hypothetical observations is as follows (to three decimal places):

$$\sqrt{0.908} = 0.953$$

Thus, a rough way to interpret this is that 0.953 represents the average distance of each observation from the mean.

Unlike the variance and standard deviation, the *interquartile range (IQR)* is not computed with respect to the sample mean. The IQR measures the width of the "middle 50 percent" of the data, that is, the range of values that lie within a 25 percent quartile on either side of the median. As such, the IQR is computed as the difference between the upper and lower quartiles of your data. Formally, where $Q_x(\cdot)$ denotes the quantile function (as defined in Section 13.2.3), the IQR is given as

$$\text{IQR}_x = Q_x(0.75) - Q_x(0.25) \tag{13.5}$$

The direct R commands for computing these measures of spread are var (variance), sd (standard deviation), and IQR (interquartile range).

```
R> var(xdata)
[1] 0.9078571
R> sd(xdata)
[1] 0.9528154
R> IQR(xdata)
[1] 1.25
```

You can confirm the relationship between the sample variance and standard deviation using the square root function sqrt on the result from var, and you can reproduce the IQR by calculating the difference between the third and first quartiles.

```
R> sqrt(var(xdata))
[1] 0.9528154
R> as.numeric(quantile(xdata,0.75)-quantile(xdata,0.25))
[1] 1.25
```

Note that as.numeric (see Section 6.2.4) strips away the percentile annotations (that label the results by default) from the returned object of quantile.

Now, do the same with the ydata observations that had the same arithmetic mean as xdata. The calculations give you the following:

```
R> sd(ydata)
[1] 2.012639
R> IQR(ydata)
[1] 1.6
```

ydata is on the same scale as xdata, so the results confirm what you can see in Figure 13-3—that the observations in the former are more spread out than in the latter.

For two quick final examples, let's return again to the chickwts and quakes data sets. In Section 13.2.1, you saw that the mean weight of all the chicks is 261.3099 grams. You can now find that the standard deviation of the weights is as follows:

```
R> sd(chickwts$weight)
[1] 78.0737
```

Informally, this implies that the weight of each chick is, on average, around 78.1 grams away from the mean weight (technically, though, remember it is merely the square root of a function of the squared distances—see the following note).

In Section 13.2.3, you used summary to obtain the five-number summary of the magnitudes of some of the earthquakes in the quakes data set. Looking at the first and third quartiles in these earlier results (4.4 and 4.9, respectively), you can quickly determine that the IQR of this subset of the events is 0.5. This can be confirmed using IQR.

```
R> IQR(quakes$mag[quakes$depth<400])
[1] 0.5
```

This gives you the width, in units of the Richter scale, of the middle 50 percent of the observations.

**NOTE**    *The definition of the variance (and hence the standard deviation) here has referred exclusively to the "sample estimator," the default in R, which uses the divisor of n − 1 in the formula. This is the formula used when the observations at hand represent a sample of an assumed larger population. In these cases, use of the divisor n − 1 is more accurate, providing what's known as an* unbiased *estimate of the true population value. Thus, you aren't exactly calculating the "average squared distance," though it can loosely be thought of as such and does indeed approach this as the sample size n increases.*

## Exercise 13.3

a.  Using the chickwts data frame, compute the 10th, 30th, and 90th percentiles of all the chick weights and then use tapply to determine which feed type is associated with the highest sample variance of weights.

b. Turn to the seismic event data in quakes and complete the following tasks:
   i. Find the IQR of the recorded depths.
   ii. Find the five-number summary of all magnitudes of seismic events that occur at a depth of 400 km *or deeper*. Compare this to the summary values found in Section 13.2.3 of those events occurring at less than 400 km and briefly comment on what you notice.
   iii. Use your knowledge of cut (Section 4.3.3) to create a new factor vector called depthcat that identifies four evenly spaced categories of quakes$depth so that when you use levels(depthcat), it gives the following:

---
```
R> levels(depthcat)
[1] "[40,200)"  "[200,360)"  "[360,520)"  "[520,680]"
```
---

   iv. Find the sample mean and standard deviation of the magnitudes of the events associated with each category of depth according to depthcat.
   v. Use tapply to compute the 0.8th quantile of the magnitudes of the seismic events in quakes, split by depthcat.

## 13.2.5 Covariance and Correlation

When analyzing data, it's often useful to be able to investigate the *relationship* between two numeric variables to assess trends. For example, you might expect height and weight observations to have a noticeable positive relationship—taller people tend to weigh more. Conversely, you might imagine that handspan and length of hair would have less of an association. One of the simplest and most common ways such associations are quantified and compared is through the idea of correlation, for which you need the covariance.

The *covariance* expresses how much two numeric variables "change together" and the nature of that relationship, whether it is positive or negative. Suppose for $n$ individuals you have a sample of observations for two variables, labeled $x = \{x_1, x_2, \ldots, x_n\}$ and $y = \{y_1, y_2, \ldots, y_n\}$, where $x_i$ corresponds to $y_i$ for $i = 1, \ldots, n$. The sample covariance $r_{xy}$ is computed with the following, where $\bar{x}$ and $\bar{y}$ represent the respective sample means of both sets of observations:

$$r_{xy} = \frac{1}{n-1} \sum_{i-1}^{n} (x_i - \bar{x})(y_i - \bar{y}) \tag{13.6}$$

When you get a positive result for $r_{xy}$, it shows that there is a positive linear relationship—as $x$ increases, $y$ increases. When you get a negative result, it shows a negative linear relationship—as $x$ increases, $y$ decreases, and vice versa. When $r_{xy} = 0$, this indicates that there is no linear relationship between the values of $x$ and $y$. It is useful to note that the order of the variables in the formula itself doesn't matter; in other words, $r_{xy} \equiv r_{yx}$.

To demonstrate, let's use the original eight illustrative observations, which I'll denote here with $x = \{2, 4.4, 3, 3, 2, 2.2, 2, 4\}$, and the additional eight observations denoted with $y = \{1, 4.4, 1, 3, 2, 2.2, 2, 7\}$. Remember that both $x$ and $y$ have sample means of 2.825. The sample covariance of these two sets of observations is as follows (rounded to three decimal places):

$$\frac{(2 - 2.825) \times (1 - 2.285) + \ldots + (4 - 2.825) \times (7 - 2.825)}{7}$$
$$= \frac{(-0.825)(-1.825) + \ldots + (1.175)(4.175)}{7}$$
$$= \frac{10.355}{7} = 1.479$$

The figure is a positive number, so this suggests there is a positive relationship based on the observations in $x$ and $y$.

*Correlation* allows you to interpret the covariance further by identifying both the direction and the strength of any association. There are different types of correlation coefficients, but the most common of these is *Pearson's product-moment correlation coefficient*, the default implemented by R (this is the estimator I will use in this chapter). Pearson's sample correlation coefficient $\rho_{xy}$ is computed by dividing the sample covariance by the product of the standard deviation of each data set. Formally, where $r_{xy}$ corresponds to Equation (13.6) and $s_x$ and $s_y$ to Equation (13.4),

$$\rho_{xy} = \frac{r_{xy}}{s_x s_y}, \tag{13.7}$$

which ensures that $-1 \leq \rho_{xy} \leq 1$.

When $\rho_{xy} = -1$, a perfect negative linear relationship exists. Any result less than zero shows a negative relationship, and the relationship gets weaker the nearer to zero the coefficient gets, until $\rho_{xy} = 0$, showing no relationship at all. As the coefficient increases above zero, a positive relationship is shown, until $\rho_{xy} = 1$, which is a perfect positive linear relationship.

If you take the standard deviations already computed for $x$ and $y$ in Section 13.2.4 ($s_x = 0.953$ and $s_y = 2.013$ to three decimal places), you find the following to three decimal places:

$$\frac{1.479}{0.953 \times 2.013} = 0.771$$

$\rho_{xy}$ is positive just like $r_{xy}$; the value of 0.771 indicates a moderate-to-strong positive association between the observations in $x$ and $y$. Again, $\rho_{xy} \equiv \rho_{yx}$.

The R commands cov and cor are used for the sample covariance and correlation; you need only to supply the two corresponding vectors of data.

```
R> xdata <- c(2,4.4,3,3,2,2.2,2,4)
R> ydata <- c(1,4.4,1,3,2,2.2,2,7)
```

```
R> cov(xdata,ydata)
[1] 1.479286
R> cov(xdata,ydata)/(sd(xdata)*sd(ydata))
[1] 0.7713962
R> cor(xdata,ydata)
[1] 0.7713962
```

You can plot these bivariate observations as a coordinate-based plot (a *scatterplot*—see more examples in Section 14.4). Executing the following gives you Figure 13-4:

```
R> plot(xdata,ydata,pch=13,cex=1.5)
```

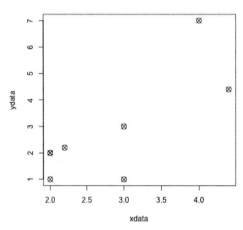

Figure 13-4: Plotting the xdata and ydata observations as bivariate data points to illustrate the interpretation of the correlation coefficient

As discussed earlier, the correlation coefficient estimates the nature of the *linear* relationship between two sets of observations, so if you look at the pattern formed by the points in Figure 13-4 and imagine drawing a perfectly straight line that best represents all the points, you can determine the strength of the linear association by how close those points are to your line. Points closer to a perfect straight line will have a value of $\rho_{xy}$ closer to either $-1$ or 1. The direction is determined by how the line is sloped—an increasing trend, with the line sloping upward toward the right, indicates positive correlation; a negative trend would be shown by the line sloping downward toward the right. Considering this, you can see that the estimated correlation coefficient for the data plotted in Figure 13-4 makes sense according to the previous calculations. The points do appear to increase together as a rough straight line in terms of the values in xdata and ydata, but this linear association is by no means perfect. How you can compute the "ideal" or "best" straight line to fit such data is discussed in Chapter 20.

To aid your understanding of the idea of correlation, Figure 13-5 displays different scatterplots, each showing 100 points. These observations have been randomly and artificially generated to follow preset "true" values of $\rho_{xy}$, labeled above each plot.

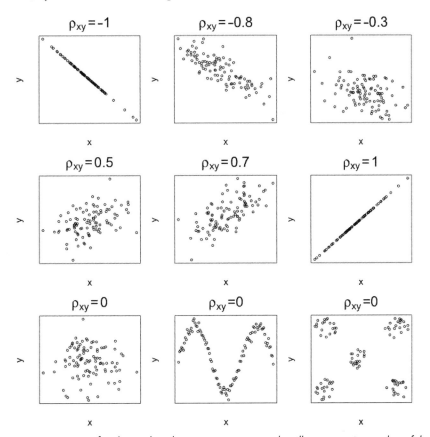

Figure 13-5: Artificial x and y observations, generated to illustrate a given value of the correlation coefficient

The first row of scatterplots shows negatively correlated data; the second shows positively correlated data. These match what you would expect to see—the direction of the line shows the negative or positive correlation of the trend, and the extremity of the coefficient corresponds to the closeness to a "perfect line."

The third and final row shows data sets generated with a correlation coefficient set to zero, implying no linear relationship between the observations in $x$ and $y$. The middle and rightmost plots are particularly important because they highlight the fact that Pearson's correlation coefficient identifies only "straight-line" relationships; these last two plots clearly show some kind of trend or pattern, but this particular statistic cannot be used to detect such a trend.

To wrap up this section, look again at the quakes data. Two of the variables are mag (the magnitude of each event) and stations (the number of stations that reported detection of the event). A plot of stations on the y-axis against mag on the x-axis can be produced with the following:

```
R> plot(quakes$mag,quakes$stations,xlab="Magnitude",ylab="No. of stations")
```

Figure 13-6 shows this image.

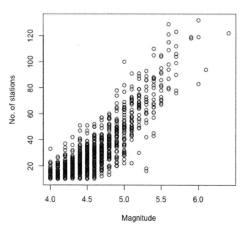

Figure 13-6: Plotting the number of stations reporting
the event (y) and the magnitude (x) of each event
in the quakes data frame

You can see by the vertical patterning that the magnitudes appear to have been recorded to a certain specific level of precision (this is owed to the difficulty associated with measuring earthquake magnitudes exactly). Nevertheless, a positive relationship (more stations tend to detect events of higher magnitude) is clearly visible in the scatterplot, a feature that is confirmed by a positive covariance.

```
R> cov(quakes$mag,quakes$stations)
[1] 7.508181
```

As you might expect from examining the pattern, Pearson's correlation coefficient confirms that the linear association is quite strong.

```
R> cor(quakes$mag,quakes$stations)
[1] 0.8511824
```

**NOTE**    *It is important to remember that correlation does not imply causation. When you detect a high correlative effect between two variables, this does not mean that one causes the other. Causation is difficult to prove in even the most controlled situations. Correlation merely allows you to measure association.*

As mentioned earlier, there are other representations of correlation that can be used; *rank* coefficients, such as Spearman's and Kendall's correlation coefficients, differ from Pearson's estimate in that they do not require the relationship to be linear. These are also available through the cor function by accessing the optional method argument (see ?cor for details). Pearson's correlation coefficient is the most commonly used, however, and is related to linear regression methods, which you'll start to examine in Chapter 20.

## 13.2.6   Outliers

An *outlier* is an observation that does not appear to "fit" with the rest of the data. It is a noticeably extreme value when compared with the bulk of the data, in other words, an anomaly. In some cases, you might suspect that such an extreme observation has not actually come from the same mechanism that generated the other observations, but there is no hard-and-fast numeric rule as to what constitutes an outlier. For example, consider the 10 hypothetical data points in foo.

```
R> foo <- c(0.6,-0.6,0.1,-0.2,-1.0,0.4,0.3,-1.8,1.1,6.0)
```

Using skills from Chapter 7 (and from creating Figure 13-3), you can plot foo on a line as follows.

```
R> plot(foo,rep(0,10),yaxt="n",ylab="",bty="n",cex=2,cex.axis=1.5,cex.lab=1.5)
R> abline(h=0,col="gray",lty=2)
R> arrows(5,0.5,5.9,0.1,lwd=2)
R> text(5,0.7,labels="outlier?",cex=3)
```

The result is given on the left of Figure 13-7.

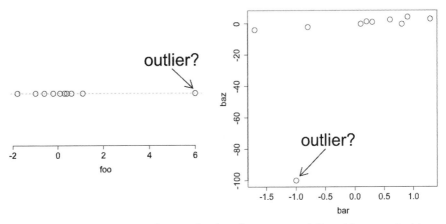

*Figure 13-7: Illustrating the definition of outliers for univariate (left) and bivariate (right) data. Should you include such values in your statistical analysis? The answer can be difficult to determine.*

From this plot, you see that most of the observations are centered around zero, but one value is way out at 6. To give a bivariate example, I'll use two further vectors, bar and baz, shown here:

```
R> bar <- c(0.1,0.3,1.3,0.6,0.2,-1.7,0.8,0.9,-0.8,-1.0)
R> baz <- c(-0.3,0.9,2.8,2.3,1.2,-4.1,-0.4,4.1,-2.3,-100.0)
```

I'll plot these data using the following code; the result is on the right of Figure 13-7.

```
R> plot(bar,baz,axes=T,cex=2,cex.axis=1.5,cex.lab=1.5)
R> arrows(-0.5,-80,-0.94,-97,lwd=2)
R> text(-0.45,-74,labels="outlier?",cex=3)
```

It's important to identify outliers because of the potential impact they can have on any statistical calculations or model fitting. For this reason, many researchers will try to identify possible outliers before computing results by conducting an "exploratory" analysis of their data using basic summary statistics and data visualization tools (like those you'll look at in Chapter 14).

Outliers can occur naturally, where the outlier is a "true" or accurate observation recorded from the population, or unnaturally, where something has "contaminated" that particular contribution to the sample, such as incorrectly inputting data. As such, it is common to omit any outliers occurring through unnatural sources prior to analysis, but in practice this is not always easy because the cause of an outlier can be difficult to determine. In some cases, researchers conduct their analysis both ways—presenting results including and excluding any perceived outliers.

With this in mind, if you return to the example shown on the left in Figure 13-7, you can see that when you include all observations, you get the following:

```
R> mean(foo)
[1] 0.49
```

However, when the possible outlier of 6 (the 10th observation) is deleted, you get the following:

```
R> mean(foo[-10])
[1] -0.1222222
```

This highlights the impact a single extreme observation can have. Without any additional information about the sample, it would be difficult to say whether it's sensible to exclude the outlier 6. The same kind of effect is noticeable if you compute, say, the correlation coefficient of bar with baz, shown on the right in Figure 13-7 (again, it's the 10th observation that is the possible outlier).

```
R> cor(bar,baz)
[1] 0.4566361
R> cor(bar[-10],baz[-10])
[1] 0.8898639
```

You see the correlation becomes much stronger without that outlier.

Again, knowing whether to delete the outlier can be hard to correctly gauge in practice. At this stage, it's important simply to be aware of the impact outliers can have on an analysis and to perform at least a cursory inspection of the raw data before beginning more rigorous statistical investigations.

**NOTE** *The extent of the effect that extreme observations have on your data analysis depends not only on their extremity but on the statistics you intend to calculate. The sample mean, for example, is highly sensitive to outliers and will differ greatly when including or excluding them, so any statistic that depends on the mean, like the variance or covariance, will be affected too. Quantiles and related statistics, such as the median or IQR, are relatively unaffected by outliers. In statistical parlance this property is referred to as* robustness.

---

### Exercise 13.4

a. In Exercise 7.1 (b) on page 139, you plotted height against weight measurements. Compute the correlation coefficient based on the observed data of these two variables.

b. Another of R's built-in, ready-to-use data sets is mtcars, containing a number of descriptive details on performance aspects of 32 automobiles.
   i. Ensure you can access this data frame by entering mtcars at the prompt. Then inspect its help file to get an idea of the types of data present.
   ii. Two of the variables describe a vehicle's horsepower and shortest time taken to travel a quarter-mile distance. Using base R graphics, plot these two data vectors with horsepower on the *x*-axis and compute the correlation coefficient.
   iii. Identify the variable in mtcars that corresponds to transmission type. Use your knowledge of factors in R to create a new factor from this variable called tranfac, where manual cars should be labeled "manual" and automatic cars "auto".
   iv. Now, use qplot from ggplot2 in conjunction with tranfac to produce the same scatterplot as in (ii) so that you're able to visually differentiate between manual and automatic cars.

v. Finally, compute separate correlation coefficients for horse-power and quarter-mile time based on the transmission of the vehicles and, comparing these estimates with the overall value from (ii), briefly comment on what you note.

c. Return to chickwts to complete the following tasks:
i. Produce a plot like the left panel of Figure 13-7, based on the weights of chicks on the sunflower diet only. Note that one of the sunflower-fed chicks has a far lower weight than the others.
ii. Compute the standard deviation and IQR of the weights of the sunflower-fed chicks.
iii. Now, suppose you're told that the lowest weight of the sunflower-fed chicks was caused by a certain illness, irrelevant to your research. Delete this observation and recalculate the standard deviation and IQR of the remaining sunflower chicks. Briefly comment on the difference in calculated values.

## Important Code in This Chapter

| Function/operator | Brief description | First occurrence |
|---|---|---|
| mean | Arithmetic mean | Section 13.2.1, p. 268 |
| median | Median | Section 13.2.1, p. 268 |
| table | Tabulate frequencies | Section 13.2.1, p. 268 |
| min, max, range | Minimum and maximum | Section 13.2.1, p. 268 |
| round | Round numeric values | Section 13.2.2, p. 272 |
| quantile | Quantiles/percentiles | Section 13.2.3, p. 274 |
| summary | Five-number summary | Section 13.2.3, p. 275 |
| jitter | Jitter points in plotting | Section 13.2.4, p. 276 |
| var, sd | Variance, standard deviation | Section 13.2.4, p. 278 |
| IQR | Interquartile range | Section 13.2.4, p. 278 |
| cov, cor | Covariance, correlation | Section 13.2.5, p. 281 |

# 14

## BASIC DATA VISUALIZATION

Data visualization is an important part of
a statistical analysis. The visualization tools
appropriate for a given data set are dependent upon the types of variables (as per the definitions in Sections 13.1.1 and 13.1.2) for which you've
made observations. In this chapter, you'll look at the
most commonly used data plots in statistical analyses and see examples using both base R graphics and
ggplot2 functionality.

## 14.1 Barplots and Pie Charts

Barplots and pie charts are commonly used to visualize qualitative data
by category frequency. In this section you'll learn how to generate both
using R.

### 14.1.1 Building a Barplot

A *barplot* draws either vertical or horizontal bars, typically separated by white
space, to visualize frequencies according to the relevant categories. Though

the raw frequencies themselves are usually displayed, a barplot can also be used to visualize other quantities, such as means or proportions, which directly depend upon these frequencies.

As an example, let's use the mtcars data set from Exercise 13.4 (b) on page 287. Detailing various characteristics of 32 classic performance cars in the mid-1970s, the first five records can be viewed directly from the prompt.

```
R> mtcars[1:5,]
                   mpg cyl disp  hp drat    wt  qsec vs am gear carb
Mazda RX4         21.0   6  160 110 3.90 2.620 16.46  0  1    4    4
Mazda RX4 Wag     21.0   6  160 110 3.90 2.875 17.02  0  1    4    4
Datsun 710        22.8   4  108  93 3.85 2.320 18.61  1  1    4    1
Hornet 4 Drive    21.4   6  258 110 3.08 3.215 19.44  1  0    3    1
Hornet Sportabout 18.7   8  360 175 3.15 3.440 17.02  0  0    3    2
```

The documentation in ?mtcars explains the variables that have been recorded. Of these, cyl provides the number of cylinders in each engine—four, six, or eight. To find out how many cars were observed with each number of cylinders, you can use table, as shown here:

```
R> cyl.freq <- table(mtcars$cyl)
R> cyl.freq

 4  6  8
11  7 14
```

The result is easily displayed as a barplot, as shown here:

```
R> barplot(cyl.freq)
```

You can find the resulting barplot on the left of Figure 14-1.

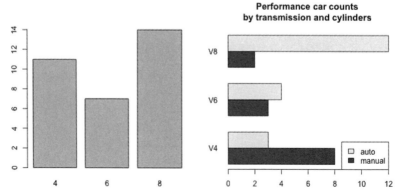

Figure 14-1: Two examples of barplots of data from mtcars using base R graphics. Left: The simplest, default version, using one categorical variable. Right: A "dodged" barplot illustrating various visual options and using two categorical variables.

This plot displays the number of four-, six-, and eight-cylinder cars in the data set but is admittedly rather uninteresting, and without annotations it's not clear what's being summarized. Fortunately, it's easy to annotate such plots and further split up the frequencies of each bar according to an additional categorical variable. Consider the following code where, this time, you're finding the counts associated with cyl by transmission (am):

```
R> table(mtcars$cyl[mtcars$am==0])

 4  6  8
 3  4 12

R> table(mtcars$cyl[mtcars$am==1])
4 6 8
8 3 2
```

If you aim to produce a barplot that's *stacked* (where bars are split up vertically) or *dodged* (where bars are broken up and placed beside each other), barplot requests its first argument as a suitably arranged matrix. You could construct it from the previous two vectors using matrix, but it's easier to just continue using table.

```
R> cyl.freq.matrix <- table(mtcars$am,mtcars$cyl)
R> cyl.freq.matrix

    4  6  8
  0 3  4 12
  1 8  3  2
```

As you can see, you can cross-tabulate counts by supplying two categorical or discrete vectors of equal length to table; the first vector stipulates row counts, and the second defines the columns. The outcome is a matrix object; here it's a 2 × 3 structure providing the quantities of the four-, six-, and eight-cylinder automatic cars in the first row and the quantities of the manual cars in the second. The rule is that each column of the barplot will correspond to a column of the supplied matrix; these will be further split with respect to each row of the supplied matrix. The plot on the right of Figure 14-1 is the result of the following code:

```
R> barplot(cyl.freq.matrix,beside=TRUE,horiz=TRUE,las=1,
           main="Performance car counts\nby transmission and cylinders",
           names.arg=c("V4","V6","V8"),legend.text=c("auto","manual"),
           args.legend=list(x="bottomright"))
```

The help file ?barplot explains the options here in detail. To label the bars according to the categories of the column variable of the matrix that was initially passed to barplot, you use a character vector of the appropriate length passed to names.arg. The options beside=TRUE and horiz=TRUE select a dodged, horizontal barplot. If both options were FALSE, a stacked, vertical barplot would be selected. The argument las=1 forces the labels on the vertical axis to appear horizontally, rather than parallel to it. The final two arguments, legend.text and args.legend, are used for the legend—you could have drawn a legend separately as in Section 7.3 via legend, but this way automates the color assignment to ensure the reference keys match the precise shading of the bars themselves.

Similar plots may be produced using ggplot2. If you load the installed package with library("ggplot2") and enter the following, it will produce the most basic barplot, given on the left of Figure 14-2:

```
R> qplot(factor(mtcars$cyl),geom="bar")
```

Note here that the relevant geom is "bar" (or geom_bar if used separately, as you'll see in a moment) and that the default mapping variable in qplot must be supplied as a factor (in mtcars the vector mtcars$cyl is just numeric, which is fine for barplot, but ggplot2 functionality is a bit more strict).

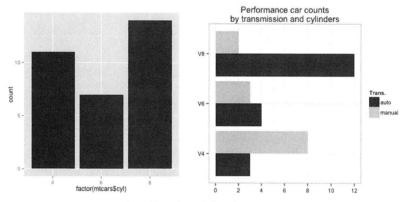

Figure 14-2: Two examples of barplots of data from mtcars using ggplot2 functionality. Left: The most simple qplot version, using one categorical variable. Right: A "dodged" barplot, the same as in Figure 14-1, based on the supply of various additional geoms and scaling options.

Again, you can create far more complicated images depending upon what you want to display. To produce a ggplot2 version of the dodged barplot from 14-1, call the following:

```
R> qplot(factor(mtcars$cyl),geom="blank",fill=factor(mtcars$am),xlab="",
         ylab="",main="Performance car counts\nby transmission and cylinders")
    + geom_bar(position="dodge")
    + scale_x_discrete(labels=c("V4","V6","V8"))
    + scale_y_continuous(breaks=seq(0,12,2))
    + theme_bw() + coord_flip()
    + scale_fill_grey(name="Trans.",labels=c("auto","manual"))
```

You can find the result on the right in Figure 14-2.

Note a number of new additions to the basic qplot setup. The default mapping, by cyl, remains the same as earlier. You further specify that the bars should be filled according to a factor created by using the transmission variable am; so, each cyl bar is instructed to split according to that variable. The initial call to qplot was "empty," in the sense that geom="blank", and therefore drawing begins with the addition of geom_bar to the ggplot2 object. It becomes a dodged barplot through position="dodge"; as in base R graphics, the default behavior is to generate a stacked plot. The scale_x_discrete modifier specifies labels for each category of the default cyl mapping; the scale_y_continuous modifier is employed to control the axis labels for the frequencies.

Further, adding theme_bw() to the object changes the visual theme of the image; in the current example, I've chosen to remove the gray background because it's too similar in color to the manual car bars. Adding coord_flip to the object flips the axes and provides horizontal bars rather than the default vertical style (note that the calls to the scale_ functions are used with respect to the unflipped image). The default behavior of fill is to use colors, so you use the scale_fill_grey modifier to force this to be grayscale and to alter the labels of the automatically generated legend to match at the same time.

The most prominent advantage of using ggplot2 over base R graphics in this case lies in the fact that you don't need to manually tabulate counts or design specific matrix structures of these frequencies—the variable mappings do this automatically. For practice, I encourage you to experiment with this code example, omitting or modifying some of the additions to the qplot object to assess the impact on the resulting image.

### 14.1.2   A Quick Pie Chart

The venerable *pie chart* is an alternative option for visualizing frequency-based quantities across levels of categorical variables, with appropriately sized "slices" representing the relative counts of each categorical variable.

```
R> pie(table(mtcars$cyl),labels=c("V4","V6","V8"),
       col=c("white","gray","black"),main="Performance cars by cylinders")
```

You can find the resulting plot in Figure 14-3.

Though it's possible to achieve with some effort, there is no direct "pie" geom in ggplot2. This may, at least in part, be due to the general preference of statisticians for barplots over pie charts. That fact itself is even summarized in the help file ?pie!

> Pie charts are a bad way of displaying information. The eye is good at judging linear measures and bad at judging relative areas.

Furthermore, barplots are of greater value than pie charts if you want frequencies split by a *second* categorical variable or if the levels of a factor are ordered.

**Performance cars by cylinders**

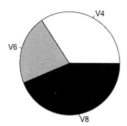

*Figure 14-3: A pie chart of the frequencies of total cylinders of the cars in the* mtcars *data frame*

## 14.2  Histograms

The barplot is intuitively sensible for counting observations in relation to categorical variables but is of virtually no use if the variable you're interested in is numeric-continuous. To visualize the distribution of continuous measurements, you can use a *histogram*—a tool that's sometimes confused with a barplot owing to its similar appearance. A histogram also measures frequencies, but in targeting a numeric-continuous variable, it's first necessary to "bin" the observed data, meaning to define intervals and then count the number of continuous observations that fall within each one. The size of this interval is known as the *binwidth*.

For a simple example of a histogram, consider the horsepower data of the 32 cars in mtcars, given in the fourth column, named hp.

```
R> mtcars$hp
 [1] 110 110  93 110 175 105 245  62  95 123 123 180 180 180 205 215 230  66
[19]  52  65  97 150 150 245 175  66  91 113 264 175 335 109
```

For this section, define horsepowers of all performance cars from that era as your population and assume that these observations represent a sample from that population. Using base R graphics, the hist command takes a vector of numeric-continuous observations and produces a histogram, as shown on the left in Figure 14-4.

```
R> hist(mtcars$hp)
```

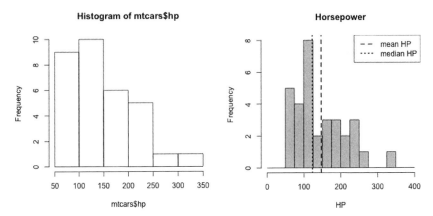

*Figure 14-4: Illustrating the default behavior of* hist *on the* mtcars *horsepower data (left); customizing binwidth, color, and title options, as well as adding markers of centrality (right)*

You can immediately see that the histogram on the left has used binwidths of 50 units spanning the range of the data, providing you with a quick and useful first impression of the distribution of horsepower measurements. It seems to be centered roughly in the range of 75 to 150, tapering off on the right (this is known as a *right* or *positive skew*; more terminology will be covered in Section 15.2.4).

The accuracy of a histogram as a representation of the shape of a distribution of measurements depends solely upon the widths of the intervals used to bin the data. Binwidths are controlled in hist by the breaks argument. You can manually set these by supplying a vector, giving each breakpoint, to breaks. This is done in the following code by halving the width of each bin from 50 to 25 and widening the overall range somewhat, using an evenly spaced sequence.

```
R> hist(mtcars$hp,breaks=seq(0,400,25),col="gray",main="Horsepower",xlab="HP")
R> abline(v=c(mean(mtcars$hp),median(mtcars$hp)),lty=c(2,3),lwd=2)
R> legend("topright",legend=c("mean HP","median HP"),lty=c(2,3),lwd=2)
```

This plot, given on the right in Figure 14-4, shows the result of using the narrower bins, as well as making the bars gray and adding a more readable title. It also includes vertical lines denoting the mean and median, using abline, and a legend (refer back to Section 7.3).

With the smaller binwidth, more detail is visible in the distribution. However, using narrower bins risks highlighting "unimportant features" (in other words, features of the histogram that represent natural variation as a consequence of the finite-sized sample). These typically occur at locations on the scale where data are scarce. For example, the single 335 horsepower car has produced an isolated bar on the right of the scale, but you might reasonably conclude that this is not a precise, accurate reflection of a "true bump" at that location in terms of the overall population. It's therefore important to note that choosing the interval widths is a balancing act of sorts.

You want to choose a width that gives you a good idea of the distribution of measurements without emphasizing unimportant detail by using too small a binwidth. Equivalently, you also want to avoid hiding important features by using too *large* a binwidth. To address this, there are data-driven algorithms that use the scale of the recorded observations to try to calculate an appropriately balanced binwidth. You can supply a character string to breaks, giving the name of the algorithm that you want to employ. The default breaks="Sturges" often works well, though it's worth trying a small number of alternative widths when exploring data in this way. For further details on this and other ways to use breaks, the documentation ?hist provides clear and concise instruction.

The issues surrounding intervals and their widths is emphasized in a different way in ggplot2. By default, the qplot function produces a histogram when you supply it with a single numeric vector but no value for the geom argument:

```
R> qplot(mtcars$hp)
`stat_bin()` using `bins = 30`. Pick better value with `binwidth`.
```

You can find the result on the left in Figure 14-5. Note, however, that a notification from qplot concerning the binwidths is printed to the console.

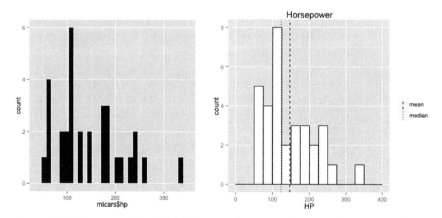

Figure 14-5: Illustrating the default behavior of qplot on the mtcars horsepower data (left); customizing binwidth, color, and title options, as well as adding markers of centrality (right)

If you don't explicitly specify the bins, exactly 30 intervals will be used to span the range of the data. Inspecting the relevant geom documentation given with a call to ?geom_histogram tells you the following:

> By default, stat_bin uses 30 bins. This is not a good default, but the idea is to get you experimenting with different binwidths. You may need to look at a few to uncover the full story behind your data.

So, rather than defaulting to a data-driven algorithm such as hist, ggplot2 encourages users to become aware of the issue and actively set their own binwidths. You can see that 30 bins yields inappropriately narrow intervals for this example—there are many gaps where no observations have fallen. There are a number of ways to choose histogram intervals in qplot, one of which is to use breaks as earlier, supplying it with an appropriate numeric vector of interval endpoints. To re-create the plot on the right of Figure 14-4 using ggplot2 functionality, use the following code, which produces the right-hand plot in Figure 14-5:

```
R> qplot(mtcars$hp,geom="blank",main="Horsepower",xlab="HP")
   + geom_histogram(color="black",fill="white",breaks=seq(0,400,25),
                    closed="right")
   + geom_vline(mapping=aes(xintercept=c(mean(mtcars$hp),median(mtcars$hp)),
                linetype=factor(c("mean","median"))),show.legend=TRUE)
   + scale_linetype_manual(values=c(2,3)) + labs(linetype="")
```

Starting with a "blank" geom, geom_histogram completes most of the work, with color governing the bar outline color and fill the internal color of the bars. The argument closed="right" determines that each interval is "closed" (in other words, exclusive) on the right and "open" (in other words, inclusive) on the left, the same as the default noted in ?hist. The geom_vline function is used to add the vertical mean and median lines; here, the mapping must be instructed to change using aes and the locations of these lines. To ensure a correctly labeled legend is created for the mean and median, you must also instruct linetype in aes to be mapped to the desired values. In this case, this is simply a factor comprised of the two desired "levels."

Since you're manually adding these lines and the associated mapping to the ggplot2 object, the legend itself must be instructed to appear with show.legend=TRUE. By default, the two lines will be drawn lty=1 (solid) and lty=2 (dashed), but to match the earlier plot, you want lty=2 and lty=3 (dotted). You can add the scale_linetype_manual modifier to make this change; the desired line type numbers are passed as a vector to values. Finally, to suppress the automatic inclusion of a title for your manually added legend, the labs(linetype="") addition instructs the scale associated with the variable mapped to linetype in the aes call to be displayed without this title.

The choice between using ggplot2 and base R graphics often comes down to your intended goal. For automated handling of graphics, especially where categorical variables are used to separate subsets of the data set, ggplot2 is particularly powerful. On the other hand, if you require manual control over the creation of a given image, traditional R graphics can be easier to handle, and you don't need to keep track of multiple aesthetic variable mappings.

## 14.3   Box-and-Whisker Plots

An especially popular alternative to the histogram is the *box-and-whisker plot*, or simply *boxplot* for short. This is merely a visual representation of the five-number summary discussed in Section 13.2.3.

### 14.3.1   Stand-Alone Boxplots

Let's return to the built-in quakes data frame of the 1,000 seismic events near Fiji. For the sake of comparison, you can examine both a histogram and a boxplot of the magnitudes of these events using default base R behavior. The following code produces the images given in Figure 14-6:

```
R> hist(quakes$mag)
R> boxplot(quakes$mag)
```

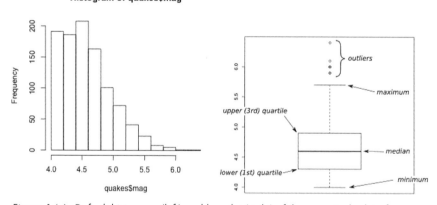

*Figure 14-6: Default histogram (left) and boxplot (right) of the magnitude data from quakes. On the boxplot, commentary (superimposed externally) points out the key information displayed.*

Like the histogram, a boxplot shows important features of the distribution, such as global (in other words, overall) centrality, spread, and skewness. It's not really possible to see local features, such as multiple significant peaks in the distribution, however. As the labeling arrows point out, the line in the middle of the box represents the median, the lower and upper edges of the box show the respective quartiles, perpendicular lines (the *whiskers*) extending from the box indicate the minimum and the maximum, and any dots drawn beyond the whiskers are deemed to be points of extremity or outliers. By default, boxplot defines an outlier as an observation that lies more than 1.5 times the IQR below the lower quartile or above the upper quartile. This is done to prevent the whiskers from extending too far and overemphasizing any skew. Thus, the "maximum" and "minimum" values marked by the whiskers are not always the raw, overall maximum or minimum values in the data set because a value that has been deemed an "outlier" might actually represent the highest or lowest value. You can control the nature of this classification via the range argument in boxplot, though the default of range=1.5 is usually sensible for basic data exploration.

### 14.3.2   Side-by-Side Boxplots

One particularly pleasing aspect of these plots is the ease with which you can compare the five-number summary distributions of different groups with *side-by-side* boxplots. Again using the quakes data, define the following corresponding factor and inspect the first five elements (review use of the cut command from Section 4.3.3 if necessary):

```
R> stations.fac <- cut(quakes$stations,breaks=c(0,50,100,150))
R> stations.fac[1:5]
[1] (0,50] (0,50] (0,50] (0,50] (0,50]
Levels: (0,50] (50,100] (100,150]
```

Recall that the stations variable records how many monitoring stations detected each event. This code has produced a factor breaking up these observations into one of three groups—events detected by 50 stations or fewer, between 51 and 100 stations, and between 101 and 150 stations. Thus, you can compare the distributions of the magnitudes of the events according to these three groups. The following line produces the left image of Figure 14-7:

```
R> boxplot(quakes$mag~stations.fac,
         xlab="# stations detected",ylab="Magnitude",col="gray")
```

With this line of code, you should note new syntax in the form of a tilde, ~, shown here in quakes$mag~stations.fac. You can read the ~ as "on," "by," or "according to" (you'll use the tilde notation frequently in Chapters 20 through 22). Here you're instructing boxplot to plot quakes$mag *according to* station.fac so that a separate boxplot is produced for each group, naturally given in the order listed in the grouping factor. Optional arguments are also employed to control axis labeling and box color. Your interpretation of this plot mirrors what you can see in Figure 13-6, in that the higher the recorded magnitude, the more stations that detected a given seismic event.

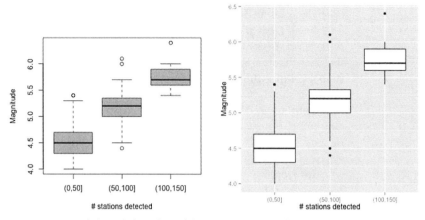

*Figure 14-7: Side-by-side boxplots of the* quakes *magnitudes, split by the three groups identified by* station.fac, *using base R graphics (left) and* ggplot2 *functionality (right)*

Turning to ggplot2 functionality, qplot can produce the same type of plot easily, with the following producing the image on the right in Figure 14-7:

```
R> qplot(stations.fac,quakes$mag,geom="boxplot",
        xlab="# stations detected",ylab="Magnitude")
```

The default boxplots look a little different, though you can make the same interpretations. In this use of qplot, you supply the boxplot grouping factor as the *x*-axis variable (first argument) and the continuous variable for which you require boxplots as the *y*-axis variable (second argument). Here I've explicitly set geom="boxplot" to ensure a boxplot display, and I've added axis labels.

## 14.4   Scatterplots

A *scatterplot* is most frequently used to identify a relationship between the observed values of two different numeric-continuous variables, displayed as *x-y* coordinate plots. The coordinate-wise nature of base R graphics lends itself naturally to the creation of scatterplots, so you've seen several examples already in this book. However, not every *x-y* coordinate-based plot is always called a scatterplot; a scatterplot usually assumes there is some "relationship of interest" present. For example, a plot of spatial coordinates like Figure 13-1 might not be regarded as a scatterplot, but a plot of the earthquake magnitude against the number of stations detecting the event, like Figure 13-6, would be.

I'll finish this chapter by expanding on how you can use scatterplots to explore more than two continuous variables. To do this, let's access another ready-to-use R data set, namely, the famous iris data. Collected in the mid-1930s, this data frame of 150 rows and 5 columns consists of petal and sepal measurements for three species of perennial iris flowers—*Iris setosa, Iris virginica,* and *Iris versicolor* (Anderson, 1935; Fisher, 1936). You can view the first five records here:

```
R> iris[1:5,]
  Sepal.Length Sepal.Width Petal.Length Petal.Width Species
1          5.1         3.5          1.4         0.2  setosa
2          4.9         3.0          1.4         0.2  setosa
3          4.7         3.2          1.3         0.2  setosa
4          4.6         3.1          1.5         0.2  setosa
5          5.0         3.6          1.4         0.2  setosa
```

Looking through ?iris, you can see that there are 50 observations for each variable, recorded in centimeters (cm), for each species.

### 14.4.1   Single Plot

You can modify a simple scatterplot to split the plotted points according to a categorical variable, exposing potential differences between any visible relationships with respect to the continuous variables. For example, using base R graphics, you can examine the petal measurements according to the three species. Using the "stepping-stone" approach first explained in Chapter 7, you can manually build up this plot by first using type="n" to generate an empty plotting region of the correct dimensions and subsequently adding the points corresponding to each species, altering point character and color as desired.

```
R> plot(iris[,4],iris[,3],type="n",xlab="Petal Width (cm)",
        ylab="Petal Length (cm)")
R> points(iris[iris$Species=="setosa",4],
          iris[iris$Species=="setosa",3],pch=19,col="black")
R> points(iris[iris$Species=="virginica",4],
          iris[iris$Species=="virginica",3],pch=19,col="gray")
R> points(iris[iris$Species=="versicolor",4],
          iris[iris$Species=="versicolor",3],pch=1,col="black")
R> legend("topleft",legend=c("setosa","virginica","versicolor"),
          col=c("black","gray","black"),pch=c(19,19,1))
```

You can find the plot in Figure 14-8. Note that the *Iris virginica* species has the largest petals, followed by *Iris versicolor,* and the smallest petals belong to *Iris setosa.* However, this code, while functional, is fairly cumbersome. You can generate the same image more simply by first setting up vectors that specify the desired point character and color *for each individual observation.*

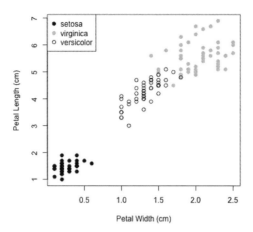

*Figure 14-8: A scatterplot of petal measurements split by species from the built-in* iris *data frame*

Consider the two objects created here:

```
R> iris_pch <- rep(19,nrow(iris))
R> iris_pch[iris$Species=="versicolor"] <- 1
R> iris_col <- rep("black",nrow(iris))
R> iris_col[iris$Species=="virginica"] <- "gray"
```

The first line creates a vector iris_pch of equal length to the number of observations in iris, with every entry being 19. Vector subsetting then overwrites the entries corresponding to *Iris versicolor* and sets the point character to 1. The same steps are followed to create iris_col; first an appropriately sized vector is filled with the character strings "black", and then those entries corresponding to *Iris virginica* are overwritten and set to "gray". With that, note that the single line shown next, when followed by the same call to legend as earlier, will produce an identical plot:

```
R> plot(iris[,4],iris[,3],col=iris_col,pch=iris_pch,
        xlab="Petal Width (cm)",ylab="Petal Length (cm)")
```

## 14.4.2   *Matrix of Plots*

The "single" type of planar scatterplot is really useful only when comparing *two* numeric-continuous variables. When there are more continuous variables of interest, it isn't possible to display this information satisfactorily on a single plot. A simple and common solution is to generate a two-variable scatterplot for each pair of variables and show them together in a structured way; this is referred to as a *scatterplot matrix*. Making use of the iris_pch and iris_col vectors created earlier, you can generate a scatterplot matrix for all four continuous variables in iris, retaining the distinction between species. Working with base R graphics, use the pairs function.

```
R> pairs(iris[,1:4],pch=iris_pch,col=iris_col,cex=0.75)
```

You can find the result of this line in Figure 14-9.

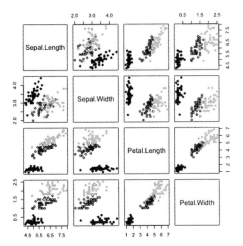

Figure 14-9: A scatterplot matrix with respect to all four continuous measurements in the data frame

The easiest way to use pairs is to supply a matrix or data frame of the raw observations as its first argument, done here by selecting all columns of iris except the Species column (iris[,1:4]). The interpretation of the plots depends upon the labeling of the diagonal panels, running from the top left to the bottom right. They will appear in the same order as the columns given as the first argument. These "label panels" allow you to determine which individual plot in the matrix corresponds to each pair of variables. For example, the first column of the scatterplot matrix in Figure 14-9 corresponds to an *x*-axis variable of Sepal Length; the third row of the matrix corresponds to a *y*-axis variable of Petal Length, and each row and column displays a scale that is constant moving left/right or up/down, respectively. This means that plots above the diagonal are mirrored in those below it— the plot of Petal Width (*y*) on Sepal Width (*x*) at position row 4, column 2 displays the same data as the scatterplot at position row 2, column 4 but flipped on its axes. As such, pairs includes an option to display only those scatterplots above *or* below the diagonal, by setting either lower.panel=NULL or upper.panel=NULL to suppress one or the other.

Scatterplot matrices therefore allow for an easier comparison of the collection of pairwise relationships formed by observations made on multiple continuous variables. In this matrix, note there's a strong positive linear association between petal measurements but a weaker relationship between the sepal measurements. Furthermore, although *Iris setosa* may reasonably be considered the smallest flower in terms of its petals, the same can't be said in terms of its sepals.

For those working with ggplot2, you know that it's natural to split the points according to a categorical variable, as in the following example.

```
R> qplot(iris[,4],iris[,3],xlab="Petal width",ylab="Petal length",
        shape=iris$Species)
    + scale_shape_manual(values=4:6) + labs(shape="Species")
```

You can find the result in Figure 14-10.

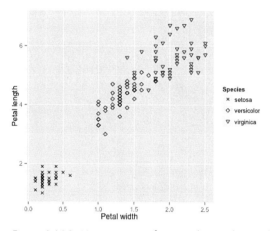

*Figure 14-10: Using* ggplot2 *functionality to plot petal dimensions for the three* iris *species, with point* shape *as the aesthetic modifier*

Here, I've split up the points using the Species variable mapped to shape (the equivalent of the base R terminology pch), and I've modified the point types using the scale_shape_manual modifier. I've also simplified the title of the automatically generated legend with labs, as done in Section 14.2. Scatterplot matrices, however, are not easily achievable using only ggplot2. To generate a matrix in a ggplot2 style, it's recommended that you download the GGally package (Schloerke et al., 2014) to access the ggpairs function. This package is designed to be an extension or add-on of ggplot2. It's installed from CRAN as per usual—for example, by running install.packages("GGally")—and must be loaded via library("GGally") prior to use. After this is done, as a quick example, the following code produces the plot in Figure 14-11:

```
R> ggpairs(iris,mapping=aes(col=Species),axisLabels="internal")
```

Though you might see familiar warnings related to the histogram binwidths, ggpairs offers an impressive visual array for such a short line of code. The output not only gives you the lower half of the scatterplot matrix produced with pairs but also provides equivalent histograms along the bottom

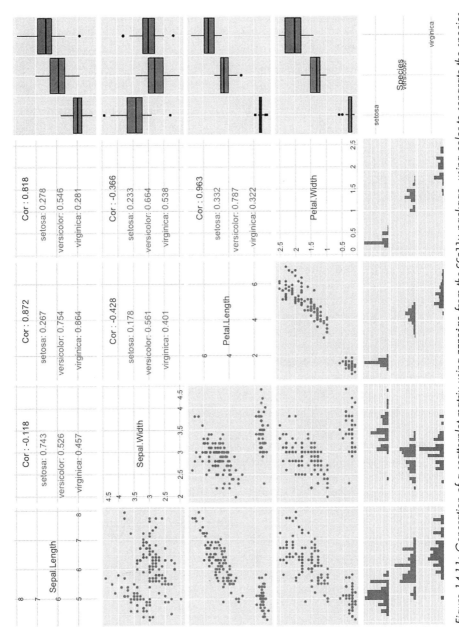

Figure 14-11: Generation of a scatterplot matrix using ggpairs from the GGally package, using color to separate the species. Note the useful addition of estimated correlation coefficients and distributional plots.

and boxplots along the right. It also displays the estimates of the correlation coefficients. As shown, you can map a variable to an aesthetic modifier to split up the plotted observations based on factor levels. In Figure 14-11, this is done by color, and you instruct ggpairs to operate on the Species variable. Documentation found in ?ggpairs gives concise information on the various options that control the presence and appearance of the individual plots.

## Exercise 14.1

Recall the built-in InsectSprays data frame, containing counts of insects on various agricultural units treated with one of six sprays.

a. Produce a histogram of the counts of insects using base R graphics.

b. Obtain the total number of insects found according to each spray (this was also asked in Exercise 13.2 (f) on page 273). Then, use base R graphics to produce a vertical barplot and a pie chart of these totals, labeling each plot appropriately.

c. Use ggplot2 functionality to generate side-by-side boxplots of the counts of insects according to each spray type and include appropriate axis labels and a title.

Yet another of R's useful ready-to-use data sets is USArrests, containing data on the number of arrests for murder, rape, and assault per 100,000 individuals in each of the 50 states of the United States, recorded in 1973 (see, for example, McNeil, 1977). It also includes a variable giving the percentage of urban-based population in each state. Briefly inspect the data frame object and the accompanying documentation ?USArrests. Then complete the following:

d. Use ggplot2 functionality to generate a right-exclusive histogram of the proportion of urban population for the states. Set your breaks to be 10 units each, between 0 and 100. Have the histogram show the first quartile, the median, and the third quartile; then provide a matching legend. Use colors as you like and include appropriate axis annotation.

e. The code t(as.matrix(USArrests[,-3])) creates a matrix of the USArrests data without the urban population column, and the built-in R object state.abb provides the two-letter state abbreviations, in alphabetical order, as a character vector. Use these two structures and base R graphics to produce a horizontal, stacked barplot with the horizontal bars labeled with state abbreviations and with each bar split according to the type of crime (murder, rape, and assault). Include a legend.

f.  Define a new factor vector urbancat that is set to 1 if the corresponding state has an urban population percentage greater than the median percentage and is set to 0 otherwise.

g.  Create a new copy of USArrests in your workspace, after deleting the UrbanPop column, leaving just the three crime rate variables. Then insert a new, fourth column in this object with urbancat.

h.  Use the data frame from (g) to produce a scatterplot matrix and other associated plots of the three crime rates against one another via GGally functionality. Use color to split the crime rates according to the two levels of urbancat.

Return to the built-in quakes data set.

i.  Create a factor vector corresponding to the magnitudes. Each entry should assume one of three categories based on breaks marked by the minimum magnitude, the $\frac{1}{3}$th quantile, the $\frac{2}{3}$th quantile, and the maximum magnitude.

j.  Re-create the plot shown next, where low-, medium-, and high-magnitude events, according to your factor vector from (i), are plotted with pch being assigned 1, 2, and 3, respectively.

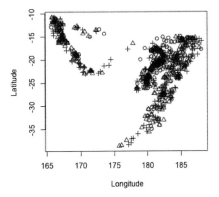

k.  Add a legend to the plot from (j) to reference the three pch values.

## Important Code in This Chapter

| Function/operator | Brief description | First occurrence |
|---|---|---|
| barplot | Create a barplot | Section 14.1.1, p. 290 |
| geom_bar | Barplot geom | Section 14.1.1, p. 292 |
| scale_x_discrete | Modify discrete x-axis (ggplot2) | Section 14.1.1, p. 292 |
| scale_y_continuous | Modify continuous y-axis | Section 14.1.1, p. 292 |
| theme_bw | Black-and-white color theme | Section 14.1.1, p. 292 |
| coord_flip | Switch x- and y-axes | Section 14.1.1, p. 292 |
| scale_fill_grey | Filled colors as grayscale | Section 14.1.1, p. 292 |
| pie | Create a pie chart | Section 14.1.2, p. 293 |
| hist | Create a histogram | Section 14.2, p. 294 |
| geom_histogram | Histogram geom | Section 14.2, p. 297 |
| geom_vline | Add vertical lines geom | Section 14.2, p. 297 |
| scale_linetype_manual | Alter ggplot2 line types | Section 14.2, p. 297 |
| labs | ggplot2 legend labels | Section 14.2, p. 297 |
| boxplot | Create boxplots | Section 14.3.1, p. 298 |
| ~ | Plot "according to" | Section 14.3.2, p. 299 |
| pairs | Scatterplot matrix | Section 14.4.2, p. 303 |
| scale_shape_manual | Alter ggplot2 point characters | Section 14.4.2, p. 304 |
| ggpairs | Scatterplot matrix (GGally) | Section 14.4.2, p. 304 |

# 15

## PROBABILITY

The concept of *probability* is central to statistical reasoning. Even the most complicated statistical techniques and models usually have the ultimate goal of making a probabilistic statement about a phenomenon. In this chapter, I'll use simple, everyday examples to illustrate this key idea in preparation for the remaining chapters. If you're already familiar with the basics of probability and random variables and the associated terminology, you may want to skip ahead to Chapter 16, where R functionality begins to feature more prominently.

## 15.1   What Is a Probability?

A *probability* is a number that describes the "magnitude of chance" associated with making a particular observation or statement. It's always a number between 0 and 1 (inclusive) and is often expressed as a fraction. Exactly how you calculate a probability depends on the definition of an *event*.

### 15.1.1　Events and Probability

In statistics, an *event* typically refers to a specific outcome that can occur. To describe the chance of event $A$ actually occurring, you use a probability, denoted by $\Pr(A)$. At the extremes, $\Pr(A) = 0$ suggests $A$ cannot occur, and $\Pr(A) = 1$ suggests that $A$ occurs with complete certainty.

Let's say you roll a six-sided, fair die. Let $A$ be the event "you roll a 5 or a 6." You can assume that each outcome on a standard die has a probability of occurring 1/6 in any given roll. Under these conditions, you have this:

$$\Pr(A) = \frac{1}{6} + \frac{1}{6} = \frac{1}{3}$$

This is what's known as a *frequentist*, or *classical*, probability, and it is assumed to be the relative frequency with which an event occurs over many identical, objective trials.

As another example, say you're married and arrive home much later than usual. Let $B$ be the event "your significant other is angry" because of your tardiness. It's a relatively straightforward process to observe $A$ in a mathematical sense, but $B$ isn't so objectively observed, and the quantity can't be easily computed. Instead, you might assign a number to $\Pr(B)$ given your own past experience. For example, you might say "I think $\Pr(B) = 0.5$" if you think there's a 50-50 chance your partner will be mad, but this would be based on your personal impressions of the situation and knowledge of your spouse's temperament or mood, not on an impartial experiment that could be easily reproduced for any two individuals. This is known as a *Bayesian* probability, which uses prior knowledge or subjective belief to inform the calculations.

Owing to its naturally implied objectivity, the frequentist interpretation is the generally assumed definition of probability; you'll focus on this kind of probability in this book. If you are interested in getting to grips with Bayesian analyses using R, Kruschke (2010) represents a well-received text on the subject.

**NOTE**　*Though it is tempting to define the concept of probability in terms of* likelihood *(and colloquially, many do), likelihood is taken to mean something slightly different in statistical theory, so I'll avoid this term for now.*

The way in which you compute probabilities when considering multiple events is determined by several important rules. These are similar in nature to the concepts of AND and OR that are key to comparing the logical values TRUE and FALSE in R via && and || (refer to Section 4.1.3). Just like these logical comparisons, calculation of probabilities based on several defined events can usually be broken down into a specific calculation concerning two distinct events. To serve as a simple running example over the next few sections, assume you roll a standard die and define event $A$ to be "you roll a 4 or more" and event $B$ to be "you roll an even number." You can therefore conclude that both $\Pr(A) = \frac{1}{2}$ and $\Pr(B) = \frac{1}{2}$.

### 15.1.2 Conditional Probability

A *conditional* probability is the probability of one event occurring after taking into account the occurrence of another event. The quantity $\Pr(A|B)$ represents "the probability that $A$ occurs, *given* that $B$ has already occurred," and vice versa if you write $\Pr(B|A)$.

If $\Pr(A|B) = \Pr(A)$, then the two events are *independent*; if $\Pr(A|B) \neq \Pr(A)$, then the two events are *dependent*. Generally, you can't assume that $\Pr(A|B)$ is equal to $\Pr(B|A)$.

Turn to $A$ and $B$ as defined previously for a roll of a die. You already know that $\Pr(A) = \frac{1}{2}$. Now think of $\Pr(A|B)$. What is the probability your outcome is a 4 or more, *given* an even number has occurred? Since there are three even numbers, 2, 4, and 6, the probability that you roll a 4 or more, assuming an even number had occurred, is $\frac{2}{3}$. Thus, $\Pr(A|B) \neq \Pr(A)$ in this context, and the two events are therefore not independent.

### 15.1.3 Intersection

The *intersection* of two events is written as $\Pr(A \cap B)$ and is read as "the probability that both $A$ and $B$ occur simultaneously." It is common to represent this as a Venn diagram, as shown here:

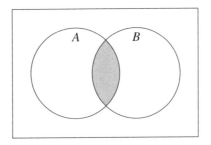

Here, the disc labeled $A$ represents the outcome (or outcomes) that satisfies $A$, and disc $B$ represents the outcomes for $B$. The shaded area represents the specific outcome (or outcomes) that satisfies both $A$ and $B$, and the area outside both discs represents the outcome (or outcomes) that satisfies neither $A$ nor $B$. Theoretically, you have this:

$$\Pr(A \cap B) = \Pr(A|B) \times \Pr(B) \quad \text{or} \quad \Pr(B|A) \times \Pr(A) \qquad (15.1)$$

If $\Pr(A \cap B) = 0$, then you say the two events are *mutually exclusive*. In other words, they cannot occur simultaneously. Also note that if the two events are independent, then Equation (15.1) simplifies to $\Pr(A \cap B) = \Pr(A) \times \Pr(B)$.

Returning to the die example, what is the probability that on a single toss you roll an even number *and* it's a 4 or more? Using the fact that $\Pr(A|B) = \frac{2}{3}$ and that $\Pr(B) = \frac{1}{2}$, it is easy to compute $\Pr(A \cap B) = \frac{2}{3} \times \frac{1}{2} = \frac{1}{3}$ and confirm this in R if you really want to.

```
R> (2/3)*(1/2)
[1] 0.3333333
```

You can see that the two events are not mutually exclusive because $\Pr(A \cap B) \neq 0$. This makes sense—it's perfectly possible in a die roll to observe a number that's both even and at least 4.

### 15.1.4   Union

The *union* of two events is written as $\Pr(A \cup B)$ and is read as "the probability that *A* or *B* occurs." Here is the representation of a union as a Venn diagram:

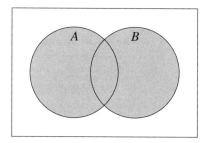

Theoretically, you have this:

$$\Pr(A \cup B) = \Pr(A) + \Pr(B) - \Pr(A \cap B) \tag{15.2}$$

The reason you need to subtract the intersection in this diagram is that in summing $\Pr(A)$ and $\Pr(B)$ alone, you'd be incorrectly counting $\Pr(A \cap B)$ twice. Note, though, that if the two events are mutually exclusive, then Equation (15.2) does simplify to $\Pr(A \cup B) = \Pr(A) + \Pr(B)$.

So, in rolling the die, what's the probability that you observe an even number *or* one that's at least 4? Using (15.2), it's easy to find that $\Pr(A \cup B) = \frac{1}{2} + \frac{1}{2} - \frac{1}{3} = \frac{2}{3}$. The following confirms this in R:

```
R> (1/2)+(1/2)-(1/3)
[1] 0.6666667
```

### 15.1.5   Complement

Lastly, the probability of the *complement* of an event is written as $\Pr(\bar{A})$ and is read as "the probability that *A* does *not* occur."

Here it is as a Venn diagram:

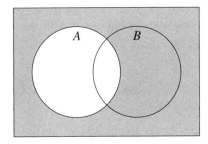

From this diagram, you can see the following:

$$\Pr(\bar{A}) = 1 - \Pr(A)$$

Wrapping up the running example, it's straightforward to find the probability that you do not roll a 4 or greater: $\Pr(\bar{A}) = 1 - \frac{1}{2} = \frac{1}{2}$. Naturally, if a 4, 5, or 6 is not obtained, then you must've rolled a 1, 2, or 3, so there are three possible outcomes left out of the six.

Sure, the die-rolling example may not represent the most pressing need facing statistical researchers today, but it has provided some clear illustrations of the behavior and terminology associated with the very real rules of probability. These rules apply across the board and play an important role in the interpretation of arguably more pressing endeavors in statistical modeling.

## Exercise 15.1

You have a standard deck of 52 playing cards. There are two colors (black and red) and four suits (spades are black, clubs are black, hearts are red, and diamonds are red). Each suit has 13 cards, in which there is an ace, numbered cards from 2 to 10, and three face cards (jack, queen, and king).

a. You randomly draw and then replace a card. What's the probability it's an ace? What's the probability it's the 4 of spades?

b. You randomly draw a card, and after replacing it, you draw another. Let $A$ be the event that the card is a club; let $B$ be the event that the card is red. What is $\Pr(A|B)$? That is, what is the probability the second card is a club, *given* the first one was a red card? Are the two events independent?

c. Repeat (b), this time assuming that when the first (club) card is drawn, it is not replaced. Would this change your answer to (b) in terms of independence?

d. Let $C$ be the event a card is a face card, and let $D$ be the event a card is black. You draw a single card. Evaluate $\Pr(C \cap D)$. Are the two events mutually exclusive?

## 15.2   Random Variables and Probability Distributions

A *random variable* is a variable whose specific outcomes are assumed to arise by chance or according to some random or *stochastic* mechanism.

You've already encountered *variables*—characteristics that describe an individual entity based on data you've observed (Section 13.1). When you're considering random variables, however, assume you have not yet made an observation. The chances of observing a specific value, or one within a specific interval, for that random variable has associated with it a probability.

It therefore makes sense to think of random variables as being tied to a function that defines these probabilities, which is referred to as a *probability distribution*. In this section, you'll look at some elementary ways in which random variables are summarized and how their corresponding probability distributions are dealt with statistically.

### 15.2.1   Realizations

So, the concept of a random variable revolves around the consideration of the possible outcomes of a variable in a probabilistic fashion. When you've actually made observations of a random variable, these are referred to as *realizations*.

Consider the following—suppose you roll your beloved die. Define the random variable $Y$ to be the result. The possible realizations are $Y = 1$, $Y = 2$, $Y = 3$, $Y = 4$, $Y = 5$, and $Y = 6$.

Now, let's say you're planning to go on a picnic and monitor the maximum daily temperature at your preferred spot. Let the random variable $W$ be the temperature in degrees Fahrenheit you observe there. Technically, you might say that the possible realizations of $W$ lie in the interval $-\infty < W < \infty$.

These examples serve to illustrate two types of random variables. $Y$ is a *discrete random variable*; $W$ is a *continuous random variable*. Whether any given random variable is discrete or continuous has consequences for the way in which you think about, and may utilize, the probabilities associated with making realizations.

## 15.2.2 Discrete Random Variables

A discrete random variable follows the same definitions as the variables covered in Chapter 13. Its realizations can take on only certain precise values, for which no other degree of measurement accuracy is possible or interpretable. Rolling a standard die can result in only those six distinct possibilities described previously by $Y$, and it would make no sense to observe, for example, "5.91."

From Section 15.1.1, you know a probability is directly tied to defined outcomes known as *events*. When discussing a discrete random variable, events are therefore defined with respect to the distinct possible values the variable can take, and the corresponding probability distribution is formed when you consider the range of all the probabilities associated with all possible realizations.

Probability distributions tied to discrete random variables are called *probability mass functions*. Since these define the probabilities of all possible outcomes, the sum of the probabilities in any complete probability mass function must always equal exactly 1.

For example, suppose you go into a casino and play a simple gambling game. At each turn, you can either lose $4 with probability 0.32, break even (win or lose nothing) with probability 0.48, win $1 with probability 0.15, or win $8 with probability 0.05. Because these are the only four possible outcomes, the probabilities sum to 1. Let the discrete random variable $X$ be defined as the "amount earned" at each turn you have. The distribution of these probabilities is expressed in Table 15-1; note that the loss of $4 is represented as a negative earning as per the definition of $X$.

**Table 15-1:** Probabilities and Cumulative Probabilities for the Amount Won, $X$, in a Hypothetical Gambling Game

| $x$ | −4 | 0 | 1 | 8 |
|---|---|---|---|---|
| $\Pr(X = x)$ | 0.32 | 0.48 | 0.15 | 0.05 |
| $\Pr(X \leq x)$ | 0.32 | 0.80 | 0.95 | 1.00 |

### Cumulative Probability Distributions of Discrete Random Variables

The *cumulative probability* is also an important part of the general idea of a probability distribution. A cumulative probability for a random variable $X$ is "the probability of observing less than or equal to $x$" and written as $\Pr(X \leq x)$. In the discrete case, you obtain the distribution of cumulative probabilities by summing the individual probabilities of the mass function up to and including any given value of $x$. This is shown in the bottom row of Table 15-1. For example, though $\Pr(X = 0)$ is 0.48, $\Pr(X \leq 0) = 0.32 + 0.48 = 0.80$.

Visualizing probability distributions is always useful, and because of the discrete nature of $X$, it's easy to use the barplot function for this. Using skills from Section 14.1, the following code first stores vectors of the possible outcomes and corresponding probabilities (X.outcomes and X.prob respectively) and then produces the left image in Figure 15-1:

```
R> X.outcomes <- c(-4,0,1,8)
R> X.prob <- c(0.32,0.48,0.15,0.05)
R> barplot(X.prob,ylim=c(0,0.5),names.arg=X.outcomes,space=0,
            xlab="x",ylab="Pr(X = x)")
```

The optional argument space=0 eliminates the gaps between the bars.

Next, you can use the built-in cumsum function to progressively sum the entries in X.prob, as shown next, giving you the cumulative probabilities:

```
R> X.cumul <- cumsum(X.prob)
R> X.cumul
[1] 0.32 0.80 0.95 1.00
```

Lastly, using X.cumul, the cumulative probability distribution can be plotted in the same way as earlier; the following line generates the right panel of Figure 15-1:

```
R> barplot(X.cumul,names.arg=X.outcomes,space=0,xlab="x",ylab="Pr(X <= x)")
```

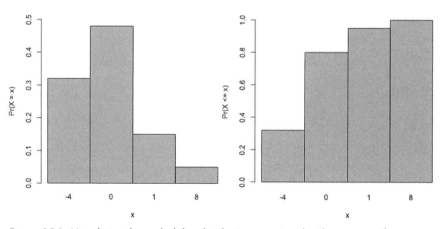

Figure 15-1: Visualizing the probability distribution associated with event-specific probabilities of a hypothetical gambling game (left) and the corresponding cumulative probability distribution (right)

Generally, it's important to remember the following for any probability mass function based on a discrete random variable $X$:

- There are $k$ distinct outcomes $x_1, \ldots, x_k$.
- For each $x_i$, where $i = \{1, \ldots, k\}$, $0 \leq \Pr(X = x_i) \leq 1$.
- $\sum_{i=1}^{k} \Pr(X = x_i) = 1$.

### Mean and Variance of a Discrete Random Variable

It's useful to be able describe or summarize properties of a random variable of interest as you would for raw data. The most useful two properties are the mean and variance, both of which depend upon the relevant distribution of probabilities associated with that random variable.

For some discrete random variable $X$, the *mean* $\mu_X$ (also referred to as the *expectation* or the *expected value* $\mathbb{E}[X]$) is the "average outcome" that you can expect over many realizations. Say $X$ has $k$ possible outcomes, labeled $x_1, x_2, \ldots, x_k$. Then, you have the following:

$$\mu_X = \mathbb{E}[X] = x_1 \times \Pr(X = x_1) + \ldots + x_k \times \Pr(X = x_k)$$

$$= \sum_{i=1}^{k} x_i \Pr(X = x_i) \qquad (15.3)$$

To find the mean, simply multiply the numeric value of each outcome by its corresponding probability and sum the results.

For a discrete random variable $X$, the *variance* $\sigma_X^2$, also written as $\mathrm{Var}[X]$, quantifies the variability in the possible realizations of $X$. Theoretically, in terms of expectations, it can be shown that $\sigma_X^2 = \mathrm{Var}[X] = \mathbb{E}[X^2] - \mathbb{E}[X]^2 = \mathbb{E}[X^2] - \mu_X^2$. As you can see, calculation of the discrete random variable variance depends upon its mean $\mu_X$ and is given as follows:

$$\sigma_X^2 = \mathrm{Var}[X] = (x_1 - \mu_X)^2 \times \Pr(X = x_1)$$

$$+ \ldots + (x_k - \mu_X)^2 \times \Pr(X = x_k)$$

$$= \sum_{i=1}^{k} (x_i - \mu_X)^2 \Pr(X = x_i) \qquad (15.4)$$

Again, the procedure is straightforward—the variance is computed by squaring the differences between each realization and mean and then multiplying by the corresponding probability of the occurrence before summing these products.

In practice, the probabilities associated with each outcome are often unknown and are estimated from observed data. Following that step, you apply the formulas in (15.3) and (15.4) to obtain *estimates* of the respective properties. Also, note that the general descriptions of mean and variance are the same as in Section 13.2—only you're now quantifying centrality and spread with respect to a random phenomenon.

Let's consider the gambling game with the possible realizations of the amount earned, $X$, and the associated probabilities as specified in Table 15-1. With vector-oriented behavior (refer to Section 2.3.4), using R to calculate the mean and variance of $X$ is easy. With the objects X.outcomes and X.prob from earlier, you can get the mean of $X$ from the element-wise multiplication in the following:

```
R> mu.X <- sum(X.outcomes*X.prob)
R> mu.X
[1] -0.73
```

So, $\mu_X = -0.73$. By the same token, the following provides the variance of $X$:

```
R> var.X <- sum((X.outcomes-mu.X)^2*X.prob)
R> var.X
[1] 7.9371
```

You can also compute the standard deviation by taking the square root of the variance (recall the definitions in Section 13.2.4). This is done with the built-in sqrt command.

```
R> sd.X <- sqrt(var.X)
R> sd.X
[1] 2.817286
```

Based on these results, you can make several comments on the gambling game and its outcomes. The expected outcome of $-0.73$ suggests that, on average, you'll lose $0.73 per turn, with a standard deviation of about $2.82. These quantities are not, and need not be, one of the specifically defined outcomes. They describe the behavior of the random mechanism over the long run.

### 15.2.3   Continuous Random Variables

Again following from the definitions of variables from Chapter 13, a continuous random variable has no limit to the number of possible realizations. For a discrete random variable, it is natural to think of a specific outcome as an event and assign it a corresponding probability. Things are a little different when you're dealing with a continuous random variable, however. If you take the picnic example in Section 15.2.1, you can see that even if you restrict the range of possible values of temperature measurement that you assume $W$ could take, say, to between 40 and 90 degrees Fahrenheit (or, expressed more formally, $40 \leq W \leq 90$), there are still an infinite number of distinct values on that continuum. Measuring 59.1 degrees makes as much sense as observing something like 59.16742 degrees.

As such, it isn't possible to assign probabilities to specific, single temperatures; instead, you assign a probability to *intervals* of values. For example, based on $W$, asking $\Pr(W = 55.2)$—"What is the probability the temperature is *exactly* 55.2 degrees Fahrenheit?"—is not a valid question. However, asking $\Pr(W \leq 55.2)$—"What is the probability it's *less than or equal to* 55.2 degrees?"—is answerable because it defines an interval.

This is easier to understand if you again think about precisely how the probabilities will be distributed. With a discrete random variable, you can straightforwardly envisage its mass function as discrete, namely, something like Table 15-1, which can be plotted like Figure 15-1. However, with continuous random variables, the function that describes the distribution of probabilities must now therefore be *continuous* on the range of possible values. Probabilities are computed as "areas underneath" that continuous function, and, as with discrete random variables, the "total area" underneath a continuous probability distribution must evaluate to exactly 1. A probability distribution tied to a continuous random variable is called a *probability density function.*

These facts will become clearer when considering the following example. Suppose you're told the probabilities associated with the picnic temperature random variable $40 \leq W \leq 90$ follow the density function $f(w)$, where the following is true:

$$f(w) = \begin{cases} \frac{w-40}{625} & \text{if } 40 \leq w \leq 65; \\ \frac{90-w}{625} & \text{if } 65 < w \leq 90; \\ 0 & \text{otherwise.} \end{cases} \quad (15.5)$$

The division by 625 is needed in this particular function to ensure a total probability of 1. This will make more sense in a visualization. To plot this density function, first consider the following code:

```
R> w <- seq(35,95,by=5)
R> w
 [1] 35 40 45 50 55 60 65 70 75 80 85 90 95
R> lower.w <- w>=40 & w<=65
R> lower.w
 [1] FALSE  TRUE  TRUE  TRUE  TRUE  TRUE  TRUE FALSE FALSE FALSE
[11] FALSE FALSE FALSE
R> upper.w <- w>65 & w<=90
R> upper.w
 [1] FALSE FALSE FALSE FALSE FALSE FALSE FALSE  TRUE  TRUE  TRUE
[11]  TRUE  TRUE FALSE
```

The first assignment sets up an even sequence of values to represent certain realizations of $w$ simply called w; the second assignment uses relational operators and the element-wise logical operator & to create a logical flag vector identifying those elements of w that form the "lower half" of values for $f(w)$ as defined by Equation (15.5); the third assignment does the same for the "upper half" of values.

The next lines make use of `lower.w` and `upper.w` to evaluate the correct result of $f(w)$ for the entries in `w`.

```
R> fw <- rep(0,length(w))
R> fw[lower.w] <- (w[lower.w]-40)/625
R> fw[upper.w] <- (90-w[upper.w])/625
R> fw
 [1] 0.000 0.000 0.008 0.016 0.024 0.032 0.040 0.032 0.024 0.016
[11] 0.008 0.000 0.000
```

This doesn't mean you've just written an R-coded function to return $f(w)$ for any $w$. You've merely created the vector `w` and obtained the corresponding values of the *mathematical* function as the vector `fw`. However, these two vectors are sufficient for plotting. Using skills from Chapter 7, you can plot a line representing the continuous density function $f(w)$ for $35 \le w \le 95$.

```
R> plot(w,fw,type="l",ylab="f(w)")
R> abline(h=0,col="gray",lty=2)
```

The plot is given in Figure 15-2; note the addition of a dashed horizontal line at $f(w) = 0$ using `abline`.

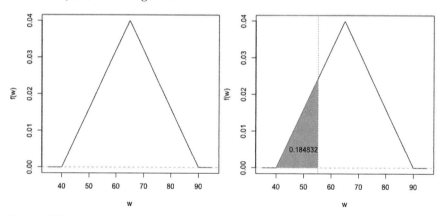

Figure 15-2: Visualizing the probability density function as defined by Equation (15.5) for the picnic temperature random variable W (left) and illustrating the computation of a specific probability from the text (right)

You can see that the continuous function defined by Equation (15.5) yields a triangular shape, with an apex at $w = 65$. The increasing line from $w = 40$ to $w = 65$ represents the first of the three components of (15.5), the decreasing line represents the second component, and for all $w < 40$ and $w > 90$, the line sits at zero, which is the third and final component of (15.5).

Generally, any function $f(w)$ that defines a probability density for a random variable $W$ must possess the following properties:

- $f(w) \geq 0$ for all $-\infty < w < \infty$; and
- $\int_{-\infty}^{\infty} f(w)\, dw = 1$ (the total area underneath the function must be 1).

In terms of the temperature example, you can see from (15.5) that $f(w) \geq 0$ for any value of $w$. To calculate the total area underneath the function, you need be concerned only with the function evaluated at $40 \leq w \leq 90$ since it's zero everywhere else.

You can do this geometrically by working out the area of the triangle formed by the function and the horizontal line at zero. For this triangle, you can use the standard "half base times height" rule. The base of the triangle is $90 - 40 = 50$, and the apex is at a value of 0.04. So, in R, half the base width times the height can be given with the following:

```
R> 0.5*50*0.04
[1] 1
```

This confirms it is indeed equal to 1; you can now see the reason behind my specific definition in (15.5).

Let's return to the question of obtaining the probability that the temperature is less than or equal to 55.2 degrees Fahrenheit. For this you must find the area underneath $f(w)$, the probability density function, bounded by the horizontal line at zero and an imaginary vertical line at 55.2. This particular area forms another triangle, for which it is again appropriate to use the "half base times height rule." In Cartesian coordinates, this is the triangle formed by the vertices at $(40, 0)$, $(55.2, 0)$, and $(55.2, f(55.2))$, as shown in the right panel of Figure 15-2—you'll see how this is plotted in a moment.

Therefore, you should first work out the value $f(55.2)$. From Equation (15.5), this is provided by creating the following object:

```
R> fw.specific <- (55.2-40)/625
R> fw.specific
[1] 0.02432
```

Note that this isn't a probability; it cannot be assigned to specific realizations. It's just the height value of the triangle on the continuous density function that you're going to need in order to calculate the interval-based probability $\Pr(W \leq 55.2)$.

You can easily determine that the base of the triangle of interest in this particular setting is $55.2 - 40 = 15.2$. Then, along with fw.specific, note that "half base times height" gives the following:

```
R> fw.specific.area <- 0.5*15.2*fw.specific
R> fw.specific.area
[1] 0.184832
```

The answer is reached. You've shown geometrically, using $f(w)$, that $\Pr(W \le 55.2) = 0.185$ (when rounded to three decimal places). In other words, you can say that there is roughly an 18.5 percent chance that the maximum temperature at the picnic spot will be less than or equal to 55.2 degrees Fahrenheit.

Again, all this is easier to digest visually. The following R code replots the density function $f(w)$ and marks off and shades the area of interest:

```
R> fw.specific.vertices <- rbind(c(40,0),c(55.2,0),c(55.2,fw.specific))
R> fw.specific.vertices
        [,1]    [,2]
[1,] 40.0 0.00000
[2,] 55.2 0.00000
[3,] 55.2 0.02432
R> plot(w,fw,type="l",ylab="f(w)")
R> abline(h=0,col="gray",lty=2)
R> polygon(fw.specific.vertices,col="gray",border=NA)
R> abline(v=55.2,lty=3)
R> text(50,0.005,labels=fw.specific.area)
```

The result is the right panel of Figure 15-2. The plotting commands should be familiar from Chapter 7, barring polygon. The built-in polygon function allows you to supply custom vertices in order to draw or shade a polygon upon an existing plot. Here, a matrix with two columns is defined using rbind, providing the $x$ and $y$ locations (first and second columns, respectively) of the three corners of the triangle to shade. Note that creation of fw.specific.vertices has made use of fw.specific, the value of $f(w)$ at $w = 55.2$; this is the topmost vertex of the shaded triangle. Further arguments to polygon control the shading (col="gray") and whether to draw a border around the defined polygon (border=NA requests no border).

Not all density functions can be appraised in this simple geometric fashion. Formally, *integration* is the mathematical operation used to find areas under a continuous function, denoted with the $\int$ symbol. That is, the mathematically inclined familiar with this technique should find it straightforward to show that "the area under $f(w)$ from $w = 40$ to $w = 55.2$, providing $\Pr(W \le 55.2)$" is yielded by the following:

$$
\begin{aligned}
\int_{40}^{55.2} f(w)\,\mathrm{d}w &= \int_{40}^{55.2} \frac{w-40}{625}\,\mathrm{d}w \\
&= \frac{w^2 - 80w}{1250}\Big|_{40}^{55.2} \\
&= \frac{55.2^2 - 80 \times 55.2 - 40^2 + 80 \times 40}{1250} \\
&= 0.185 \qquad \text{(rounded to 3 d.p.)}
\end{aligned}
$$

In R, the third line of this calculation looks like this:

```
R> (55.2^2-80*55.2-40^2+80*40)/1250
[1] 0.184832
```

R finds the correct result. I'll leave the mathematical details aside, but it's nice to confirm that the more general integral does match the intuitive geometric solution based on the area of the triangle, computed earlier as fw.specific.area.

It's now becoming clearer why it doesn't make sense to assign probabilities to single, specific realizations associated with continuous random variables. For example, evaluating the "area under the function $f(w)$" at a single value is the same as finding the area of a polygon with a base width of *zero*, and hence, the probability itself is technically zero for *any* $\Pr(W = w)$. Furthermore, in the continuous setting, it makes no difference to your calculations if you use < or ≤, or > or ≥. So although you found $\Pr(W \leq 55.2)$ earlier, if you had been tasked to find $\Pr(W < 55.2)$, you would have gotten the same answer of 0.185. It may all seem a little unnatural at first, but it all comes down to the idea of an infinite number of possible realizations so that there's no meaningful interpretation of "equality" to a specific value.

### Cumulative Probability Distributions of Continuous Random Variables

The cumulative probability distribution for a continuous variable is interpreted in the same way as for a discrete variable. Given a certain value $w$, the cumulative distribution function provides the probability of observing $w$ or less. This may seem familiar; the probability you worked out earlier, $\Pr(W \leq 55.2)$, based on the shaded triangle on the right of Figure 15-2 or using analytical methods, is itself a cumulative probability. More generally, you find a cumulative probability for a continuous random variable by calculating the area under the density function of interest from $-\infty$ to $w$. This general treatment therefore requires mathematical integration of the relevant probability density function. Looking at Figure 15-2, you should imagine a vertical line moving from left to right of the density plot and, at every location, evaluating the area under the density function to the left of that line.

For the picnic temperature example, it can be shown that the cumulative distribution function $F$ is given with the following:

$$F(w) = \int_{-\infty}^{w} f(u)\, du$$

$$= \begin{cases} 0 & \text{if } w < 40; \\ \frac{w^2 - 80w + 1600}{1250} & \text{if } 40 \leq w \leq 65; \\ \frac{180w - w^2 - 6850}{1250} & \text{if } 65 < w \leq 90; \\ 1 & \text{otherwise.} \end{cases} \quad (15.6)$$

Making use of the sequence w and the logical flag vectors `lower.w` and `upper.w` from earlier, you can use the same vector subset-and-overwrite approach to plot $F(w)$; the following code creates the required vector `Fw` and produces Figure 15-3:

```
R> Fw <- rep(0,length(w))
R> Fw[lower.w] <- (w[lower.w]^2-80*w[lower.w]+1600)/1250
R> Fw[upper.w] <- (180*w[upper.w]-w[upper.w]^2-6850)/1250
R> Fw[w>90] <- 1
R> plot(w,Fw,type="l",ylab="F(w)")
R> abline(h=c(0,1),col="gray",lty=2)
```

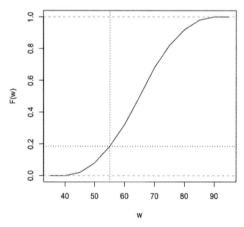

Figure 15-3: Plotting the cumulative distribution function of the picnic temperature example, which is given as Equation (15.6). The cumulative probability of observing a temperature less than (or equal to) 55.2 is marked off.

Including these extra two lines following creation of this plot clearly identifies the fact that at $w = 55.2$, the cumulative probability is located precisely on the curve of $F$:

```
R> abline(v=55.2,lty=3)
R> abline(h=fw.specific.area,lty=3)
```

## Mean and Variance of a Continuous Random Variable

Naturally, it's also possible, and useful, to determine the mean and variance of a continuous random variable.

For a continuous random variable $W$ with density $f$, the mean $\mu_W$ (or *expectation* or *expected value* $\mathbb{E}[W]$) is again interpreted as the "average outcome" that you can expect over many realizations. This is expressed mathematically as follows:

$$\mu_W = \mathbb{E}[W] = \int_{-\infty}^{\infty} w f(w) \, dw \qquad (15.7)$$

This equation represents the continuous analogue of Equation (15.3) and can be read as "the total area underneath the function given by multiplication of the density $f(w)$ with the value of $w$ itself."

For $W$, the variance $\sigma_W^2$, also written as Var[$W$], quantifies the variability inherent in realizations of $W$. Calculation of the continuous random variable variance depends upon its mean $\mu_W$ and is given as follows:

$$\mu_W = \text{Var}[W] = \int_{-\infty}^{\infty} (w - \mu_W)^2 f(w) \, dw \qquad (15.8)$$

Again, the procedure is to find the area under the density function multiplied by a certain quantity—in this case, the squared difference of the value of $w$ with the overall expected value $\mu_W$.

Evaluation of the mean and variance of the picnic temperature random variable must follow (15.7) and (15.8), respectively. These calculations become rather complex, so I won't reproduce them here. However, Figure 15-2 shows that the mean of $W$ must be $\mu_W = 65$; it is the perfect center of the symmetric density function $f(w)$.

In terms of the required integrals, you can therefore use the previously stored w and fw objects to view the two functions, $w f(w)$ and $(w - \mu_W)^2 f(w)$, by executing the following, which produces the two images in Figure 15-4:

```
R> plot(w,w*fw,type="l",ylab="wf(w)")
R> plot(w,(w-65)^2*fw,type="l",ylab="(w-65)^2 f(w)")
```

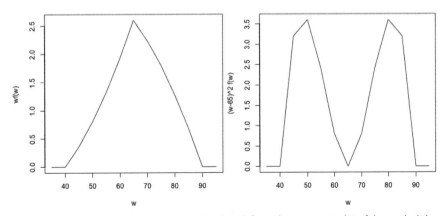

Figure 15-4: Integrands for the expected value (left) and variance (right) of the probability density function for the temperature example

Rest assured that the following can be shown mathematically using Equations (15.7) and (15.8).

$$\mu_W = 65 \quad \text{and} \quad \text{Var}[W] = 104.1667$$

By visually approximating the area underneath each of the images in Figure 15-4, you find these results are consistent. As earlier, the standard deviation of the distribution of $W$ is given with the square root of the variance, and the following readily provides this value:

```
R> sqrt(104.1667)
[1] 10.20621
```

### 15.2.4   Shape, Skew, and Modality

At this point, you're familiar with both continuous and discrete random variables and their natural pairings with a distribution of probabilities, and you've had a look at visualizations of distributions of probability mass and density functions. In this section, I'll define some terminology used to describe the appearance of these distributions—being able to describe your visual impressions is just as important as being able to readily compute them.

You'll often hear or read about the following descriptors:

**Symmetry**   A distribution is *symmetric* if you can draw a vertical line down the center, and it is equally reflected with 0.5 probability falling on either side of this center line (see Figure 15-2). A symmetric probability distribution implies that the mean and the median of the distribution are identical.

**Skew**   If a distribution is *asymmetric*, you can qualify your description further by discussing *skew*. When the "tail" of a distribution (in other words, moving away from its measures of centrality) tapers off longer in one direction than the other, it is in this direction that the distribution is said to be skewed. *Positive* or *right* skew indicates a tail extending longer to the right of center; *negative* or *left* skew refers to a tail extending longer to the left of center. You could also qualify the strength or prominence of the skew.

**Modality**   A probability distribution doesn't always necessarily have a single peak. *Modality* describes the number of easily identifiable peaks in the distribution of interest. *Unimodal*, *bimodal*, and *trimodal*, for example, are the terms used to describe distributions with one, two, and three peaks, respectively.

Figure 15-5 provides some visual interpretations of symmetry, asymmetry, skew, and modality. (Note that although they are drawn with a continuous line, you can assume they represent the general shape of either a discrete probability mass function *or* a continuous density function.)

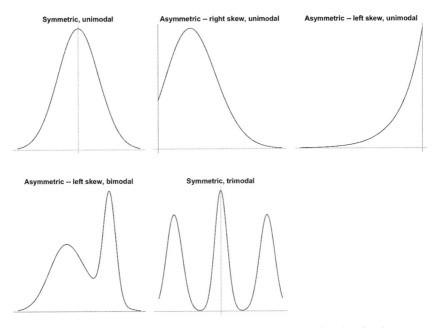

*Figure 15-5: General examples of the terms used to describe probability distributions. The top three images are unimodal and highlight the notion of symmetry versus asymmetric skew; the bottom two images emphasize the reference to modality.*

You can use these descriptors when discussing the probability distributions for the gambling game and picnic temperature examples. The mass function for *X*, on the left in Figure 15-1, is unimodal and asymmetric—it appears to have a mild but noticeable right skew. The density function for *W*, given in Figure 15-2, is also unimodal, though as noted earlier is perfectly symmetric.

---

## Exercise 15.2

a.  For each of the following definitions, identify whether it's best described as a random variable or as a *realization* of a random variable. Furthermore, identify whether each statement describes a continuous or a discrete quantity.

    i.   The number of coffees *x* made by your local shop on June 3, 2016

ii. The number of coffees $X$ made by your local shop on any given day
iii. $Y$, whether or not it rains tomorrow
iv. $Z$, the amount of rain that falls tomorrow
v. How many crumbs $k$ on your desk right now
vi. Total collective weight $W$ of the crumbs on your desk at any specified time

b. Suppose you construct the following table providing probabilities associated with the random variable $S$, the total stars given to any movie in a particular genre by a certain critic:

| $s$ | 1 | 2 | 3 | 4 | 5 |
|---|---|---|---|---|---|
| $Pr(S = s)$ | 0.10 | 0.13 | 0.21 | ??? | 0.15 |

i. Assuming this table describes the complete set of outcomes, evaluate the missing probability $Pr(S = 4)$.
ii. Obtain the cumulative probabilities.
iii. What is the mean of $S$, the expected number of stars this critic will award any given movie in this genre?
iv. What is the standard deviation of $S$?
v. What is the probability that any given movie in this genre will be given at least three stars?
vi. Visualize, and briefly comment on the appearance of, the probability mass function.

c. Return to the picnic temperature example based on the random variable $W$ defined in Section 15.2.3.
i. Write an R function to return $f(w)$ as per Equation (15.5) for any numeric vector of values supplied as $w$. Try to avoid using a loop in favor of vector-oriented operations.
ii. Write an R function to return $F(w)$ as per Equation (15.6) for any numeric vector of values supplied as $w$. Again, try to avoid using a loop, either explicit or implicit.
iii. Use your functions from (i) and (ii) to confirm the results from the text, in other words, that $f(55.2) = 0.02432$ and that $F(55.2) = 0.184832$.
iv. Make use of your function for $F(w)$ to compute $Pr(W > 60)$. Hint: Note that because the total area underneath $f(w)$ is one, $Pr(W > 60) = 1 - Pr(W \leq 60)$.
v. Find $Pr(60.3 < W < 76.89)$.

d. Assume each of the following plots labeled (i)–(iv) shows the general appearance of a probability distribution. Use terminology from Section 15.2.4 to describe the shape of each.

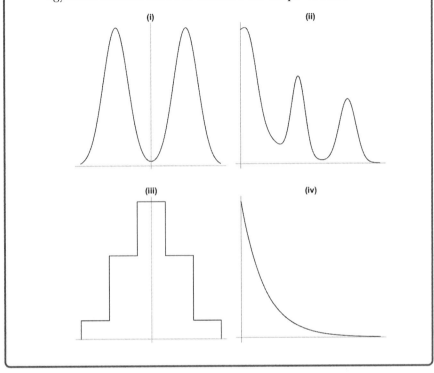

## Important Code in This Chapter

| Function/operator | Brief description | First occurrence |
|---|---|---|
| polygon | Add shaded polygon to plot | Section 15.2.3, p. 322 |

# 16

## COMMON PROBABILITY DISTRIBUTIONS

In this chapter, you'll look at a number of standard probability distributions that exist for dealing with commonly occurring random phenomena in statistical modeling. These distributions follow the same natural rules as the examples presented in Chapter 15, and they're useful because their properties are well understood and documented. In fact, they are so ubiquitous that most statistical software packages have corresponding built-in functionality for their evaluation, and R is no exception. Several of these distributions represent an essential ingredient in traditional statistical hypothesis testing, which is explored in Chapters 17 and 18.

Just like the random variables they model, the common distributions you'll examine here are broadly categorized as either discrete or continuous. Each distribution has four core R functions tied to it—a d-function, providing specific mass or density function values; a p-function, providing cumulative distribution probabilities; a q-function, providing quantiles; and an r-function, providing random variate generation.

## 16.1  Common Probability Mass Functions

You'll start here by looking at definitions and examples of some common probability mass functions for discrete random variables. Continuous distributions will be explored in in Section 16.2.

### 16.1.1  Bernoulli Distribution

The *Bernoulli* distribution is the probability distribution of a discrete random variable that has only two possible outcomes, such as success or failure. This type of variable can be referred to as *binary* or *dichotomous.*

Let's say you've defined a binary random variable $X$ for the success or failure of an event, where $X = 0$ is failure, $X = 1$ is success, and $p$ is the known probability of success. Table 16-1 shows the probability mass function for $X$.

**Table 16-1:** The Bernoulli
Probability Mass Function

| $x$ | 0 | 1 |
|---|---|---|
| $\Pr(X = x)$ | $1 - p$ | $p$ |

From Section 15.2.2 you know that the probabilities associated with all possible outcomes must sum to 1. Therefore, if the probability of success is $p$ for a binary random variable, the only other alternative outcome, failure, must occur with probability $1 - p$.

In mathematical terms, for a discrete random variable $X = x$, the Bernoulli mass function $f$ is

$$f(x) = p^x(1 - p)^{1-x}; \quad x = \{0, 1\} \tag{16.1}$$

where $p$ is a parameter of the distribution. The notation

$$X \sim \text{BERN}(p)$$

is often used to indicate that "$X$ follows a Bernoulli distribution with parameter $p$."

The following are the key points to remember:

- $X$ is dichotomous and can take only the values 1 ("success") or 0 ("failure").

- $p$ should be interpreted as "the probability of success," and therefore $0 \le p \le 1$.

The mean and variance are defined as follows, respectively:

$$\mu_X = p \quad \text{and} \quad \sigma_X^2 = p(1 - p)$$

Say you use the common example of rolling a die, with success defined as getting a 4, and you roll once. You therefore have a binary random variable $X$ that can be modeled using the Bernoulli distribution, with the probability of success $p = \frac{1}{6}$. For this example, $X \sim \text{BERN}(\frac{1}{6})$. You can easily determine, using (16.1), that

$$\Pr(\text{rolling a 4}) = \Pr(X = 1) = f(1) = \left(\frac{1}{6}\right)^1 \left(\frac{5}{6}\right)^0 = \frac{1}{6}$$

and, in much the same way, that $f(0) = \frac{5}{6}$. Furthermore, you'd have $\mu_X = \frac{1}{6}$ and $\sigma_X^2 = \frac{1}{6} \times \frac{5}{6} = \frac{5}{36}$.

## 16.1.2  Binomial Distribution

The *binomial distribution* is the distribution of successes in $n$ number of trials involving binary discrete random variables. The role of the Bernoulli distribution is typically one of a "building block" for more complicated distributions, like the binomial, that give you more interesting results.

For example, suppose you define a random variable $X = \sum_{i=1}^{n} Y_i$, where $Y_1, Y_2, \ldots, Y_n$ are each Bernoulli random variables corresponding to the same event, in other words, the die roll with success defined as rolling a 4. The new random variable $X$, a sum of Bernoulli random variables, now describes *the number of successes in $n$ trials* of the defined action. Providing that certain reasonable assumptions are satisfied, the probability distribution that describes this success count is the binomial distribution.

In mathematical terms, for a discrete random variable and a realization $X = x$, the binomial mass function $f$ is

$$f(x) = \binom{n}{x} p^x (1-p)^{n-x}; \quad x = \{0, 1, \ldots, n\} \tag{16.2}$$

where

$$\binom{n}{x} = \frac{n!}{x!(n-x)!} \tag{16.3}$$

is known as the *binomial coefficient*. (Recall the use of the integer factorial operator !, as first discussed in Exercise 10.4 on page 203.) This coefficient, also referred to as a *combination*, accounts for all different orders in which you might observe $x$ successes throughout $n$ trials.

The parameters of the binomial distribution are $n$ and $p$, and the notation

$$X \sim \text{BIN}(n, p)$$

is often used to indicate that $X$ follows a binomial distribution for $n$ trials with parameter $p$.

The following are the key points to remember:

- $X$ can take only the values 0, 1, ..., $n$ and represents the total number of successes.

- $p$ should be interpreted as "the probability of success at each trial." Therefore, $0 \leq p \leq 1$, and $n > 0$ is an integer interpreted as "the number of trials."

- Each of the *n* trials is a Bernoulli success and failure event, the trials are independent (in other words, the outcome of one doesn't affect the outcome of any other), and *p* is constant.

The mean and variance are defined as follows:

$$\mu_X = np \quad \text{and} \quad \sigma_X^2 = np(1-p)$$

Counting the number of successes of repeated trials of a binary-valued test is one of the common random phenomena mentioned at the start of this section. Consider the specific situation in which there's only one "trial," that is, $n = 1$. Examining Equations (16.2) and (16.3), it should become clear that (16.2) simplifies to (16.1). In other words, the Bernoulli distribution is just a special case of the binomial. Clearly, this makes sense with respect to the definition of a binomial random variable as a sum of *n* Bernoulli random variables. In turn, R provides functionality for the binomial distribution though not explicitly for the Bernoulli.

To illustrate this, I'll return to the example of rolling a die with success defined as getting a 4. If you roll the die independently eight times, what is the probability of observing exactly five successes (five 4s) in total? Well, you'd have $X \sim \text{BIN}(8, \frac{1}{6})$, and this probability can be worked through mathematically using (16.2).

$$\Pr(\text{get five 4s}) = \Pr(X = 5) = f(5)$$

$$= \frac{8!}{5! \times 3!} \left(\frac{1}{6}\right)^5 \left(\frac{5}{6}\right)^3$$

$$= 0.004 \quad \text{(rounded to 3 d.p.)}$$

The result tells you there is approximately a 0.4 percent chance that you'll observe exactly five 4s in eight rolls of the die. This is small and makes sense—it's far more probable that you might observe zero to two 4s in eight rolls of a die.

Fortunately, R functions will handle the arithmetic in these situations. The built-in functions dbinom, pbinom, qbinom, and rbinom are all relevant to the binomial and Bernoulli distributions and are summarized in one help file indexed by each of these function names.

- dbinom directly provides the mass function probabilities $\Pr(X = x)$ for any valid *x*—that is, $0 \le x \le n$.

- pbinom provides the cumulative probability distribution—given a valid *x*, it yields $\Pr(X \le x)$.

- qbinom provides the *inverse* cumulative probability distribution (also known as the *quantile function* of the distribution)—given a valid probability $0 \le p \le 1$, it yields the value of *x* that satisfies $\Pr(X \le x) = p$.

- rbinom is used to generate any number of realizations of *X* given a specific binomial distribution.

## The dbinom Function

With this knowledge, you can use R to confirm the result of $\Pr(X = 5)$ for the die-roll example described a moment ago.

```
R> dbinom(x=5,size=8,prob=1/6)
[1] 0.004167619
```

To the dbinom function, you provide the specific value of interest as x; the total number of trials, $n$, as size; and the probability of success at each trial, $p$, as prob. True to R, a vector argument is possible for x. If you want the full probability mass function table for $X$ for this example, you can supply the vector 0:8 to x.

```
R> X.prob <- dbinom(x=0:8,size=8,prob=1/6)
R> X.prob
[1] 2.325680e-01 3.721089e-01 2.604762e-01 1.041905e-01 2.604762e-02
[6] 4.167619e-03 4.167619e-04 2.381497e-05 5.953742e-07
```

These can be confirmed to sum to 1.

```
R> sum(X.prob)
[1] 1
```

The resulting vector of probabilities, which corresponds to the specific outcomes $x = \{0, 1, \ldots, 8\}$, is returned using e-notation (refer to Section 2.1.3). You can tidy this up by rounding the results using the round function introduced in Section 13.2.2. Rounding to three decimal places, the results are easier to read.

```
R> round(X.prob,3)
[1] 0.233 0.372 0.260 0.104 0.026 0.004 0.000 0.000 0.000
```

The achievement of one success in eight trials has the highest probability, at approximately 0.372. Furthermore, the mean (expected value) and variance of $X$ in this example are $\mu_X = np = 8 \times \frac{1}{6} = 8/6$ and $\sigma_X^2 = np(1 - p) = 8 \times \frac{1}{6} \times \frac{5}{6}$.

```
R> 8/6
[1] 1.333333
R> 8*(1/6)*(5/6)
[1] 1.111111
```

You can plot the corresponding probability mass function in the same way as for the example in Section 15.2.2; the following line produces Figure 16-1:

```
R> barplot(X.prob,names.arg=0:8,space=0,xlab="x",ylab="Pr(X = x)")
```

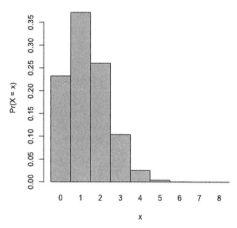

Figure 16-1: The probability mass function associated with the binomial distribution of the die-rolling example

## The pbinom Function

The other R functions for the binomial distribution work in much the same way. The first argument is always the value (or values) of interest; $n$ is supplied as size and $p$ as prob. To find, for example, the probability that you observe three or fewer 4s, $\Pr(X \leq 3)$, you either sum the relevant individual entries from dbinom as earlier or use pbinom.

```
R> sum(dbinom(x=0:3,size=8,prob=1/6))
[1] 0.9693436
R> pbinom(q=3,size=8,prob=1/6)
[1] 0.9693436
```

Note that the pivotal argument to pbinom is tagged q, not x; this is because, in a cumulative sense, you are searching for a probability based on a quantile. The cumulative distribution results from pbinom can be used in the same way to search for "upper-tail" probabilities (probabilities to the right of a given value) since you know that the total probability mass is always 1. To find the probability that you observe *at least* three 4s in eight rolls of the die, $\Pr(X \geq 3)$ (which is equivalent to $\Pr(X > 2)$ in the context of this discrete random variable), note that the following finds the correct result because it's the complement of $\Pr(X \leq 2)$ that you're looking for:

```
R> 1-pbinom(q=2,size=8,prob=1/6)
[1] 0.1348469
```

## The qbinom Function

Less frequently used is the qbinom function, which is the inverse of pbinom. Where pbinom provides a cumulative probability when given a quantile value q, the function qbinom provides a quantile value when given a cumulative probability p. The discrete nature of a binomial random variable means qbinom will return the nearest value of $x$ below which p lies. For example, note that

```
R> qbinom(p=0.95,size=8,prob=1/6)
[1] 3
```

provides 3 as the quantile value, even though you know from earlier that the exact probability that lies at or below 3, $\Pr(X \leq 3)$, is 0.9693436. You'll look at p- and q-functions more when dealing with continuous probability distributions; see Section 16.2.

## The rbinom Function

Lastly, the random generation of realizations of a binomially distributed variable is retrieved using the rbinom function. Again, going with the $\mathrm{BIN}(8, \frac{1}{6})$ distribution, note the following:

```
R> rbinom(n=1,size=8,prob=1/6)
[1] 0
R> rbinom(n=1,size=8,prob=1/6)
[1] 2
R> rbinom(n=1,size=8,prob=1/6)
[1] 2
R> rbinom(n=3,size=8,prob=1/6)
[1] 2 1 1
```

The initial argument n doesn't refer to the number of trials. The number of trials is still provided to size with $p$ given to prob. Here, n requests the number of realizations you want to generate for the random variable $X \sim \mathrm{BIN}(8, \frac{1}{6})$. The first three lines each request a single realization—in the first eight rolls, you observe zero successes (4s), and in the second and third sets of eight rolls, you observe two and two 4s, respectively. The fourth line highlights the fact that multiple realizations of $X$ are easily obtained and stored as a vector by increasing n. As these are *randomly generated realizations*, if you run these lines now, you'll probably observe some different values.

Though not used often in standard statistical testing methods, the r-functions for probability distributions, either discrete or continuous, play an important role when it comes to simulation and various advanced numeric algorithms in computational statistics.

## Exercise 16.1

A forested nature reserve has 13 bird-viewing platforms scattered throughout a large block of land. The naturalists claim that at any point in time, there is a 75 percent chance of seeing birds at each platform. Suppose you walk through the reserve and visit every platform. If you assume that all relevant conditions are satisfied, let $X$ be a binomial random variable representing the total number of platforms at which you see birds.

a. Visualize the probability mass function of the binomial distribution of interest.

b. What is the probability you see birds at all sites?

c. What is the probability you see birds at more than 9 platforms?

d. What is the probability of seeing birds at between 8 and 11 platforms (inclusive)? Confirm your answer by using only the d-function and then again using only the p-function.

e. Say that, before your visit, you decide that if you see birds at fewer than 9 sites, you'll make a scene and demand your entry fee back. What's the probability of your embarrassing yourself in this way?

f. Simulate realizations of $X$ that represent 10 different visits to the reserve; store your resulting vector as an object.

g. Compute the mean and standard deviation of the distribution of interest.

### 16.1.3 Poisson Distribution

In this section, you'll use the *Poisson* distribution to model a slightly more general, but just as important, discrete random variable—a *count*. For example, the variable of interest might be the number of seismic tremors detected by a certain station in a given year or the number of imperfections found on square-foot pieces of sheet metal coming off a factory production line.

Importantly, the events or items being counted are assumed to manifest independently of one another. In mathematical terms, for a discrete random variable and a realization $X = x$, the Poisson mass function $f$ is given as follows, where $\lambda_p$ is a parameter of the distribution (this will be explained further momentarily):

$$f(x) = \frac{\lambda_p^x \exp(-\lambda_p)}{x!}; \quad x = \{0, 1, \ldots\} \tag{16.4}$$

The notation

$$X \sim \text{POIS}(\lambda_p)$$

is often used to indicate that "$X$ follows a Poisson distribution with parameter $\lambda_p$."

The following are the keys points to remember:

- The entities, features, or events being counted occur independently in a well-defined interval at a constant rate.

- $X$ can take only non-negative integers: $0, 1, \ldots$.

- $\lambda_p$ should be interpreted as the "mean number of occurrences" and must therefore be finite and strictly positive; that is, $0 < \lambda_p < \infty$.

The mean and variance are as follows:

$$\mu_X = \lambda_p \quad \text{and} \quad \sigma_X^2 = \lambda_p$$

Like the binomial random variable, the values taken by a Poisson random variable are discrete, non-negative integers. Unlike the binomial, however, there's typically no upper limit on a Poisson count. While this implies that an "infinite count" is allowed to occur, it's a distinct feature of the Poisson distribution that the probability mass associated with some value $x$ goes to zero as $x$ itself goes to infinity.

As noted in Equation (16.4), any Poisson distribution depends upon the specification of a single parameter, denoted here with $\lambda_p$. This parameter describes the mean number of occurrences, which impacts the overall shape of the mass function, as shown in Figure 16-2.

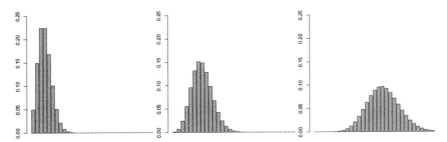

Figure 16-2: Three examples of the Poisson probability mass function, plotted for $0 \leq x \leq 30$. The "expected count" parameter $\lambda_p$ is altered from 3.00 (left) to 6.89 (middle) and to 17.20 (right).

Again, it's worth noting that the total probability mass over all possible outcomes is 1, no matter what the value of $\lambda_p$ is and regardless of the fact that possible outcomes can, technically, range from 0 to infinity.

By definition, it's easy to understand why the mean of $X$, $\mu_X$, is equal to $\lambda_p$; in fact, it turns out that the variance of a Poisson distributed random variable is also equal to $\lambda_p$.

Consider the example of blemishes on 1-foot-square sheets of metal coming off a production line, mentioned in the opening of this section. Suppose you're told that the number of blemishes found, $X$, is thought to follow a Poisson distribution with $\lambda_p = 3.22$, as in $X \sim \text{POIS}(3.22)$. In other words, you'd expect to see an average of 3.22 blemishes on your 1-foot sheets.

## The dpois and ppois Functions

The R dpois function provides the individual Poisson mass function probabilities $\Pr(X = x)$ for the Poisson distribution. The ppois function provides the left cumulative probabilities, as in $\Pr(X \leq x)$. Consider the following lines of code:

```
R> dpois(x=3,lambda=3.22)
[1] 0.2223249
R> dpois(x=0,lambda=3.22)
[1] 0.03995506
R> round(dpois(0:10,3.22),3)
[1] 0.040 0.129 0.207 0.222 0.179 0.115 0.062 0.028 0.011 0.004 0.001
```

The first call finds that $\Pr(X = 3) = 0.22$ (to 2 d.p.); in other words, the probability that you observe exactly three blemishes on a randomly selected piece of sheet metal is equal to about 0.22. The second call indicates a less than 4 percent chance that the piece is flawless. The third line returns a rounded version of the relevant mass function for the values $0 \leq x \leq 10$. By hand you can confirm the first result like this:

```
R> (3.22^3*exp(-3.22))/prod(3:1)
[1] 0.2223249
```

You create a visualization of the mass function with the following:

```
R> barplot(dpois(x=0:10,lambda=3.22),ylim=c(0,0.25),space=0,
           names.arg=0:10,ylab="Pr(X=x)",xlab="x")
```

This is shown on the left of Figure 16-3.

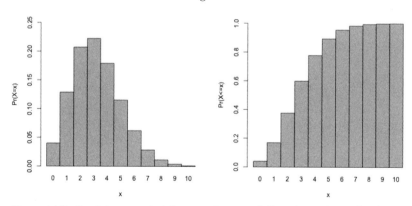

Figure 16-3: The Poisson probability mass function (left) and cumulative distribution function (right) for $\lambda_p = 3.22$ plotted for the integers $0 \leq x \leq 10$, with reference to the sheet metal example

To calculate cumulative results, you use ppois.

```
R> ppois(q=2,lambda=3.22)
[1] 0.3757454
R> 1-ppois(q=5,lambda=3.22)
[1] 0.1077005
```

These lines find that the probability you observe at most two imperfections, $\Pr(X \leq 2)$, is about 0.38, and the probability you observe strictly more than five blemishes, $\Pr(X \geq 6)$, is roughly 0.11.

A visualization of the cumulative mass function is given on the right of Figure 16-3, created with the following:

```
R> barplot(ppois(q=0:10,lambda=3.22),ylim=0:1,space=0,
            names.arg=0:10,ylab="Pr(X<=x)",xlab="x")
```

### The qpois Function

The q-function for the Poisson distribution, qpois, provides the inverse of ppois, in the same way as qbinom in Section 16.1.2 provides the inverse of pbinom.

### The rpois Function

To produce random variates, you use rpois; you supply the number of variates you want as n and supply the all-important parameter as lambda. You can imagine

```
R> rpois(n=15,lambda=3.22)
[1] 0 2 9 1 3 1 9 3 4 3 2 2 6 3 5
```

as selecting fifteen 1-foot-square metal sheets from the production line at random and counting the number of blemishes on each. Note again that this is random generation; your specific results are likely to vary.

---

### Exercise 16.2

Every Saturday, at the same time, an individual stands by the side of a road and tallies the number of cars going by within a 120-minute window. Based on previous knowledge, she believes that the mean number of cars going by during this time is exactly 107. Let $X$ represent the appropriate Poisson random variable of the number of cars passing her position in each Saturday session.

a. What is the probability that more than 100 cars pass her on any given Saturday?

b. Determine the probability that no cars pass.

c. Plot the relevant Poisson mass function over the values in $60 \leq x \leq 150$.

d. Simulate 260 results from this distribution (about five years of weekly Saturday monitoring sessions). Plot the simulated results using hist; use xlim to set the horizontal limits from 60 to 150. Compare your histogram to the shape of your mass function from (c).

### 16.1.4    Other Mass Functions

There are many other well-defined probability mass functions in R's built-in suite of statistical calculations. All model a discrete random variable in a certain way under certain conditions and are defined with at least one parameter, and most are represented by their own set of d-, p-, q-, and r-functions. Here I summarize a few more:

- The *geometric* distribution counts the number of failures before a success is recorded and is dependent on a "probability of success parameter" prob. Its functions are dgeom, pgeom, qgeom, and rgeom.

- The *negative binomial* distribution is a generalization of the geometric distribution, dependent upon parameters size (number of trials) and prob. Its functions are dnbinom, pnbinom, qnbinom, and rnbinom.

- The *hypergeometric* distribution is used to model sampling without replacement (in other words, a "success" can change the probabilities associated with further successes), dependent upon parameters m, n, and k describing the nature of sampled items. Its functions are dhyper, phyper, qhyper, and rhyper.

- The *multinomial* distribution is a generalization of the binomial, where a success can occur in one of multiple categories at each trial, with parameters size and prob (this time, prob must be a vector of probabilities corresponding to the multiple categories). Its built-in functions are limited to dmultinom and rmultinom.

As noted earlier, some familiar probability distributions are just simplifications or special cases of functions that describe a more general class of distributions.

## 16.2    Common Probability Density Functions

When considering continuous random variables, you need to deal with probability density functions. There are a number of common continuous probability distributions frequently used over many different types of problems. In this section, you'll be familiarized with some of these and R's accompanying d-, p-, q-, and r-functions.

## 16.2.1  Uniform

The *uniform* distribution is a simple density function that describes a continuous random variable whose interval of possible values offers no fluctuations in probability. This will become clear in a moment when you plot Figure 16-4.

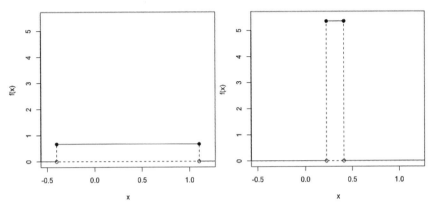

Figure 16-4: Two uniform distributions plotted on the same scale for comparability. Left: X ~ UNIF(−0.4,1.1); right: X ~ UNIF(0.223,0.410). The total area underneath each density function is, as always, 1.

For a continuous random variable $a \leq X \leq b$, the uniform density function $f$ is

$$f(x) = \begin{cases} \frac{1}{b-a} & \text{if } a \leq x \leq b; \\ 0 & \text{otherwise} \end{cases} \qquad (16.5)$$

where $a$ and $b$ are parameters of the distribution defining the limits of the possible values $X$ can take. The notation

$$X \sim \text{UNIF}(a,b)$$

is often used to indicate that "$X$ follows a uniform distribution with limits $a$ and $b$."

The following are the key points to remember:

- $X$ can take any value in the interval bounded by $a$ and $b$.

- $a$ and $b$ can be any values, provided that $a < b$, and they represent the lower and upper limits, respectively, of the interval of possible values.

The mean and variance are as follows:

$$\mu_X = \frac{a+b}{2} \qquad \text{and} \qquad \sigma_X^2 = \frac{(b-a)^2}{12}$$

For the more complicated densities in this section, it's especially useful to visualize the functions in order to understand the probabilistic structure associated with a continuous random variable. For the uniform distribution, given Equation (16.5), you can recognize the two different uniform distributions shown in Figure 16-4. I'll provide the code to produce these types of plots shortly.

For the left plot in Figure 16-4, you can confirm the exact height of the $X \sim \mathrm{UNIF}(-0.4, 1.1)$ density by hand: $1/\{1.1 - (-0.4)\} = 1/1.5 = \frac{2}{3}$. For the plot on the right, based on $X \sim \mathrm{UNIF}(0.223, 0.410)$, you can use R to find that its height is roughly 5.35.

```
R> 1/(0.41-0.223)
[1] 5.347594
```

### The dunif Function

You can use the built-in d-function for the uniform distribution, dunif, to return these heights for any value within the defined interval. The dunif command returns zero for values outside of the interval. The parameters of the distribution, $a$ and $b$, are provided as the arguments min and max, respectively. For example, the line

```
R> dunif(x=c(-2,-0.33,0,0.5,1.05,1.2),min=-0.4,max=1.1)
[1] 0.0000000 0.6666667 0.6666667 0.6666667 0.6666667 0.0000000
```

evaluates the uniform density function of $X \sim \mathrm{UNIF}(-0.4, 1.1)$ at the values given in the vector passed to x. You'll notice that the first and last values fall outside the bounds defined by min and max, and so they are zero. All others evaluate to the height value of $\frac{2}{3}$, as previously calculated.

As a second example, the line

```
R> dunif(x=c(0.3,0,0.41),min=0.223,max=0.41)
[1] 5.347594 0.000000 5.347594
```

confirms the correct density values for the $X \sim \mathrm{UNIF}(0.223, 0.410)$ distribution, with the second value, zero, falling outside the defined interval.

This most recent example in particular should remind you that probability density functions for continuous random variables, unlike mass functions for discrete variables, *do not* directly provide probabilities, as mentioned in Section 15.2.3. In other words, the results just returned by dunif represent the respective density functions themselves and not any notion of chance attached to the specific values of $x$ at which they were evaluated.

To calculate some probabilities based on the uniform density function, use the example of a faulty drill press. In a woodworker's shop, imagine there's a drill press that cannot keep to a constant alignment when in use; instead, it randomly hits the intended target at up to 0.4 cm to the left or 1.1 cm to the right. Let the random variable $X \sim \mathrm{UNIF}(-0.4, 1.1)$ represent where the drill hits the material relative to the target at 0. Figure 16-5 replots the left image of Figure 16-4 on a more detailed scale. You have three versions, each marking off a different area under the density function: $\Pr(X \leq -0.21)$, $\Pr(-0.21 \leq X \leq 0.6)$, and $\Pr(X \geq 0.6)$.

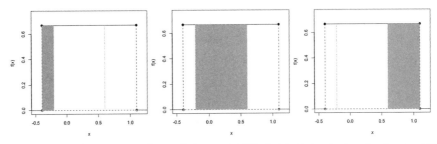

Figure 16-5: Three areas underneath the $X \sim UNIF(-0.4, 1.1)$ density function for the drill press example. Left: $Pr(X \leq -0.21)$; middle: $Pr(-0.21 \leq X \leq 0.6)$; right: $Pr(X \geq 0.6)$.

These plots are created using the coordinate-based plotting skills covered in Chapter 7. The density itself is presented with the following:

```
R> a1 <- -4/10
R> b1 <- 11/10
R> unif1 <- 1/(b1-a1)
R> plot(c(a1,b1),rep(unif1,2),type="o",pch=19,xlim=c(a1-1/10,b1+1/10),
      ylim=c(0,0.75),ylab="f(x)",xlab="x")
R> abline(h=0,lty=2)
R> segments(c(a1-2,b1+2,a1,b1),rep(0,4),rep(c(a1,b1),2),rep(c(0,unif1),each=2),
      lty=rep(1:2,each=2))
R> points(c(a1,b1),c(0,0))
```

You can use much of the same code to produce the plots in Figure 16-4 by modifying the xlim and ylim arguments to adjust the scale of the axes.

You add the vertical lines denoting $f(-0.21)$ and $f(0.6)$ in Figure 16-5 with another call to segments.

```
R> segments(c(-0.21,0.6),c(0,0),c(-0.21,0.6),rep(unif1,2),lty=3)
```

Finally, you can shade the areas using the polygon function, which was first explored in Section 15.2.3. For example, in the leftmost plot in Figure 16-5, use the previous plotting code followed by this:

```
R> polygon(rbind(c(a1,0),c(a1,unif1),c(-0.21,unif1),c(-0.21,0)),col="gray",
      border=NA)
```

As mentioned earlier, the three shaded areas in Figure 16-5 represent, from left to right, $Pr(X < -0.21)$, $Pr(-0.21 < X < 0.6)$, and $Pr(X > 0.6)$, respectively. In terms of the drill press example, you can interpret these as the probability that the drill hits the target 0.21 cm to the left or more, the probability that the drill hits the target between 0.21 cm to the left and 0.6 cm to the right, and the probability that the drill hits the target 0.6 cm to the right or more, respectively. (Remember from Section 15.2.3 that it makes no difference if you use $\leq$ or $<$ (or $\geq$ or $>$) for probabilities associated with continuous random variables.) Though you could evaluate these probabilities geometrically for such a simple density function, it's still faster to use R.

## The punif Function

Remember that probabilities associated with continuous random variables are defined as *areas under the function,* and therefore your study focuses on the appropriate intervals of $X$ rather than any specific value. The p-function for densities, just like the p-function for discrete random variables, provides the cumulative probability distribution $\Pr(X \leq x)$. In the context of the uniform density, this means that given a specific value of $x$ (supplied as a "quantile" argument q), punif will provide the left-directional area underneath the function from that specific value.

Accessing punif, the line

```
R> punif(-0.21,min=a1,max=b1)
[1] 0.1266667
```

tells you that the leftmost area in Figure 16-5 represents a probability of about 0.127. The line

```
R> 1-punif(q=0.6,min=a1,max=b1)
[1] 0.3333333
```

tells you that $\Pr(X > 0.6) = \frac{1}{3}$. The final result for $\Pr(-0.21 < X < 0.6)$, giving a 54 percent chance, is found with

```
R> punif(q=0.6,min=a1,max=b1) - punif(q=-0.21,min=a1,max=b1)
[1] 0.54
```

since the first call provides the area under the density from 0.6 all the way left and the second call provides the area from $-0.21$ all the way left. Therefore, this difference is the middle area as defined.

It's essential to be able to manipulate cumulative probability results like this when working with probability distributions in R, and the beginner might find it useful to sketch out the desired area before using p-functions, especially with respect to density functions.

## The qunif Function

The q-functions for densities are used more often than they are for mass functions because the continuous nature of the variable means that a unique quantile value can be found for any valid probability p.

The qunif function is the inverse of punif:

```
R> qunif(p=0.1266667,min=a1,max=b1)
[1] -0.21
R> qunif(p=1-1/3,min=a1,max=b1)
[1] 0.6
```

These lines confirm the values of $X$ used earlier to get the lower- and upper-tail probabilities $\Pr(X < -0.21)$ and $\Pr(X > 0.6)$, respectively. Any q-function expects a *cumulative* (in other words, left-hand) probability as its first argument, which is why you need to supply 1-1/3 in the second line to recover 0.6. (The total area is 1. You know that you want the area to the *right* of 0.6 to be $\frac{1}{3}$; thus, the area on the left must be $1 - \frac{1}{3}$.)

## The runif Function

Lastly, to generate random realizations of a specific uniform distribution, you use runif. Let's say the woodworker drills 10 separate holes using the faulty press; you can simulate one instance of the position of each of these holes relative to their target with the following call.

```
R> runif(n=10,min=a1,max=b1)
 [1] -0.2429272 -0.1226586  0.9318365  0.4829028  0.5963365
 [6]  0.2009347  0.3073956 -0.1416678  0.5551469  0.4033372
```

Again, note that the specific values of r-function calls like runif will be different each time they are run.

---

### Exercise 16.3

You visit a national park and are informed that the height of a certain species of tree found in the forest is uniformly distributed between 3 and 70 feet.

a. What is the probability you encounter a tree shorter than $5\frac{1}{2}$ feet?

b. For this probability density function, what is the height that marks the cutoff point of the tallest 15 percent of trees?

c. Evaluate the mean and standard deviation of the tree height distribution.

d. Using (c), confirm that the chance that you encounter a tree with a height that is within half a standard deviation (that is, below or above) of the mean height is roughly 28.9 percent.

e. At what height is the density function itself? Show it in a plot.

f. Simulate 10 observed tree heights. Based on these data, use quantile (refer to Section 13.2.3) to estimate the answer you arrived at in (b). Repeat your simulation, this time generating 1,000 variates, and estimate (b) again. Do this a handful of times, taking a mental note of your two estimates each time. Overall, what do you notice of your two estimates (one based on 10 variates at a time and the other based on 1,000) with respect to the "true" value in (b)?

---

## 16.2.2 Normal

The *normal distribution* is one of the most well-known and commonly applied probability distributions in modeling continuous random variables. Characterized by a distinctive "bell-shaped" curve, it's also referred to as the *Gaussian* distribution.

For a continuous random variable $-\infty < X < \infty$, the normal density function $f$ is

$$f(x) = \frac{1}{\sigma\sqrt{2\pi}} \exp\left\{-\frac{(x-\mu)^2}{2\sigma^2}\right\} \tag{16.6}$$

where $\mu$ and $\sigma$ are parameters of the distribution, $\pi$ is the familiar geometric value $3.1415\ldots$, and $\exp\{\cdot\}$ is the exponential function (refer to Section 2.1.2). The notation

$$X \sim \mathrm{N}(\mu,\sigma)$$

is often used to indicate that "$X$ follows a normal distribution with mean $\mu$ and standard deviation $\sigma$."

The following are the key points to remember:

- Theoretically, $X$ can take any value from $-\infty$ to $\infty$.

- As hinted at earlier, the parameters $\mu$ and $\sigma$ directly describe the mean and the standard deviation of the distribution, with the square of the latter, $\sigma^2$, being the variance.

- In practice, the mean parameter is finite $-\infty < \mu < \infty$, and the standard deviation parameter is strictly positive and finite $0 < \sigma < \infty$.

- If you have a random variable $X \sim \mathrm{N}(\mu,\sigma)$, then you can create a new random variable $Z = (X - \mu)/\sigma$, which means $Z \sim \mathrm{N}(0,1)$. This is known as *standardization* of $X$.

The two parameters noted earlier fully define a particular normal distribution. These are always perfectly symmetric, unimodal, and centered on the mean $\mu$, and they have a degree of "spread" defined using the standard deviation $\sigma$.

The top image of Figure 16-6 provides the density functions for four specific normal distributions. You can see that altering the mean results in a translation, where the center of the distribution is simply shifted to sit on the specific value of $\mu$. The effect of a smaller standard deviation is to reduce the spread, resulting in a taller, skinnier appearance of the density. Increasing $\sigma$ flattens the density out around the mean.

The bottom image zooms in on the $\mathrm{N}(0,1)$ distribution when you have a normal density centered on $\mu = 0$ and with a standard deviation of $\sigma = 1$. This distribution, known as the *standard normal*, is frequently used as a standard reference to compare different normally distributed random variables with one another on the same scale of probabilities. It's common practice to rescale, or *standardize*, some variable $X \sim \mathrm{N}(\mu_X, \sigma_X)$ to a new variable $Z$ such that $Z \sim \mathrm{N}(0,1)$ (you'll see this practiced in Chapter 18). Vertical lines in the plot show plus or minus one, two, and three times the standard deviation away from the mean of zero. This serves to highlight the fact that for *any* normal distribution, a probability of exactly 0.5 lies either above or below the mean. Furthermore, note that there's a probability of approximately 0.683 of a value falling within one standard deviation of the mean, approximately 0.954 probability under the curve from $-2\sigma$ to $+2\sigma$, and approximately 0.997 probability between $-3\sigma$ and $+3\sigma$.

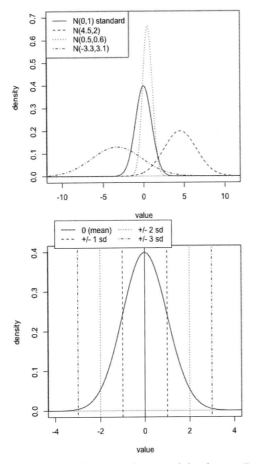

Figure 16-6: Illustrating the normal distribution. Top: Four different instances of the density achieved by varying the mean $\mu$ and standard deviation $\sigma$. Bottom: The "standard normal" distribution, N(0,1), marking off the mean $\pm 1\sigma$, $\pm 2\sigma$, and $\pm 3\sigma$.

**NOTE**    *The mathematical definition of the normal density means that as you move further away from the mean, the value of the density function itself will approach zero. In actual fact, any normal density function never actually touches the horizontal line at zero; it just gets closer and closer as you move to positive or negative infinity. This behavior is formally referred to as* asymptotic; *in this case, you'd say that the normal distribution $f(x)$ has a* horizontal asymptote *at $f(x) = 0$. In discussing probabilities as areas under the curve, you'd refer to the fact that "the total area under the curve from negative to positive infinity" is 1, in other words, $\int_{-\infty}^{\infty} f(x)\, dx = 1$.*

## The dnorm Function

Being a probability density function, the dnorm function itself doesn't provide probabilities—merely the value of the desired normal function curve $f(x)$ at any $x$. To plot a normal density, you'd therefore be able to use seq (refer to Section 2.3.2) to create a fine sequence of values for $x$, evaluate the density at these values with dnorm, and then plot the result as a line. For example, to produce an image of the standard normal distribution curve similar to that in the bottom image of Figure 16-6, the following code will create the desired $x$ values as xvals.

```
R> xvals <- seq(-4,4,length=50)
R> fx <- dnorm(xvals,mean=0,sd=1)
R> fx
 [1] 0.0001338302 0.0002537388 0.0004684284 0.0008420216 0.0014737603
 [6] 0.0025116210 0.0041677820 0.0067340995 0.0105944324 0.0162292891
[11] 0.0242072211 0.0351571786 0.0497172078 0.0684578227 0.0917831740
[16] 0.1198192782 0.1523049307 0.1885058641 0.2271744074 0.2665738719
[21] 0.3045786052 0.3388479358 0.3670573564 0.3871565916 0.3976152387
[26] 0.3976152387 0.3871565916 0.3670573564 0.3388479358 0.3045786052
[31] 0.2665738719 0.2271744074 0.1885058641 0.1523049307 0.1198192782
[36] 0.0917831740 0.0684578227 0.0497172078 0.0351571786 0.0242072211
[41] 0.0162292891 0.0105944324 0.0067340995 0.0041677820 0.0025116210
[46] 0.0014737603 0.0008420216 0.0004684284 0.0002537388 0.0001338302
```

Then dnorm, which includes specification of $\mu$ as mean and $\sigma$ as sigma, produces the precise values of $f(x)$ at those xvals. Finally, a call such as plot(xvals,fx,type="l") achieves the bare bones of a density plot, which you can easily enhance by adding titles and using commands such as abline and segments to mark locations off (I'll produce another plot in a moment, so this basic one isn't shown here).

Note that if you don't supply any values to mean and sd, the default behavior of R is to implement the standard normal distribution; the object fx shown earlier could have been created with an even shorter call using just dnorm(xvals).

## The pnorm Function

The pnorm function obtains left-side probabilities under the specified normal density. As with dnorm, if no parameter values are supplied, R automatically sets mean=0 and sd=1. In the same way you used punif in Section 16.2.1, you can find differences of results from pnorm to find any areas you want when you provide the function with the desired values in the argument q.

For example, it was mentioned earlier that a probability of approximately 0.683 lies within one standard deviation of the mean. You can confirm this using pnorm for the standard normal.

```
R> pnorm(q=1)-pnorm(q=-1)
[1] 0.6826895
```

The first call to pnorm evaluates the area under the curve from positive 1 left (in other words, all the way to −∞) and then finds the difference between that and the area from −1 left. The result reflects the proportion between the two dashed lines in the bottom of Figure 16-6. These kinds of probabilities will be the same for *any* normal distribution. Consider the distribution where $\mu = -3.42$ and $\sigma = 0.2$. Then the following provides the same value:

```
R> mu <- -3.42
R> sigma <- 0.2
R> mu.minus.1sig <- mu-sigma
R> mu.minus.1sig
[1] -3.62
R> mu.plus.1sig <- mu+sigma
R> mu.plus.1sig
[1] -3.22
R> pnorm(q=mu.plus.1sig,mean=mu,sd=sigma) -
     pnorm(q=mu.minus.1sig,mean=mu,sd=sigma)
[1] 0.6826895
```

It takes a little more work to specify the distribution of interest since it's not standard, but the principle is the same: plus and minus one standard deviation away from the mean.

The symmetry of the normal distribution is also useful when it comes to calculating probabilities. Sticking with the N(3.42, 0.2) distribution, you can see that the probability you make an observation greater than $\mu + \sigma = -3.42 + 0.2 = -3.22$ (an *upper-tail* probability) is identical to the probability of making an observation less than $\mu - \sigma = -3.42 - 0.2 = -3.62$ (a *lower-tail* probability).

```
R> 1-pnorm(mu.plus.1sig,mu,sigma)
[1] 0.1586553
R> pnorm(mu.minus.1sig,mu,sigma)
[1] 0.1586553
```

You can also evaluate these values by hand, given the result you computed earlier that says $\Pr(\mu - \sigma < X < \mu + \sigma) = 0.6826895$. The remaining probability *outside* of this middle area must be as follows:

```
R> 1-0.6826895
[1] 0.3173105
```

So, in each of the lower- and upper-tail areas marked off by $\mu - \sigma$ and $\mu + \sigma$, respectively, there must be a probability of the following:

```
R> 0.3173105/2
[1] 0.1586552
```

This is what was just found using pnorm (note that there can be some minor rounding errors in these types of calculations). You can see this in Figure 16-7, which is, initially, plotted with the following:

```
R> xvals <- seq(-5,-2,length=300)
R> fx <- dnorm(xvals,mean=mu,sd=sigma)
R> plot(xvals,fx,type="l",xlim=c(-4.4,-2.5),main="N(-3.42,0.2) distribution",
        xlab="x",ylab="f(x)")
R> abline(h=0,col="gray")
R> abline(v=c(mu.plus.1sig,mu.minus.1sig),lty=3:2)
R> legend("topleft",legend=c("-3.62\n(mean - 1 sd)","\n-3.22\n(mean + 1 sd)"),
        lty=2:3,bty="n")
```

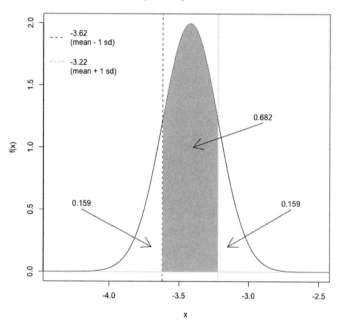

*Figure 16-7: Illustrating the example in the text, where the symmetry of the normal distribution is used to point out features of probabilities under the curve. Note that the total area under the density is 1, which in conjunction with symmetry is useful for making calculations.*

To add the shaded area between $\mu \pm \sigma$, you can use polygon, for which you need the vertices of the shape of interest. To get a smooth curve, make use of the fine sequence xvals and corresponding fx as defined in the code, and use logical vector subsetting to restrict attention to those locations of $x$ such that $-3.62 \leq x \leq -3.22$.

```
R> xvals.sub <- xvals[xvals>=mu.minus.1sig & xvals<=mu.plus.1sig]
R> fx.sub <- fx[xvals>=mu.minus.1sig & xvals<=mu.plus.1sig]
```

You can then sandwich these between the two corners at the bottom of the shaded polygon using the matrix structure that the polygon function expects.

```
R> polygon(rbind(c(mu.minus.1sig,0),cbind(xvals.sub,fx.sub),c(mu.plus.1sig,0)),
           border=NA,col="gray")
```

Finally, arrows and text indicate the areas discussed in the text.

```
R> arrows(c(-4.2,-2.7,-2.9),c(0.5,0.5,1.2),c(-3.7,-3.15,-3.4),c(0.2,0.2,1))
R> text(c(-4.2,-2.7,-2.9),c(0.5,0.5,1.2)+0.05,
        labels=c("0.159","0.159","0.682"))
```

### The qnorm Function

Let's turn to qnorm. To find the quantile value that will give you a lower-tail probability of 0.159, you use the following:

```
R> qnorm(p=0.159,mean=mu,sd=sigma)
[1] -3.619715
```

Given the earlier results and what you already know about previous q-functions, it should be clear why the result is a value of approximately −3.62. You find the upper quartile (the value *above which* you'd find a probability of 0.25) with this:

```
R> qnorm(p=1-0.25,mean=mu,sd=sigma)
[1] -3.285102
```

Remember that the q-function will operate based on the (left) lower-tail probability, so to find a quantile based on an upper-tail probability, you must first subtract it from the total probability of 1.

In some methods and models used in frequentist statistics, it's common to assume that your observed data are normal. You can test the validity of this assumption by using your knowledge of the theoretical quantiles of the normal distribution, found in the results of qnorm: calculate a range of sample quantile values for your observed data and plot these against the same quantiles for a correspondingly standardized normal distribution. This visual tool is referred to as a normal *quantile-quantile* or *QQ* plot and is useful when viewed alongside a histogram. If the plotted points don't lie on a straight line, then the quantiles from your data do not match the appearance of those from a normal curve, and the assumption that your data are normal may not be valid.

The built-in qqnorm function takes in your raw data and produces the corresponding plot. Go back once more to the ready-to-use chickwts data set. Let's say you want to find out whether it's reasonable to assume the weights are normally distributed. To that end, you use

```
R> hist(chickwts$weight,main="",xlab="weight")
R> qqnorm(chickwts$weight,main="Normal QQ plot of weights")
R> qqline(chickwts$weight,col="gray")
```

to produce the histogram of the 71 weights and the normal QQ plot given in Figure 16-8. The additional qqline command adds the "optimal" line that the coordinates would lie along if the data were perfectly normal.

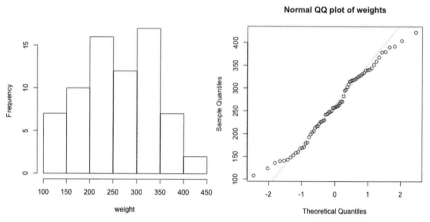

Figure 16-8: Histogram (left) and normal QQ plot (right) of the weights of chicks in the chickwts data set. Are the data normally distributed?

If you inspect the histogram of the weights, you can see that the data match the general appearance of a normal distribution, with a roughly symmetric unimodal appearance. That said, it doesn't quite achieve the smoothness and naturally decaying height that produces the familiar normal bell shape. This is reflected in the QQ plot on the right; the central quantile values appear to lie on the line relatively well, except for some relatively minor "wiggles." There are some clear discrepancies in the outer tails, but note that it is typical to observe discrepancies in these extreme quantiles in any QQ plot because fewer data points naturally occur there. Taking all of this into consideration, for this example the assumption of normality isn't completely unreasonable.

NOTE    *It's important to consider the sample size when assessing the validity of these kinds of assumptions; the larger the sample size, the less random variability will creep into the histogram and QQ plot, and you can more confidently reach a conclusion about whether your data are normal. For instance, the assumption of normality in this example may be complicated by the fact there's a relatively small sample size of only 71.*

### The rnorm Function

Random variates of any given normal distribution are generated with rnorm; for example, the line

```
R> rnorm(n=7,mu,sigma)
[1] -3.764532 -3.231154 -3.124965 -3.490482 -3.884633 -3.192205 -3.475835
```

produces seven normally distributed values arising from $N(-3.42, 0.2)$. In contrast to the QQ plot produced for the chick weights in Figure 16-8, you can use rnorm, qqnorm, and qqline to examine the degree to which hypothetically observed data sets that are truly normal vary in the context of a QQ plot.

The following code generates 71 standard normal values and produces a corresponding normal QQ plot and then does the same for a separate data set of $n = 710$; these are displayed in Figure 16-9.

```
R> fakedata1 <- rnorm(n=71)
R> fakedata2 <- rnorm(n=710)
R> qqnorm(fakedata1,main="Normal QQ plot of generated N(0,1) data; n=71")
R> qqline(fakedata1,col="gray")
R> qqnorm(fakedata2,main="Normal QQ plot of generated N(0,1) data; n=710")
R> qqline(fakedata2,col="gray")
```

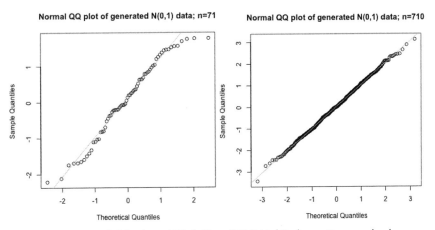

Figure 16-9: Normal QQ plots of 71 (left) and 710 (right) observations randomly generated from the standard normal distribution

You can see that the QQ plot for the simulated data set of size $n = 71$ shows similar deviation from the optimal line as does the chick weights data set. Bumping the sample size up by a factor of 10 shows that the QQ plot for the $n = 710$ normal observations offers up far less random variation, although visible discrepancies in the tails do still occur. A good way to get used to assessing these effects is to rerun these lines of code several times (in other words, generating new data sets each time) and examine how each new QQ plot varies.

### Normal Functions in Action: A Quick Example

Let's finish this section with one more working problem. Assume the manufacturer of a certain type of snack knows that the total net weight of the snacks in its 80-gram advertised package, $X$, is normally distributed with a mean of 80.2 grams and a standard deviation of 1.1 grams. The manufacturer weighs the contents of randomly selected individual packets. The probability a randomly selected packet is less than 78 grams (that is, $\Pr(X < 78)$) is as follows:

```
R> pnorm(78,80.2,1.1)
[1] 0.02275013
```

The probability a packet is found to weigh between 80.5 and 81.5 grams is as follows:

```
R> pnorm(81.5,80.2,1.1)-pnorm(80.5,80.2,1.1)
[1] 0.2738925
```

The weight below which the lightest 20 percent of packets lie is as follows:

```
R> qnorm(0.2,80.2,1.1)
[1] 79.27422
```

A simulation of five randomly selected packets can be found with the following:

```
R> round(rnorm(5,80.2,1.1),1)
[1] 78.6 77.9 78.6 80.2 80.8
```

## Exercise 16.4

a.  A tutor knows that the length of time taken to complete a certain statistics question by first-year undergraduate students, $X$, is normally distributed with a mean of 17 minutes and a standard deviation of 4.5 minutes.

  i.   What is the probability a randomly selected undergraduate takes more than 20 minutes to complete the question?
  ii.  What's the chance that a student takes between 5 and 10 minutes to finish the question?
  iii. Find the time that marks off the slowest 10 percent of students.

   iv. Plot the normal distribution of interest between $\pm 4\sigma$ and shade in the probability area of (iii), the slowest 10 percent of students.

   v. Generate a realization of times based on a class of 10 students completing the question.

b. A meticulous gardener is interested in the length of blades of grass on his lawn. He believes that blade length $X$ follows a normal distribution centered on 10 mm with a variance of 2 mm.

   i. Find the probability that a blade of grass is between 9.5 and 11 mm long.

   ii. What are the standardized values of 9.5 and 11 in the context of this distribution? Using the standardized values, confirm that you can obtain the same probability you found in (i) with the standard normal density.

   iii. Below which value are the shortest 2.5 percent of blade lengths found?

   iv. Standardize your answer from (iii).

## 16.2.3 Student's t-distribution

The *Student's t-distribution* is a continuous probability distribution generally used when dealing with statistics estimated from a sample of data. It will become particularly relevant in the next two chapters, so I'll briefly explain it here first.

Any particular *t*-distribution looks a lot like the standard normal distribution—it's bell-shaped, symmetric, and unimodal, and it's centered on zero. The difference is that while a normal distribution is typically used to deal with a population, the *t*-distribution deals with *sample* from a population.

For the *t*-distribution you don't have to define any parameters per se, but you must choose the appropriate *t*-distribution by way of a strictly positive integer $v > 0$; this is referred to as the *degrees of freedom* (df), called so because it represents the number of individual components in the calculation of a given statistic that are "free to change." You'll see in the upcoming chapters that this quantity is usually directly related to sample sizes.

For the moment, though, you should just loosely think of the *t*-distribution as the representation of a family of curves and think of the degrees of freedom as the "selector" you use to tell you which particular version of the density to use. The precise equation for the density of the *t*-distribution is also not especially useful in an introductory setting, though it is useful to remember that the total probability underneath any *t* curve is naturally 1.

For a *t*-distribution, the dt, pt, qt, and rt functions represent the R implementation of the density, the cumulative distribution (left probabilities), the quantile, and the random variate generation functions, respectively. The first arguments, x, q, p, and n, respectively, provide the relevant value (or values) of interest to these functions; the second argument in all of these is df, to which you must specify the degrees of freedom $v$.

The best way to get an impression of the $t$ family is through a visualization. Figure 16-10 plots the standard normal distribution, as well as the $t$-distribution curve with $v = 1$, $v = 6$, and $v = 20$ df.

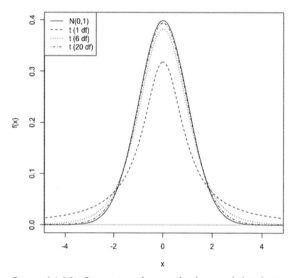

Figure 16-10: Comparing the standard normal distribution with three instances of the t-distribution. Note that the higher the degrees of freedom, the closer the t-distribution approximation becomes to the normal.

The one important note to take away from Figure 16-10, and indeed from this section, is the way in which the $t$ density function changes with respect to the $N(0, 1)$ distribution as you increase the df. For small values of $v$ close to 1, the $t$-distribution is shorter, in terms of its mode, with more probability occurring in noticeably fatter tails. It turns out that the $t$ density approaches the standard normal density as $v \to \infty$. As a case in point, note that the upper 5 percent tail of the standard normal distribution is delineated by the following value:

```
R> qnorm(1-0.05)
[1] 1.644854
```

The same upper tail of the $t$-distribution is provided with df values of $v = 1$, $v = 6$, and $v = 20$, respectively.

```
R> qt(1-0.05,df=1)
[1] 6.313752
R> qt(1-0.05,df=6)
[1] 1.94318
```

```
R> qt(1-0.05,df=20)
[1] 1.724718
```

In direct comparison with the standard normal, the heavier weight in the tails of the $t$ density leads naturally to more extreme quantile values given a specific probability. Notice that this extremity, however, is reduced as the df is increased—fitting in with the aforementioned fact that the $t$-distribution continues to improve in terms of its approximation to the standard normal as you raise the df.

### 16.2.4 Exponential

Of course, probability density functions don't have to be symmetrical like those you've encountered so far, nor do they need to allow for the random variable to be able to take values from negative infinity to positive infinity (like the normal or $t$-distributions). A good example of this is the *exponential distribution*, for which realizations of a random variable $X$ are valid only on a $0 \leq X < \infty$ domain.

For a continuous random variable $0 \leq X < \infty$, the exponential density function $f$ is

$$f(x) = \lambda_e \exp\{-\lambda_e x\}; \quad 0 \leq x < \infty \tag{16.7}$$

where $\lambda_e$ is a parameter of the distribution and $\exp\{\cdot\}$ is the exponential function. The notation

$$X \sim \text{EXP}(\lambda_e)$$

is often used to indicate that "$X$ follows an exponential distribution with rate $\lambda_e$."

The following are the key points to note:

- Theoretically, $X$ can take any value in the range 0 to $\infty$, and $f(x)$ decreases as $x$ increases.

- The "rate" parameter must be strictly positive; in other words, $\lambda_e > 0$. It defines $f(0)$ and the rate of decay of the function to the horizontal asymptote at zero.

The mean and variance are as follows, respectively:

$$\mu_X = \frac{1}{\lambda_e} \quad \text{and} \quad \sigma_X^2 = \frac{1}{\lambda_e^2}$$

#### The dexp Function

The density function for the exponential distribution is a steadily decreasing line beginning at $f(0) = \lambda$; the rate of this decay ensures a total area of 1 underneath the curve. You create Figure 16-11 with the relevant d-function in the following code.

```
R> xvals <- seq(0,10,length=200)
R> plot(xvals,dexp(x=xvals,rate=1.65),xlim=c(0,8),ylim=c(0,1.65),type="l",
      xlab="x",ylab="f(x)")
```

```
R> lines(xvals,dexp(x=xvals,rate=1),lty=2)
R> lines(xvals,dexp(x=xvals,rate=0.4),lty=3)
R> abline(v=0,col="gray")
R> abline(h=0,col="gray")
R> legend("topright",legend=c("EXP(1.65)","EXP(1)","EXP(0.4)"),lty=1:3)
```

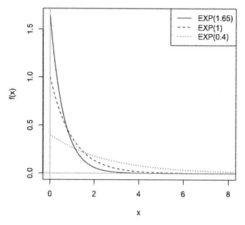

Figure 16-11: Three different exponential density functions.
Decreasing $\lambda_e$ lowers the mode and extends the tail.

The parameter $\lambda_e$ is provided to rate in dexp, which is evaluated at $x$, provided to the first argument x (via the xvals object in this example). You can see that a distinct feature of the exponential density function is that aforementioned decay to zero, with larger values of $\lambda_e$ translating to a taller (yet sharper and more rapid) drop.

This naturally decreasing behavior helps identify the role the exponential distribution often plays in applications—one of a "time-until-event" nature. In fact, there's a special relationship between the exponential distribution and the Poisson distribution introduced in Section 16.1.3. When the Poisson distribution is used to model the count of a certain event through time, you use the exponential distribution to model the time between these events. In such a setting, the exponential parameter $\lambda_e$ defines the mean *rate* at which the events occur over time.

### The pexp Function

Let's revisit the example from Exercise 16.2, where the average number of cars passing an individual within a 120-minute window was said to be 107. Define the random variable $X$ to be the waiting time between two consecutive cars passing and, using an exponential distribution for $X$ on a minute time scale, set $\lambda_e = 107/120 \approx 0.89$ (rounded to 2 d.p.). If 107 cars are typically observed in a two-hour window, then you see cars at an average rate of 0.89 per minute.

Thus, you interpret $\lambda_e$ as the "per-unit-time" measure of the $\lambda_p$ parameter from the Poisson mass function. The interpretation of the mean as the reciprocal of the rate, $\mu_X = 1/\lambda_e$, is also intuitive. For example, when observing cars at a rate of about 0.89 per minute, note that the average waiting time between cars is roughly $1/0.89 \approx 1.12$ minutes.

So, in the current example, you want to examine the density $X \sim \text{EXP}\left(\frac{107}{120}\right)$.

```
R> lambda.e <- 107/120
R> lambda.e
[1] 0.8916667
```

Say a car has just passed the individual's location and you want to find the probability that they must wait more than two and a half minutes before seeing another car, in other words, $\Pr(X > 2.5)$. You can do so using pexp.

```
R> 1-pexp(q=2.5,rate=lambda.e)
[1] 0.1076181
```

This indicates that you have just over a 10 percent chance of observing at least a 2-minute 30-second gap before the next car appears. Remember that the default behavior of the p-function is to find the cumulative, left-hand probability from the provided value, so you need to subtract the result from 1 to find an upper-tail probability. You find the probability of waiting less than 25 seconds with the following, which gives a result of roughly 0.31:

```
R> pexp(25/60,lambda.e)
[1] 0.3103202
```

Note the need to first convert the value of interest from seconds to minutes since you've defined $f(x)$ via $\lambda_e \approx 0.89$ on the scale of the latter.

### The qexp Function

Use the appropriate quantile function qexp to find, say, the cutoff point for the shortest 15 percent of waits.

```
R> qexp(p=0.15,lambda.e)
[1] 0.1822642
```

This indicates that the value of interest is about 0.182 minutes, in other words, roughly $0.182 \times 60 = 10.9$ seconds.

As usual, you can use rexp to generate random variates of any specific exponential distribution.

NOTE     *It is important to distinguish between the "exponential distribution," the "exponential family* of distributions," *and the "exponential function." The first refers to the density function that's just been studied, whereas the second refers to a general class of probability distributions, including the Poisson, the normal, and the exponential itself. The third is just the standard mathematical exponential function upon which the exponential family members depend and is directly accessible in R via exp.*

a. Situated in the central north island of New Zealand, the Pohutu geyser is said to be the largest active geyser in the southern hemisphere. Suppose that it erupts an average of 3,500 times every year.

    i. With the intention of modeling a random variable $X$ as the time between consecutive eruptions, evaluate the parameter value $\lambda_e$ with respect to a time scale in days (assume 365.25 days per year to account for leap years).

    ii. Plot the density function of interest. What's the mean wait in days between eruptions?

    iii. What's the probability of waiting less than 30 minutes for the next eruption?

    iv. What waiting time defines the longest 10 percent of waits? Convert your answer to hours.

b. You can also use the exponential distribution to model certain product survival times, or "time-to-failure" type of variables. Say a manufacturer of a particular air conditioning unit knows that the product has an average life of 11 years before it needs any type of repair callout. Let the random variable $X$ represent the time until the necessary repair of one of these units and assume $X$ follows an exponential distribution with $\lambda_e = 1/11$.

    i. The company offers a five-year full repair warranty on this unit. What's the probability that a randomly selected air conditioner owner makes use of the warranty?

    ii. A rival company offers a six-year guarantee on its competing air conditioning unit but knows that its units last, on average, only nine years before requiring some kind of repair. What are the chances of making use of that warranty?

    iii. Determine the probabilies that the units in (i) and the units in (ii) last more than 15 years.

## 16.2.5 Other Density Functions

There are a number of other common probability density functions used for a wide variety of tasks involving continuous random variables. I'll summarize a few here:

- The *chi-squared distribution* models sums of squared normal variates and is thus often related to operations concerning sample variances of normally distributed data. Its functions are dchisq, pchisq, qchisq, and rchisq, and like the *t*-distribution (Section 16.2.3), it's dependent upon specification of a degrees of freedom provided as the argument df.

- The *F-distribution* is used to model ratios of two chi-squared random variables and is useful in, for example, regression problems (see Chapter 20). Its functions are df, pf, qf, and rf, and as it involves two chi-squared values, it's dependent upon the specification of a *pair* of degrees of freedom values supplied as the arguments df1 and df2.

- The *gamma distribution* is a generalization of both the exponential and chi-squared distributions. Its functions are dgamma, pgamma, qgamma, and rgamma, and it's dependent upon "shape" and "scale" parameters provided as the arguments shape and scale, respectively.

- The *beta distribution* is often used in Bayesian modeling, and it has implemented functions dbeta, pbeta, qbeta, and rbeta. It's defined by two "shape" parameters $\alpha$ and $\beta$, supplied as shape1 and shape2, respectively.

In particular, you'll encounter the chi-squared and *F*-distributions over the next couple of chapters.

**NOTE** *In all of the common probability distributions you've examined over the past couple of sections, I've emphasized the need to perform "one-minus" operations to find probabilities or quantiles with respect to an upper- or right-tailed area. This is because of the cumulative nature of the p- and q-functions—by definition, it's the lower tail that's dealt with. However, most p- and q-functions in R include an optional logical argument, lower.tail, which defaults to TRUE. Therefore, an alternative is to set lower.tail=FALSE in any relevant function call, in which case R will expect or return upper-tail areas specifically.*

## Important Code in This Chapter

| Function/operator | Brief description | First occurrence |
|---|---|---|
| dbinom | Binomial mass function | Section 16.1.2, p. 335 |
| pbinom | Binomial cumulative probabilities | Section 16.1.2, p. 336 |
| qbinom | Binomial quantiles function | Section 16.1.2, p. 337 |
| rbinom | Binomial random realizations | Section 16.1.2, p. 337 |
| dpois | Poisson mass function | Section 16.1.3, p. 340 |
| ppois | Poisson cumulative probabilities | Section 16.1.3, p. 341 |
| rpois | Poisson random realizations | Section 16.1.3, p. 341 |
| dunif | Uniform density function | Section 16.2.1, p. 344 |
| punif | Uniform cumulative probabilities | Section 16.2.1, p. 346 |
| qunif | Uniform quantiles | Section 16.2.1, p. 346 |
| runif | Uniform random realizations | Section 16.2.1, p. 347 |
| dnorm | Normal density function | Section 16.2.2, p. 350 |
| pnorm | Normal cumulative probabilities | Section 16.2.2, p. 350 |
| qnorm | Normal quantiles | Section 16.2.2, p. 353 |
| rnorm | Normal random realizations | Section 16.2.2, p. 355 |
| qt | Student's *t* quantiles | Section 16.2.3, p. 358 |
| dexp | Exponential density function | Section 16.2.4, p. 359 |
| pexp | Exponential cumulative probabilities | Section 16.2.4, p. 361 |
| qexp | Exponential quantiles | Section 16.2.4, p. 361 |

# PART IV

## STATISTICAL TESTING AND MODELING

# 17

## SAMPLING DISTRIBUTIONS AND CONFIDENCE

In Chapters 15 and 16, you applied the idea of a probability distribution to examples where a random variable is defined as some measurement or observation of interest. In this chapter, you'll consider sample statistics themselves as random variables to introduce the concept of a *sampling distribution*—a probability distribution that is used to account for the variability naturally present when you estimate population parameters using sample statistics. I'll then introduce the idea of a *confidence interval*, which is a direct reflection of the variability in a sampling distribution, used in a way that results in an interval estimate of a population parameter. This will form the foundation for formal hypothesis testing in Chapter 18.

## 17.1   Sampling Distributions

A sampling distribution is just like any other probability distribution, but it is specifically associated with a random variable that is a sample statistic. In Chapters 15 and 16, we assumed we knew the parameters of the relevant example distribution (for example, the mean and the standard deviation of a normal distribution or the probability of success in a binomial distribution), but in practice these kinds of quantities are often unknown. In these cases, you'd typically estimate the quantities from a sample (see Figure 13-2 on page 266 for a visual illustration of this). Any statistic estimated from a sample can be treated as a random variable, with the estimated value itself as the realization of that random variable. It's therefore entirely possible that different samples from the same population will provide a different value for the same statistic—realizations of random variables are naturally subject to variability. Being able to understand and model this natural variability inherent in estimated sample statistics (using relevant sampling distributions) is a key part of many statistical analyses.

Like any other probability distribution, the central "balance" point of a sampling distribution is its mean, but the standard deviation of a sampling distribution is referred to as a *standard error*. The slight change in terminology reflects the fact that the probabilities of interest are no longer tied to raw measurements or observations per se, but rather to a quantity calculated from a *sample* of such observations. The theoretical formulas for various sampling distributions therefore depend upon (a) the original probability distributions that are assumed to have generated the raw data and (b) the size of the sample itself.

This section will explain the key ideas and provide some examples, and I'll focus on two simple and easily recognized statistics: a single sample mean and a single sample proportion. I'll then expand on this in Chapter 18 when covering hypothesis testing, and you'll need to understand the role of sampling distributions in assessing important model parameters when you look at regression methods in Chapters 20 to 22.

NOTE *The validity of the theory of sampling distributions as discussed in this chapter makes an important assumption. Whenever I talk about a sample of data from which a given statistic is calculated, I assume those observations are independent of one another and that they are identically distributed. You'll see this notion—independent, identically distributed observations—frequently abbreviated as* iid *in statistical material.*

### 17.1.1  Distribution for a Sample Mean

The arithmetic mean is arguably the most common measure of centrality (Section 13.2.1) used when summarizing a data set.

Mathematically, the variability inherent in an estimated sample mean is described as follows: Formally, denote the random variable of interest as $\bar{X}$. This represents the mean of a sample of $n$ observations from the "raw observation" random variable $X$, as in $x_1, x_2, \ldots, x_n$. Those observations are assumed to have a true finite mean $-\infty < \mu_X < \infty$ and a true finite standard deviation $0 < \sigma_X < \infty$. The conditions for finding the probability distribution of a sample mean vary depending on whether you know the value of the standard deviation.

## Situation 1: Standard Deviation Known

When the true value of the standard deviation $\sigma_X$ is known, then the following are true:

- If $X$ itself is normal, the sampling distribution of $\bar{X}$ is a normal distribution, with mean $\mu_X$ and standard error $\sigma_X/\sqrt{n}$.

- If $X$ is not normal, the sampling distribution of $\bar{X}$ is still approximately normal, with mean $\mu_X$ and standard error $\sigma_X/\sqrt{n}$, and this approximation improves arbitrarily as $n \to \infty$. This is known as the *central limit theorem (CLT)*.

## Situation 2: Standard Deviation Unknown

In practice, you commonly won't know the true value of the standard deviation of the raw measurement distribution that generated your sample data. In this eventuality, it's usual to just replace $\sigma_X$ with $s_X$, which is the standard deviation of the sampled data. However, this substitution introduces additional variability that affects the distribution associated with the sample mean random variable.

- Standardized values (Section 16.2.2) of the sampling distribution of $\bar{X}$ follow a *t*-distribution with $v = n - 1$ degrees of freedom; standardization is performed using the standard error $s_X/\sqrt{n}$.

- If, additionally, $n$ is small, then it is necessary to assume the distribution of $X$ is normal for the validity of this *t*-based sampling distribution of $\bar{X}$.

The nature of the sampling distribution of $\bar{X}$ therefore depends upon whether the true standard deviation of the observations is known, as well as the sample size $n$. The CLT states that normality occurs even if the raw observation distribution is itself not normal, but this approximation is less reliable if $n$ is small. It's a common rule of thumb to rely on the CLT only if $n \geq 30$. If $s_X$, the sample standard deviation, is used to calculate the standard error of $\bar{X}$, then the sampling distribution is the *t*-distribution (following standardization). Again, this is generally taken to be reliable only if $n \geq 30$.

## Example: Dunedin Temperatures

As an example, suppose that the daily maximum temperature in the month of January in Dunedin, New Zealand, follows a normal distribution, with a mean of 22 degrees Celsius and a standard deviation of 1.5 degrees. Then, in line with the comments for situation 1, for samples of size $n = 5$, the sampling distribution of $\bar{X}$ will be normal, with mean 22 and standard error $1.5/\sqrt{5} \approx 0.671$.

The top image of Figure 17-1 shows the raw measurement distribution along with this sampling distribution. You can produce this with code that's familiar from Chapter 16.

```
R> xvals <- seq(16,28,by=0.1)
R> fx.samp <- dnorm(xvals,22,1.5/sqrt(5))
R> plot(xvals,fx.samp,type="l",lty=2,lwd=2,xlab="",ylab="")
R> abline(h=0,col="gray")
R> fx <- dnorm(xvals,22,1.5)
R> lines(xvals,fx,lwd=2)
R> legend("topright",legend=c("raw obs. distbn.","sampling distbn. (mean)"),
          lty=1:2,lwd=c(2,2),bty="n")
```

In this example, the sampling distribution of $\bar{X}$ is clearly a taller, skinnier normal distribution than the one tied to the observations. This makes sense—you expect less variability in an *average* of several measurements as opposed to the raw, individual measurements. Furthermore, the presence of $n$ in the denominator of the standard error dictates a more precise distribution around the mean if you increase the sample size. Again, this makes sense—means will "vary less" between samples of a larger size.

You can now ask various probability questions; note that distinguishing between the measurement distribution and the sampling distribution is important. For example, the following code provides $Pr(X < 21.5)$, the probability that a randomly chosen day in January has a maximum temperature of less than 21.5 degrees:

```
R> pnorm(21.5,mean=22,sd=1.5)
[1] 0.3694413
```

The next bit of code provides the probability that the sample mean will be less than 21.5 degrees, $Pr(\bar{X} < 21.5)$, based on a sample of five random days in January:

```
R> pnorm(21.5,mean=22,sd=1.5/sqrt(5))
[1] 0.2280283
```

The line-shaded areas on the top of Figure 17-1 show these two probabilities. In R, these shaded areas can be added to that plot by running the following lines directly after the earlier code:

```
R> abline(v=21.5,col="gray")
R> xvals.sub <- xvals[xvals<=21.5]
R> fx.sub <- fx[xvals<=21.5]
R> fx.samp.sub <- fx.samp[xvals<=21.5]
R> polygon(cbind(c(21.5,xvals.sub),c(0,fx.sub)),density=10)
R> polygon(cbind(c(21.5,xvals.sub),c(0,fx.samp.sub)),density=10,
           angle=120,lty=2)
```

Note that in previous uses of polygon, you've simply specified a col; in this example, I implemented shading lines instead, using the arguments density (number of lines per inch) and angle (slope of lines in degrees; defaults to angle=45).

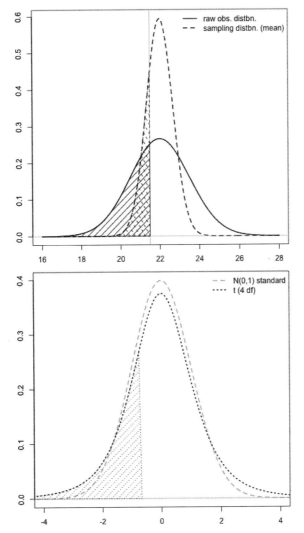

Figure 17-1: Illustrating the sampling distribution of a sample mean for n = 5, based on an N(22, 1.5) raw observation distribution. Top: the normal-based version of the sampling distribution (assuming $\sigma_X$ is known) compared to the observation distribution. Bottom: the t-based version of the sampling distribution, using 4 degrees of freedom (in other words, assuming s has been used to calculate the standard error), compared to a standard normal. Shaded areas represent $\Pr(X < 21.5)$, $\Pr(\bar{X} < 21.5)$ (solid and dashed, topmost plot) and $\Pr(T < (21.5 - \bar{x})/(s/\sqrt{5}))$ (dotted, bottom plot).

To evaluate the probabilities, note that you've required knowledge of the parameters governing $X$. In practice, you'll rarely have these quantities (as noted in situation 2). Instead, you obtain a sample of data and calculate summary statistics.

Running the following line produces five randomly generated Dunedin temperatures from the $X \sim N(22, 1.5)$ distribution:

```
R> obs <- rnorm(5,mean=22,sd=1.5)
R> obs
[1] 22.92233 23.09505 20.98653 20.10941 22.33888
```

Now, for the sake of the example, say these five values constitute all the data you have for this particular problem; in other words, pretend you don't know that $\mu_X = 22$ and $\sigma_X = 1.5$. Your best guesses of the true values of $\mu_X$ and $\sigma_X$, denoted $\bar{x}$ and $s$, respectively, are therefore as follows:

```
R> obs.mean <- mean(obs)
R> obs.mean
[1] 21.89044
R> obs.sd <- sd(obs)
R> obs.sd
[1] 1.294806
```

The estimated standard error can be calculated:

```
R> obs.mean.se <- obs.sd/sqrt(5)
R> obs.mean.se
[1] 0.5790549
```

Because $n = 5$ is relatively small, you must assume the values in obs are realizations from a normal distribution, in line with the points made for situation 2. This allows you to handle the sampling distribution of $\bar{X}$ using the $t$-distribution with 4 degrees of freedom. Recall from Section 16.2.3, though, that any $t$-distribution is typically placed on a standardized scale. Therefore, for you to find the probability that the mean temperature (in a sample of five days) is less than 21.5 based on your calculated sample statistics, you must first standardize this value using the rules outlined in Section 16.2.2. Label the corresponding random variable as $T$ and the specific value as $t_4$, stored as the object t4 in R.

```
R> t4 <- (21.5-obs.mean)/obs.mean.se
R> t4
[1] -0.6742706
```

This has placed the value of interest, 21.5, on the standardized scale, making it interpretable with respect to a standard normal distribution or, as is correct in this setting (because you are using the estimate $s$ rather than the unknown $\sigma_X$ in calculating the standard error), $t_4$ follows the aforementioned $t$-distribution with 4 degrees of freedom. The estimated probability is as follows.

```
R> pt(t4,df=4)
[1] 0.26855
```

Note that when you calculated the "true" theoretical probability from the sampling distribution of $\Pr(\bar{X} < 21.5)$, you got a result of about 0.23 (see page 370), but the same probability based on standardization using sample statistics of the data obs (in other words, *estimates* of the true theoretical values $\Pr(T < t_4)$) has been computed as 0.27 (2 d.p.).

The bottom image of Figure 17-1 provides the *t*-distribution with $v = 4$, marking off the probability described. The $N(0, 1)$ density is also plotted for comparison; this represents the standardized version of the $N(22, 1.5/\sqrt{5})$ sampling distribution from earlier, in situation 1. You can produce this image with the following lines:

```
R> xvals <- seq(-5,5,length=100)
R> fx.samp.t <- dt(xvals,df=4)
R> plot(xvals,dnorm(xvals),type="l",lty=2,lwd=2,col="gray",xlim=c(-4,4),
         xlab="",ylab="")
R> abline(h=0,col="gray")
R> lines(xvals,fx.samp.t,lty=3,lwd=2)
R> polygon(cbind(c(t4,-5,xvals[xvals<=t4]),c(0,0,fx.samp.t[xvals<=t4])),
           density=10,lty=3)
R> legend("topright",legend=c("N(0,1) standard","t (4 df)"),
           col=c("gray","black"),lty=2:3,lwd=c(2,2),bty="n")
```

Consideration of probability distributions associated with sample means is clearly not a trivial exercise. Using sample statistics governs the nature of the sampling distribution; in particular, it will be *t* based if you use the sample standard deviation to calculate the standard error. However, as the examples here have shown, once that's been established, the calculation of various probabilities is easy and follows the same general rules and R functionality detailed in Section 16.2.

### 17.1.2   Distribution for a Sample Proportion

Sampling distributions for sample proportions are interpreted in much the same way. If *n* trials of a success/failure event are performed, you can obtain an estimate of the proportion of successes; if another *n* trials are performed, the new estimate could vary. It's this variability that you're investigating.

The random variable of interest, $\hat{P}$, represents the estimated proportions of successes over any *n* trials, each resulting in some defined binary outcome. It is estimated as $\hat{p} = \frac{x}{n}$, where *x* is the number of successes in a sample of size *n*. Let the corresponding true proportion of successes (often unknown) simply be denoted with $\pi$.

**NOTE**    *Note that $\pi$ as used in this setting doesn't refer to the common geometric value 3.14 (2 d.p.). Rather, it's simply standard notation to refer to a true population proportion using the $\pi$ symbol.*

The sampling distribution of $\hat{P}$ is approximately normal with mean $\pi$ and standard error $\sqrt{\pi(1-\pi)/n}$. The following are the key things to note:

- This approximation is valid if $n$ is large and/or $\pi$ is not too close to either 0 or 1.

- There are rules of thumb to determine this validity; one such rule is to assume the normal approximation is satisfactory if both $n\pi$ and $n(1-\pi)$ are greater than 5.

- When the true $\pi$ is unknown or is unassumed to be a certain value, it is typically replaced by $\hat{p}$ in all of the previous formulas.

As long as you can deem the approximation to the normal distribution valid, this is the only probability distribution that you need to be concerned with. However, it's worth noting that the standard error of the sampling distribution of a sample proportion depends directly upon the proportion $\pi$. This becomes important when constructing confidence intervals and carrying out hypothesis tests, which you'll begin to explore in Chapter 18.

Let's look at a practical example. Suppose a political commentator in the United States is interested in the proportion of voting-age citizens in her home city that already know how they will vote in the next presidential election. She obtains a yes or no answer from 118 suitable randomly selected individuals. Of these individuals, 80 say they know how they'll vote. To investigate the variability associated with the proportion of interest, you'll therefore need to consider

$$\hat{P} \sim \mathrm{N}\left(\hat{p},\ \sqrt{\frac{\hat{p}(1-\hat{p})}{n}}\right), \tag{17.1}$$

where $\hat{p} = \frac{80}{118}$. In R, the following gives you the estimate of interest:

```
R> p.hat <- 80/118
R> p.hat
[1] 0.6779661
```

In the sample, about 68 percent of the surveyed individuals know how they will vote in the next election. Note also that, according to the aforementioned rule of thumb, the approximation to the normal distribution is valid because both values are greater than 5.

```
R> 118*p.hat
[1] 80
R> 118*(1-p.hat)
[1] 38
```

Estimate the standard error with the following:

```
R> p.se <- sqrt(p.hat*(1-p.hat)/118)
R> p.se
[1] 0.04301439
```

Then, you can plot the corresponding sampling distribution using this code:

```
R> pvals <- seq(p.hat-5*p.se,p.hat+5*p.se,length=100)
R> p.samp <- dnorm(pvals,mean=p.hat,sd=p.se)
R> plot(pvals,p.samp,type="l",xlab="",ylab="",
        xlim=p.hat+c(-4,4)*p.se,ylim=c(0,max(p.samp)))
R> abline(h=0,col="gray")
```

Figure 17-2 gives the result.

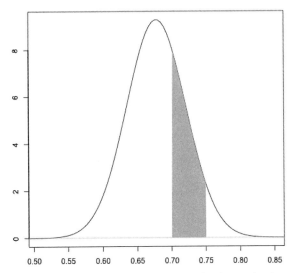

*Figure 17-2: Visualizing the sampling distribution for the voting example as per Equation (17.1). The shaded area represents $\Pr(0.7 < \hat{P} < 0.75)$, which is the probability that the true sample proportion for samples of size n = 118 lies between 0.7 and 0.75.*

Now you can use this distribution to describe the variability in the sample proportion of voters who already know how they will vote, for other samples of this size.

For example, the shaded area in Figure 17-2 highlights the probability that in another sample of the same size, the sample proportion of voters in the given city who already know how they're going to vote is somewhere between 0.7 and 0.75. This shaded area can be added with the following code:

```
R> pvals.sub <- pvals[pvals>=0.7 & pvals<=0.75]
R> p.samp.sub <- p.samp[pvals>=0.7 & pvals<=0.75]
R> polygon(cbind(c(0.7,pvals.sub,0.75),c(0,p.samp.sub,0)),
        border=NA,col="gray")
```

And with knowledge of pnorm, introduced in Section 16.2.2, you can use the following code to calculate the probability of interest:

```
R> pnorm(0.75,mean=p.hat,sd=p.se) - pnorm(0.7,mean=p.hat,sd=p.se)
[1] 0.257238
```

This sampling distribution suggests that the chance of another sample proportion, based on the same sample size, lying somewhere between these two values is about 25.7 percent.

---

### Exercise 17.1

A teacher wants to test all of the 10th-grade students at his school to gauge their basic mathematical understanding, but the photocopier breaks after making only six copies of the test. With no other choice, he chooses six students at random to take the test. Their results, recorded as a score out of 65, have a sample mean of 41.1. The standard deviation of the marks of this test is known to be 11.3.

a. Find the standard error associated with the mean test score.

b. Assuming the scores themselves are normally distributed, evaluate the probability that the mean score lies between 45 and 55 if the teacher took another sample of the same size.

c. A student who gets less than half the questions correct receives a failing grade (F). Find the probability that the average score is an F based on another sample of the same size.

A marketing company wants to find out which of two energy drinks teenagers prefer—drink A or drink B. It surveys 140 teens, and the results indicate that only 35 percent prefer drink A.

d. Use a quick check to decide whether it is valid to use the normal distribution to represent the sampling distribution of this proportion.

e. What is the probability that in another sample of the same size, the proportion of teenagers who prefer drink A is greater than 0.4?

f. Find the two values of this sampling distribution that identify the central 80 percent of values of the proportion of interest.

In Section 16.2.4, the time between cars passing an individual's location was modeled using an exponential distribution. Say that on the other side of town, her friend is curious about a similar problem. Standing outside her house, she records 63 individual times between cars passing. These sampled times have a mean of $\bar{x} = 37.8$ seconds with a standard deviation of $s = 34.51$ seconds.

g. The friend inspects a histogram of her raw measurements and notices that her raw data are heavily right-skewed. Briefly identify and describe the nature of the sampling distribution with respect to the sample mean and calculate the appropriate standard error.

h. Using the standard error from (g) and the appropriate probability distribution, calculate the probability that in another sample of the same size, the sample mean time between cars passing is as follows:
    i. More than 40 seconds
    ii. Less than half a minute
    iii. Between the given sample mean and 40 seconds

## *17.1.3  Sampling Distributions for Other Statistics*

So far you've looked at sampling distributions in cases dealing with a single sample mean or sample proportion, though it's important to note that many problems require more complicated measures. Nevertheless, you can apply the ideas explored in this section to any statistic estimated from a finite-sized sample. The key, always, is to be able to understand the variability associated with your point estimates.

In some settings, such as those covered so far, the sampling distribution is parametric, meaning that the functional (mathematical) form of the probability distribution itself is known and depends only on the provision of specific parameter values. This is sometimes contingent upon the satisfaction of certain conditions, as you've seen with the application of the normal distribution covered in this chapter. For other statistics, it may be the case that you do not know the form of the appropriate sampling distribution—in these cases, you could use computer simulation to obtain the required probabilities.

In the remainder of this chapter and over the next few chapters, you'll continue to explore statistics that are tied to parametric sampling distributions for common tests and models.

**NOTE**  *The variability of an estimated quantity is actually only one side of the coin. Just as important is the issue of statistical bias. Where "natural variability" should be associated with random error, bias is associated with systematic error, in the sense that a biased statistic does not settle on the corresponding true parameter value as the sample size increases. Bias can be caused by flaws in a study design or collection of data or can be the result of a poor estimator of the statistic of interest. Bias is an undesirable trait of any given estimator and/or statistical analysis unless it can be quantified and removed, which is often difficult if not impossible in practice. I've therefore dealt so far only with unbiased statistical estimators, many of which are those you may already be familiar with (for example, the arithmetic mean), and I'll continue to assume unbiasedness moving forward.*

## 17.2 Confidence Intervals

A *confidence interval (CI)* is an interval defined by a lower limit $l$ and an upper limit $u$, used to describe possible values of a corresponding true population parameter in light of observed sample data. Interpretation of a confidence interval therefore allows you to state a "level of confidence" that a true parameter of interest falls between this upper and lower limit, often expressed as a percentage. As such, it is a common and useful tool built directly from the sampling distribution of the statistic of interest.

The following are the important points to note:

- The level of confidence is usually expressed as a percentage, such that you'd construct a $100 \times (1 - \alpha)$ percent confidence interval, where $0 < \alpha < 1$ is an "amount of tail probability."

- The three most common intervals are defined with either $\alpha = 0.1$ (a 90 percent interval), $\alpha = 0.05$ (a 95 percent interval), or $\alpha = 0.01$ (a 99 percent interval).

- Colloquially, you'd state the interpretation of a confidence interval $(l,u)$ as "I am $100 \times (1 - \alpha)$ percent confident that the true parameter value lies somewhere between $l$ and $u$."

Confidence intervals may be constructed in different ways, depending on the type of statistic and therefore the shape of the corresponding sampling distribution. For symmetrically distributed sample statistics, like those involving means and proportions that will be used in this chapter, a general formula is

$$statistic \pm critical\ value \times standard\ error, \qquad (17.2)$$

where *statistic* is the sample statistic under scrutiny, *critical value* is a value from the standardized version of the sampling distribution that corresponds to $\alpha$, and *standard error* is the standard deviation of the sampling distribution. The product of the critical value and standard error is referred to as the *error component* of the interval; subtraction of the error component from the value of the statistic provides $l$, and addition provides $u$.

With reference to the appropriate sampling distribution, all that a CI yields are the two values of the distribution that mark off the central $100 \times (1 - \alpha)$ percent of the area under the density. (This is the process that was briefly mentioned in Exercise 17.1 (f).) You then use the CI to make further interpretations concerning the true (typically unknown) parameter value that's being estimated by the statistic of interest.

### 17.2.1 An Interval for a Mean

You know from Section 17.1.1 that the sampling distribution of a single sample mean depends primarily on whether you know the true standard deviation of the raw measurements, $\sigma_X$. Then, provided the sample size for this sample mean is roughly $n \geq 30$, the CLT ensures a symmetric sampling distribution—which will be normal if you know the true value of $\sigma_X$, or $t$ based with $\nu = n - 1$ df if you must use the sample standard deviation, $s$, to estimate $\sigma_X$ (as is more common in practice). You've seen that the standard error is defined as the standard deviation divided by the square root of $n$. For a small $n$, you must also assume that the raw observations are normally distributed, since the CLT will not apply.

To construct an appropriate interval, you must first find the critical value corresponding to $\alpha$. By definition the CI is symmetric, so this translates to a central probability of $(1 - \alpha)$ around the mean, which is exactly $\alpha/2$ in the lower tail and the same in the upper tail.

Return to the example from Section 17.1.1, dealing with the mean daily maximum temperatures (degrees Celsius) in January for Dunedin, New Zealand. Suppose you know the observations are normally distributed but you don't know the true mean $\mu_X$ (which is set at 22) or the true standard deviation $\sigma_X$ (which is set at 1.5). Setting it up in the same way as earlier, assume you've made the following five independent observations:

```
R> temp.sample <- rnorm(n=5,mean=22,sd=1.5)
R> temp.sample
[1] 20.46097 21.45658 21.06410 20.49367 24.92843
```

As you're interested in the sample mean and its sampling distribution, you must calculate the sample mean $\bar{x}$, the sample standard deviation $s$, and the appropriate standard error of the sample mean, $s/\sqrt{n}$.

```
R> temp.mean <- mean(temp.sample)
R> temp.mean
[1] 21.68075
R> temp.sd <- sd(temp.sample)
R> temp.sd
[1] 1.862456
R> temp.se <- temp.sd/sqrt(5)
R> temp.se
[1] 0.8329155
```

Now, let's say the aim is to construct a 95 percent confidence interval for the true, unknown mean $\mu_X$. This implies $\alpha = 0.05$ (the total amount of tail probability) for the relevant sampling distribution. Given the fact that you know the raw observations are normal and that you're using $s$ (not $\sigma_X$), the appropriate distribution is the $t$-distribution with $n - 1 = 4$ degrees of freedom. For a central area of 0.95 under this curve, $\alpha/2 = 0.025$ must be in either tail. Knowing that R's q functions operate based on a total lower tail area, the (positive) critical value is therefore found by supplying a probability of $1 - \alpha/2 = 0.975$ to the appropriate function.

```
R> 1-0.05/2
[1] 0.975
R> critval <- qt(0.975,df=4)
R> critval
[1] 2.776445
```

Figure 17-3 shows why the qt function is used in this way (since I used similar code throughout Chapter 16, I haven't reproduced the code for Figure 17-3 here).

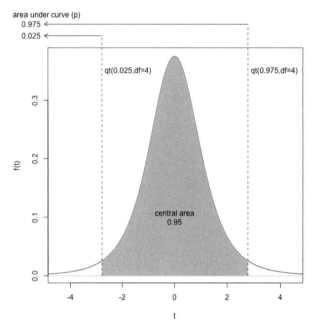

Figure 17-3: Illustrating the role of the critical value in a confidence interval for a sample mean, using the Dunedin temperature example. The sampling distribution is t with 4 df, and the use of qt with respect to symmetric tail probabilities related to α/2 = 0.025 yields a central area of 0.95.

Note that when viewed with respect to the negative version of the same critical value ("reflected" around the mean and obtained by using qt(0.025,4)), the central, symmetric area under the curve must be 0.95. You can confirm this using pt.

```
R> pt(critval,4)-pt(-critval,4)
[1] 0.95
```

So, all the ingredients are present. You find the 95 percent confidence interval for the true mean $\mu_X$ via Equation (17.2) with the following lines, which give $l$ and $u$, respectively:

```
R> temp.mean-critval*temp.se
[1] 19.36821
R> temp.mean+critval*temp.se
[1] 23.99329
```

The CI given by (19.37, 23.99) is therefore interpreted as follows: you are 95 percent confident that the true mean maximum temperature in Dunedin in January lies somewhere between 19.37 and 23.99 degrees Celsius.

With this result, you've combined knowledge of the estimate of the mean itself with the inherent variability of a sample to define an interval of values in which you're fairly sure the true mean will lie. As you know, the true mean in this case is 22, which is indeed included in the calculated CI.

From this, it's easy to alter the intervals to change the confidence levels. You need to change only the critical value, which, as always, must define $\alpha/2$ in each tail. For example, an 80 percent CI ($\alpha = 0.2$) and a 99 percent CI ($\alpha = 0.01$) for the same example value given here can be found with these two lines, respectively:

```
R> temp.mean+c(-1,1)*qt(p=0.9,df=4)*temp.se
[1] 20.40372 22.95778
R> temp.mean+c(-1,1)*qt(p=0.995,df=4)*temp.se
[1] 17.84593 25.51557
```

Note here the use of multiplication by the vector c(-1,1) so that the lower and upper limits can be obtained at once and the result returned as a vector of length 2. As usual, the qt function is used with respect to a complete lower-tail area, so p is set at $1 - \alpha/2$.

These most recent intervals highlight the natural consequence of moving to a higher confidence level for a given CI. A higher probability in the central area translates directly to a more extreme critical value, resulting in a wider interval. This makes sense—in order to be "more confident" about the true parameter value, you'd need to take into account a larger range of possible values.

## 17.2.2   An Interval for a Proportion

Establishing a CI for a sample proportion follows the same rules as for the mean. With knowledge of the sampling distribution as per Section 17.1.2, you obtain critical values from the standard normal distribution, and for an estimate of $\hat{p}$ from a sample of size $n$, the interval itself is constructed with the standard error $\sqrt{\hat{p}(1 - \hat{p})/n}$.

Let's return to the example from Section 17.1.2, where 80 of 118 surveyed individuals said that they knew how they were going to vote in the next US presidential election. Recall you have the following:

```
R> p.hat <- 80/118
R> p.hat
[1] 0.6779661
R> p.se <- sqrt(p.hat*(1-p.hat)/118)
R> p.se
[1] 0.04301439
```

To construct a 90 percent CI ($\alpha = 0.1$), the appropriate critical value from the standardized sampling distribution of interest is as follows, implying $\Pr(-1.644854 < Z < 1.644854) = 0.9$ for $Z \sim \mathrm{N}(0,1)$:

```
R> qnorm(0.95)
[1] 1.644854
```

Now you again follow Equation (17.2):

```
R> p.hat+c(-1,1)*qnorm(0.95)*p.se
[1] 0.6072137 0.7487185
```

You can conclude that you're 90 percent confident that the true proportion of voters who know how they will vote in the next election lies somewhere between 0.61 and 0.75 (rounded to two decimal places).

### 17.2.3 Other Intervals

The two simple situations presented in Sections 17.2.1 and 17.2.2 serve to highlight the importance of associating any point estimate (in other words, a sample statistic) with the idea of its variability. Confidence intervals can of course be constructed for other quantities, and over the following sections (as part of testing hypotheses), I'll expand on the discussion of confidence intervals to investigate differences between two means and two proportions, as well as ratios of categorical counts. These more complicated statistics come with their own standard error formulas, though the corresponding sampling distributions are still symmetric via the normal and $t$-curves (if, again, some standard assumptions are met), which means that the now familiar formulation of Equation (17.2) still applies.

Generally, a confidence interval seeks to mark off a central area of $1 - \alpha$ from the sampling distribution of interest, including sampling distributions that are asymmetric. In those cases, however, it doesn't make much sense to have a symmetric CI based on a single, standardized critical value as per Equation (17.2). Similarly, you might not know the functional, parametric form of the sampling distribution and so may not be willing to make any distributional assumptions, such as symmetry. In these cases, you can take an alternative path based on the raw quantiles (or estimated raw quantiles; see Section 13.2.3) of the supposed asymmetric sampling distribution. Using specific quantile values to mark off identical $\alpha/2$ upper- and lower-tail areas is a valid method that remains sensitive to the shape of the sampling distribution of interest, while still allowing you to construct a useful interval that describes potential true parameter values.

### 17.2.4 Comments on Interpretation of a CI

The typical statement about the interpretation of any CI references a degree of confidence in where the true parameter value lies, but a more formally correct interpretation should consider and clarify the probabilistic nature of the construction. Technically, given a $100(1 - \alpha)$ percent confidence level, the more accurate interpretation is as follows: over many samples of the same size and from the same population where a CI, of the same confidence level, is constructed with respect to the same statistic from each sample, you would expect the true corresponding parameter value to fall within the limits of $100(1 - \alpha)$ percent of those intervals.

This comes from the fact that the theory of a sampling distribution describes the variability in multiple samples, not just the sample that has been taken. At first glance it may be difficult to fully appreciate the difference between this and the colloquially used "confidence statement," but it is important to remain aware of the technically correct definition, particularly given that a CI is typically estimated based on only one sample.

## Exercise 17.2

A casual runner records the average time it takes him to sprint 100 meters. He completes the dash 34 times under identical conditions and finds that the mean of these is 14.22 seconds. Assume that he knows the standard deviation of his runs is $\sigma_X = 2.9$ seconds.

a. Construct and interpret a 90 percent confidence interval for the true mean time.

b. Repeat (a), but this time, assume that the standard deviation is not known and that $s = 2.9$ is estimated from the sample. How, if at all, does this change the interval?

In a particular country, the true proportion of citizens who are left handed or ambidextrous is unknown. A random sample of 400 people is taken, and each individual is asked to identify with one of three options: right-handed only, left-handed only, or ambidextrous. The results show that 37 selected left-handed and 11 selected ambidextrous.

c. Calculate and interpret a 99 percent CI for the true proportion of left-handed-only citizens.

d. Calculate and interpret a 99 percent CI for the true proportion of citizens who are either left-handed *or* ambidextrous.

In Section 17.2.4, the technical interpretation of a CI with respect to its confidence level was described as the proportion of many similar intervals (that is, when calculated for samples of the same size from the same population) that contain the true value of the parameter of interest.

e. Your task is to write an example to demonstrate this behavior of confidence intervals using simulation. To do so, follow these instructions:

- Set up a matrix (see Chapter 3) filled with NAs (Chapter 6) that has 5,000 rows and 3 columns.
- Use skills from Chapter 10 to write a for loop that, at each of 5,000 iterations, generates a random sample of size 300 from an exponential distribution with rate parameter $\lambda_e = 0.1$ (Section 16.2.4).
- Evaluate the sample mean and sample standard deviation of each sample, and use these quantities with the critical values from the appropriate sampling distribution to calculate a 95 percent CI for the true mean of the distribution.
- Within the for loop, the matrix should now be filled, row by row, with your results. The first column will contain the lower limit, the second will contain the upper limit, and the third column will be a logical value that is TRUE if the corresponding interval contains the true mean of $1/\lambda_e$ and that is FALSE otherwise.
- When the loop is completed, compute the proportion of TRUEs in the third column of the filled matrix. You should find that this proportion is close to 0.95; this will vary randomly each time you rerun the loop.

f. Create a plot that draws the first 100 of your estimated confidence intervals as separate horizontal lines drawn from $l$ to $u$, one on top of another. One way to do this is to first create an empty plot with preset $x$- and $y$-limits (the latter as c(1,100)) and then progressively add each line using lines with appropriate coordinates (this could be done using another for loop). As a final touch, add to the plot a red vertical line that denotes the true mean. Confidence intervals that do not include the true mean will not intersect that vertical line.

The following shows an example of this plot:

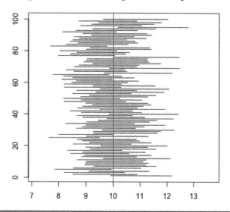

# 18

## HYPOTHESIS TESTING

In this chapter, you'll build on your experience with confidence intervals and sampling distributions to make more formal statements about the value of a true, unknown parameter of interest. For this, you'll learn about frequentist *hypothesis testing*, where a probability from a relevant sampling distribution is used as evidence against some claim about the true value. When a probability is used in this way, it is referred to as a *p*-value. In this chapter, I talk about interpreting results for relatively basic statistics, but you can apply the same concepts to statistics arising from more complicated methods (such as regression modeling in Chapter 19).

## 18.1   Components of a Hypothesis Test

To give you an example of hypothesis testing, suppose I told you that 7 percent of a certain population was allergic to peanuts. You then randomly selected 20 individuals from that population and found that 18 of them were allergic to peanuts. Assuming your sample was unbiased and truly reflective of the population, what would you then think about my claim that the true proportion of allergic individuals is 7 percent?

Naturally, you would doubt the correctness of my claim. In other words, there is such a small probability of observing 18 or more successes out of 20 trials for a set success rate of 0.07 that you can state that you have statistical evidence against the claim that the true rate is 0.07. Indeed, when defining $X$ as the number of allergic individuals out of 20 by assuming $X \sim \mathrm{BIN}(20, 0.07)$, evaluating $\Pr(X \geq 18)$ gives you the precise $p$-value, which is tiny.

```
R> dbinom(18,size=20,prob=0.07) + dbinom(19,size=20,prob=0.07) +
    dbinom(20,size=20,prob=0.07)
[1] 2.69727e-19
```

This $p$-value represents the probability of observing the results in your sample, $X = 18$, or a more extreme outcome ($X = 19$ or $X = 20$), if the chance of success was truly 7 percent.

Before looking at specific hypothesis tests and their implementation in R, this section will introduce terminology that you'll come across often in the reporting of such tests.

### 18.1.1  Hypotheses

As the name would suggest, in hypothesis testing, formally stating a claim and the subsequent hypothesis test is done with a *null* and an *alternative* hypothesis. The null hypothesis is interpreted as the *baseline* or *no-change* hypothesis and is the claim that is assumed to be true. The alternative hypothesis is the conjecture that you're testing for, against the null hypothesis.

In general, null and alternative hypotheses are denoted $H_0$ and $H_A$, respectively, and they are written as follows:

$$H_0 : \ldots$$
$$H_A : \ldots$$

The null hypothesis is often (but not always) defined as an equality, =, to a null value. Conversely, the alternative hypothesis (the situation you're testing for) is often defined in terms of an inequality to the null value.

- When $H_A$ is defined in terms of a less-than statement, with <, it is *one-sided*; this is also called a *lower-tailed test*.

- When $H_A$ is defined in terms of a greater-than statement, with >, it is *one-sided*; this is also called an *upper-tailed test*.

- When $H_A$ is merely defined in terms of a different-to statement, with ≠, it is *two-sided*; this is also called a *two-tailed test*.

These test variants are entirely situation specific and depend upon the problem at hand.

### 18.1.2 Test Statistic

Once the hypotheses are formed, sample data are collected, and statistics are calculated according to the parameters detailed in the hypotheses. The *test statistic* is the statistic that's compared to the appropriate standardized sampling distribution to yield the *p*-value.

A test statistic is typically a standardized or rescaled version of the sample statistic of interest. The distribution and extremity (that is, distance from zero) of the test statistic are the sole drivers of the smallness of the *p*-value (which indicates the strength of the evidence against the null hypothesis—see Section 18.1.3). Specifically, the test statistic is determined by both the difference between the original sample statistic and the null value and the standard error of the sample statistic.

### 18.1.3 *p-value*

The *p*-value is the probability value that's used to quantify the amount of evidence, if any, against the null hypothesis. More formally, the *p*-value is found to be the probability of observing the test statistic, or something more extreme, assuming the null hypothesis is true.

The exact nature of calculating a *p*-value is dictated by the type of statistics being tested and the nature of $H_A$. In reference to this, you'll see the following terms:

- A lower-tailed test implies the *p*-value is a left-hand tail probability from the sampling distribution of interest.

- For an upper-tailed test, the *p*-value is a right-hand tail probability.

- For a two-sided test, the *p*-value is the sum of a left-hand tail probability and right-hand tail probability. When the sampling distribution is symmetric (for example, normal or *t*, as in all examples coming up in Sections 18.2 and 18.3), this is equivalent to two times the area in one of those tails.

Put simply, the more extreme the test statistic, the smaller the *p*-value. The smaller the *p*-value, the greater the amount of statistical evidence against the assumed truth of $H_0$.

### 18.1.4 Significance Level

For every hypothesis test, a *significance level*, denoted $\alpha$, is assumed. This is used to qualify the result of the test. The significance level defines a cutoff point, at which you decide whether there is sufficient evidence to view $H_0$ as incorrect and favor $H_A$ instead.

- If the *p*-value is greater than or equal to $\alpha$, then you conclude there is insufficient evidence against the null hypothesis, and therefore you *retain* $H_0$ when compared to $H_A$.

- If the *p*-value is less than $\alpha$, then the result of the test is *statistically significant*. This implies there is sufficient evidence against the null hypothesis, and therefore you *reject* $H_0$ in favor of $H_A$.

Common or conventional values of $\alpha$ are $\alpha = 0.1$, $\alpha = 0.05$, and $\alpha = 0.01$.

### 18.1.5   Criticisms of Hypothesis Testing

The terminology just presented becomes easier to understand once you look at some examples in the upcoming sections. However, even at this early stage, it's important to recognize that hypothesis testing is susceptible to justifiable criticism. The end result of any hypothesis test is to either retain or reject the null hypothesis, a decision that is solely dependent upon the rather arbitrary choice of significance level $\alpha$; this is most often simply set at one of the conventionally used values.

Before you begin looking at examples, it is also important to note that a *p*-value never provides "proof" of either $H_0$ or $H_A$ being truly correct. It can only ever quantify evidence against the null hypothesis, which one rejects given a sufficiently small *p*-value $< \alpha$. In other words, rejecting a null hypothesis is not the same as disproving it. Rejecting $H_0$ merely implies that the sample data suggest $H_A$ ought to be preferred, and the *p*-value merely indicates the strength of this preference.

In recent years, there has been a push against emphasizing these aspects of statistical inference in some introductory statistics courses owing at least in part to the overuse, and even misuse, of *p*-values in some areas of applied research. A particularly good article by Sterne and Smith (2001) discusses the role of, and problems surrounding, hypothesis testing from the point of view of medical research. Another good reference is Reinhart (2015), which discusses common misinterpretations of *p*-values in statistics.

That being said, probabilistic inference with respect to sampling distributions is, and will always remain, a cornerstone of frequentist statistical practice. The best way to improve the use and interpretation of statistical tests and modeling is with a sound introduction to the relevant ideas and methods so that, from the outset, you understand statistical significance and what it can and cannot tell you.

## 18.2   Testing Means

The validity of hypothesis tests involving sample means is dependent upon the same assumptions and conditions mentioned in Section 17.1.1. In particular, throughout this section you should assume that the central limit theorem holds, and if the sample sizes are small (in other words, roughly less than 30), the raw data are normally distributed. You'll also focus on examples where the sample standard deviation $s$ is used to estimate the true standard deviation, $\sigma_X$, because this is the most common situation you'll encounter in practice. Again, mirroring Section 17.1.1, this means you need to use the *t*-distribution instead of the normal distribution when calculating the critical values and *p*-values.

## 18.2.1 Single Mean

As you've already met the standard error formula, $s/\sqrt{n}$, and the R functionality needed to obtain quantiles and probabilities from the $t$-distribution (qt and pt), the only new concepts to introduce here are related to the definition of the hypotheses themselves and the interpretation of the result.

### Calculation: One-Sample t-Test

Let's dive straight into an example—a one-sample $t$-test. Recall the problem in Section 16.2.2 where a manufacturer of a snack was interested in the mean net weight of contents in an advertised 80-gram pack. Say that a consumer calls in with a complaint—over time they have bought and precisely weighed the contents of 44 randomly selected 80-gram packs from different stores and recorded the weights as follows:

```
R> snacks <- c(87.7,80.01,77.28,78.76,81.52,74.2,80.71,79.5,77.87,81.94,80.7,
              82.32,75.78,80.19,83.91,79.4,77.52,77.62,81.4,74.89,82.95,
              73.59,77.92,77.18,79.83,81.23,79.28,78.44,79.01,80.47,76.23,
              78.89,77.14,69.94,78.54,79.7,82.45,77.29,75.52,77.21,75.99,
              81.94,80.41,77.7)
```

The customer claims that they've been shortchanged because their data cannot have arisen from a distribution with mean $\mu = 80$, so the true mean weight must be less than 80. To investigate this claim, the manufacturer conducts a hypothesis test using a significance level of $\alpha = 0.05$.

First, the hypotheses must be defined, with a null value of 80 grams. Remember, the alternative hypothesis is "what you're testing for"; in this case, $H_A$ is that $\mu$ is smaller than 80. The null hypothesis, interpreted as "no change," will be defined as $\mu = 80$: that the true mean is in fact 80 grams. These hypotheses are formalized like this:

$$H_0 : \ \mu = 80$$
$$H_A : \ \mu < 80 \tag{18.1}$$

Second, the mean and standard deviation must be estimated from the sample.

```
R> n <- length(snacks)
R> snack.mean <- mean(snacks)
R> snack.mean
[1] 78.91068
R> snack.sd <- sd(snacks)
R> snack.sd
[1] 3.056023
```

The question your hypotheses seek to answer is this: given the estimated standard deviation, what's the probability of observing a sample mean (when $n = 44$) of 78.91 grams or less if the true mean is 80 grams? To answer this, you need to calculate the relevant test statistic.

Formally, the test statistic $T$ in a hypothesis test for a single mean with respect to a null value of $\mu_0$ is given as

$$T = \frac{\bar{x} - \mu_0}{(s/\sqrt{n})} \qquad (18.2)$$

based on a sample of size $n$, a sample mean of $\bar{x}$, and a sample standard deviation of $s$ (the denominator is the estimated standard error of the mean). Assuming the relevant conditions have been met, $T$ follows a $t$-distribution with $v = n - 1$ degrees of freedom.

In R, the following provides you with the standard error of the sample mean for the snacks data:

```
R> snack.se <- snack.sd/sqrt(n)
R> snack.se
[1] 0.4607128
```

Then, $T$ can be calculated as follows:

```
R> snack.T <- (snack.mean-80)/snack.se
R> snack.T
[1] -2.364419
```

Finally, the test statistic is used to obtain the $p$-value. Recall that the $p$-value is the probability that you observe $T$ or something more extreme. The nature of "more extreme" is determined by the alternative hypothesis $H_A$, which, as a less-than statement, directs you to find a left-hand, lower-tail probability as the $p$-value. In other words, the $p$-value is provided as the area under the sampling distribution (a $t$-distribution with 43 df in the current example) to the left of a vertical line at $T$. From Section 16.2.3, this is easily done, as shown here:

```
R> pt(snack.T,df=n-1)
[1] 0.01132175
```

Your result states that if the $H_0$ were true, there would be only a little more than a 1 percent chance that you'd observe the customer's sample mean of $\bar{x} = 78.91$, or less, as a random phenomenon. Since this $p$-value is smaller than the predefined significance level of $\alpha = 0.05$, the manufacturer concludes that there is sufficient evidence to reject the null hypothesis in favor of the alternative, suggesting the true value of $\mu$ is in fact less than 80 grams.

Note that if you find the corresponding 95 percent CI for the single sample mean, as described in Section 17.2.1 and given by

```
R> snack.mean+c(-1,1)*qt(0.975,n-1)*snack.se
[1] 77.98157 79.83980
```

it does *not* include the null value of 80, mirroring the result of the hypothesis test at the 0.05 level.

## R Function: t.test

The result of the one-sample *t*-test can also be found with the built-in t.test function.

```
R> t.test(x=snacks,mu=80,alternative="less")

        One Sample t-test

data:  snacks
t = -2.3644, df = 43, p-value = 0.01132
alternative hypothesis: true mean is less than 80
95 percent confidence interval:
     -Inf 79.68517
sample estimates:
mean of x
 78.91068
```

The function takes the raw data vector as x, the null value for the mean as mu, and the direction of the test (in other words, how to find the *p*-value under the appropriate *t*-curve) as alternative. The alternative argument has three available options: "less" for $H_A$ with <; "greater" for $H_A$ with >; and "two.sided" for $H_A$ with ≠. The default value of $\alpha$ is 0.05. If you want a different significance level than 0.05, this must be provided to t.test as $1 - \alpha$, passed to the argument conf.level.

Note that the value of $T$ is reported in the output of t.test, as are the degrees of freedom and the *p*-value. You also get a 95 percent "interval," but its values of -Inf and 79.68517 do not match the interval calculated just a moment ago. The manually calculated interval is in fact a two-sided interval—a bounded interval formed by using an error component that's equal on both sides.

The CI in the t.test output, on the other hand, takes instruction from the alternative argument. It provides a *one-sided confidence bound*. For a lower-tailed test, it provides an upper bound on the statistic such that the entire lower-tail area of the sampling distribution of interest is 0.95, as opposed to a *central* area as the traditional two-sided interval does. One-sided bounds are less frequently used than the fully bounded two-sided interval, which can be obtained (as the component conf.int) from a relevant call to t.test by setting alternative="two.sided".

```
R> t.test(x=snacks,mu=80,alternative="two.sided")$conf.int
[1] 77.98157 79.83980
attr(,"conf.level")
[1] 0.95
```

This result matches your manually computed version from earlier. Note also that the corresponding confidence level, $1 - \alpha$, is stored alongside this component as an attribute (refer to Section 6.2.1).

In examining the result for the snack example, with a *p*-value of around 0.011, remember to be careful when interpreting hypothesis tests. With $\alpha$ set at 0.05 for this particular test, $H_0$ is rejected. But what if the test were carried out with $\alpha = 0.01$? The *p*-value is greater than 0.01, so in that case, $H_0$ would be retained, for no reason other than the arbitrary movement of the value of $\alpha$. In these situations, it's helpful to comment on the perceived strength of the evidence against the null hypothesis. For the current example, you could reasonably state that there exists some evidence to support $H_A$ but that this evidence is not especially strong.

---

### Exercise 18.1

a. Adult domestic cats of a certain breed are said to have an average weight of 3.5 kilograms. A feline enthusiast disagrees and collects a sample of 73 weights of cats of this breed. From her sample, she calculates a mean of 3.97 kilograms and a standard deviation of 2.21 kilograms. Perform a hypothesis test to test her claim that the true mean weight $\mu$ is *not* 3.5 kilograms by setting up the appropriate hypothesis, carrying out the analysis, and interpreting the *p*-value (assume the significance level is $\alpha = 0.05$).

b. Suppose it was previously believed that the mean magnitude of seismic events off the coast of Fiji is 4.3 on the Richter scale. Use the data in the `mag` variable of the ready-to-use `quakes` data set, providing 1,000 sampled seismic events in that area, to test the claim that the true mean magnitude is in fact *greater* than 4.3. Set up appropriate hypotheses, use `t.test` (conduct the test at a significance level of $\alpha = 0.01$), and draw a conclusion.

c. Manually compute a two-sided confidence interval for the true mean of (b).

---

## 18.2.2    Two Means

Often, testing a single sample mean isn't enough to answer the question you're interested in. In many settings, a researcher wants to directly compare the means of two distinct groups of measurements, which boils down to a hypothesis test for the true difference between two means; call them $\mu_1$ and $\mu_2$.

The way in which two groups of data relate to each other affects the specific form of standard error for the difference between two sample means and therefore the test statistic itself. The actual comparison of the two means, however, is often of the same nature—the typical null hypothesis is usually defined as $\mu_1$ and $\mu_2$ being equal. In other words, the null value of the difference between the two means is often zero.

## Unpaired/Independent Samples: Unpooled Variances

The most general case is where the two sets of measurements are based on two independent, separate groups (also referred to as *unpaired* samples). You compute the sample means and sample standard deviations of both data sets, define the hypotheses of interest, and then calculate the test statistic.

When you cannot assume the variances of the two populations are equal, then you perform the *unpooled* version of the two-sample $t$-test; this will be discussed first. If, however, you can safely assume equal variances, then you can perform a *pooled* two-sample $t$-test, which improves the precision of the results. You'll look at the pooled version of the test in a moment.

For an unpooled example, return to the 80-gram snack packet example from Section 18.2.1. After collecting a sample of 44 packs from the original manufacturer (label this sample size $n_1$), the disgruntled consumer goes out and collects $n_2 = 31$ randomly selected 80-gram packs from a rival snack manufacturer. This second set of measurements is stored as snacks2.

```
R> snacks2 <- c(80.22,79.73,81.1,78.76,82.03,81.66,80.97,81.32,80.12,78.98,
                79.21,81.48,79.86,81.06,77.96,80.73,80.34,80.01,81.82,79.3,
                79.08,79.47,78.98,80.87,82.24,77.22,80.03,79.2,80.95,79.17,81)
```

From Section 18.2.1, you already know the mean and standard deviation of the first sample of size $n_1 = 44$—these are stored as snack.mean (around 78.91) and snack.sd (around 3.06), respectively—think of these as $\bar{x}_1$ and $s_1$. Compute the same quantities, $\bar{x}_2$ and $s_2$, respectively, for the new data.

```
R> snack2.mean <- mean(snacks2)
R> snack2.mean
[1] 80.1571
R> snack2.sd <- sd(snacks2)
R> snack2.sd
[1] 1.213695
```

Let the true mean of the original sample be denoted with $\mu_1$ and the true mean of the new sample from the rival company packs be denoted with $\mu_2$. You're now interested in testing whether there is statistical evidence to support the claim that $\mu_2$ is greater than $\mu_1$. This suggests the hypotheses of $H_0 : \mu_1 = \mu_2$ and $H_A : \mu_1 < \mu_2$, which can be written as follows:

$$H_0 : \mu_2 - \mu_1 = 0$$
$$H_A : \mu_2 - \mu_1 > 0$$

That is, the difference between the true mean of the rival company packs and the original manufacturer's packs, when the original is subtracted from the rival, is bigger than zero. The "no-change" scenario, the null hypothesis, is that the two means are the same, so their difference is truly zero.

Now that you've constructed the hypotheses, let's look at how to actually test them. The difference between two means is the quantity of interest. For two independent samples arising from populations with true means $\mu_1$ and $\mu_2$, sample means $\bar{x}_1$ and $\bar{x}_2$, and sample standard deviations $s_1$ and $s_2$, respectively (and that meet the relevant conditions for the validity of the $t$-distribution), the standardized test statistic $T$ for testing the difference between $\mu_2$ and $\mu_1$, in that order, is given as

$$T = \frac{\bar{x}_2 - \bar{x}_1 - \mu_0}{\sqrt{s_1^2/n_1 + s_2^2/n_2}}, \tag{18.3}$$

whose distribution is approximated by a $t$-distribution with $\nu$ degrees of freedom, where

$$\nu = \left\lfloor \frac{(s_1^2/n_1 + s_2^2/n_2)^2}{(s_1^2/n_1)^2/(n_1 - 1) + (s_2^2/n_2)^2/(n_2 - 1)} \right\rfloor \tag{18.4}$$

In (18.3), $\mu_0$ is the null value of interest—typically zero in tests concerned with "difference" statistics. This term would therefore disappear from the numerator of the test statistic. The denominator of $T$ is the standard error of the difference between two means in this setting.

The $\lfloor \cdot \rfloor$ on the right of Equation (18.4) denotes a *floor* operation—rounding strictly down to the nearest integer.

**NOTE**    *This two-sample* t-test, *conducted using Equation (18.3), is also called* Welch's t-test. *This refers to use of Equation (18.4), called the* Welch-Satterthwaite *equation. Crucially, it assumes that the two samples have different true variances, which is why it's called the* unpooled variance *version of the test.*

It's important to be consistent when defining your two sets of parameters and when constructing the hypotheses. In this example, since the test aims to find evidence for $\mu_2$ being greater than $\mu_1$, a difference of $\mu_2 - \mu_1 > 0$ forms $H_A$ (a greater-than, upper-tailed test), and this order of subtraction is mirrored when calculating $T$. The same test could be carried out if you defined the difference the other way around. In that case, your alternative hypothesis would suggest a lower-tailed test because if you're testing for $\mu_2$ being bigger than $\mu_1$, $H_A$ would correctly be written as $\mu_1 - \mu_2 < 0$. Again, this would modify the order of subtraction in the numerator of Equation (18.3) accordingly.

The same care must apply to the use of t.test for two-sample comparisons. The two samples must be supplied as the arguments x and y, but the function interprets x as greater than y when doing an upper-tailed test and interprets x as less than y when doing a lower-tailed test. Therefore, when performing the test with alternative="greater" for the snack pack example, it's snacks2 that must be supplied to x:

```
R> t.test(x=snacks2,y=snacks,alternative="greater",conf.level=0.9)

        Welch Two Sample t-test

data:  snacks2 and snacks
t = 2.4455, df = 60.091, p-value = 0.008706
alternative hypothesis: true difference in means is greater than 0
90 percent confidence interval:
 0.5859714      Inf
sample estimates:
mean of x mean of y
 80.15710  78.91068
```

With a small $p$-value of 0.008706, you'd conclude that there is sufficient evidence to reject $H_0$ in favor of $H_A$ (indeed, the $p$-value is certainly smaller than the stipulated $\alpha = 0.1$ significance level as implied by conf.level=0.9). The evidence suggests that the mean net weight of snacks from the rival manufacturer's 80-gram packs is greater than the mean net weight for the original manufacturer.

Note that the output from t.test has reported a df value of 60.091, which is the unfloored result of (18.4). You also receive a one-sided confidence bound (based on the aforementioned confidence level), triggered by the one-sided nature of this test. Again, the more common two-sided 90 percent interval is also useful; knowing that $v = \lfloor 60.091 \rfloor = 60$ and using the statistic and the standard error of interest (numerator and denominator of Equation (18.3), respectively), you can calculate it.

```
R> (snack2.mean-snack.mean) +
      c(-1,1)*qt(0.95,df=60)*sqrt(snack.sd^2/44+snack2.sd^2/31)
[1] 0.3949179 2.0979120
```

Here, you've used the previously stored sample statistics snack.mean, snack.sd (the mean and standard deviation of the 44 raw measurements from the original manufacturer's sample), snack2.mean, and snack2.sd (the same quantities for the 31 observations corresponding to the rival manufacturer). Note that the CI takes the same form as detailed by Equation (17.2) on page 378 and that to provide the correct $1 - \alpha$ central area, the q-function for the appropriate $t$-distribution requires $1 - \alpha/2$ as its supplied probability value. You can interpret this as being "90 percent confident that the true difference in mean net weight between the rival and the original manufacturer (in that order) is somewhere between 0.395 and 2.098 grams." The fact that zero isn't included in the interval, and that the interval is wholly positive, supports the conclusion from the hypothesis test.

## Unpaired/Independent Samples: Pooled Variance

In the unpooled variance example just passed, there was no assumption that the variances of the two populations whose means were being compared were equal. This is an important note to make because it leads to the use of (18.3) for the test statistic calculation and (18.4) for the associated degrees of freedom in the corresponding $t$-distribution. However, if you *can* assume equivalence of variances, the precision of the test is improved—you use a different formula for the standard error of the difference and for calculating the associated df.

Again, the quantity of interest is the difference between two means, written as $\mu_2 - \mu_1$. Assume you have two independent samples of sizes $n_1$ and $n_2$ arising from populations with true means $\mu_1$ and $\mu_2$, sample means $\bar{x}_1$ and $\bar{x}_2$, and sample standard deviations $s_1$ and $s_2$, respectively, and assume that the relevant conditions for the validity of the $t$-distribution have been met. Additionally, assume that the true variances of the samples, $\sigma_1^2$ and $\sigma_2^2$, are equal such that $\sigma_p^2 = \sigma_1^2 = \sigma_2^2$.

**NOTE** *There is a simple rule of thumb to check the validity of the "equal variance" assumption. If the ratio of the larger sample standard deviation to the smaller sample standard deviation is less than 2, you can assume equal variances. For example, if $s_1 > s_2$, then if $\frac{s_1}{s_2} < 2$, you can use the pooled variance test statistic that follows.*

The standardized test statistic $T$ for this scenario is given as

$$T = \frac{\bar{x}_2 - \bar{x}_1 - \mu_0}{\sqrt{s_p^2 \left(1/n_1 + 1/n_2\right)}}, \tag{18.5}$$

whose distribution is a $t$-distribution with $\nu = n_1 + n_2 - 2$ degrees of freedom, where

$$s_p^2 = \frac{(n_1 - 1)s_1^2 + (n_2 - 1)s_2^2}{n_1 + n_2 - 2} \tag{18.6}$$

is the *pooled estimate of the variance* of all the raw measurements. This is substituted in place of $s_1$ and $s_2$ in the denominator of Equation (18.3), resulting in Equation (18.5).

All other aspects of the two-sample $t$-test remain as earlier, including the construction of appropriate hypotheses, the typical null value of $\mu_0$, and the calculation and interpretation of the $p$-value.

For the comparison of the two means in the snack pack example, you'd find it difficult to justify using the pooled version of the $t$-test. Applying the rule of thumb, the two estimated standard deviations ($s_1 \approx 3.06$ and $s_2 \approx 1.21$ for the original and rival manufacturer's samples, respectively) have a large-to-small ratio that is greater than 2.

```
R> snack.sd/snack2.sd
[1] 2.51795
```

Though this is rather informal, if the assumption cannot reasonably be made, it's best to stick with the unpooled version of the test.

To illustrate this, let's consider a new example. The intelligence quotient (IQ) is a quantity commonly used to measure how clever a person is. IQ scores are reasonably assumed to be normally distributed, and the average IQ of the population is said to be 100. Say that you're interested in assessing whether there is a difference in mean IQ scores between men and women, suggesting the following hypotheses where you have $n_{\text{men}} = 12$ and $n_{\text{women}} = 20$:

$$H_0 : \mu_{\text{men}} - \mu_{\text{women}} = 0$$
$$H_A : \mu_{\text{men}} - \mu_{\text{women}} \neq 0$$

You randomly sample the following data:

```
R> men <- c(102,87,101,96,107,101,91,85,108,67,85,82)
R> women <- c(73,81,111,109,143,95,92,120,93,89,119,79,90,126,62,92,77,106,
        105,111)
```

As usual, let's calculate the basic statistics required.

```
R> mean(men)
[1] 92.66667
R> sd(men)
[1] 12.0705
R> mean(women)
[1] 98.65
R> sd(women)
[1] 19.94802
```

These give the sample averages $\bar{x}_{\text{men}}$ and $\bar{x}_{\text{women}}$, as well as their respective sample standard deviations $s_{\text{men}}$ and $s_{\text{women}}$. Enter the following to quickly check the ratio of the standard deviations:

```
R> sd(women)/sd(men)
[1] 1.652626
```

You can see that the ratio of the larger sample standard deviation to the smaller is less than 2, so you could assume equal variances in carrying out the hypothesis test.

The t.test command also enables the pooled two-sample $t$-test as per Equations (18.5) and (18.6). To execute it, you provide the optional argument var.equal=TRUE (as opposed to the default var.equal=FALSE, which triggers Welch's $t$-test).

```
R> t.test(x=men,y=women,alternative="two.sided",conf.level=0.95,var.equal=TRUE)

        Two Sample t-test
```

```
data:  men and women
t = -0.9376, df = 30, p-value = 0.3559
alternative hypothesis: true difference in means is not equal to 0
95 percent confidence interval:
 -19.016393   7.049727
sample estimates:
mean of x mean of y
 92.66667  98.65000
```

Note also that $H_A$ for this example implies a two-tailed test, hence the provision of alternative="two.sided".

The resulting $p$-value of this test, 0.3559, is certainly larger than the conventional cutoff level of 0.05. Thus, your conclusion here is that there is no evidence to reject $H_0$—there is insufficient evidence to support a true difference in the mean IQ scores of men compared to women.

### Paired/Dependent Samples

Finally, we'll look comparing two means in *paired* data. This setting is distinctly different from that of both unpaired $t$-tests because it concerns the way the data have been collected. The issue concerns *dependence* between pairs of observations across the two groups of interest—previously, the measurements in each group have been defined as independent. This notion has important consequences for how the test can be carried out.

Paired data occur if the measurements forming the two sets of observations are recorded on the same individual or if they are related in some other important or obvious way. A classic example of this is "before" and "after" observations, such as two measurements made on each person before and after some kind of intervention treatment. These situations still focus on the difference between the mean outcomes in each group, but rather than working with the two data sets separately, a paired $t$-test works with a single mean—the true mean of the individual paired differences $\mu_d$.

As an example, consider a company interested in the efficacy of a drug designed to reduce resting heart rates in beats per minute (bpm). The resting heart rates of 16 individuals are measured. The individuals are then administered a course of the treatment, and their resting heart rates are again measured. The data are provided in the two vectors rate.before and rate.after as follows:

```
R> rate.before <- c(52,66,89,87,89,72,66,65,49,62,70,52,75,63,65,61)
R> rate.after <- c(51,66,71,73,70,68,60,51,40,57,65,53,64,56,60,59)
```

It quickly becomes clear why any test comparing these two groups must take dependence into account. Heart rate is affected by an individual's age, build, and level of physical fitness. An unfit individual older than 60 is likely to have a higher baseline resting heart rate than a fit 20-year-old, and if both are given the same drug to lower their heart rate, their final heart rates are still likely to reflect the baselines. Any true effect of the drug therefore has the potential to be hidden if you approached the analysis using either of the unpaired *t*-tests.

To overcome this problem, the paired two-sample *t*-test considers the difference between each pair of values. Labeling one set of $n$ measurements as $x_1, \ldots, x_n$ and the other set of $n$ observations as $y_1, \ldots, y_n$, the difference, $d$, is defined as $d_i = y_i - x_i; i = 1, \ldots, n$. In R, you can easily compute the pairwise differences:

```
R> rate.d <- rate.after-rate.before
R> rate.d
 [1]  -1   0 -18 -14 -19  -4  -6 -14  -9  -5  -5   1 -11  -7  -5  -2
```

The following code calculates the sample mean $\bar{d}$ and standard deviation $s_d$ of these differences:

```
R> rate.dbar <- mean(rate.d)
R> rate.dbar
[1] -7.4375
R> rate.sd <- sd(rate.d)
R> rate.sd
[1] 6.196437
```

You want to see how much the heart rate is reduced by, so the test at hand will be concerned with the following hypotheses:

$$H_0 : \mu_d = 0$$
$$H_A : \mu_d < 0$$

Given the order or subtraction used to obtain the differences, detection of a successful reduction in heart rate will be represented by an "after" mean that is smaller than the "before" mean.

Expressing all this mathematically, the value of interest is the true mean difference, $\mu_d$, between two means of dependent pairs of measurements. There are two sets of $n$ measurements, $x_1, \ldots, x_n$ and $y_1, \ldots, y_n$, with pairwise differences $d_1, \ldots, d_n$. The relevant conditions for the validity of the *t*-distribution must be met; in this case, if the number of pairs $n$ is less than 30, then you must be able to assume the raw data are normally distributed. The test statistic $T$ is given as

$$T = \frac{\bar{d} - \mu_0}{s_d / \sqrt{n}}, \tag{18.7}$$

where $\bar{d}$ is the mean of the pairwise differences, $s_d$ is the sample standard deviation of the pairwise differences, and $\mu_0$ is the null value (usually zero). The statistic $T$ follows a *t*-distribution with $n - 1$ df.

The form of Equation (18.7) is actually the same as the form of the test statistic in (18.2), once the sample statistics for the individual paired differences have been calculated. Furthermore, it's important to note that *n* represents the total number of *pairs*, not the total number of individual observations.

For the current example hypotheses, you can find the test statistic and *p*-value with rate.dbar and rate.sd.

```
R> rate.T <- rate.dbar/(rate.sd/sqrt(16))
R> rate.T
[1] -4.801146
R> pt(rate.T,df=15)
[1] 0.000116681
```

These results suggest evidence to reject $H_0$. In t.test, the optional logical argument paired must be set to TRUE.

```
R> t.test(x=rate.after,y=rate.before,alternative="less",conf.level=0.95,
          paired=TRUE)


        Paired t-test

data:  rate.after and rate.before
t = -4.8011, df = 15, p-value = 0.0001167
alternative hypothesis: true difference in means is less than 0
95 percent confidence interval:
      -Inf -4.721833
sample estimates:
mean of the differences
              -7.4375
```

Note that the order you supply your data vectors to the x and y arguments follows the same rules as for the unpaired tests, given the desired value of alternative. The same *p*-value as was calculated manually is confirmed through the use of t.test, and since this is less than an assumed conventional significance level of $\alpha = 0.05$, a valid conclusion would be to state that there is statistical evidence that the medication does reduce the mean resting heart rate. You could go on to say you're 95 percent confident that the true mean difference in heart rate after taking the course of medication lies somewhere between

```
R> rate.dbar-qt(0.975,df=15)*(rate.sd/sqrt(16))
[1] -10.73935
```

and

```
R> rate.dbar+qt(0.975,df=15)*(rate.sd/sqrt(16))
[1] -4.135652
```

**NOTE**

*On some occasions, such as when your data strongly indicate non-normality, you may not be comfortable assuming the validity of the CLT (refer back to Section 17.1.1). An alternative approach to the tests discussed here is to employ a* nonparametric *technique that relaxes these distributional requirements. In the two-sample case, you could employ the* Mann-Whitney U test *(also known as the* Wilcoxon rank-sum test*). This is a hypothesis test that compares two medians, as opposed to two means. You can use the R function* `wilcox.test` *to access this methodology; its help page provides useful commentary and references on the particulars of the technique.*

---

### Exercise 18.2

In the package `MASS` you'll find the data set `anorexia`, which contains data on pre- and post-treatment weights (in pounds) of 72 young women suffering from the disease, obtained from Hand et al. (1994). One group of women is the control group (in other words, no intervention), and the other two groups are the cognitive behavioral program and family support intervention program groups. Load the library and ensure you can access the data frame and understand its contents. Let $\mu_d$ denote the mean difference in weight, computed as (*post-weight − pre-weight*).

a. Regardless of which treatment group the participants fall into, conduct and conclude an appropriate hypothesis test with $\alpha = 0.05$ for the entire set of weights for the following hypotheses:

$$H_0 : \mu_d = 0$$
$$H_A : \mu_d > 0$$

b. Next, conduct three separate hypothesis tests using the same defined hypotheses, based on which treatment group the participants fall into. What do you notice?

Another ready-to-use data set in R is `PlantGrowth` (Dobson, 1983), which records a continuous measure of the yields of a certain plant, looking at the potential effect of two supplements administered during growth to increase the yield when compared to a control group with no supplement.

c. Set up hypotheses to test whether the mean yield for the control group is less than the mean yield from a plant given either of the treatments. Determine whether this test should proceed using a pooled estimate of the variance or whether Welch's *t*-test would be more appropriate.

d. Conduct the test and make a conclusion (assuming normality of the raw observations).

As discussed, there is a rule of thumb for deciding whether to use a pooled estimate of the variance in an unpaired *t*-test.

e. Your task is to write a *wrapper* function that calls t.test after deciding whether it should be executed with var.equal=FALSE according to the rule of thumb. Use the following guidelines:
  – Your function should take four defined arguments: x and y with no defaults, to be treated in the same way as the same arguments in t.test; and var.equal and paired, with defaults that are the same as the defaults of t.test.
  – An ellipsis (Section 9.2.5) should be included to represent any additional arguments to be passed to t.test.
  – Upon execution, the function should determine whether paired=FALSE.
    * If paired is TRUE, then there is no need to proceed with the check of a pooled variance.
    * If paired is FALSE, then the function should determine the value for var.equal automatically by using the rule of thumb.
  – If the value of var.equal was set automatically, you can assume it will override any value of this argument initially supplied by the user.
  – Then, call t.test appropriately.

f. Try your new function on all three examples in the text of Section 18.2.2, ensuring you reach identical results.

## 18.3   Testing Proportions

A focus on means is especially common in statistical modeling and hypothesis testing, and therefore you must also consider sample proportions, interpreted as the mean of a series of $n$ binary trials, in which the results are success (1) or failure (0). This section focuses on the parametric tests of proportions, which assume normality of the target sampling distributions (otherwise referred to as Z-tests).

The general rules regarding the setup and interpretation of hypothesis tests for sample proportions remain the same as for sample means. In this introduction to Z-tests, you can consider these as tests regarding the true value of a single proportion or the difference between two proportions.

### 18.3.1   Single Proportion

Section 17.1.2 introduced the sampling distribution of a single sample proportion to be normally distributed, with a mean centered on the true proportion $\pi$ and with a standard error of $\sqrt{\pi(1-\pi)/n}$. Provided the trials are independent and that $n$ isn't "too small" and $\pi$ isn't "too close" to 0 or 1, those formulas are applicable here.

*A rule of thumb to check the latter condition on n and π simply involves checking that $n\hat{p}$ and $n(1 - \hat{p})$ are both greater than 5, where $\hat{p}$ is the sample estimate of π.*

It's worth noting that the standard error in the case of hypothesis tests involving proportions is itself dependent upon π. This is important—remember that any hypothesis test assumes satisfaction of $H_0$ in the relevant calculations. In dealing with proportions, that means when computing the test statistic, the standard error must make use of the null value $\pi_0$ rather than the estimated sample proportion $\hat{p}$.

I'll clarify this in an example. Suppose an individual fond of a particular fast-food chain notices that he tends to have an upset stomach within a certain amount of time after having his usual lunch. He comes across the website of a blogger who believes that the chance of getting an upset stomach shortly after eating that particular food is 20 percent. The individual is curious to determine whether his true rate of stomach upset π is any different from the blogger's quoted value and, over time, visits these fast-food outlets for lunch on $n = 29$ separate occasions, recording the success (TRUE) or failure (FALSE) of experiencing an upset stomach. This suggests the following pair of hypotheses:

$$H_0 : \pi = 0.2$$
$$H_A : \pi \neq 0.2$$

These may be tested according to the general rules discussed in the following sections.

### Calculation: One-Sample Z-Test

In testing for the true value of some proportion of success, π, let $\hat{p}$ be the sample proportion over $n$ trials, and let the null value be denoted with $\pi_0$. You find the test statistic with the following:

$$Z = \frac{\hat{p} - \pi_0}{\sqrt{\frac{\pi_0(1-\pi_0)}{n}}} \tag{18.8}$$

Provided the aforementioned conditions on the size of $n$ and the value of π can be assumed, $Z \sim N(0, 1)$.

The denominator of Equation (18.8), the standard error of the proportion, is calculated with respect to the null value $\pi_0$, not $\hat{p}$. As mentioned just a moment ago, this is to satisfy the assumption of "truth" of $H_0$ as the test is carried out, so it allows interpretation of the resulting *p*-value as usual. The standard normal distribution is used to find the *p*-value with respect to Z; the direction underneath this curve is governed by the nature of $H_A$ just as before.

Getting back to the fast-food example, suppose these are the observed data, where 1 is recorded for an upset stomach and is 0 otherwise.

```
sick <- c(0,0,1,1,0,0,0,0,0,1,0,0,0,0,0,0,0,0,1,0,0,0,1,1,1,0,0,0,1)
```

The number of successes and probability of success in this sample are as follows:

```
R> sum(sick)
[1] 8
R> p.hat <- mean(sick)
R> p.hat
[1] 0.2758621
```

A quick check indicates that as per the rule of thumb, the test is reasonable to carry out:

```
R> 29*0.2
[1] 5.8
R> 29*0.8
[1] 23.2
```

Following Equation (18.8), the test statistic $Z$ for this example is as follows:

```
R> Z <- (p.hat-0.2)/sqrt(0.2*0.8/29)
R> Z
[1] 1.021324
```

The alternative hypothesis is two-sided, so you compute the corresponding $p$-value as a two-tailed area under the standard normal curve. With a positive test statistic, this can be evaluated by doubling the upper-tailed area from $Z$.

```
R> 2*(1-pnorm(Z))
[1] 0.3071008
```

Assume a conventional $\alpha$ level of 0.05. The high $p$-value given as 0.307 suggests the results in the sample of size 29 are not unusual enough, under the assumption that the null hypothesis is true, to reject $H_0$. There is insufficient evidence to suggest that the proportion of instances of an upset stomach that this individual experiences is any different from 0.2 as noted by the blogger.

You can support this conclusion with a confidence interval. At the level of 95 percent, you calculate the CI:

```
R> p.hat+c(-1,1)*qnorm(0.975)*sqrt(p.hat*(1-p.hat)/29)
[1] 0.1131927 0.4385314
```

This interval easily includes the null value of 0.2.

### R Function: prop.test

Once more, R rescues you from tedious step-by-step calculation. The ready-to-use prop.test function allows you to perform, among other things, a single sample proportion test. The function actually performs the test in a slightly different way, using the chi-squared distribution (which will be explored more in Section 18.4). However, the test is equivalent, and the resulting *p*-value from prop.test is identical to the one reached using the Z-based test.

To the prop.test function, as used for a single sample test of a proportion, you provide the number of successes observed as x, the total number of trials as n, and the null value as p. The two further arguments, alternative (defining the nature of $H_A$) and conf.level (defining $1 - \alpha$), are identical to the same arguments in t.test and have defaults of "two.sided" and 0.95, respectively. Lastly, it is recommended to explicitly set the optional argument correct=FALSE if your data satisfy the $n\hat{p}$ and $n(1 - \hat{p})$ rule of thumb.

For the current example, you perform the test with this code:

```
R> prop.test(x=sum(sick),n=length(sick),p=0.2,correct=FALSE)

        1-sample proportions test without continuity correction

data:  sum(sick) out of length(sick), null probability 0.2
X-squared = 1.0431, df = 1, p-value = 0.3071
alternative hypothesis: true p is not equal to 0.2
95 percent confidence interval:
 0.1469876 0.4571713
sample estimates:
        p
0.2758621
```

The *p*-value is the same as you got earlier. Note, however, that the reported CI is not quite the same (the normal-based interval, dependent upon the CLT). The CI produced by prop.test is referred to as the *Wilson score interval*, which takes into account the direct association that a "probability of success" has with the binomial distribution. For simplicity, you'll continue to work with normal-based intervals when performing hypothesis tests involving proportions here.

Note also that, just like t.test, any one-sided test performed with prop.test will provide only a single-limit confidence bound; you'll see this in the following example.

## 18.3.2 Two Proportions

With a basic extension to the previous procedure, by way of a modification to the standard error, you can compare *two* estimated proportions from independent populations. As with the difference between two means, you're often testing whether the two proportions are the same and thus have a difference of zero. Therefore, the typical null value is zero.

For an example, consider a group of students taking a statistics exam. In this group are $n_1 = 233$ students majoring in psychology, of whom $x_1 = 180$ pass, and $n_2 = 197$ students majoring in geography, of whom 175 pass. Suppose it is claimed that the geography students have a higher pass rate in statistics than the psychology students.

Representing the true pass rates for psychology students as $\pi_1$ and geography students as $\pi_2$, this claim can be statistically tested using a pair of hypotheses defined as follows:

$$H_0 : \pi_2 - \pi_1 = 0$$
$$H_A : \pi_2 - \pi_1 > 0$$

Just as with a comparison between two means, it's important to keep the order of differencing consistent throughout the test calculations. This example shows an upper-tailed test.

### Calculation: Two-Sample Z-Test

In testing for the true difference between two proportions mathematically, $\pi_1$ and $\pi_2$, let $\hat{p}_1 = x_1/n_1$ be the sample proportion for $x_1$ successes in $n_1$ trials corresponding to $\pi_1$, and the same quantities as $\hat{p}_2 = x_2/n_2$ for $\pi_2$. With a null value of the difference denoted $\pi_0$, the test statistic is given by the following:

$$Z = \frac{\hat{p}_2 - \hat{p}_1 - \pi_0}{\sqrt{p^*(1 - p^*)\left(\frac{1}{n_1} + \frac{1}{n_2}\right)}} \tag{18.9}$$

Provided you can assume to apply the aforementioned conditions for a proportion with respect to $n_1$, $n_2$ and $\pi_1$, $\pi_2$, you can treat $Z \sim N(0,1)$.

There is a new quantity, $p^*$, present in the denominator of (18.9). This is a *pooled* proportion, given as follows:

$$p^* = \frac{x_1 + x_2}{n_1 + n_2} \tag{18.10}$$

As noted, in this kind of test it is common for the null value, the true difference in proportions, to be set to zero (in other words, $\pi_0 = 0$).

The denominator of Equation (18.9) is itself the standard error of the difference between two proportions as used in a hypothesis test. The need to use $p^*$ lies once more in the fact that $H_0$ is assumed to be true. Using $\hat{p}_1$ and $\hat{p}_2$ separately in the denominator of (18.9), in the form of $\sqrt{\hat{p}_1(1 - \hat{p}_1)/n_1 + \hat{p}_2(1 - \hat{p}_2)/n_2}$ (the standard error of the difference between two proportions outside the confines of a hypothesis test), would violate the assumed "truth" of $H_0$.

So, returning to the statistics exam taken by the psychology and geography students, you can evaluate the required quantities as such:

```
R> x1 <- 180
R> n1 <- 233
```

```
R> p.hat1 <- x1/n1
R> p.hat1
[1] 0.7725322
R> x2 <- 175
R> n2 <- 197
R> p.hat2 <- x2/n2
R> p.hat2
[1] 0.8883249
```

The results indicate sample pass rates of around 77.2 percent for the psychology students and 88.8 percent for the geography students; this is a difference of roughly 11.6 percent. From examining the values of $\hat{p}_1$, $n_1$ and $\hat{p}_2$, $n_2$, you can see that the rule of thumb is satisfied for this test; again, assume a standard significance level of $\alpha = 0.05$.

The pooled proportion $p^*$, following (18.10), is as follows:

```
R> p.star <- (x1+x2)/(n1+n2)
R> p.star
[1] 0.8255814
```

With that you calculate the test statistic $Z$ as per Equation (18.9) with the following:

```
R> Z <- (p.hat2-p.hat1)/sqrt(p.star*(1-p.star)*(1/n1+1/n2))
R> Z
[1] 3.152693
```

In light of the hypotheses, you find the corresponding $p$-value as a right-hand, upper-tail area from $Z$ underneath the standard normal curve as follows:

```
R> 1-pnorm(Z)
[1] 0.0008088606
```

You observe a $p$-value that's substantially smaller than $\alpha$, so the formal decision is of course to reject the null hypothesis in favor of the alternative. The sample data provide sufficient evidence against $H_0$ such that you can conclude that evidence exists to support the pass rate for geography students being higher than the pass rate for psychology students.

### R Function: prop.test

Once more, R allows you to perform the test with one line of code using prop.test. For comparisons of two proportions, you pass the number of successes in each group as a vector of length 2 to x and the respective sample sizes as another vector of length 2 to n. Note that the order of the entries must reflect the order of alternative if this is one-sided (in other words, here, the proportion that is to be tested as "greater" corresponds to the first elements of x and n). Once more, correct is set to FALSE.

```
R> prop.test(x=c(x2,x1),n=c(n2,n1),alternative="greater",correct=FALSE)

        2-sample test for equality of proportions without continuity correction

data:  c(x2, x1) out of c(n2, n1)
X-squared = 9.9395, df = 1, p-value = 0.0008089
alternative hypothesis: greater
95 percent confidence interval:
 0.05745804 1.00000000
sample estimates:
   prop 1    prop 2
0.8883249 0.7725322
```

The $p$-value is identical to the one generated by the previous series of calculations, suggesting a rejection of $H_0$. Since prop.test was called as a one-sided test, the confidence interval returned provides a single bound. To provide a two-sided CI for the true difference, it makes sense, considering the outcome of the test, to construct this using the separate $\hat{p}_1$ and $\hat{p}_2$ instead of using the denominator of (18.9) specifically (which assumes truth of $H_0$). The "separate-estimate" version of the standard error of the difference between two proportions was given earlier (in the text beneath Equation (18.10)), and a 95 percent CI is therefore calculated with the following:

```
R> (p.hat2-p.hat1) +
     c(-1,1)*qnorm(0.975)*sqrt(p.hat1*(1-p.hat1)/n1+p.hat2*(1-p.hat2)/n2)
[1] 0.04628267 0.18530270
```

With that, you're 95 percent confident that the true difference between the proportion of geography students passing the exam and the proportion of psychology students passing the exam lies somewhere between 0.046 and 0.185. Naturally, the interval also reflects the result of the hypothesis test—it doesn't include the null value of zero and is wholly positive.

## Exercise 18.3

An advertisement for a skin cream claims nine out of ten women who use it would recommend it to a friend. A skeptical salesperson in a department store believes the true proportion of women users who'd recommend it, $\pi$, is much smaller than 0.9. She follows up with 89 random customers who had purchased the skin cream and asks if they would recommend it to others, to which 71 answer yes.

a. Set up an appropriate pair of hypotheses for this test and determine whether it will be valid to carry out using the normal distribution.

b. Compute the test statistic and the *p*-value and state your conclusion for the test using a significance level of $\alpha = 0.1$.

c. Using your estimated sample proportion, construct a two-sided 90 percent confidence interval for the true proportion of women who would recommend the skin cream.

The political leaders of a particular country are curious as to the proportion of citizens in two of its states that support the decriminalization of marijuana. A small pilot survey taken by officials reveals that 97 out of 445 randomly sampled voting-age citizens residing in state 1 support the decriminalization and that 90 out of 419 voting-age citizens residing in state 2 support the same notion.

d. Letting $\pi_1$ denote the true proportion of citizens in support of decriminalization in state 1, and $\pi_2$ the same measure in state 2, conduct and conclude a hypothesis test under a significance level of $\alpha = 0.05$ with reference to the following hypotheses:

$$H_0 : \pi_2 - \pi_1 = 0$$
$$H_A : \pi_2 - \pi_1 \neq 0$$

e. Compute and interpret a corresponding CI.

Though there is standard, ready-to-use R functionality for the *t*-test, at the time of this writing, there is no similar function for the Z-test (in other words, the normal-based test of proportions described here) except in contributed packages.

f. Your task is to write a relatively simple R function, `Z.test`, that can perform a one- or two-sample Z-test, using the following guidelines:
   - The function should take the following arguments: `p1` and `n1` (no default) to pose as the estimated proportion and sample size; `p2` and `n2` (both defaulting to `NULL`) that contain the second sample proportion and sample size in the event of a two-sample test; `p0` (no default) as the null value; and `alternative` (default `"two.sided"`) and `conf.level` (default `0.95`), to be used in the same way as in `t.test`.
   - When conducting a two-sample test, it should be `p1` that is tested as being smaller or larger than `p2` when `alternative="less"` or `alternative="greater"`, the same as in the use of x and y in `t.test`.
   - The function should perform a one-sample Z-test using `p1`, `n1`, and `p0` if either `p2` or `n2` (or both) is `NULL`.

- The function should contain a check for the rule of thumb to ensure the validity of the normal distribution in both one- and two-sample settings. If this is violated, the function should still complete but should issue an appropriate warning message (see Section 12.1.1).
- All that need be returned is a list containing the members Z (test statistic), P (appropriate *p*-value—this can be determined by alternative; for a two-sided test, determining whether Z is positive or not can help), and CI (two-sided CI with respect to conf.level).

g. Replicate the two examples in the text of Sections 18.3.1 and 18.3.2 using Z.test; ensure you reach identical results.

h. Call Z.test(p1=0.11,n1=10,p0=0.1) to try your warning message in the one-sample setting.

## 18.4 Testing Categorical Variables

The normal-based *Z*-test is particular to data that are binary in nature. To statistically test claims regarding more general categorical variables, with more than two distinct levels, you use the ubiquitous *chi-squared test*. Pronounced *kai*, "chi" refers to the Greek symbol $\chi$ and is sometimes noted in shorthand as the $\chi^2$ *test*.

There are two common variants of the chi-squared test. The first—a chi-squared test of distribution, also called a *goodness of fit (GOF)* test—is used when assessing the frequencies in the levels of a single categorical variable. The second—a chi-squared test of *independence*—is employed when you're investigating the relationship between frequencies in the levels of two such variables.

### 18.4.1 Single Categorical Variable

Like the *Z*-test, the one-dimensional chi-squared test is also concerned with comparing proportions but in a setting where there are more than two proportions. A chi-squared test is used when you have *k* levels (or categories) of a categorical variable and want to hypothesize about their relative frequencies to find out what proportion of *n* observations fall into each defined category. In the following examples, it must be assumed that the categories are *mutually exclusive* (in other words, an observation cannot take more than one of the possible categories) and *exhaustive* (in other words, the *k* categories cover all possible outcomes).

I'll illustrate how hypotheses are constructed and introduce the relevant ideas and methods with the following example. Suppose a researcher in sociology is interested in the dispersion of rates of facial hair in men of his local city and whether they are uniformly represented in the male population. He defines a categorical variable with three levels: clean shaven (1), beard only *or* moustache only (2), and beard *and* moustache (3). He collects data on 53 randomly selected men and finds the following outcomes:

```
R> hairy <- c(2,3,2,3,2,1,3,3,3,2,2,3,2,2,2,3,3,3,3,2,3,2,2,2,1,3,2,2,2,1,2,2,3,
              2,2,2,2,1,2,1,1,1,2,2,2,3,1,2,1,2,1,2,1,3,3)
```

Now, the research question asks whether the proportions in each category are equally represented. Let $\pi_1$, $\pi_2$, and $\pi_3$ represent the true proportion of men in the city who fall into groups 1, 2, and 3, respectively. You therefore seek to test these hypotheses:

$$H_0 : \pi_1 = \pi_2 = \pi_3 = \frac{1}{3}$$

$$H_A : H_0 \text{ is incorrect}$$

For this test, use a standard significance level of 0.05.

The appearance of the alternative hypothesis is a little different from what you've seen so far but is an accurate reflection of the interpretation of a chi-squared goodness of fit test. In these types of problems, $H_0$ is always that the proportions in each group are equal to the stated values, and $H_A$ is that the data, as a whole, do not match the proportions defined in the null. The test is conducted assuming the null hypothesis is true, and evidence against the no-change, baseline setting will be represented as a small *p*-value.

### Calculation: Chi-Squared Test of Distribution

The quantities of interest are the proportion of $n$ observations in each of $k$ categories, $\pi_1, \ldots, \pi_k$, for a single mutually exclusive and exhaustive categorical variable. The null hypothesis defines hypothesized null values for each proportion; label these respectively as $\pi_{0(1)}, \ldots, \pi_{0(k)}$. The test statistic $\chi^2$ is given as

$$\chi^2 = \sum_{i=1}^{k} \frac{(O_i - E_i)^2}{E_i}, \tag{18.11}$$

where $O_i$ is the *observed* count and $E_i$ is the *expected* count in the $i$th category; $i = 1, \ldots, k$. The $O_i$ are obtained directly from the raw data, and the expected counts, $E_i = n\pi_{0(i)}$, are merely the product of the overall sample size $n$ with the respective null proportion for each category. The result of $\chi^2$ follows a *chi-squared distribution* (explained further momentarily) with $\nu = k - 1$ degrees of freedom. You usually consider the test to be valid based on an informal rule of thumb stating that at least 80 percent of the expected counts $E_i$ should be at least 5.

In this type of chi-squared test, it is important to note the following:

- The term *goodness of fit* refers to the proximity of the observed data to the distribution hypothesized in $H_0$.

- Positive extremity of the result of (18.11) provides evidence against $H_0$. As such, the corresponding *p-value is always computed as an upper-tail area.*

- As in the current example, a test for uniformity simplifies the null hypothesis slightly by having equivalent null proportions $\pi_0 = \pi_{0(1)} = \ldots = \pi_{0(k)}$.

- A rejected $H_0$ doesn't tell you about the true values of $\pi_i$. It merely suggests that they do not follow $H_0$ specifically.

The chi-squared distribution relies on specification of a degree of freedom, much like the *t*-distribution. Unlike a *t* curve, however, a chi-squared curve is unidirectional in nature, being defined for non-negative values and with a positive (right-hand) horizontal asymptote (tail going to zero).

It's this unidirectional distribution that leads to *p*-values being defined as upper-tail areas only; decisions like one- or two-tailed areas have no relevance in these types of chi-squared tests. To get an idea of what the density functions actually look like, Figure 18-1 shows three particular curves defined with $v = 1$, $v = 5$, and $v = 10$ degrees of freedom.

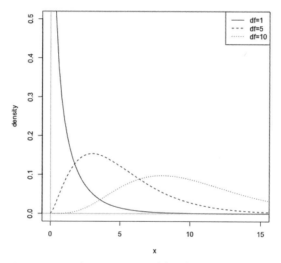

*Figure 18-1: Three instances of the chi-squared density function using differing degrees of freedom values. Note the positive domain of the function and the "flattening" and "right-extending" behavior as v is increased.*

This image was produced using the relevant d-function, dchisq, with $v$ passed to the argument df.

```
R> x <- seq(0,20,length=100)
R> plot(x,dchisq(x,df=1),type="l",xlim=c(0,15),ylim=c(0,0.5),ylab="density")
R> lines(x,dchisq(x,df=5),lty=2)
R> lines(x,dchisq(x,df=10),lty=3)
R> abline(h=0,col="gray")
```

```
R> abline(v=0,col="gray")
R> legend("topright",legend=c("df=1","df=5","df=10"),lty=1:3)
```

The current facial hair example is a test for the uniformity of the distribution of frequencies in the three categories. You can obtain the observed counts and corresponding proportions with table.

```
R> n <- length(hairy)
R> n
[1] 53
R> hairy.tab <- table(hairy)
R> hairy.tab
hairy
 1  2  3
11 28 14
R> hairy.tab/n
hairy
        1         2         3
0.2075472 0.5283019 0.2641509
```

For computation of the test statistic $\chi^2$, you have the observed counts $O_i$ in hairy.tab. The expected count $E_i$ is a straightforward arithmetic calculation of the total number of observations multiplied by the null proportion 1/3 (the result stored as expected), giving you the same value for each category.

These, as well as the contribution of each category to the test statistic, are nicely presented in a matrix constructed with cbind (Section 3.1.2).

```
R> expected <- 1/3*n
R> expected
[1] 17.66667
R> hairy.matrix <- cbind(1:3,hairy.tab,expected,
                         (hairy.tab-expected)^2/expected)
R> dimnames(hairy.matrix) <- list(c("clean","beard OR mous.",
                                    "beard AND mous."),
                                  c("i","Oi","Ei","(Oi-Ei)^2/Ei"))
R> hairy.matrix
                i Oi       Ei (Oi-Ei)^2/Ei
clean           1 11 17.66667    2.5157233
beard OR mous.  2 28 17.66667    6.0440252
beard AND mous. 3 14 17.66667    0.7610063
```

Note that all the expected counts are comfortably greater than 5, which satisfies the informal rule of thumb mentioned earlier. In terms of R coding, note also that the single number expected is implicitly recycled to match the length of the other vectors supplied to cbind and that you've used the dimnames attribute (refer to Section 6.2.1) to annotate the rows and columns.

The test statistic, as per (18.11), is given as the sum of the $(O_i - E_i)^2/E_i$ contributions in the fourth column of hairy.matrix.

```
R> X2 <- sum(hairy.matrix[,4])
R> X2
[1] 9.320755
```

The corresponding $p$-value is the appropriate upper-tail area from the chi-squared distribution with $v = 3 - 1 = 2$ degrees of freedom.

```
R> 1-pchisq(X2,df=2)
[1] 0.009462891
```

This small $p$-value provides evidence to suggest that the true frequencies in the defined categories of male facial hair are not uniformly distributed in a $1/3, 1/3, 1/3$ fashion. Remember that the test result doesn't give you the true proportions but only suggests that they do not follow those in $H_0$.

### R Function: chisq.test

Like t.test and prop.test, R provides a quick-use function for performing a chi-squared GOF test. The chisq.test function takes the vector of observed frequencies as its first argument x. For the facial hair example, this simple line therefore provides the same results as found previously:

```
R> chisq.test(x=hairy.tab)

        Chi-squared test for given probabilities

data:  hairy.tab
X-squared = 9.3208, df = 2, p-value = 0.009463
```

By default, the function performs a test for uniformity, taking the number of categories as the length of the vector supplied to x. However, suppose that the researcher collecting the facial hair data realizes that he was doing so in November, a month during which many men grow mustaches in support of "Mo-vember" to raise awareness of men's health. This changes thoughts on the true rates in terms of his clean-shaven (1), beard-only *or* moustache-only (2), and beard *and* moustache (3) categories. He now wants to test the following:

$$H_0 : \pi_{0(1)} = 0.25; \pi_{0(2)} = 0.5; \pi_{0(3)} = 0.25$$
$$H_A : H_0 \text{ is incorrect.}$$

If a GOF test of uniformity is not desired, when the "true" rates across the categories are not all the same, the chisq.test function requires you to supply the null proportions as a vector of the same length as x to the p argument. Naturally, each entry in p must correspond to the categories tabulated in x.

```
R> chisq.test(x=hairy.tab,p=c(0.25,0.5,0.25))

        Chi-squared test for given probabilities

data:  hairy.tab
X-squared = 0.5094, df = 2, p-value = 0.7751
```

With a very high *p*-value, there is no evidence to reject $H_0$ in this scenario. In other words, there is no evidence to suggest that the proportions hypothesized in $H_0$ are incorrect.

### 18.4.2   Two Categorical Variables

The chi-squared test can also apply to the situation in which you have *two* mutually exclusive and exhaustive categorical variables at hand—call them variable *A* and variable *B*. It is used to detect whether there might be some influential relationship (in other words, *dependence*) between *A* and *B* by looking at the way in which the distribution of frequencies change together with respect to their categories. If there is no relationship, the distribution of frequencies in variable *A* will have nothing to do with the distribution of frequencies in variable *B*. As such, this particular variant of the chi-squared test is called a *test of independence* and is always performed with the following hypotheses:

> $H_0$ : Variables *A* and *B* are independent.
>
> (*or*, There is no relationship between *A* and *B*.)
>
> $H_A$ : Variables *A* and *B* are *not* independent.
>
> (*or*, There *is* a relationship between *A* and *B*.)

To carry out the test, therefore, you compare the observed data to the counts you'd expect to see if the distributions were completely unrelated (satisfying the assumption that $H_0$ is true). An overall large departure from the expected frequencies will result in a small *p*-value and thus provide evidence against the null.

So, how are such data best presented? For two categorical variables, a two-dimensional structure is appropriate; in R, this is a standard matrix. For example, suppose some dermatologists at a certain clinical practice are interested in their successes in treating a common skin affliction. Their records show $N = 355$ patients have been treated at their clinic using one of four possible treatments—a course of tablets, a series of injections, a laser treatment, and an herbal-based remedy. The level of success in curing the affliction is also recorded—none, partial success, and full success. The data are given in the constructed matrix skin.

```
R> skin <- matrix(c(20,32,8,52,9,72,8,32,16,64,30,12),4,3,
                  dimnames=list(c("Injection","Tablet","Laser","Herbal"),
                  c("None","Partial","Full")))
```

```
R> skin
          None Partial Full
Injection   20       9   16
Tablet      32      72   64
Laser        8       8   30
Herbal      52      32   12
```

A two-dimensional table presenting frequencies in this fashion is called a *contingency table*.

### Calculation: Chi-Squared Test of Independence

To compute the test statistic, presume data are presented as a $k_r \times k_c$ contingency table, in other words, a matrix of counts, based on two categorical variables (both mutually exclusive and exhaustive). The focus of the test is the way in which the frequencies of $N$ observations between the $k_r$ levels of the "row" variable and the $k_c$ levels of the "column" variable are jointly distributed. The test statistic $\chi^2$ is given with

$$\chi^2 = \sum_{i=1}^{k_r} \sum_{j=1}^{k_c} \frac{(O_{[i,j]} - E_{[i,j]})^2}{E_{[i,j]}}, \tag{18.12}$$

where $O_{[i,j]}$ is the observed count and $E_{[i,j]}$ is the expected count at row position $i$ and column position $j$. Each $E_{[i,j]}$ is found as the sum total of row $i$ multiplied by the sum total of column $j$, all divided by $N$.

$$E_{[i,j]} = \frac{\left(\sum_{u=1}^{k_r} O_{[u,j]}\right) \times \left(\sum_{v=1}^{k_c} O_{[i,v]}\right)}{N} \tag{18.13}$$

The result, $\chi^2$, follows a chi-squared distribution with $v = (k_r - 1) \times (k_c - 1)$ degrees of freedom. Again, the *p*-value is always an upper-tailed area, and you can consider the test valid with the satisfaction of the condition that at least 80 percent of the $E_{[i,j]}$ are at least 5.

For this calculation, it's important to note the following:

- It's not necessary to assume that $k_r = k_c$.

- The functionality of Equation (18.12) is the same as that of (18.11)—an overall sum involving the squared differences between the observed and expected values of each cell.

- The double-sum in (18.12) just represents the total sum over all the cells, in the sense that you can compute the total sample size $N$ with $\sum_{i=1}^{k_r} \sum_{j=1}^{k_c} O_{[i,j]}$.

- A rejected $H_0$ doesn't tell you about the nature of *how* the frequencies depend on one another, just that there is evidence to suggest that some kind of dependency between the two categorical variables exists.

Continuing with the example, the dermatologists want to determine whether their records suggest there is statistical evidence to indicate some relationship between the type of treatment and the level of success in curing the skin affliction. For convenience, store the total number of categories $k_r$ and $k_c$ for the row and column variables, respectively.

```
R> kr <- nrow(skin)
R> kc <- ncol(skin)
```

You have the $O_{[i,j]}$ in skin, so now you must now compute the $E_{[i,j]}$. In light of Equation (18.13), which deals with row and column sums, you can evaluate these using the built-in rowSums and colSums functions.

```
R> rowSums(skin)
Injection    Tablet     Laser    Herbal
       45       168        46        96
R> colSums(skin)
  None Partial     Full
   112     121      122
```

These results indicate the totals in each group, regardless of the other variable. To get the expected counts for all cells of the matrix, Equation (18.13) requires each row sum to be multiplied by each column sum once. You could write a for loop, but this would be inefficient and rather inelegant. It is better to use rep with the optional each argument (refer to Section 2.3.2). By repeating each element of the column totals (level of success) four times, you can then use vector-oriented behavior to multiply that repeated vector by the shorter vector produced by rowSums. You can then call sum(skin) to divide this by $N$ and rearrange it into a matrix. The following lines show how this example works step-by-step:

```
R> rep(colSums(skin),each=kr)
    None     None     None     None Partial Partial Partial Partial     Full
     112      112      112      112     121     121     121     121      122
    Full     Full     Full
     122      122      122
R> rep(colSums(skin),each=kr)*rowSums(skin)
    None     None     None     None Partial Partial Partial Partial     Full
    5040    18816     5152    10752    5445   20328    5566   11616     5490
    Full     Full     Full
   20496     5612    11712
R> rep(colSums(skin),each=kr)*rowSums(skin)/sum(skin)
     None     None     None     None  Partial  Partial  Partial  Partial
 14.19718 53.00282 14.51268 30.28732 15.33803 57.26197 15.67887 32.72113
     Full     Full     Full     Full
 15.46479 57.73521 15.80845 32.99155
R> skin.expected <- matrix(rep(colSums(skin),each=kr)*rowSums(skin)/sum(skin),
                    nrow=kr,ncol=kc,dimnames=dimnames(skin))
```

```
R> skin.expected
            None  Partial     Full
Injection 14.19718 15.33803 15.46479
Tablet    53.00282 57.26197 57.73521
Laser     14.51268 15.67887 15.80845
Herbal    30.28732 32.72113 32.99155
```

Note that all the expected values are greater than 5, as preferred.

It's best to construct a single object to hold the results of the different stages of calculations leading to the test statistic, as you did for the one-dimensional example. Since each stage is a matrix, you can bind the relevant matrices together with cbind and produce an array of the appropriate dimensions (refer to Section 3.4 for a refresher).

```
R> skin.array <- array(data=cbind(skin,skin.expected,
                                (skin-skin.expected)^2/skin.expected),
                    dim=c(kr,kc,3),
                    dimnames=list(dimnames(skin)[[1]],dimnames(skin)[[2]],
                            c("O[i,j]","E[i,j]",
                                "(O[i,j]-E[i,j])^2/E[i,j]")))
R> skin.array
, , O[i,j]

          None Partial Full
Injection   20       9   16
Tablet      32      72   64
Laser        8       8   30
Herbal      52      32   12

, , E[i,j]

            None  Partial     Full
Injection 14.19718 15.33803 15.46479
Tablet    53.00282 57.26197 57.73521
Laser     14.51268 15.67887 15.80845
Herbal    30.28732 32.72113 32.99155

, , (O[i,j]-E[i,j])^2/E[i,j]

             None   Partial       Full
Injection 2.371786 2.6190199  0.01852279
Tablet    8.322545 3.7932587  0.67978582
Laser     2.922614 3.7607992 12.74002590
Herbal   15.565598 0.0158926 13.35630339
```

The final steps are easy—the test statistic given by (18.12) is just the grand total of all elements of the matrix that is the third layer of skin.array.

```
R> X2 <- sum(skin.array[,,3])
R> X2
[1] 66.16615
```

The corresponding *p*-value for this test of independence is as follows:

```
R> 1-pchisq(X2,df=(kr-1)*(kc-1))
[1] 2.492451e-12
```

Recall that the relevant degrees of freedom are defined as $v = (k_r - 1) \times (k_c - 1)$.

The extremely small *p*-value provides strong evidence against the null hypothesis. The appropriate conclusion would be to reject $H_0$ and state that there does appear to be a relationship between the type of treatment for the skin affliction and the level of success in curing it.

### R Function: chisq.test

Yet once more, no section in this chapter would be complete without showcasing the built-in functionality R possesses for these fundamental procedures. The default behavior of chisq.test, when supplied a matrix as x, is to perform a chi-squared test of independence with respect to the row and column frequencies—just as performed manually here for the skin affliction example. The following result easily confirms your previous calculations:

```
R> chisq.test(x=skin)

        Pearson's Chi-squared test

data:  skin
X-squared = 66.1662, df = 6, p-value = 2.492e-12
```

## Exercise 18.4

HairEyeColor is a ready-to-use data set in R that you haven't yet come across. This $4 \times 4 \times 2$ array provides frequencies of hair and eye colors of 592 statistics students, split by sex (Snee, 1974).

a.  Perform and interpret, at a significance level of $\alpha = 0.01$, a chi-squared test of independence for hair against eye color for all students, regardless of their sex.

In Exercise 8.1 on page 161, you accessed the Duncan data set of the contributed package car, which contains markers of job prestige collected in 1950. Install the package if you haven't already and load the data frame.

b. The first column of Duncan is the variable type, recording the type of job as a factor with three levels: prof (professional or managerial), bc (blue collar), and wc (white collar). Construct appropriate hypotheses and perform a chi-squared GOF test to determine whether the three job types are equally represented in the data set.

    i. Interpret the resulting $p$-value with respect to a significance level of $\alpha = 0.05$.

    ii. What conclusion would you reach if you used a significance level of $\alpha = 0.01$?

## 18.5   Errors and Power

In discussing all these forms of statistical hypothesis testing, there has been one common thread: the interpretation of a $p$-value and what it tells you about your problem in terms of the hypotheses. Frequentist statistical hypothesis testing is ubiquitous in many fields of research, so it is important to at least briefly explore directly related concepts.

### 18.5.1   Hypothesis Test Errors

Hypothesis testing is performed with the objective of obtaining a $p$-value in order to quantify evidence against the null statement $H_0$. This is rejected in favor of the alternative, $H_A$, if the $p$-value is itself less than a predefined significance level $\alpha$, which is conventionally 0.05 or 0.01. As touched upon, this approach is justifiably criticized since the choice of $\alpha$ is essentially arbitrary; a decision to reject or retain $H_0$ can change depending solely upon the $\alpha$ value.

Consider for the moment, given a specific test, what the *correct* outcome is. If $H_0$ is really true, then you'd want to retain it. If $H_A$ is really true, you'd want to reject the null. This "truth," one way or another, is impossible to know in practice. That being said, it's useful to consider in a theoretical sense just how good (or bad) a given hypothesis test is at yielding a result that leads to the correct conclusion.

To be able to test the validity of your rejection or retention of the null hypothesis, you must be able to identify two kinds of errors:

- A *Type I error* occurs when you incorrectly reject a true $H_0$. In any given hypothesis test, the probability of a Type I error is equivalent to the significance level $\alpha$.

- A *Type II error* occurs when you incorrectly retain a false $H_0$ (in other words, fail to accept a true $H_A$). Since this depends upon what the true $H_A$ actually is, the probability of committing such an error, labeled $\beta$, is not usually known in practice.

## 18.5.2 Type I Errors

If your *p*-value is less than $\alpha$, you reject the null statement. If the null is really true, though, the $\alpha$ directly defines the probability that you *incorrectly* reject it. This is referred to as a *Type I error*.

Figure 18-2 provides a conceptual illustration of a Type I error probability for a supposed hypothesis test of a sample mean, where the hypotheses are set up as $H_0 : \mu = \mu_0$ and $H_A : \mu > \mu_0$.

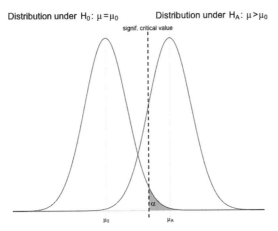

Figure 18-2: A conceptual diagram of the Type I error probability $\alpha$

The null hypothesis distribution is centered on the null value $\mu_0$; the alternative hypothesis distribution is centered to its right at some mean $\mu_A$ in Figure 18-2. As you can see, if the null hypothesis is really true, then the probability it is incorrectly rejected for this test will be equal to the significance level $\alpha$, located in the upper tail of the null distribution.

### Simulating Type I Errors

To demonstrate the Type I error rate via *numerical simulation* (here, this refers to randomly generating hypothetical data samples), you can write code that does the equivalent of repeating a hypothesis test under known conditions. So that you can use this code multiple times, in the R script editor define the following function:

```
typeI.tester <- function(mu0,sigma,n,alpha,ITERATIONS=10000){
    pvals <- rep(NA,ITERATIONS)
    for(i in 1:ITERATIONS){
        temporary.sample <- rnorm(n=n,mean=mu0,sd=sigma)
```

```
        temporary.mean <- mean(temporary.sample)
        temporary.sd <- sd(temporary.sample)
        pvals[i] <- 1-pt((temporary.mean-mu0)/(temporary.sd/sqrt(n)),df=n-1)
    }
    return(mean(pvals<alpha))
}
```

The typeI.tester function is designed to generate ITERATIONS samples from a particular normal distribution. With each sample, you'll perform an upper-tailed test of the mean (refer to Section 18.2.1) in the spirit of Figure 18-2, assuming the hypotheses of $H_0 : \mu = \mu_0$ and $H_A : \mu > \mu_0$.

You can decrease ITERATIONS to generate fewer entire samples, and this will speed up computation time but will result in simulated rates that are more variable. Each entire sample of size n of hypothetical raw measurements is generated using rnorm with the mean equal to the mu0 argument (and standard deviation equal to the sigma argument). The desired significance level is set by alpha. In the for loop, the sample mean and sample standard deviation are calculated for each generated sample.

Were each sample subjected to a "real" hypothesis test, the *p*-value would be taken from the right-hand area of the *t*-distribution with n-1 degrees of freedom (using pt), with respect to the standardized test statistic given earlier in Equation (18.2).

The calculated *p*-value, at each iteration, is stored in a predefined vector pvals. The logical vector pvals<alpha therefore contains corresponding TRUE/FALSE values; the former logical value flags rejection of the null hypothesis, and the latter flags retention. The Type I error rate is determined by calling mean on that logical vector, which yields the proportion of TRUEs (in other words, the overall proportion of "null hypothesis rejections") arising from the simulated samples. Remember, the samples are generated randomly, so your results are liable to change slightly each time you run the function.

This function works because, by definition of the problem, the samples that are being generated come from a distribution that truly has the mean set at the null value, in other words, $\mu_A = \mu_0$. Therefore, any statistical rejection of this statement, obtained with a *p*-value less than the significance level $\alpha$, is clearly incorrect and is purely a result of random variation.

To try this, import the function and execute it generating the default ITERATIONS=10000 samples. Use the standard normal as the null (and "true" in this case!) distribution; make each sample of size 40 and set the significance level at the conventional $\alpha = 0.05$. Here's an example:

```
R> typeI.tester(mu0=0,sigma=1,n=40,alpha=0.05)
[1] 0.0489
```

This indicates that $10{,}000 \times 0.0489 = 489$ of the samples taken yielded a corresponding test statistic that provided a $p$-value, which would incorrectly result in rejection of $H_0$. This simulated Type I error rate lies close to the preset alpha=0.05.

Here's another example, this time for nonstandard normal data samples with $\alpha = 0.01$:

```
R> typeI.tester(mu0=-4,sigma=0.3,n=60,alpha=0.01)
[1] 0.0108
```

Note that again, the numerically simulated rate of Type I error reflects the significance level.

These results are not difficult to understand theoretically—if the true distribution does indeed have a mean equal to the null value, you'll naturally observe those "extreme" test statistic values in practice at a rate equal to $\alpha$. The catch, of course, is that in practice the true distribution is unknown, highlighting once more the fact that a rejection of any $H_0$ can never be interpreted as proof of the truth of $H_A$. It might simply be that the sample you observed followed the null hypothesis but produced an extreme test statistic value by chance, however small that chance might be.

## Bonferroni Correction

The fact that Type I errors naturally occur because of random variation is particularly important and leads us to consider the *multiple testing problem*. If you're conducting many hypothesis tests, you should be cautious in simply reporting the "number of statistically significant outcomes"—as you increase the number of hypothesis tests, you increase the chance of receiving an erroneous result. In, say, 20 tests conducted under $\alpha = 0.05$, on average one will be a so-called false positive; if you conduct 40 or 60 tests, you are inevitably more likely to find more false positives.

When several hypothesis tests are conducted, you can curb the multiple testing problem with respect to committing a Type I error by using the *Bonferroni correction*. The Bonferroni correction suggests that when performing a total of $N$ independent hypothesis tests, each under a significance level of $\alpha$, you should instead use $\alpha_B = \alpha/N$ for any interpretation of statistical significance. Be aware, however, that this correction to the level of significance represents the simplest solution to the multiple testing problem and can be criticized for its conservative nature, which is potentially problematic when $N$ is large.

The Bonferroni and other corrective measures were developed in an attempt to formalize remedies to making a Type I error in multiple tests. In general, though, it suffices to be aware of the possibility that $H_0$ may be true, even if the $p$-value is considered small.

### 18.5.3    Type II Errors

The issues with Type I errors might suggest that it's desirable to perform a hypothesis test with a smaller $\alpha$ value. Unfortunately, it's not quite so simple; reducing the significance level for any given test leads directly to an increase in the chance of committing a Type II error.

A Type II error refers to incorrect retention of the null hypothesis—in other words, obtaining a $p$-value greater than the significance level when it's the alternative hypothesis that's actually true. For the same scenario you've been looking at so far (an upper-tailed test for a single sample mean), Figure 18-3 illustrates the probability of a Type II error, shaded and denoted $\beta$.

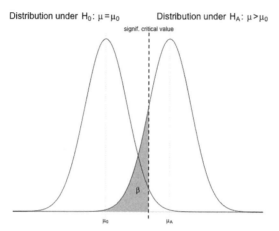

*Figure 18-3: A conceptual diagram of the Type II error probability $\beta$*

It's not as easy to find $\beta$ as it is to find the probability of making a Type I error because $\beta$ depends, among other things, on what the true value of $\mu_A$ is (which in general you won't know). If $\mu_A$ is closer to the hypothesized null value of $\mu_0$, you can imagine the alternative distribution in Figure 18-3 translating (shifting) to the left, resulting in an increase in $\beta$. Similarly, staying with Figure 18-3, imagine decreasing the significance level $\alpha$. Doing so means the vertical dashed line (denoting the corresponding critical value) moves to the right, also increasing the shaded area of $\beta$. Intuitively, this makes sense—the closer the true alternative value is to the null and/or the smaller the significance level, the harder $H_A$ is to detect by rejection of $H_0$.

As noted, $\beta$ usually can't be calculated in practice because of the need to know what the true distribution actually is. This quantity is, however, useful in giving you an idea of how prone a test is to the incorrect retention of a null hypothesis under particular conditions. Suppose, for example, you're performing a one-sample $t$-test for $H_0 : \mu = \mu_0$ and $H_A : \mu > \mu_0$ with $\mu_0 = 0$ but that the (true) alternative distribution of the raw measurements has mean $\mu_A = 0.5$ and standard deviation $\sigma = 1$. Given a random sample of size $n = 30$ and using $\alpha = 0.05$, what is the probability of committing a Type II error in any given hypothesis test (using the same standard deviation for the null distribution)? To answer this, look again at Figure 18-3; you need the critical value marked off by the significance level (the dashed vertical line). If you assume $\sigma$ is known, then the sampling distribution of interest will be normal with mean $\mu_0 = 0$ and a standard error of $1/\sqrt{30}$ (see Section 17.1.1). Therefore, with an upper-tail area of 0.05, you can find the critical value with the following:

```
R> critval <- qnorm(1-0.05,mean=0,sd=1/sqrt(30))
R> critval
[1] 0.3003078
```

This represents the vertical dashed line in this specific setting (see Section 16.2.2 for a refresher on use of qnorm). The Type II error in this example is found as the left-hand tail area under the alternative, "true" distribution, from that critical value:

```
R> pnorm(critval,mean=0.5,sd=1/sqrt(30))
[1] 0.1370303
```

From this, you can see that a hypothesis test under these conditions has roughly a 13.7 percent chance of incorrect retention of the null.

## Simulating Type II Errors

Simulation is especially useful here. In the editor, consider the function typeII.tester defined as follows:

```
typeII.tester <- function(mu0,muA,sigma,n,alpha,ITERATIONS=10000){
    pvals <- rep(NA,ITERATIONS)
    for(i in 1:ITERATIONS){
        temporary.sample <- rnorm(n=n,mean=muA,sd=sigma)
        temporary.mean <- mean(temporary.sample)
        temporary.sd <- sd(temporary.sample)
        pvals[i] <- 1-pt((temporary.mean-mu0)/(temporary.sd/sqrt(n)),df=n-1)
    }
    return(mean(pvals>=alpha))
}
```

This function is similar to typeI.tester. The null value, standard deviation of raw measurements, sample size, significance level, and number of iterations are all as before. Additionally, you now have muA, providing the "true" mean $\mu_A$ under which to generate the samples. Again, at each iteration, a random sample of size n is generated, its mean and standard deviation are calculated, and the appropriate *p*-value for the test is computed using pt from the usual standardized test statistic with df=n-1. (Remember, since you're estimating the true standard deviation of the measurements $\sigma$ with the sample standard deviation $s$, it's technically correct to use the *t*-distribution.) Following completion of the for loop, the proportion of *p*-values that were greater than or equal to the significance level alpha is returned.

After importing the function into the workspace, you can simulate $\beta$ for this test.

```
R> typeII.tester(mu0=0,muA=0.5,sigma=1,n=30,alpha=0.05)
[1] 0.1471
```

My result indicates something close to the theoretical $\beta$ evaluated previously, albeit slightly larger because of the additional uncertainty that is naturally present when using a *t*-based sampling distribution instead of a normal. Again, each time you run typeII.tester, the results will vary slightly since everything is based on randomly generated hypothetical data samples.

Turning your attention to Figure 18-3, you can see (in line with a comment made earlier) that if, in an effort to decrease the chance of a Type I error, you use $\alpha = 0.01$ instead of 0.05, the vertical line moves to the right, thereby increasing the probability of a Type II error, with all other conditions being held constant.

```
R> typeII.tester(mu0=0,muA=0.5,sigma=1,n=30,alpha=0.01)
[1] 0.3891
```

### Other Influences on the Type II Error Rate

The significance level isn't the only contributing factor in driving $\beta$. Keeping $\alpha$ at 0.01, this time see what happens if the standard deviation of the raw measurements is increased from $\sigma = 1$ to $\sigma = 1.1$ and then $\sigma = 1.2$.

```
R> typeII.tester(mu0=0,muA=0.5,sigma=1.1,n=30,alpha=0.01)
[1] 0.4815
R> typeII.tester(mu0=0,muA=0.5,sigma=1.2,n=30,alpha=0.01)
[1] 0.5501
```

Increasing the variability of the measurements, without touching anything else in the scenario, also increases the chance of a Type II error. You can imagine the curves in Figure 18-3 becoming flatter and more widely dispersed owing to a larger standard error of the mean, which would result in more probability weight in the left-hand tail marked off by the critical value. Conversely, if the variability of the raw measurements is smaller, then the sampling distributions of the sample mean will be taller and skinnier, meaning a reduction in $\beta$.

A smaller or larger sample size will have a similar impact. Located in the denominator of the standard error formula, a smaller *n* will result in a larger standard error and hence that flatter curve and an increased $\beta$; a larger sample size will have the opposite effect. If you remain with the latest values of $\mu_0 = 0$, $\mu_A = 0.5$, $\sigma = 1.2$, and $\alpha = 0.01$, note that reducing the sample size to 20 (from 30) results in an increased simulated Type II error rate compared with the most recent result of 0.5501, but increasing the sample size to 40 improves the rate.

```
R> typeII.tester(mu0=0,muA=0.5,sigma=1.2,n=20,alpha=0.01)
[1] 0.7319
R> typeII.tester(mu0=0,muA=0.5,sigma=1.2,n=40,alpha=0.01)
[1] 0.4219
```

Finally, as noted at the beginning of the discussion, the specific value of $\mu_A$ itself affects $\beta$ just as you'd expect. Again, keeping the latest values for all other components, which resulted in my case in $\beta = 0.4219$, note that shifting the "true" mean closer to $\mu_0$ by changing from $\mu_A = 0.5$ to $\mu_A = 0.4$ means the probability of committing a Type II error is increased; the opposite is true if the difference is increased to $\mu_A = 0.6$.

```
R> typeII.tester(mu0=0,muA=0.4,sigma=1.2,n=40,alpha=0.01)
[1] 0.6147
R> typeII.tester(mu0=0,muA=0.6,sigma=1.2,n=40,alpha=0.01)
[1] 0.2287
```

To summarize, although these simulated rates have been applied to the specific situation in which the hypothesis test is an upper-tailed test for a single mean, the general concepts and ideas discussed here hold for any hypothesis test. It's easy to establish that the Type I error rate matches the predefined significance level and so can be decreased by reducing $\alpha$. In contrast, controlling the Type II error rate is a complex balancing act that can involve sample size, significance level, observation variability, and magnitude of the difference between the true value and the null. This problem is largely academic since the "truth" is typically unknown in practice. However, the Type II error rate's direct relationship to statistical power often plays a critical role in preparing for data collection, especially when you're considering sample size requirements, as you'll see in the next section.

### Exercise 18.5

a. Write a new version of typeI.tester called typeI.mean. The new function should be able to simulate the Type I error rate for tests of a single mean in any direction (in other words, one- or two-sided). The new function should take an additional argument, test, which takes a character string "less", "greater", or "two.sided" depending on the type of desired test. You can achieve this by modifying typeI.tester as follows:
   – Instead of calculating and storing the *p*-values directly in the for loop, simply store the test statistic.

- When the loop is complete, set up stacked if-else statements that cater to each of the three types of test, calculating the *p*-value as appropriate.
- For the two-sided test, remember that the *p*-value is defined as twice the area "more extreme" than the null. Computationally, this means you must use the upper-tail area if the test statistic is positive and the lower-tail area otherwise. If this area is less than half of $\alpha$ (since it is subsequently multiplied by 2 in a "real" hypothesis test), then a rejection of the null should be flagged.
- If the value of test is not one of the three possibilities, the function should throw an appropriate error using stop.

i. Experiment with your function using the first example setting in the text with $\mu_0 = 0$, $\sigma = 1$, $n = 40$, and $\alpha = 0.05$. Call typeI.mean three times, using each of the three possible options for test. You should find that all simulated results sit close to 0.05.

ii. Repeat (i) using the second example setting in the text with $\mu_0 = -4$, $\sigma = 0.3$, $n = 60$, and $\alpha = 0.01$. Again, you should find that all simulated results sit close to the value of $\alpha$.

b. Modify typeII.tester in the same way as you did typeI.tester; call the new function typeII.mean. Simulate the Type II error rates for the following hypothesis tests. As per the text, assume $\mu_A$, $\sigma$, $\alpha$, and $n$ denote the true mean, standard deviation of raw observations, significance level, and sample size, respectively.

i. $H_0 : \mu = -3.2$;  $H_A : \mu \neq -3.2$
with $\mu_A = -3.3$, $\sigma = 0.1$, $\alpha = 0.05$, and $n = 25$.

ii. $H_0 : \mu = 8994$;  $H_A : \mu < 8994$
with $\mu_A = 5600$, $\sigma = 3888$, $\alpha = 0.01$, and $n = 9$.

iii. $H_0 : \mu = 0.44$;  $H_A : \mu > 0.44$
with $\mu_A = 0.4$, $\sigma = 2.4$, $\alpha = 0.05$, and $n = 68$.

### 18.5.4  Statistical Power

For any hypothesis test, it is useful to consider its potential statistical power. *Power* is the probability of correctly rejecting a null hypothesis that is untrue. For a test that has a Type II error rate of $\beta$, the statistical power is found simply with $1 - \beta$. It's desirable for a test to have a power that's as high as possible. The simple relationship with the Type II error probability means that all factors impacting the value of $\beta$ also directly affect power.

For the same one-sided $H_0 : \mu = \mu_0$ and $H_A : \mu > \mu_0$ example discussed in the previous section, Figure 18-4 shades the power of the test—the complement to the Type II error rate. By convention, a hypothesis test that has a power greater than 0.8 is considered *statistically powerful*.

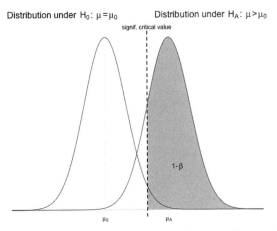

Distribution under $H_0$: $\mu = \mu_0$     Distribution under $H_A$: $\mu > \mu_0$

signif. critical value

$1-\beta$

$\mu_0$          $\mu_A$

*Figure 18-4: A conceptual diagram of statistical power $1 - \beta$*

You can numerically evaluate power under specific testing conditions using simulation. For the previous discussion on Type II errors, you're able to subtract all simulated results of $\beta$ from 1 to evaluate the power of that particular test. For example, the power of detection of $\mu_A = 0.6$ when $\mu_0 = 0$, taking samples of size $n = 40$ and with $\sigma = 1.2$ and $\alpha = 0.01$, is simulated as $1 - 0.2287 = 0.7713$ (using my most recent result of $\beta$ from earlier). This means there's roughly a 77 percent chance of correctly detecting the true mean of 0.6 in a hypothesis test based on a sample of measurements generated under those conditions.

Researchers are often interested in the relationship between power and sample size (though it is important to bear in mind that this is only one of the influential ingredients in the determination of power). Before you begin to collect data to examine a particular hypothesis, you might have an idea of the potential true value of the parameter of interest from past research or pilot studies. This is useful in helping to determine your sample size, such as in helping to answer questions like "How big does my sample need to be in order to be able to conduct a statistically powerful test to correctly reject $H_0$, if the true mean is actually $\mu_A$?"

### Simulating Power

For the most recent testing conditions, with a sample size of $n = 40$, you've seen that there's a power of around 0.77 of detecting $\mu_A = 0.6$. For the purposes of this example, let's say you want to find how much you should increase $n$ by in order to conduct a statistically powerful test. To answer this, define the following function power.tester in the editor:

```
power.tester <- function(nvec,...){
    nlen <- length(nvec)
    result <- rep(NA,nlen)
    pbar <- txtProgressBar(min=0,max=nlen,style=3)
    for(i in 1:nlen){
```

```
                result[i] <- 1-typeII.tester(n=nvec[i],...)
                setTxtProgressBar(pbar,i)
        }
        close(pbar)
        return(result)
}
```

The `power.tester` function uses the `typeII.tester` function defined in Section 18.5.3 to evaluate the power of a given upper-tailed hypothesis test of a single sample mean. It takes a vector of sample sizes supplied as the `nvec` argument (you pass all other arguments to `typeII.tester` using an ellipsis—refer to Section 11.2.4). A for loop defined in `power.tester` cycles through the entries of `nvec` one at a time, simulates the power for each sample size, and stores them in a corresponding vector that's returned to the user. Remember, through `typeII.tester`, this function is using random generation of hypothetical data samples, so there may be some fluctuation in the results you observe each time you run `power.tester`.

There can be a slight delay when evaluating the power for many individual sample sizes, so this function also provides a good opportunity to showcase a progress bar in a practical implementation (refer to Section 12.2.1 for details).

Set up the following vector, which uses the colon operator (see Section 2.3.2) to construct a sequence of integers between 5 and 100 inclusive for the sample sizes to be examined:

```
R> sample.sizes <- 5:100
```

Importing the `power.tester` function, you can then simulate the power for each of these integers for this particular test (`ITERATIONS` is halved to 5000 to reduce the overall completion time).

```
R> pow <- power.tester(nvec=sample.sizes,
                   mu0=0,muA=0.6,sigma=1.2,alpha=0.01,ITERATIONS=5000)
 |==================================================================| 100%
R> pow
 [1] 0.0630 0.0752 0.1018 0.1226 0.1432 0.1588 0.1834 0.2162 0.2440 0.2638
[11] 0.2904 0.3122 0.3278 0.3504 0.3664 0.3976 0.4232 0.4478 0.4680 0.4920
[21] 0.5258 0.5452 0.5552 0.5616 0.5916 0.6174 0.6326 0.6438 0.6638 0.6844
[31] 0.6910 0.7058 0.7288 0.7412 0.7552 0.7718 0.7792 0.7950 0.8050 0.8078
[41] 0.8148 0.8316 0.8480 0.8524 0.8600 0.8702 0.8724 0.8800 0.8968 0.8942
[51] 0.8976 0.9086 0.9116 0.9234 0.9188 0.9288 0.9320 0.9378 0.9370 0.9448
[61] 0.9436 0.9510 0.9534 0.9580 0.9552 0.9648 0.9656 0.9658 0.9684 0.9756
[71] 0.9742 0.9770 0.9774 0.9804 0.9806 0.9804 0.9806 0.9854 0.9848 0.9844
[81] 0.9864 0.9886 0.9890 0.9884 0.9910 0.9894 0.9906 0.9930 0.9926 0.9938
[91] 0.9930 0.9946 0.9948 0.9942 0.9942 0.9956
```

As expected, the power of detection rises steadily as $n$ increases; the conventional cutoff of 80 percent is visible in these results as lying between

0.7950 and 0.8050. If you don't want to identify the value visually, you can find which entry of `sample.sizes` corresponds to the 80 percent cutoff by first using `which` to identify the indexes of `pow` that are at least 0.8 and then returning the lowest value in that category with `min`. The identified index may then be specified in square brackets to `sample.sizes` to give you the value of *n* that corresponds to that simulated power (0.8050 in this case). These commands can be nested as follows:

```
R> minimum.n <- sample.sizes[min(which(pow>=0.8))]
R> minimum.n
[1] 43
```

The result indicates that if your sample size is at least 43, a hypothesis test under these particular conditions should be statistically powerful (based on the randomly simulated output in `pow` in this instance).

What if the significance level for this test were relaxed? Say you wanted to conduct the test (still upper-tailed under the condition of $\mu_0 = 0$, $\mu_A = 0.6$, and $\sigma = 1.2$) using a significance level of $\alpha = 0.05$ rather than 0.01. If you look again at Figure 18-4, this alteration means the vertical line (critical value) moves to the left, decreasing $\beta$ and so increasing power. That would therefore suggest you'd require a smaller sample size than earlier, in other words, $n < 43$, in order to perform a statistically powerful test when $\alpha$ is increased.

To simulate this situation for the same range of sample sizes and store the resulting powers in `pow2`, examine the following:

```
R> pow2 <- power.tester(nvec=sample.sizes,
               muO=0,muA=0.6,sigma=1.2,alpha=0.05,ITERATIONS=5000)
  |==================================================================| 100%
R> minimum.n2 <- sample.sizes[min(which(pow2>0.8))]
R> minimum.n2
[1] 27
```

This result indicates a sample size of at least 27 is required, which is a noticeable reduction from the 43 noted if $\alpha = 0.01$. However, relaxing $\alpha$ means an increased risk of committing a Type I error!

### Power Curves

For comparison, you can plot your simulated powers as a kind of power curve using both `pow` and `pow2` with the following code:

```
R> plot(sample.sizes,pow,xlab="sample size n",ylab="simulated power")
R> points(sample.sizes,pow2,col="gray")
R> abline(h=0.8,lty=2)
R> abline(v=c(minimum.n,minimum.n2),lty=3,col=c("black","gray"))
R> legend("bottomright",legend=c("alpha=0.01","alpha=0.05"),
          col=c("black","gray"),pch=1)
```

My particular image is given in Figure 18-5. A horizontal line marks off the power of 0.8, and the vertical line marks the minimum sample size values identified and stored in `minimum.n` and `minimum.n2`. As a final touch, a legend is added to reference the $\alpha$ values of each curve.

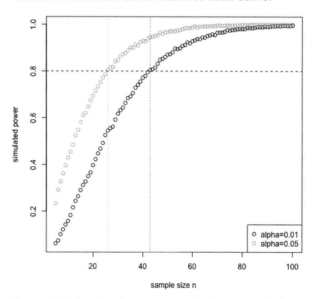

Figure 18-5: Simulated power curves for the upper-tailed hypothesis test of a single sample mean

The curves themselves indicate exactly what you'd expect—the power of detection increases as the sample size is incremented. You can also note the flattening off occurring as the power rises ever closer to the "perfect" rate of 1, which is typical of a power curve. For $\alpha = 0.05$, the curve sits almost consistently above the curve for $\alpha = 0.01$, though the difference looks negligible as $n$ rises above 75 or so.

The preceding discussion of errors and power highlights the need for care in interpreting the results of even the most basic of statistical tests. A $p$-value is merely a probability, and as such, no matter how small it may be in any circumstance, it can never prove or disprove a claim on its own. Issues surrounding the quality of a hypothesis test (parametric or otherwise) should be considered, though this is arguably difficult in practice. Nevertheless, an awareness of Type I and Type II errors, as well as the concept of statistical power, is extremely useful in the implementation and appraisal of any formal statistical testing procedure.

## Exercise 18.6

a. For this exercise you'll need to have written typeII.mean from Exercise 18.5 (b). Using this function, modify power.tester so that a new function, power.mean, calls typeII.mean instead of calling typeII.tester.

   i. Confirm that the power of the test given by $H_0 : \mu = 10$; $H_A : \mu \neq 10$, with $\mu_A = 10.5$, $\sigma = 0.9$, $\alpha = 0.01$, and $n = 50$, is roughly 88 percent.

   ii. Remember the hypothesis test in Section 18.2.1 for the mean net weight of an 80-gram pack of snacks, based on the $n = 44$ observations provided in the snack vector. The hypotheses were as follows:

   $$H_0 : \mu = 80$$
   $$H_A : \mu < 80$$

   If the true mean is $\mu_A = 78.5$ g and the true standard deviation of the weights is $\sigma = 3.1$ g, use power.mean to determine whether the test is statistically powerful, assuming $\alpha = 0.05$. Does your answer to this change if $\alpha = 0.01$?

b. Staying with the snacks hypothesis test, using the sample.sizes vector from the text, determine the minimum sample size required for a statistically powerful test using both $\alpha = 0.05$ and $\alpha = 0.01$. Produce a plot showing the two power curves.

## Important Code in This Chapter

| Function/operator | Brief description | First occurrence |
|---|---|---|
| t.test | One- and two-sample *t*-test | Section 18.2.1, p. 391 |
| prop.test | One- and two-sample Z-test | Section 18.3.1, p. 405 |
| pchisq | $\chi^2$ cumulative problems | Section 18.4.1, p. 414 |
| chisq.test | $\chi^2$ test of distribution/independence | Section 18.4.1, p. 414 |
| rowSums | Matrix row totals | Section 18.4.2, p. 417 |
| colSums | Matrix column totals | Section 18.4.2, p. 417 |

# 19

## ANALYSIS OF VARIANCE

*Analysis of variance (ANOVA),* in its simplest form, is used to compare multiple means in a test for equivalence. In that sense, it's a straightforward extension of the hypothesis test comparing two means. There's a continuous variable from which the means of interest are calculated, and there's at least one categorical variable that tells you how to define the groups for those means. In this chapter, you'll explore the ideas surrounding ANOVA and look at comparing means first split by one categorical variable (one-way ANOVA) and then split by multiple categorical variables (multiple-factor ANOVA).

## 19.1 One-Way ANOVA

The simplest version of ANOVA is referred to as *one-way* or *one-factor* analysis. Simply put, the one-way ANOVA is used to test two or more means for equality. Those means are split by a categorical *group* or *factor* variable. ANOVA is often used to analyze experimental data to assess the impact of an intervention. You might, for example, be interested in comparing the mean weights of the chicks in the built-in `chickwts` data set, split according to the different food types they were fed.

### 19.1.1  Hypotheses and Diagnostic Checking

Say you have a categorical-nominal variable that splits a total of $N$ numeric observations into $k$ distinct groups, where $k \geq 2$. You're looking to statistically compare the $k$ groups' means, $\mu_1, \ldots, \mu_k$, to see whether they can be claimed to be equal. The standard hypotheses are as follows:

$$H_0 : \mu_1 = \mu_2 = \ldots = \mu_k$$

$$H_A : \mu_1, \mu_2, \ldots, \mu_k \text{ are not all equal}$$

(alternatively, at least one mean differs).

In fact, when $k = 2$, the two-sample $t$-test is equivalent to ANOVA; for that reason, ANOVA is most frequently employed when $k > 2$.

The following assumptions need to be satisfied in order for the results of the basic one-way ANOVA test to be considered reliable:

**Independence**  The samples making up the $k$ groups must be independent of one another, and the observations in each group must be independent and identically distributed (iid).

**Normality**  The observations in each group should be normally distributed, or at least approximately so.

**Equality of variances**  The variance of the observations in each group should be equal, or at least approximately so.

If the assumptions of equality of variances or normality are violated, it doesn't necessarily mean your results will be completely worthless, but it will impact the overall effectiveness of detecting a true difference in the means (refer to the discussion on statistical power in Section 18.5.4). It's always a good idea to assess the validity of these assumptions before using ANOVA; I'll do this informally for the upcoming example.

It's also worth noting that you don't need to have an equal number of observations in each group to perform the test (in which case it is referred to as *unbalanced*). However, having unbalanced groups does render the test more sensitive to potentially detrimental effects if your assumptions of equality of variances and normality are not sound.

Let's return to the chickwts data for the example—the weights of chicks based on $k = 6$ different feeds. You're interested in comparing the mean weights according to feed type to see whether they're all equal. Use table to summarize the six sample sizes and use tapply (see, for example, Section 13.2.1) to get each group mean, as follows:

```
R> table(chickwts$feed)

  casein horsebean   linseed  meatmeal   soybean sunflower
      12        10        12        11        14        12
R> chick.means <- tapply(chickwts$weight,INDEX=chickwts$feed,FUN=mean)
R> chick.means
  casein horsebean   linseed  meatmeal   soybean sunflower
323.5833  160.2000  218.7500  276.9091  246.4286  328.9167
```

Your skills from Section 14.3.2 allow you to produce side-by-side boxplots of the distributions of weights. The next two lines give you the plot on the left of Figure 19-1:

```
R> boxplot(chickwts$weight~chickwts$feed)
R> points(1:6,chick.means,pch=4,cex=1.5)
```

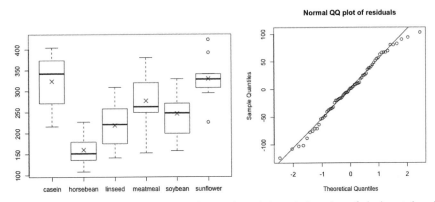

Figure 19-1: Exploring the chickwts data. Left: Side-by-side boxplots of chick weight split by feed type, with the mean marked by ×. Right: Normal QQ plot of the mean-centered data of each feed group.

Because boxplots display the median, not the mean, the second line of code adds the feed-specific means (stored in the chick.means object you just created) to each box using points.

Inspecting the left plot of Figure 19-1, it certainly looks as though there's a difference in the mean weights. Is any apparent difference statistically significant, though? To find out, the ANOVA test for this example concerns the following hypotheses:

$H_0 : \mu_{casein} = \mu_{horsebean} = \mu_{linseed} = \mu_{meatmeal} = \mu_{soybean} = \mu_{sunflower}$

$H_A :$ The means are not all equal.

Assuming independence of the data, before implementing the test, you must first check that the other assumptions are valid. To examine equality of variances, you can use the same informal rule of thumb as used in the two-sample $t$-test. That is, you can assume equality of variances if the ratio of the largest sample standard deviation to the smallest is less than 2. For the chick weights data, the following code will determine this:

```
R> chick.sds <- tapply(chickwts$weight,INDEX=chickwts$feed,FUN=sd)
R> chick.sds
   casein horsebean    linseed   meatmeal    soybean sunflower
 64.43384  38.62584   52.23570   64.90062   54.12907  48.83638
R> max(chick.sds)/min(chick.sds)
[1] 1.680238
```

This informal result indicates that it's reasonable to make the assumption.

Next, consider the assumption of normality of the raw observations. This can be difficult to determine in many real-data examples. At the least, though, it's worthwhile to inspect histograms and QQ plots for signs of non-normality. You already inspected histograms and QQ plots for all 71 weights in Section 16.2.2, but for an ANOVA, you need to do this with respect to the grouping of the observations (that is, not just "overall" for the whole set of weights regardless of groups).

To achieve this for the chickwts data, you need to first *mean-center* each weight by its respective sample mean. You can do this by taking the original vector of weights and subtracting from it the chick.means vector, but first you must rearrange and replicate the latter elements to correspond to the elements in the former. This is done by using as.numeric on the factor vector that represents feed type, giving the numeric value of the levels of chickwts$feed for each record in the original data frame. When that numeric vector is passed via the square brackets to chick.means, you get the correct group mean matched to each observation. As an exercise, you can inspect all the ingredients that go into creating the following chick.meancen object to satisfy yourself of what's going on:

```
R> chick.meancen <- chickwts$weight-chick.means[as.numeric(chickwts$feed)]
```

In the context of the current analysis, these group-wise, mean-centered values are also referred to as *residuals*, a term you'll come across frequently when you study regression methods in the next few chapters.

You can now assess normality of the observations as a whole using the residuals. To inspect a normal QQ plot, the relevant functions are qqnorm and qqline, which you first met in Section 16.2.2. The following two lines produce the image on the right of Figure 19-1.

```
R> qqnorm(chick.meancen,main="Normal QQ plot of residuals")
R> qqline(chick.meancen)
```

Based on this plot (the proximity of the plotted points to the perfect straight line), it doesn't seem unreasonable to assume normality for these data, particularly when compared to QQ plots of generated normal data of the same sample size (an example was given on the left of Figure 16-9 on page 355).

Investigating the validity of any required assumptions is referred to as *diagnostic checking*. If you wanted to perform a more rigorous diagnostic check for an ANOVA, other visual diagnostics could involve inspecting QQ plots split by group (you'll do this in an example in Section 19.3) or plotting the sample standard deviation for each group against the corresponding sample means. Indeed, there are also general hypothesis tests for normality (such as the Shapiro-Wilk test or Anderson-Darling test—you'll see the former used in Section 22.3.2), as well as tests for equality of variances (such as

Levene's test), but I'll stick with the basic rule of thumb and visual checks in this example.

## 19.1.2 One-Way ANOVA Table Construction

Turning your attention back to the left of Figure 19-1, remember that the goal is to statistically evaluate the equality of the means marked by ×. This task will therefore require you to consider not only the variability *within* each of the $k$ samples but the variability *between* the samples; this is why the test is referred to as an analysis of variance.

The test proceeds by first calculating various metrics associated with the overall variability and then calculating the within- and between-group variability. These figures involve sums of squared quantities and associated degrees of freedom values. All this culminates in a single test statistic and $p$-value targeting the aforementioned hypotheses. These ingredients are typically presented in a table, which is defined as follows.

Let $x_1, \ldots, x_N$ represent all $N$ observations, regardless of group; let $x_{1(j)}, \ldots, x_{n_j(j)}$ denote the specific group observations in group $j = 1, \ldots, k$ such that $n_1 + \ldots + n_k = N$. Let the "grand mean" of all observations be defined as $\bar{x} = \frac{1}{N} \sum_{i=1}^{N} x_i$. The ANOVA table is then constructed, where SS stands for sum-of-squares, df stands for degrees of freedom, MS stands for mean square, $F$ refers to the $F$ test statistic, and $p$ refers to the $p$-value.

| | df | SS | MS | $F$ | $p$ |
|---|---|---|---|---|---|
| Overall | 1 | (1) | | | |
| Group (or "Factor") | $k - 1$ | (2) | (5) | (5)÷(6) | $p$-value |
| Error (or "Residual") | $N - k$ | (3) | (6) | | |
| TOTAL | $N$ | (4) | | | |

You calculate the values with these formulas:

(1): $N\bar{x}^2$

(2): $\sum_{j=1}^{k} \frac{\left( \sum_{i=1}^{n_j} x_{i(j)} \right)^2}{n_j}$

(3): (4)−(2)−(1)

(4): $\sum_{i=1}^{N} x_i^2$

(5): (2)÷(k − 1)

(6): (3)÷(N − k)

There are three input sources that are assumed to make up the observed data, which, when added together, result in the TOTAL row. Let's think about these in a little more detail:

**Overall row**   This relates to the scale on which the data as a whole sit. It doesn't affect the outcome of the hypothesis test (since you're interested only in the relative differences between means) and is sometimes removed from the table, affecting the TOTAL values accordingly.

**Group row/Factor row** This relates to the data in the individual groups of interest, thereby accounting for the *between-group variability*.

**Error row/Residual row** This accounts for the random deviation from the estimated means of each group, thereby accounting for the *within-group variability*.

**TOTAL row** This represents the raw data, based on the previous three ingredients. It is used to find the Error SS by differencing.

The three input sources each have a corresponding degrees of freedom (df) value in the first column and a sum-of-squares (SS) value attached to the df in the second column. Between- and within-group variability is averaged by dividing the SS by the df, giving the mean squared (MS) component for these two items. The test statistic, $F$, is found by dividing the mean squared group (MSG) effect by the mean squared error (MSE) effect. This test statistic follows the $F$-distribution (refer to Section 16.2.5), which itself requires a pair of degrees of freedom values ordered as $df_1$ (which represents the Group df, $k-1$) and $df_2$ (which represents the Error df, $N-k$). Like the chi-squared distribution, the $F$-distribution is unidirectional in nature, and the $p$-value is obtained as the upper-tail area from the test statistic $F$.

### 19.1.3   Building ANOVA Tables with the aov Function

As you might expect, R allows you to easily construct an ANOVA table for the chick weight test using the built-in aov function as follows:

```
R> chick.anova <- aov(weight~feed,data=chickwts)
```

Then, the table is printed to the console screen using summary.

```
R> summary(chick.anova)
            Df Sum Sq Mean Sq F value   Pr(>F)
feed         5 231129   46226   15.37 5.94e-10 ***
Residuals   65 195556    3009
---
Signif. codes:  0 '***' 0.001 '**' 0.01 '*' 0.05 '.' 0.1 ' ' 1
```

There are several comments to make here. Note that you employ formula notation weight~feed to specify the measurement variable of interest, weight, as modeled by the categorical-nominal variable of interest, feed type. In this case, the variable names weight and feed are not required to be prefaced by chickwts$ since the optional data argument has been passed the data frame of interest.

Remember from Section 14.3.2 that for the notation in the expression weight~feed, the "outcome" of interest must always appear on the left of the ~ (this notation will become particularly relevant in Chapters 20 to 22).

To actually view the table, you must apply the summary command to the object resulting from the call to aov. R omits the first and last rows (Overall and TOTAL) since these are not directly involved in calculating the *p*-value. Other than that, it's easy to identify that the feed row refers to the Group row and the Residuals row refers to the Error row.

NOTE *By default, R annotates model-based summary output like this with* significance stars. *These show intervals of significance, and the number of stars increases as the p-value decreases beyond a cutoff mark of 0.1. This can be useful when you're examining more complicated analyses where multiple p-values are summarized, though not everyone likes this feature. If you want, you can turn off this feature in a given R session by entering* options(show.signif.stars=FALSE) *at the prompt. Alternatively, you can turn off the feature directly in the call to* summary *by setting the additional argument* signif.stars=FALSE. *In this book, I'll leave them be.*

From the contents of the ANOVA for this example, you can quickly confirm the calculations. Note that the MSE, 3009, was defined as the Error SS divided by the Error df. Indeed, in R, the same result is achieved manually (the table output has been rounded to the nearest integer).

```
R> 195556/65
[1] 3008.554
```

You can confirm all the other results in the table output using the relevant equations from earlier.

Interpreting a hypothesis test based on ANOVA follows the same rules as any other test. With the understanding of a *p*-value as "the probability that you observe the sample statistics at hand or something more extreme, if $H_0$ is true," a small *p*-value indicates evidence against the null hypothesis. In the current example, a tiny *p*-value provides strong evidence against the null that the mean chick weights are the same for the different diets. In other words, you reject $H_0$ in favor of $H_A$; the latter states that there is a difference.

In a similar fashion as in the chi-squared tests, rejection of the null in one-way ANOVA doesn't tell you exactly where a difference lies, merely that there's evidence one exists. Further scrutiny of the data in the individual groups is necessary to identify the offending means. At the simplest level, you could turn back to pairwise two-sample *t*-tests, in which case you could also use the MSE from the ANOVA table as an estimate of the pooled variance. The substitution is valid if the assumption of equal variance holds, and such a step is beneficial because the corresponding *t*-based sampling distribution will utilize the Error df (this is naturally higher than would otherwise be the case if the df was based on just the sample sizes of the two groups of specific interest).

Analysis of Variance **441**

## Exercise 19.1

Consider the following data:

| Site I | Site II | Site III | Site IV |
|--------|---------|----------|---------|
| 93     | 85      | 100      | 96      |
| 120    | 45      | 75       | 58      |
| 65     | 80      | 65       | 95      |
| 105    | 28      | 40       | 90      |
| 115    | 75      | 73       | 65      |
| 82     | 70      | 65       | 80      |
| 99     | 65      | 50       | 85      |
| 87     | 55      | 30       | 95      |
| 100    | 50      | 45       | 82      |
| 90     | 40      | 50       |         |
| 78     |         | 45       |         |
| 95     |         | 55       |         |
| 93     |         |          |         |
| 88     |         |          |         |
| 110    |         |          |         |

These figures provide the depths (in centimeters) at which important archaeological finds were made at four sites in New Mexico (see Woosley and Mcintyre, 1996). Store these data in your R workspace, with one vector containing depth and the other vector containing the site of each observation.

a. Produce side-by-side boxplots of the depths split by group, and use additional points to mark the locations of the sample means.

b. Assuming independence, execute diagnostic checks for normality and equality of variances.

c. Perform and conclude a one-way ANOVA test for evidence of a difference between the means.

In Section 14.4, you looked at the data set providing measurements on petal and sepal sizes for three species of iris flowers. This is available in R as iris.

d. Based on diagnostic checks for normality and equality of variances, decide which of the four outcome measurements (sepal length/width and petal length/width) would be suitable for ANOVA (using the species as the group variable).

e. Carry out one-way ANOVA for any suitable measurement variables.

## 19.2 Two-Way ANOVA

In many studies, the numeric outcome variable you're interested in will be categorized by more than just one grouping variable. In these cases, you would use the *multiple-factor* ANOVA rather than the one-way ANOVA. This technique is directly referred to by the number of grouping variables used, with two- and three-way ANOVA being the next and most common extensions.

Increasing the number of grouping variables complicates matters somewhat—performing just a one-way ANOVA for each variable separately is inadequate. In dealing with more than one categorical grouping factor, you must consider the *main effects* of each factor on the numeric outcome, while simultaneously accounting for the presence of the other grouping factor(s). That's not all, though. It's just as important to additionally investigate the idea of an *interactive effect*; if an interactive effect exists, then it suggests that the impact one of the grouping variables has on the outcome of interest, specified by its main effect, varies according to the levels of the other grouping variable(s).

### 19.2.1 A Suite of Hypotheses

For this explanation, denote your numeric outcome variable with $O$ and your two grouping variables as $G_1$ and $G_2$. In two-way ANOVA, the hypotheses should be set along the following lines:

$H_0 :$ $G_1$ has no main (marginal) effect on the mean of $O$.

$G_2$ has no main (marginal) effect on the mean of $O$.

There is no interactive effect of $G_1$ with $G_2$ on the mean of $O$.

$H_A :$ Separately, each statement in $H_0$ is incorrect.

You can see from these general hypotheses that you now have to obtain a *p*-value for each of the three components.

For the example, let's use the built-in warpbreaks data frame (Tippett, 1950), which provides the number of "warp break" imperfections (column breaks) observed in 54 pieces of yarn of equal length. Each piece of yarn is classified according to two categorical variables: wool (the type of yarn, with levels A and B) and tension (the level of tension applied to that piece—L, M, or H for low, medium, or high). Using tapply, you can inspect the mean number of warp breaks for each classification.

```
R> tapply(warpbreaks$breaks,INDEX=list(warpbreaks$wool,warpbreaks$tension),
        FUN=mean)
         L        M        H
A 44.55556 24.00000 24.55556
B 28.22222 28.77778 18.77778
```

You can supply more than one grouping variable to the INDEX argument as separate members of a list (any factor vectors given to this argument should be the same length as the first argument that specifies the data of interest). The results are returned as a matrix for two grouping variables, a 3D array for three grouping variables, and so on.

For some analyses, however, you might need the same information provided earlier in a different format. The aggregate function is similar to tapply, but it returns a data frame, the results in *stacked* format according to the specified grouping variables (as opposed to an array as returned by tapply). It's called in much the same way. The first argument takes the data vector of interest. The second argument, by, should be a list of the desired grouping variables, and in FUN, you specify the function to operate on each subset.

```
R> wb.means <- aggregate(warpbreaks$breaks,
                     by=list(warpbreaks$wool,warpbreaks$tension),FUN=mean)
R> wb.means
  Group.1 Group.2        x
1       A       L 44.55556
2       B       L 28.22222
3       A       M 24.00000
4       B       M 28.77778
5       A       H 24.55556
6       B       H 18.77778
```

Here I've stored the result of the call to aggregate as the object wb.means for later use.

### 19.2.2  Main Effects and Interactions

I mentioned earlier that you could perform just a one-way ANOVA on each grouping variable separately, but this, in general, isn't a good idea. I'll demonstrate this now with the warpbreaks data (a quick inspection of the relevant diagnostics shows no obvious cause for concern):

```
R> summary(aov(breaks~wool,data=warpbreaks))
            Df Sum Sq Mean Sq F value Pr(>F)
wool         1    451   450.7   2.668  0.108
Residuals   52   8782   168.9
R> summary(aov(breaks~tension,data=warpbreaks))
            Df Sum Sq Mean Sq F value  Pr(>F)
tension      2   2034  1017.1   7.206 0.00175 **
Residuals   51   7199   141.1
---
Signif. codes:  0 '***' 0.001 '**' 0.01 '*' 0.05 '.' 0.1 ' ' 1
```

This output tells you that if you ignore tension, there is no evidence to suggest that there is any difference in the mean number of imperfections based on the type of wool alone (*p*-value 0.108). If you ignore wool, however, there *is* evidence to suggest a difference in warp breaks according to tension only.

The problem here is that by ignoring one of the variables, you lose the ability to detect differences (or, more generally, statistical relationships) that may occur at a finer level. For example, though the wool type alone seems to have no remarkable impact on the mean number of warp breaks, you cannot tell whether this would be the case if you just looked at wool types at one particular level of tension.

Instead, you investigate this kind of question using two-way ANOVA. The following executes a two-way ANOVA for the warp breaks data based only on the main effects of the two grouping variables:

```
R> summary(aov(breaks~wool+tension,data=warpbreaks))
            Df Sum Sq Mean Sq F value  Pr(>F)
wool         1    451   450.7   3.339 0.07361 .
tension      2   2034  1017.1   7.537 0.00138 **
Residuals   50   6748   135.0
---
Signif. codes:  0 '***' 0.001 '**' 0.01 '*' 0.05 '.' 0.1 ' ' 1
```

Take a look at the formula. Specifying wool+tension to the right of the outcome variable and the ~ allows you to take both grouping variables into account at the same time. The results reveal a small drop in the size of the *p*-values now attached to each grouping variable; indeed, the *p*-value for wool is around 0.073, approaching the conventional cutoff significance level of $\alpha = 0.05$. To interpret the results, you hold one grouping variable constant—if you focus on just one type of wool, there is still statistically significant evidence to suggest a difference in the mean number of warp breaks between the different tension levels. If you focus on just one level of tension, the evidence of a difference considering the two wool types has increased a little but is still not statistically significant (assuming the aforementioned $\alpha = 0.05$).

There's still a limitation with considering only main effects. While the previous analysis shows that there's variation in the outcome between the different levels of the two categorical variables, it doesn't address the possibility that a difference in the mean number of warp breaks might vary further according to precisely *which* level of either tension or wool is being used when holding the other variable constant. This relatively subtle yet important consideration is known as an *interaction*. Specifically, if there is an interactive effect present between tension and wool with respect to warp breaks, then this would imply that the magnitude and/or direction of the difference in the mean number of warp breaks *is not the same* at different levels of the two grouping factors.

To account for interactions, you make a slight adjustment to the two-way ANOVA model code.

```
R> summary(aov(breaks~wool+tension+wool:tension,data=warpbreaks))
             Df Sum Sq Mean Sq F value   Pr(>F)
wool          1    451   450.7   3.765 0.058213 .
tension       2   2034  1017.1   8.498 0.000693 ***
wool:tension  2   1003   501.4   4.189 0.021044 *
Residuals    48   5745   119.7
---
Signif. codes:  0 '***' 0.001 '**' 0.01 '*' 0.05 '.' 0.1 ' ' 1
```

You can explicitly specify the interaction as the main effects model formula *plus* the notation wool:tension, where the two grouping variables are separated by a :. (Note, in this setting, the : operator has nothing to do with the shortcut for creating an integer sequence first discussed in Section 2.3.2.)

You can see from the ANOVA table output that, statistically, there is evidence of an interactive effect; that is, the very nature of the difference in the means is dependent upon the factor levels themselves, even though that evidence is relatively weak. Of course, the *p*-value of around 0.021 tells you only that, overall, there might be an interaction but not the precise features of the interaction.

To help with this, you can interpret such a two-way interaction effect in more detail with an *interaction plot*, provided in R with interaction.plot.

```
R> interaction.plot(x.factor=wb.means[,2],trace.factor=wb.means[,1],
                    response=wb.means$x,trace.label="wool",
                    xlab="tension",ylab="mean warp breaks")
```

When interaction.plot is called, the outcome means should be supplied to the argument response, and the vectors providing the corresponding levels of each of the two factors should be supplied to the arguments x.factor (for the variable on the horizontal axis that refers to moving between levels from the left to right) and trace.factor (each level of which will produce a different line, referenced in an automatically produced legend; the title of this legend is passed to trace.label). It doesn't matter which grouping variable is which; the appearance of the plot will change accordingly, but your interpretations will (should!) be the same. The result is shown in Figure 19-2.

The two-way interaction plot displays the outcome variable on the vertical axis and splits the recorded means by the levels of the two grouping variables. This allows you to inspect the potential effect that varying the levels of the grouping variables has on the outcome. In general, when the lines (or segments thereof) are not parallel, it suggests an interaction could be present. Vertical separations between the plotted locations indicate the individual main effects of the grouping variables.

It turns out that the columns returned by a call to aggregate are actually perfectly suited to interaction.plot. As usual, you can specify the common graphical parameters, like those you initially encountered in Section 7.2, to control specific features of the plot and axis annotation. For Figure 19-2, you've specified that x.factor should be the second column of the wb.means matrix, meaning that the tension levels vary horizontally. The trace.factor here is the type of wool, so there are only two distinct lines corresponding to the two levels A and B. The response is that third column of wb.means, extracted using $x (take a look at the wb.means object; you'll see the column containing the results of interest is labeled x by default after a call to aggregate).

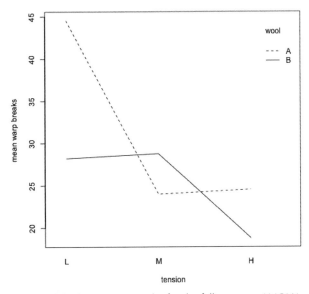

Figure 19-2: An interaction plot for the full two-way ANOVA model of the warpbreaks data set

Considering the actual appearance of the plot in Figure 19-2, it does indeed appear that the mean number of warp breaks for wool type A is higher if tension is low, but the nature of the difference changes if you move to a medium tension, where B has a higher point estimate than A. Moving to a high tension, type A again has a higher estimate of the mean number of breaks than B, though here the difference between A and B is nowhere near as big as it is at a low tension. (Note, however, that the interaction plot does not display any kind of standard error measurements, so you must remember that all point estimates of the means are subject to variability.)

Interactions are certainly not a concept unique to multiple-factor ANOVA; they form an important consideration in many different types of statistical models. For the moment, it's good just to gain a basic appreciation of interactions.

## 19.3  Kruskal-Wallis Test

When comparing multiple means, there may be situations when you're unwilling to assume normality or have even found the assumption of normality invalid in diagnostic checks. In this case, you can use the *Kruskal-Wallis test*, an alternative to the one-way ANOVA that relaxes the dependence on the necessity for normality. This method tests for "equality of distributions" of the measurements in each level of the grouping factor. If you make the usual assumption of equal variances across these groups, you can therefore think of this test as one that compares multiple medians rather than means.

The hypotheses governing the test alter accordingly.

$H_0$ : Group medians are all equal.

$H_A$ : Group medians are not all equal

(alternatively, at least one group median differs).

The Kruskal-Wallis test is a *nonparametric* approach since it does not rely on quantiles of a standardized parametric distribution (in other words, the normal distribution) or any of its functions. In the same way that the ANOVA is a generalization of the two-sample *t*-test, the Kruskal-Wallis ANOVA is a generalization of the Mann-Whitney test for two medians. It's also referred to as the Kruskal-Wallis *rank sum* test, and you use the chi-squared distribution to calculate the *p*-value.

Turn your attention to the data frame survey, located in the MASS package. These data record particular characteristics of 237 first-year undergraduate statistics students collected from a class at the University of Adelaide, South Australia. Load the required package first with a call to library("MASS") and then enter ?survey at the prompt. You can read the help file to understand which variables are present in the data frame.

Suppose you're interested to see whether the age of the students, Age, tends to differ with respect to four smoking categories reported in Smoke. An inspection of the relevant side-by-side boxplots and a normal QQ plot of the residuals (mean-centered observations with respect to each group) suggests a straightforward one-way ANOVA isn't necessarily a good idea. The following code (which mimics the steps you saw in Section 19.1.1) produces the two images in Figure 19-3, which show normality is questionable:

```
R> boxplot(Age~Smoke,data=survey)
R> age.means <- tapply(survey$Age,survey$Smoke,mean)
R> age.meancen <- survey$Age-age.means[as.numeric(survey$Smoke)]
R> qqnorm(age.meancen,main="Normal QQ plot of residuals")
R> qqline(age.meancen)
```

With this possible violation of normality, you could therefore apply the Kruskal-Wallis test instead of the parametric ANOVA. A quick check for equality of variances further supports this, with the ratio of the largest to the smallest group standard deviations clearly being less than 2.

```
R> tapply(survey$Age,survey$Smoke,sd)
   Heavy    Never    Occas    Regul
6.332628 6.675257 5.861992 5.408822
```

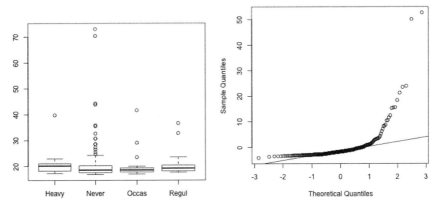

Figure 19-3: Side-by-side boxplots (left) and a normal QQ plot of the residuals (right) for the student age observations split by smoking status

In R, a Kruskal-Wallis test is performed using kruskal.test.

```
R> kruskal.test(Age~Smoke,data=survey)

        Kruskal-Wallis rank sum test

data:  Age by Smoke
Kruskal-Wallis chi-squared = 3.9262, df = 3, p-value = 0.2695
```

The syntax for this test is the same as for aov. As you might suspect from Figure 19-3, the large *p*-value suggests there's no evidence against the null hypothesis that states that the medians are all equal. In other words, there doesn't seem to be an overall age difference between the students in the four smoking categories.

---

## Exercise 19.2

Bring up the quakes data frame again, which describes the locations, magnitudes, depths, and number of observation stations that detected 1,000 seismic events off the coast of Fiji.

a.  Use cut (see Section 4.3.3) to create a new factor vector defining the depths of each event according to the following three categories: $(0, 200]$, $(200, 400]$, and $(400, 680]$.

b.  Decide whether a one-way ANOVA or a Kruskal-Wallis test is more appropriate to use to compare the distributions of the number of detecting stations, split according to the three categories in (a).

c.  Perform your choice of test in (b) (assume a $\alpha = 0.01$ level of significance) and conclude.

Load the MASS package with a call to library("MASS") if you haven't already done so in the current R session. This package includes the ready-to-use Cars93 data frame, which contains detailed data on 93 cars for sale in the United States in 1993 (Lock, 1993; Venables and Ripley, 2002).

d.  Use aggregate to compute the mean length of the 93 cars, split by two categorical variables: AirBags (type of airbags available—levels are Driver & Passenger, Driver only, and None), and Man.trans.avail (whether the car comes in a manual transmission—levels are Yes and No).

e.  Produce an interaction plot using the results in (d). Does there appear to be an interactive effect of AirBags with Man.trans.avail on the mean length of these cars (if you consider only these variables)?

f.  Fit a full two-way ANOVA model for the mean lengths according to the two grouping variables (assume satisfaction of all relevant assumptions). Is the interactive effect statistically significant? Is there evidence of any main effects?

## Important Code in This Chapter

| Function/operator | Brief description | First occurrence |
|---|---|---|
| aov | Produce ANOVA table | Section 19.1.3, p. 440 |
| aggregate | Stacked statistics by factor | Section 19.2.1, p. 444 |
| interaction.plot | Two-factor interaction plot | Section 19.2.2, p. 446 |
| kruskal.test | Kruskal-Wallis test | Section 19.3, p. 449 |

# 20

## SIMPLE LINEAR REGRESSION

Though straightforward comparative tests of individual statistics are useful in their own right, you'll often want to learn more from your data. In this chapter, you'll look at *linear regression* models: a suite of methods used to evaluate precisely *how* variables relate to each other.

Simple linear regression models describe the effect that a particular variable, called the *explanatory variable*, might have on the value of a continuous outcome variable, called the *response variable*. The explanatory variable may be continuous, discrete, or categorical, but to introduce the key concepts, I'll concentrate on continuous explanatory variables for the first several sections in this chapter. Then, I'll cover how the representation of the model changes if the explanatory variable is categorical.

## 20.1   An Example of a Linear Relationship

As an example to start with, let's continue with the data used in Section 19.3 and look at the student survey data (the survey data frame in the package MASS) a little more closely. If you haven't already done so, with the required package loaded (call library("MASS")), you can read the help file ?survey for details on the variables present.

Plot the student heights on the *y*-axis and their handspans (of their writing hand) on the *x*-axis.

```
R> plot(survey$Height~survey$Wr.Hnd,xlab="Writing handspan (cm)",
        ylab="Height (cm)")
```

Figure 20-1 shows the result.

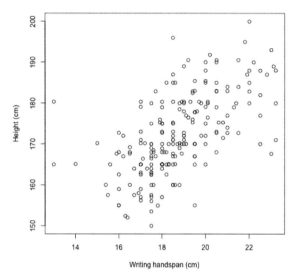

*Figure 20-1: A scatterplot of height against writing handspan for a sample of first-year statistics students*

Note that the call to plot uses formula notation (also referred to as *symbolic notation*) to specify "height *on* handspan." You can produce the same scatterplot by using the coordinate vector form of $(x, y)$, that is, plot(survey$Wr.Hnd,survey$Height,...), but I'm using the symbolic notation here because it nicely reflects how you'll fit the linear model in a moment.

As you might expect, there's a positive association between a student's handspan and their height. That relationship appears to be linear in nature. To assess the strength of the linear relationship (refer to Section 13.2.5), you can find the estimated correlation coefficient.

```
R> cor(survey$Wr.Hnd,survey$Height,use="complete.obs")
[1] 0.6009909
```

Though there are 237 records in the data frame, the plot doesn't actually show 237 points. This is because there are missing observations (coded NA; see Section 6.1.3). By default, R removes any "incomplete" pairs when producing a plot like this. To find out how many offending observations have been deleted, you can use the short-form logical operator |

(Section 4.1.3) in conjunction with is.na (Section 6.1.3) and which (Section 4.1.5). You then use length to discover there are 29 missing observation pairs.

```
R> incomplete.obs <- which(is.na(survey$Height)|is.na(survey$Wr.Hnd))
R> length(incomplete.obs)
[1] 29
```

**NOTE**     *Because there are NAs in the vectors supplied to the correlation coefficient function* cor, *you must also specify the optional argument* use="complete.obs"*. This means that the calculated statistic takes into account only those observation pairs in the* Wr.Hnd *and* Height *vectors where* neither *element is* NA*. You can think of this argument as doing much the same thing as* na.rm=TRUE *in univariate summary statistic functions such as* mean *and* sd*.*

## 20.2   General Concepts

The purpose of a linear regression model is to come up with a function that estimates the *mean* of one variable given a particular value of another variable. These variables are known as the *response variable* (the "outcome" variable whose mean you are attempting to find) and the *explanatory variable* (the "predictor" variable whose value you already have).

   In terms of the student survey example, you might ask something like "What's the expected height of a student if their handspan is 14.5 cm?" Here the response variable is the height, and the explanatory variable is the handspan.

### 20.2.1   Definition of the Model

Assume you're looking to determine the value of response variable $Y$ given the value of an explanatory variable $X$. The *simple linear regression model* states that the value of a response is expressed as the following equation:

$$Y|X = \beta_0 + \beta_1 X + \epsilon \tag{20.1}$$

   On the left side of Equation (20.1), the notation $Y|X$ reads as "the value of $Y$ conditional upon the value of $X$."

**Residual Assumptions**

The validity of the conclusions you can draw based on the model in (20.1) is critically dependent on the assumptions made about $\epsilon$, which are defined as follows:

- The value of $\epsilon$ is assumed to be normally distributed such that $\epsilon \sim N(0, \sigma)$.

- That $\epsilon$ is centered (that is, has a mean of) zero.

- The variance of $\epsilon$, $\sigma^2$, is constant.

The $\epsilon$ term represents random error. In other words, you assume that any raw value of the response is owed to a linear change in a given value of $X$, plus or minus some random, *residual* variation or normally distributed *noise*.

### Parameters

The value denoted by $\beta_0$ is called the *intercept*, and that of $\beta_1$ is called the *slope*. Together, they are also referred to as the *regression coefficients* and are interpreted as follows:

- The intercept, $\beta_0$, is interpreted as the expected value of the response variable when the predictor is zero.

- Generally, the slope, $\beta_1$, is the focus of interest. This is interpreted as the change in the mean response for each one-unit increase in the predictor. When the slope is positive, the regression line increases from left to right (the mean response is higher when the predictor is higher); when the slope is negative, the line decreases from left to right (the mean response is lower when the predictor is higher). When the slope is zero, this implies that the predictor has no effect on the value of the response. The more extreme the value of $\beta_1$ (that is, away from zero), the steeper the increasing or decreasing line becomes.

## 20.2.2  Estimating the Intercept and Slope Parameters

The goal is to use your data to estimate the regression parameters, yielding the estimates $\hat{\beta}_0$ and $\hat{\beta}_1$; this is referred to as *fitting* the linear model. In this case, the data comprise $n$ pairs of observations for each individual. The fitted model of interest concerns the mean response value, denoted $\hat{y}$, for a specific value of the predictor, $x$, and is written as follows:

$$\hat{y} = \hat{\beta}_0 + \hat{\beta}_1 x \tag{20.2}$$

Sometimes, alternative notation such as $\mathbb{E}[Y]$ or $\mathbb{E}[Y|X = x]$ is used on the left side of (20.2) to emphasize the fact that the model gives the mean (that is, the expected value) of the response. For compactness, many simply use something like $\hat{y}$, as shown here.

Let your $n$ observed data pairs be denoted $x_i$ and $y_i$ for the predictor and response variables, respectively; $i = 1, \ldots, n$. Then, the parameter estimates for the simple linear regression function are

$$\hat{\beta}_1 = \rho_{xy} \frac{s_y}{s_x} \quad \text{and} \quad \hat{\beta}_0 = \bar{y} - \hat{\beta}_1 \bar{x} \tag{20.3}$$

where

- $\bar{x}$ and $\bar{y}$ are the sample means of the $x_i$s and $y_i$s.

- $s_x$ and $s_y$ are the sample standard deviations of the $x_i$s and $y_i$s.

- $\rho_{xy}$ is the estimate of correlation between $X$ and $Y$ based on the data (see Section 13.2.5).

Estimating the model parameters in this way is referred to as *least-squares regression*; the reason for this will become clear in a moment.

### 20.2.3   Fitting Linear Models with lm

In R, the command lm performs the estimation for you. For example, the following line creates a fitted linear model object of the mean student height by handspan and stores it in your global environment as survfit:

```
R> survfit <- lm(Height~Wr.Hnd,data=survey)
```

The first argument is the now-familiar *response ~ predictor* formula, which specifies the desired model. You don't have to use the survey$ prefix to extract the vectors from the data frame because you specifically instruct lm to look in the object supplied to the data argument.

The fitted linear model object itself, survfit, has a special class in R—one of "lm". An object of class "lm" can essentially be thought of as a list containing several components that describe the model. You'll look at these in a moment.

If you simply enter the name of the "lm" object at the prompt, it will provide the most basic output: a repeat of your call and the estimates of the intercept ($\hat{\beta}_0$) and slope ($\hat{\beta}_1$).

```
R> survfit

Call:
lm(formula = Height ~ Wr.Hnd, data = survey)

Coefficients:
(Intercept)        Wr.Hnd
    113.954         3.117
```

This reveals that the linear model for this example is estimated as follows:

$$\hat{y} = 113.954 + 3.117x \tag{20.4}$$

If you evaluate the mathematical function for $\hat{y}$—Equation (20.2)—at a range of different values for $x$, you end up with a straight line when you plot the results. Considering the definition of intercept given earlier as the expected value of the response variable when the predictor is zero, in the current example, this would imply that the mean height of a student with a handspan of 0 cm is 113.954 cm (an arguably less-than-useful statement since a value of zero for the explanatory variable doesn't make sense; you'll consider these and related issues in Section 20.4). The slope, the change in the mean response for each one-unit increase in the predictor, is 3.117. This states that, on average, for every 1 cm increase in handspan, a student's height is estimated to increase by 3.117 cm.

With all this in mind, once more run the line to plot the raw data as given in Section 20.1 and shown in Figure 20-1, but now add the fitted regression line using abline. So far, you've only used the abline command to add perfectly horizontal and vertical lines to an existing plot, but when passed an object of class "lm" that represents a simple linear model, like survfit, the fitted regression line will be added instead.

```
R> abline(survfit,lwd=2)
```

This adds the slightly thickened diagonally increasing line shown in Figure 20-2.

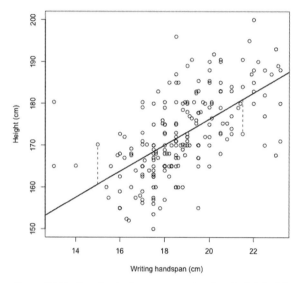

*Figure 20-2: The simple linear regression line (solid, bold) fitted to the observed data. Two dashed vertical line segments provide examples of a positive (leftmost) and negative (rightmost) residual.*

### 20.2.4   Illustrating Residuals

When the parameters are estimated as shown here, using (20.3), the fitted line is referred to as an implementation of *least-squares regression* because it's the line that minimizes the average squared difference between the observed data and itself. This concept is easier to understand by drawing the distances between the observations and the fitted line, formally called *residuals*, for a couple of individual observations within Figure 20-2.

First, let's extract two specific records from the Wr.Hnd and Height data vectors and call the resulting vectors obsA and obsB.

```
R> obsA <- c(survey$Wr.Hnd[197],survey$Height[197])
R> obsA
[1]   15.00 170.18
```

```
R> obsB <- c(survey$Wr.Hnd[154],survey$Height[154])
R> obsB
[1]   21.50 172.72
```

Next, briefly inspect the names of the members of the survfit object.

```
R> names(survfit)
 [1] "coefficients"  "residuals"     "effects"     "rank"
 [5] "fitted.values" "assign"        "qr"          "df.residual"
 [9] "na.action"     "xlevels"       "call"        "terms"
[13] "model"
```

These members are the components that automatically make up a fitted model object of class "lm", mentioned briefly earlier. Note that there's a component called "coefficients". This contains a numeric vector of the estimates of the intercept and slope.

You can extract this component (and indeed any of the other ones listed here) in the same way you would perform a member reference on a named list: by entering survfit$coefficients at the prompt. Where possible, though, it's technically preferable for programming purposes to extract such components using a "direct-access" function. For the coefficients component of an "lm" object, the function you use is coef.

```
R> mycoefs <- coef(survfit)
R> mycoefs
(Intercept)      Wr.Hnd
 113.953623    3.116617
R> beta0.hat <- mycoefs[1]
R> beta1.hat <- mycoefs[2]
```

Here, the regression coefficients are extracted from the object and then separately assigned to the objects beta0.hat and beta1.hat. Other common direct-access functions include resid and fitted; these two pertain to the "residuals" and "fitted.values" components, respectively.

Finally, I use segments to draw the vertical dashed lines present in Figure 20-2.

```
R> segments(x0=c(obsA[1],obsB[1]),y0=beta0.hat+beta1.hat*c(obsA[1],obsB[1]),
            x1=c(obsA[1],obsB[1]),y1=c(obsA[2],obsB[2]),lty=2)
```

Note that the dashed lines meet the fitted line at the vertical axis locations passed to y0, which, with the use of the regression coefficients beta0.hat and beta1.hat, reflects Equation (20.4).

Now, imagine a collection of alternative regression lines drawn through the data (achieved by altering the value of the intercept and slope). Then, for each of the alternative regression lines, imagine you calculate the residuals (vertical distances) between the response value of every observation and the *fitted value* of that line. The simple linear regression line estimated as per (20.3) is the line that lies "closest to all observations." By this, it is meant that the fitted regression model is represented by the estimated line that passes through the coordinate provided by the variable means $(\bar{x}, \bar{y})$, and it's the line that yields the smallest overall measure of the squared residual distances. For this reason, another name for a least-squares-estimated regression equation like this is the *line of best fit*.

## 20.3   Statistical Inference

The estimation of a regression equation is relatively straightforward, but this is merely the beginning. You should now think about what can be inferred from your result. In simple linear regression, there's a natural question that should always be asked: Is there statistical evidence to support the presence of a relationship between the predictor and the response? To put it another way, is there evidence that a change in the explanatory variable affects the mean outcome? You investigate this following the same ideas that were introduced in Chapter 17 when you began thinking about the variability present in estimated statistics and then continued to infer from your results using confidence intervals and, in Chapter 18, hypothesis testing.

### 20.3.1   Summarizing the Fitted Model

This kind of *model-based inference* is automatically carried out by R when lm objects are processed. Using the summary function on an object created by lm provides you with output far more detailed than simply printing the object to the console. For the moment, you'll focus on just two aspects of the information presented in summary: the significance tests associated with the regression coefficients and the interpretation of the so-called *coefficient of determination* (labeled R-squared in the output), which I'll explain shortly.

Use summary on the current model object survfit, and you'll see the following:

```
R> summary(survfit)

Call:
lm(formula = Height ~ Wr.Hnd, data = survey)

Residuals:
    Min       1Q   Median       3Q      Max
-19.7276  -5.0706  -0.8269   4.9473  25.8704

Coefficients:
            Estimate Std. Error t value Pr(>|t|)
(Intercept) 113.9536     5.4416   20.94   <2e-16 ***
Wr.Hnd        3.1166     0.2888   10.79   <2e-16 ***
---
Signif. codes:  0 '***' 0.001 '**' 0.01 '*' 0.05 '.' 0.1 ' ' 1
```

Residual standard error: 7.909 on 206 degrees of freedom
  (29 observations deleted due to missingness)
Multiple R-squared:  0.3612,    Adjusted R-squared:  0.3581
F-statistic: 116.5 on 1 and 206 DF,  p-value: < 2.2e-16

## 20.3.2  Regression Coefficient Significance Tests

Let's begin by focusing on the way the estimated regression coefficients are reported. The first column of the coefficients table contains the point estimates of the intercept and slope (the intercept is labeled as such, and the slope is labeled after the name of the predictor variable in the data frame); the table also includes estimates of the standard errors of these statistics. It can be shown that simple linear regression coefficients, when estimated using least-squares, follow a $t$-distribution with $n - 2$ degrees of freedom (when given the number of observations, $n$, used in the model fit). The standardized $t$ value and a $p$-value are reported for each parameter. These represent the results of a two-tailed hypothesis test formally defined as

$$H_0 : \beta_j = 0$$
$$H_A : \beta_j \neq 0$$

where $j = 0$ for the intercept and $j = 1$ for the slope, using the notation in Equation (20.1).

Focus on the row of results for the predictor. With a null value of zero, truth of $H_0$ implies that the predictor has no effect on the response. The claim here is interested in whether there is *any effect* of the covariate, not the direction of this effect, so $H_A$ is two-sided (via $\neq$). As with any hypothesis test, the smaller the $p$-value, the stronger the evidence against $H_0$. With a small $p$-value ($< 2 \times 10^{-16}$) attached to this particular test statistic (which you can confirm using the formula in Chapter 18: $T = (3.116 - 0)/0.2888 = 10.79$), you'd therefore conclude there is strong evidence *against* the claim that the predictor has no effect on the mean level of the response.

The same test is carried out for the intercept, but the test for the slope parameter $\beta_1$ is typically more interesting (since rejection of the null hypothesis for $\beta_0$ simply indicates evidence that the regression line does not strike the vertical axis at zero), especially when the observed data don't include $x = 0$, as is the case here.

From this, you can conclude that the fitted model suggests there is evidence that an increase in handspan is associated with an increase in height among the population being studied. For each additional centimeter of handspan, the average increase in height is approximately 3.12 cm.

You could also produce confidence intervals for your estimates using Equation (17.2) on page 378 and knowledge of the sampling distributions of the regression parameters; however, yet again, R provides a convenient function for an object of class "lm" to do this for you.

```
R> confint(survfit,level=0.95)
                2.5 %     97.5 %
(Intercept) 103.225178 124.682069
Wr.Hnd        2.547273   3.685961
```

To the confint function you pass your model object as the first argument and your desired level of confidence as level. This indicates that you should be 95 percent confident the true value of $\beta_1$ lies somewhere between 2.55 and 3.69 (to 2 d.p.). As usual, the exclusion of the null value of zero reflects the statistically significant result from earlier.

### 20.3.3   Coefficient of Determination

The output of summary also provides you with the values of Multiple R-squared and Adjusted R-squared, which are particularly interesting. Both of these are referred to as the *coefficient of determination*; they describe the proportion of the variation in the response that can be attributed to the predictor.

For simple linear regression, the first (unadjusted) measure is simply obtained as the square of the estimated correlation coefficient (refer to Section 13.2.5). For the student height example, first store the estimated correlation between Wr.Hnd and Height as rho.xy, and then square it:

```
R> rho.xy <- cor(survey$Wr.Hnd,survey$Height,use="complete.obs")
R> rho.xy^2
[1] 0.3611901
```

You get the same result as the Multiple R-squared value (usually written mathematically as $R^2$). This tells you that about 36.1 percent of the variation in the student heights can be attributed to handspan.

The adjusted measure is an alternative estimate that takes into account the number of parameters that require estimation. The adjusted measure is generally important only if you're using the coefficient of determination to assess the overall "quality" of the fitted model in terms of a balance between goodness of fit and complexity. I'll cover this in Chapter 22, so I won't go into any more detail just yet.

### 20.3.4   Other summary Output

The summary of the model object provides you with even more useful information. The "residual standard error" is the estimated standard error of the $\epsilon$ term (in other words, the square root of the estimated variance of $\epsilon$, namely, $\sigma^2$); below that it also reports any missing values. (The 29 observation pairs "deleted due to missingness" here matches the number of incomplete observations determined in Section 20.1.)

The output also provides a five-number summary (Section 13.2.3) for the residual distances—I'll cover this further in Section 22.3. As the final result, you're provided with a certain hypothesis test performed using the *F*-distribution. This is a global test of the impact of your predictor(s) on the response; this will be explored alongside multiple linear regression in Section 21.3.5.

You can access all the output provided by summary directly, as individual R objects, rather than having to read them off the screen from the entire printed summary. Just as names(survfit) provides you with an indication of the contents of the stand-alone survfit object, the following code gives you the names of all the components accessible *after* summary is used to process survfit.

```
R> names(summary(survfit))
 [1] "call"          "terms"      "residuals"    "coefficients"
 [5] "aliased"       "sigma"      "df"           "r.squared"
 [9] "adj.r.squared" "fstatistic" "cov.unscaled" "na.action"
```

It's fairly easy to match most of the components with the printed summary output, and they can be extracted using the dollar operator as usual. The residual standard error, for example, can be retrieved directly with this:

```
R> summary(survfit)$sigma
[1] 7.90878
```

There are further details on this in the ?summary.lm help file.

# 20.4   Prediction

To wrap up these preliminary details of linear regression, you'll now look at using your fitted model for predictive purposes. The ability to fit a statistical model means that you not only can understand and quantify the nature of relationships in your data (like the estimated 3.1166 cm increase in mean height per 1 cm increase in handspan for the student example) but can also *predict* values of the outcome of interest, even where you haven't actually observed the values of any explanatory variables in the original data set. As with any statistic, though, there is always a need to accompany any point estimates or predictions with a measure of spread.

## 20.4.1   Confidence Interval or Prediction Interval?

With a fitted simple linear model you're able to calculate a point estimate of the *mean response* value, conditional upon the value of an explanatory variable. To do this, you simply plug in (to the fitted model equation) the value of $x$ you're interested in. A statistic like this is always subject to variation, so just as with sample statistics explored in earlier chapters, you use a *confidence interval for the mean response (CI)* to gauge this uncertainty.

Assume a simple linear regression line has been fitted to $n$ observations such that $\hat{y} = \hat{\beta}_0 + \hat{\beta}_1 x$. A $100(1-\alpha)$ percent confidence interval for the mean response given a value of $x$ is calculated with

$$\hat{y} \pm t_{(1-\alpha/2,n-2)} s_\epsilon \sqrt{\frac{1}{n} + \frac{(x-\bar{x})^2}{(n-1)s_x^2}} \tag{20.5}$$

where you obtain the lower limit by subtraction, the upper limit by addition.

Here, $\hat{y}$ is the fitted value (from the regression line) at $x$; $t_{(1-\alpha/2,n-2)}$ is the appropriate critical value from a $t$-distribution with $n-2$ degrees of freedom (in other words, resulting in an upper-tail area of exactly $\alpha/2$); $s_\epsilon$ is the estimated residual standard error; and $\bar{x}$ and $s_x^2$ represent the sample mean and the variance of the observations of the predictor, respectively.

A *prediction interval (PI)* for an observed response is different from the confidence interval in terms of context. Where CIs are used to describe the variability of the *mean* response, a PI is used to provide the possible range of values that an *individual realization* of the response variable might take, given $x$. This distinction is subtle but important: the CI corresponds to a mean, and the PI corresponds to an individual observation.

Let's remain with the previous notation. It can be shown that $100(1-\alpha)$ percent prediction interval for an individual response given a value of $x$ is calculated with the following:

$$\hat{y} \pm t_{(1-\alpha/2,n-2)} s_\epsilon \sqrt{1 + \frac{1}{n} + \frac{(x-\bar{x})^2}{(n-1)s_x^2}} \tag{20.6}$$

It turns out that the only difference from (20.5) is the 1+ that appears in the square root. As such, a PI at $x$ is wider than a CI at $x$.

### 20.4.2    Interpreting Intervals

Continuing with our example, let's say you want to determine the mean height for students with a handspan of 14.5 cm and for students with a handspan of 24 cm. The point estimates themselves are easy—just plug the desired $x$ values into the regression equation (20.4).

```
R> as.numeric(beta0.hat+beta1.hat*14.5)
[1] 159.1446
R> as.numeric(beta0.hat+beta1.hat*24)
[1] 188.7524
```

According to the model, you can expect mean heights to be around 159.14 and 188.75 cm for handspans of 14.5 and 24 cm, respectively. The `as.numeric` coercion function (first encountered in Section 6.2.4) is used simply to strip the result of the annotative names that are otherwise present from the `beta0.hat` and `beta1.hat` objects.

## Confidence Intervals for Mean Heights

To find confidence intervals for these estimates, you could calculate them manually using (20.5), but of course R has a built-in predict command to do it for you. To use predict, you first need to store your $x$ values in a particular way: as a column in a new data frame. The name of the column must match the predictor used in the original call to create the fitted model object. In this example, I'll create a new data frame, xvals, with the column named Wr.Hnd, which contains only two values of interest—the handspans of 14.5 and 24 cm.

```
R> xvals <- data.frame(Wr.Hnd=c(14.5,24))
R> xvals
  Wr.Hnd
1   14.5
2   24.0
```

Now, when predict is called, the first argument must be the fitted model object of interest, survfit for this example. Next, in the argument newdata, you pass the specially constructed data frame containing the specified predictor values. To the interval argument you must specify "confidence" as a character string value. The confidence level, here set for 95 percent, is passed (on the scale of a probability) to level.

```
R> mypred.ci <- predict(survfit,newdata=xvals,interval="confidence",level=0.95)
R> mypred.ci
       fit      lwr      upr
1 159.1446 156.4956 161.7936
2 188.7524 185.5726 191.9323
```

This call will return a matrix with three columns, whose number (and order) of rows correspond to the predictor values you supplied in the newdata data frame. The first column, with a heading of fit, is the point estimate on the regression line; you can see that these numbers match the values you worked out earlier. The other columns provide the lower and upper CI limits as the lwr and upr columns, respectively. In this case, you'd interpret this as 95 percent confidence that the mean height of a student with a handspan of 14.5 cm lies somewhere between 156.5 cm and 161.8 cm and lies between 185.6 cm and 191.9 cm for a handspan of 24 cm (when rounded to 1 d.p.). Remember, these CIs, calculated as per (20.5) through predict, are for the *mean* response value.

## Prediction Intervals for Individual Observations

The predict function will also provide your prediction intervals. To find the prediction interval for possible individual observations with a certain probability, you simply need to change the interval argument to "prediction".

```
R> mypred.pi <- predict(survfit,newdata=xvals,interval="prediction",level=0.95)
R> mypred.pi
```

```
       fit      lwr      upr
1 159.1446 143.3286 174.9605
2 188.7524 172.8390 204.6659
```

Notice that the fitted values remain the same, as Equations (20.5) and (20.6) indicate. The widths of the PIs, however, are significantly larger than those of the corresponding CIs—this is because raw observations themselves, at a specific $x$ value, will naturally be more variable than their mean.

Interpretation changes accordingly. The intervals describe where raw student heights are predicted to lie "95 percent of the time." For a handspan of 14.5 cm, the model predicts individual observations to lie somewhere between 143.3 cm and 175.0 cm with a probability of 0.95; for a handspan of 24 cm, the same PI is estimated at 172.8 cm and 204.7 cm (when rounded to 1 d.p.).

### 20.4.3   Plotting Intervals

Both CIs and PIs are well suited to visualization for simple linear regression models. With the following code, you can start off Figure 20-3 by plotting the data and estimated regression line just as for Figure 20-2, but this time using xlim and ylim in plot to widen the $x$- and $y$-limits a little in order to accommodate the full length and breadth of the CI and PI.

```
R> plot(survey$Height~survey$Wr.Hnd,xlim=c(13,24),ylim=c(140,205),
        xlab="Writing handspan (cm)",ylab="Height (cm)")
R> abline(survfit,lwd=2)
```

To this you add the locations of the fitted values for $x = 14.5$ and $x = 24$, as well as two sets of vertical lines showing the CIs and PIs.

```
R> points(xvals[,1],mypred.ci[,1],pch=8)
R> segments(x0=c(14.5,24),y0=c(mypred.pi[1,2],mypred.pi[2,2]),
            x1=c(14.5,24),y1=c(mypred.pi[1,3],mypred.pi[2,3]),col="gray",lwd=3)
R> segments(x0=c(14.5,24),y0=c(mypred.ci[1,2],mypred.ci[2,2]),
            x1=c(14.5,24),y1=c(mypred.ci[1,3],mypred.ci[2,3]),lwd=2)
```

The call to points marks the fitted values for these two particular values of $x$. The first call to segments lays down the PIs as thickened vertical gray lines, and the second lays down the CIs as the shorter vertical black lines. The coordinates for these plotted line segments are taken directly from the mypred.pi and mypred.ci objects, respectively.

You can also produce "bands" around the fitted regression line that mark one or both of these intervals over *all* values of the predictor. From a programming standpoint, this isn't technically possible for a continuous variable, but you can achieve it practically by defining a fine sequence of values along the $x$-axis (using seq with a high length value) and evaluating the CI and PI at every point in this fine sequence. Then you just join resulting points as lines when plotting.

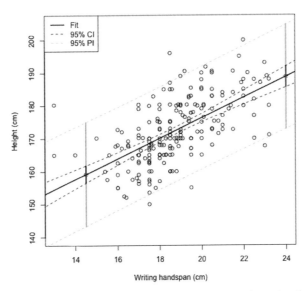

Figure 20-3: The student height regression example, with a fitted regression line and point estimates at x = 14.5 and x = 24 and with corresponding 95 percent CIs (black vertical lines) and PIs (gray vertical lines). The dashed black and dashed gray lines provide 95 percent confidence and prediction bands for the response variable over the visible range of x values.

In R, this requires you to rerun the predict command as follows:

```
R> xseq <- data.frame(Wr.Hnd=seq(12,25,length=100))
R> ci.band <- predict(survfit,newdata=xseq,interval="confidence",level=0.95)
R> pi.band <- predict(survfit,newdata=xseq,interval="prediction",level=0.95)
```

The first line in this code creates the fine sequence of predictor values and stores it in the format required by the newdata argument. The y-axis coordinates for CI and PI bands are stored as the second and third columns of the matrix objects ci.band and pi.band. Finally, lines is used to add each of the four dashed lines corresponding to the upper and lower limits of the two intervals, and a legend adds a final touch.

```
R> lines(xseq[,1],ci.band[,2],lty=2)
R> lines(xseq[,1],ci.band[,3],lty=2)
R> lines(xseq[,1],pi.band[,2],lty=2,col="gray")
R> lines(xseq[,1],pi.band[,3],lty=2,col="gray")
R> legend("topleft",legend=c("Fit","95% CI","95% PI"),lty=c(1,2,2),
        col=c("black","black","gray"),lwd=c(2,1,1))
```

Note that the black dashed CI bands meet the vertical black lines and the gray dashed PI bands meet the vertical gray lines for the two individual x values from earlier, just as you'd expect.

Figure 20-3 shows the end result of all these additions to the plot. The "bowing inwards" curvature of the intervals is characteristic of this kind of plot and is especially visible in the CI. This curve occurs because there is naturally less variation if you're predicting where there are more data. For more information on predict for linear model objects, take a look at the ?predict.lm help file.

### 20.4.4  Interpolation vs. Extrapolation

Before finishing this introduction to prediction, it's important to clarify the definitions of two key terms: *interpolation* and *extrapolation*. These terms describe the nature of a given prediction. A prediction is referred to as interpolation if the $x$ value you specify falls within the range of your observed data; extrapolation is when the $x$ value of interest lies outside this range. From the point-predictions you just made, you can see that the location $x = 14.5$ is an example of interpolation, and $x = 24$ is an example of extrapolation.

In general, interpolation is preferable to extrapolation—it makes more sense to use a fitted model for prediction in the vicinity of data that have already been observed. Extrapolation that isn't too far out of that vicinity may still be considered reliable, though. The extrapolation for the student height example at $x = 24$ is a case in point. This is outside the range of the observed data, but not by much in terms of scale, and the estimated intervals for the expected value of $\hat{y} = 188.75$ cm appear, at least visually, not unreasonable given the distribution of the other observations. In contrast, it would make less sense to use the fitted model to predict student height at a handspan of, say, 50 cm:

```
R> predict(survfit,newdata=data.frame(Wr.Hnd=50),interval="confidence",
        level=0.95)
        fit      lwr      upr
1 269.7845 251.9583 287.6106
```

Such an extreme extrapolation suggests that the mean height of an individual with a handspan of 50 cm is almost 270 cm, both being fairly unrealistic measurements. The same is true in the other direction; the intercept $\hat{\beta}_0$ doesn't have a particularly useful practical interpretation, indicating that the mean height of a student with a handspan of 0 cm is around 114 cm.

The main message here is to use common sense when making any prediction from a linear model fit. In terms of the reliability of the results, predictions made at values within an appropriate proximity of the observed data are preferable.

# Exercise 20.1

Continue to use the survey data frame from the package MASS for the next few exercises.

a. Using your fitted model of student height on writing handspan, survfit, provide point estimates and 99 percent confidence intervals for the mean student height for handspans of 12, 15.2, 17, and 19.9 cm.

b. In Section 20.1, you defined the object incomplete.obs, a numeric vector that provides the records of survey that were automatically removed from consideration when estimating the model parameters. Now, use the incomplete.obs vector along with survey and Equation (20.3) to calculate $\hat{\beta}_0$ and $\hat{\beta}_1$ in R. (Remember the functions mean, sd, and cor. Ensure your answers match the output from survfit.)

c. The survey data frame has a number of other variables present aside from Height and Wr.Hnd. For this exercise, the end aim is to fit a simple linear model to predict the mean student height, but this time from their pulse rate, given in Pulse (continue to assume the conditions listed in Section 20.2 are satisfied).

    i. Fit the regression model and produce a scatterplot with the fitted line superimposed upon the data. Make sure you can write down the fitted model equation and keep the plot open.

    ii. Identify and interpret the point estimate of the slope, as well as the outcome of the test associated with the hypotheses $H_0 : \beta_1 = 0$; $H_A : \beta_1 \neq 0$. Also find a 90 percent CI for the slope parameter.

    iii. Using your model, add lines for 90 percent confidence and prediction interval bands on the plot from (i) and add a legend to differentiate between the lines.

    iv. Create an incomplete.obs vector for the current "height on pulse" data. Use that vector to calculate the sample mean of the height observations that were used for the model fitted in (i). Then add a perfectly horizontal line to the plot at this mean (use color or line type options to avoid confusion with the other lines present). What do you notice? Does the plot support your conclusions from (ii)?

Next, examine the help file for the mtcars data set, which you first saw in Exercise 13.4 on page 287. For this exercise, the goal is to model fuel efficiency, measured in miles per gallon (MPG), in terms of the overall weight of the vehicle (in thousands of pounds).

d. Plot the data—mpg on the y-axis and wt on the x-axis.

e. Fit the simple linear regression model. Add the fitted line to the plot from (d).

f. Write down the regression equation and interpret the point estimate of the slope. Is the effect of wt on mean mpg estimated to be statistically significant?

g. Produce a point estimate and associated 95 percent PI for a car that weighs 6,000 lbs. Do you trust the model to predict observations accurately for this value of the explanatory variable? Why or why not?

## 20.5 Understanding Categorical Predictors

So far, you've looked at simple linear regression models that rely on continuous explanatory variables, but it's also possible to use a discrete or categorical explanatory variable, made up of $k$ distinct groups or levels, to model the mean response. You must be able to make the same assumptions noted in Section 20.2: that observations are all independent of one another and residuals are normally distributed with an equal variance. To begin with, you'll look at the simplest case in which $k = 2$ (a binary-valued predictor), which forms the basis of the slightly more complicated situation in which the categorical predictor has more than two levels (a multilevel predictor: $k > 2$).

### 20.5.1 Binary Variables: k = 2

Turn your attention back to Equation (20.1), where the regression model is specified as $Y|X = \beta_0 + \beta_1 X + \epsilon$ for a response variable $Y$ and predictor $X$, and $\epsilon \sim N(0, \sigma^2)$. Now, suppose your predictor variable is categorical, with only two possible levels (binary; $k = 2$) and observations coded either 0 or 1. For this case, (20.1) still holds, but the interpretation of the model parameters, $\beta_0$ and $\beta_1$, isn't really one of an "intercept" and a "slope" anymore. Instead, it's better to think of them as being something like two intercepts, where $\beta_0$ provides the *baseline* or *reference* value of the response when $X = 0$ and $\beta_1$ represents the *additive effect* on the mean response if $X = 1$. In other words, if $X = 0$, then $Y = \beta_0 + \epsilon$; if $X = 1$, then $Y = \beta_0 + \beta_1 + \epsilon$. As usual, estimation is in terms of finding the *mean* response $\hat{y} \equiv \mathbb{E}[Y|X = x]$ as per Equation (20.2), so the equation becomes $\hat{y} = \hat{\beta}_0 + \hat{\beta}_1 x$.

Go back to the survey data frame and note that you have a Sex variable, where the students recorded their gender. Look at the documentation on the help page ?survey or enter something like this:

```
R> class(survey$Sex)
[1] "factor"
R> table(survey$Sex)

Female   Male
   118    118
```

You'll see that the sex data column is a factor vector with two levels, Female and Male, and that there happens to be an equal number of the two (one of the 237 records has a missing value for this variable).

You're going to determine whether there is statistical evidence that the height of a student is affected by sex. This means that you're again interested in modeling height as the response variable, but this time, it's with the categorical sex variable as the predictor.

To visualize the data, if you make a call to plot as follows, you'll get a pair of boxplots.

```
R> plot(survey$Height~survey$Sex)
```

This is because the response variable specified to the left of the ~ is numeric and the explanatory variable to the right is a factor, and the default behavior of R in that situation is to produce side-by-side boxplots.

To further emphasize the categorical nature of the explanatory variable, you can superimpose the raw height and sex observations on top of the boxplots. To do this, just convert the factor vector to numeric with a call to as.numeric; this can be done directly in a call to points.

```
R> points(survey$Height~as.numeric(survey$Sex),cex=0.5)
```

Remember that boxplots mark off the median as the central bold line but that least-squares linear regressions are defined by the mean outcome, so it's useful to also display the mean heights according to sex.

```
R> means.sex <- tapply(survey$Height,INDEX=survey$Sex,FUN=mean,na.rm=TRUE)
R> means.sex
  Female     Male
165.6867 178.8260
R> points(1:2,means.sex,pch=4,cex=3)
```

You were introduced to tapply in Section 10.2.3; in this call, the argument na.rm=TRUE is matched to the ellipsis in the definition of tapply and is passed to mean (you need it to ensure the missing values present in the data do not end up producing NAs as the results). A further call to points adds those coordinates (as × symbols) to the image; Figure 20-4 gives the final result.

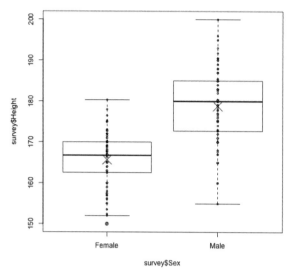

Figure 20-4: Boxplots of the student heights split by sex, with the raw observations and sample means (small ○ and large × symbols, respectively) superimposed

The plot indicates, overall, that males tend to be taller than females—but is there statistical evidence of a difference to back this up?

### Linear Regression Model of Binary Variables

To answer this with a simple linear regression model, you can use lm to produce least-squares estimates just like with every other model you've fitted so far.

```
R> survfit2 <- lm(Height~Sex,data=survey)
R> summary(survfit2)

Call:
lm(formula = Height ~ Sex, data = survey)

Residuals:
    Min      1Q  Median      3Q     Max
-23.886  -5.667   1.174   4.358  21.174

Coefficients:
             Estimate Std. Error t value Pr(>|t|)
(Intercept)  165.687      0.730  226.98   <2e-16 ***
SexMale       13.139      1.022   12.85   <2e-16 ***
---
```

```
Signif. codes:  0 '***' 0.001 '**' 0.01 '*' 0.05 '.' 0.1 ' ' 1

Residual standard error: 7.372 on 206 degrees of freedom
  (29 observations deleted due to missingness)
Multiple R-squared:  0.4449, Adjusted R-squared:  0.4422
F-statistic: 165.1 on 1 and 206 DF,  p-value: < 2.2e-16
```

However, because the predictor is a factor vector instead of a numeric vector, the reporting of the coefficients is slightly different. The estimate of $\beta_0$ is again reported as (Intercept); this is the estimate of the mean height if a student is female. The estimate of $\beta_1$ is reported as SexMale. The corresponding regression coefficient of 13.139 is the estimated difference that is imparted upon the mean height of a student if male. If you look at the corresponding regression equation

$$\hat{y} = \hat{\beta}_0 + \hat{\beta}_1 x = 165.687 + 13.139x \qquad (20.7)$$

you can see that the model has been fitted assuming the variable $x$ is defined as "the individual is male"—0 for no/false, 1 for yes/true. In other words, the level of "female" for the sex variable is assumed as a reference, and it is the effect of "being male" on mean height that is explicitly estimated. The hypothesis test for $\beta_0$ and $\beta_1$ is performed with the same hypotheses defined in Section 20.3.2:

$$H_0 : \beta_j = 0$$
$$H_A : \beta_j \neq 0$$

Again, it's the test for $\beta_1$ that's generally of the most interest since it's this value that tells you whether there is statistical evidence that the mean response variable is affected by the explanatory variable, that is, if $\beta_1$ is significantly different from zero.

### Predictions from a Binary Categorical Variable

Because there are only two possible values for $x$, prediction is straightforward here. When you evaluate the equation, the only decision that needs to be made is whether $\hat{\beta}_1$ needs to be used (in other words, if an individual is male) or not (if an individual is female). For example, you can enter the following code to create a factor of five extra observations with the same level names as the original data and store the new data in extra.obs:

```
R> extra.obs <- factor(c("Female","Male","Male","Male","Female"))
R> extra.obs
[1] Female Male   Male   Male   Female
Levels: Female Male
```

Then, use predict in the now-familiar fashion to find the mean heights at those extra values of the predictor. (Remember that when you pass in new data to predict using the newdata argument, the predictors must be in the same form as the data that were used to fit the model in the first place.)

```
R> predict(survfit2,newdata=data.frame(Sex=extra.obs),interval="confidence",
        level=0.9)
       fit      lwr      upr
1 165.6867 164.4806 166.8928
2 178.8260 177.6429 180.0092
3 178.8260 177.6429 180.0092
4 178.8260 177.6429 180.0092
5 165.6867 164.4806 166.8928
```

You can see from the output that the predictions are different only between the two sets of values—the point estimates of the two instances of Female are identical, simply $\hat{\beta}_0$ with 90 percent CIs. The point estimates and CIs for the instances of Male are also all the same as each other, based on a point estimate of $\hat{\beta}_0 + \hat{\beta}_1$.

On its own, admittedly, this example isn't too exciting. However, it's critical to understand how R presents regression results when using categorical predictors, especially when considering multiple regression in Chapter 21.

### 20.5.2    Multilevel Variables: k > 2

It's common to work with data where the categorical predictor variables have more than two levels so that ($k > 2$). These can also be referred to as *multilevel* categorical variables. To deal with this more complicated situation while retaining interpretability of your parameters, you must first dummy code your predictor into $k - 1$ binary variables.

#### Dummy Coding Multilevel Variables

To see how this is done, assume that you want to find the value of response variable $Y$ when given the value of a categorical variable $X$, where $X$ has $k > 2$ levels (also assume the conditions for validity of the linear regression model—Section 20.2—are satisfied).

In regression modeling, *dummy coding* is the procedure used to create several binary variables from a categorical variable like $X$. Instead of the single categorical variable with possible realizations

$$X = 1, 2, 3, \ldots, k$$

you recode it into several yes/no variables—one for each level—with possible realizations:

$$X_{(1)} = 0, 1; \quad X_{(2)} = 0, 1; \quad X_{(3)} = 0, 1; \quad \ldots; \quad X_{(k)} = 0, 1$$

As you can see, $X_{(i)}$ represents a binary variable for the $i$th level of the original $X$. For example, if an individual has $X = 2$ in the original categorical variable, then $X_{(2)} = 1$ (yes) and all of the others ($X_{(1)}, X_{(3)}, \ldots, X_{(k)}$) will be zero (no).

Suppose $X$ is a variable that can take any one of the $k = 4$ values 1, 2, 3, or 4, and you've made six observations of this variable: 1, 2, 2, 1, 4, 3. Table 20-1 shows these observations and their dummy-coded equivalents $X_{(1)}$, $X_{(2)}$, $X_{(3)}$, and $X_{(4)}$.

**Table 20-1:** Illustrative Example of Dummy Coding for Six Observations of a Categorical Variable with $k = 4$ Groups

| $X$ | $X_{(1)}$ | $X_{(2)}$ | $X_{(3)}$ | $X_{(4)}$ |
|---|---|---|---|---|
| 1 | 1 | 0 | 0 | 0 |
| 2 | 0 | 1 | 0 | 0 |
| 2 | 0 | 1 | 0 | 0 |
| 1 | 1 | 0 | 0 | 0 |
| 4 | 0 | 0 | 0 | 1 |
| 3 | 0 | 0 | 1 | 0 |

In fitting the subsequent model, you usually only use $k - 1$ of the dummy binary variables—one of the variables acts as a *reference* or *baseline* level, and it's incorporated into the overall intercept of the model. In practice, you would end up with an estimated model like this,

$$\hat{y} = \hat{\beta}_0 + \hat{\beta}_1 X_{(2)} + \hat{\beta}_2 X_{(3)} + \ldots + \hat{\beta}_{k-1} X_{(k)} \tag{20.8}$$

assuming 1 is the reference level. As you can see, in addition to the overall intercept term $\hat{\beta}_0$, you have $k - 1$ other estimated intercept terms that modify the baseline coefficient $\hat{\beta}_0$, depending on which of the original categories an observation takes on. For example, in light of the coding imposed in (20.8), if an observation has $X_{(3)} = 1$ and all other binary values are therefore zero (so that observation would've had a value of $X = 3$ for the original categorical variable), the predicted mean value of the response would be $\hat{y} = \hat{\beta}_0 + \hat{\beta}_2$. On the other hand, because the reference level is defined as 1, if an observation has values of zero for all the binary variables, it implies the observation originally had $X = 1$, and the prediction would be simply $\hat{y} = \hat{\beta}_0$.

The reason it's necessary to dummy code for categorical variables of this nature is that, in general, categories cannot be related to each other in the same numeric sense as continuous variables. It's often not appropriate, for example, to think that an observation in category 4 is "twice as much" as one in category 2, which is what the estimation methods would assume. Binary presence/absence variables are valid, however, and can be easily incorporated into the modeling framework. Choosing the reference level is generally of secondary importance—the specific values of the estimated coefficients will change accordingly, but any overall interpretations you make based on the fitted model will be the same regardless.

**NOTE**    *Implementing this dummy-coding approach is technically a form of multiple regression since you're now including several binary variables in the model. It's important, however, to be aware of the somewhat artificial nature of dummy coding—you should still think of the multiple coefficients as representing a single categorical variable since the binary variables $X_{(1)}, \ldots, X_{(k)}$ are not independent of one another. This is why I've chosen to define these models in this chapter; multiple regression will be formally discussed in Chapter 21.*

## Linear Regression Model of Multilevel Variables

R makes working with categorical predictors in this way quite simple since it automatically dummy codes for any such explanatory variable when you call lm. There are two things you should check before fitting your model, though.

1.  The categorical variable of interest should be stored as a (formally unordered) factor.

2.  You should check that you're happy with the category assigned as the reference level (for interpretative purposes—see Section 20.5.3).

You must also of course be happy with the validity of the familiar assumptions of normality and independence of $\epsilon$.

To demonstrate all these definitions and ideas, let's return to the student survey data from the MASS package and keep "student height" as the response variable of interest. Among the data is the variable Smoke. This variable describes the kind of smoker each student reports themselves as, defined by frequency and split into four categories: "heavy," "never," "occasional," and "regular."

```
R> is.factor(survey$Smoke)
[1] TRUE
R> table(survey$Smoke)

Heavy Never Occas Regul
   11   189    19    17
R> levels(survey$Smoke)
[1] "Heavy" "Never" "Occas" "Regul"
```

Here, the result from is.factor(survey$Smoke) indicates that you do indeed have a factor vector at hand, the call to table yields the number of students in each of the four categories, and as per Chapter 5, you can explicitly request the levels attribute of any R factor via levels.

Let's ask whether there's statistical evidence to support a difference in mean student height according to smoking frequency. You can create a set of boxplots of these data with the following two lines; Figure 20-5 shows the result.

```
R> boxplot(Height~Smoke,data=survey)
R> points(1:4,tapply(survey$Height,survey$Smoke,mean,na.rm=TRUE),pch=4)
```

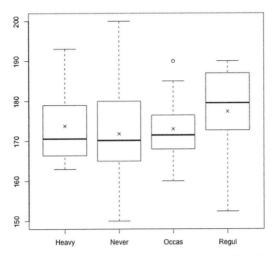

Figure 20-5: Boxplots of the observed student heights split by smoking frequency; respective sample means marked with ×

Note from earlier R output that unless explicitly defined at creation, the levels of a factor appear in alphabetical order by default—as is the case for Smoke—and R will automatically set the first one (as shown in the output of a call to levels) as the reference level when that factor is used as a predictor in subsequent model fitting. Fitting the linear model in mind using lm, you can see from a subsequent call to summary that indeed the first level of Smoke, for "heavy", has been used as the reference:

```
R> survfit3 <- lm(Height~Smoke,data=survey)
R> summary(survfit3)

Call:
lm(formula = Height ~ Smoke, data = survey)

Residuals:
   Min     1Q Median     3Q    Max
-25.02  -6.82  -1.64   8.18  28.18

Coefficients:
            Estimate Std. Error t value Pr(>|t|)
(Intercept) 173.7720     3.1028  56.005   <2e-16 ***
SmokeNever   -1.9520     3.1933  -0.611    0.542
SmokeOccas   -0.7433     3.9553  -0.188    0.851
SmokeRegul    3.6451     4.0625   0.897    0.371
---
Signif. codes:  0 '***' 0.001 '**' 0.01 '*' 0.05 '.' 0.1 ' ' 1

Residual standard error: 9.812 on 205 degrees of freedom
  (28 observations deleted due to missingness)
```

```
Multiple R-squared:  0.02153, Adjusted R-squared:  0.007214
F-statistic: 1.504 on 3 and 205 DF,  p-value: 0.2147
```

As outlined in Equation (20.8), you get estimates of coefficients corresponding to the dummy binary variables for three of the four possible categories in this example—the three nonreference levels. The observation in the reference category Heavy is represented solely by $\hat{\beta}_0$, designated first as the overall (Intercept), with the other coefficients providing the effects associated with an observation in one of the other categories.

### Predictions from a Multilevel Categorical Variable

You find point estimates through prediction, as usual.

```
R> one.of.each <- factor(levels(survey$Smoke))
R> one.of.each
[1] Heavy Never Occas Regul
Levels: Heavy Never Occas Regul
R> predict(survfit3,newdata=data.frame(Smoke=one.of.each),
          interval="confidence",level=0.95)
       fit      lwr      upr
1 173.7720 167.6545 179.8895
2 171.8200 170.3319 173.3081
3 173.0287 168.1924 177.8651
4 177.4171 172.2469 182.5874
```

Here, I've created the object one.of.each for illustrative purposes; it represents one observation in each of the four categories, stored as an object matching the class (and levels) of the original Smoke data. A student in the Occas category, for example, is predicted to have a mean height of $173.772 - 0.7433 = 173.0287$.

The output from the model summary earlier, however, shows that none of the binary dummy variable coefficients are considered statistically significant from zero (because all the $p$-values are too large). The results indicate, as you might have suspected, that there's no evidence that smoking frequency (or more specifically, having a smoking frequency that's different from the reference level) affects mean student heights based on this sample of individuals. As is common, the baseline coefficient $\hat{\beta}_0$ is highly statistically significant—but that only suggests that the overall intercept probably isn't zero. (Because your response variable is a measurement of height and will clearly not be centered anywhere near 0 cm, that result makes sense.) The confidence intervals supplied are calculated in the usual $t$-based fashion.

The small R-Squared value reinforces this conclusion, indicating that barely any of the variation in the response can be explained by changing the category of smoking frequency. Furthermore, the overall $F$-test $p$-value is rather large at around 0.215, suggesting an overall nonsignificant effect of the predictor on the response; you'll look at this in more detail in a moment in Section 20.5.5 and later on in Section 21.3.5.

As noted earlier, it's important that you interpret these results—indeed any based on a $k$-level categorical variable in regression—in a *collective* fashion. You can claim only that there is *no* discernible effect of smoking on height because *all* the $p$-values for the binary dummy coefficients are nonsignificant. If one of the levels was in fact highly significant (through a small $p$-value), it would imply that the smoking factor as defined here, as a whole, *does* have a statistically detectable effect on the response (even if the other two levels were still associated with very high $p$-values). This will be discussed further in several more examples in Chapter 21.

### 20.5.3 Changing the Reference Level

Sometimes you might decide to change the automatically selected reference level, compared to which the effects of taking on any of the other levels are estimated. Changing the baseline will result in the estimation of different coefficients, meaning that individual $p$-values are subject to change, but the overall result (in terms of global significance of the factor) will not be affected. Because of this, altering the reference level is only done for interpretative purposes—sometimes there's an intuitively natural baseline of the predictor (for example, "Placebo" versus "Drug A" and "Drug B" as a treatment variable in the analysis of some clinical trial) from which you want to estimate deviation in the mean response with respect to the other possible categories.

Redefining the reference level can be achieved quickly using the built-in relevel function in R. This function allows you to choose which level comes first in the definition of a given factor vector object and will therefore be designated as the reference level in subsequent model fitting. In the current example, let's say you'd rather have the nonsmokers as the reference level.

```
R> SmokeReordered <- relevel(survey$Smoke,ref="Never")
R> levels(SmokeReordered)
[1] "Never" "Heavy" "Occas" "Regul"
```

The relevel function has moved the Never category into the first position in the new factor vector. If you go ahead fit the model again using SmokeReordered instead of the original Smoke column of survey, it'll provide estimates of coefficients associated with the three different levels of smokers.

It's worth noting the differences in the treatment of unordered versus ordered factor vectors in regression applications. It might seem sensible to formally order the smoking variable by, for example, increasing the frequency of smoking when creating a new factor vector. However, when an ordered factor vector is supplied in a call to lm, R reacts in a different way—it doesn't perform the relatively simple dummy coding discussed here, where an effect is associated with each optional level to the baseline (technically referred to as *orthogonal contrasts*). Instead, the default behavior is to fit the model based on something called *polynomial contrasts*, where the effect of the ordered categorical variable on the response is defined in a more complicated functional form. That discussion is beyond the scope of this text, but it suffices to say that this approach can be beneficial when your interest lies in the specific functional nature of "moving up" through an ordered set of categories. For more on the technical details, see Kuhn and Johnson (2013). For all relevant regression examples in this book, we'll work exclusively with unordered factor vectors.

### 20.5.4   Treating Categorical Variables as Numeric

The way in which lm decides to define the parameters of the fitted model depends primarily on the kind of data you pass to the function. As discussed, lm imposes dummy coding only if the explanatory variable is an unordered factor vector.

Sometimes the categorical data you want to analyze haven't been stored as a factor in your data object. If the categorical variable is a character vector, lm will implicitly coerce it into a factor. If, however, the intended categorical variable is numeric, then lm performs linear regression exactly as if it were a continuous numeric predictor; it estimates a single regression coefficient, which is interpreted as a "per-one-unit-change" in the mean response.

This may seem inappropriate if the original explanatory variable is supposed to be made up of distinct groups. In some settings, however, especially when the variable can be naturally treated as numeric-discrete, this treatment is not only valid statistically but also helps with interpretation.

Let's take a break from the survey data and go back to the ready-to-use mtcars data set. Say you're interested in the variables mileage, mpg (continuous), and number of cylinders, cyl (discrete; the data set contains cars with either 4, 6, or 8 cylinders). Now, it's perfectly sensible to automatically think of cyl as a categorical variable. Taking mpg to be the response variable, boxplots are well suited to reflect the grouped nature of cyl as a predictor; the result of the following line is given on the left of Figure 20-6:

```
R> boxplot(mtcars$mpg~mtcars$cyl,xlab="Cylinders",ylab="MPG")
```

When fitting the associated regression model, you must be aware of what you're instructing R to do. Since the cyl column of mtcars is numeric, and not a factor vector per se, lm will treat it as continuous if you just directly access the data frame.

```
R> class(mtcars$cyl)
[1] "numeric"
R> carfit <- lm(mpg~cyl,data=mtcars)
R> summary(carfit)

Call:
lm(formula = mpg ~ cyl, data = mtcars)

Residuals:
    Min      1Q  Median      3Q     Max
-4.9814 -2.1185  0.2217  1.0717  7.5186
```

```
Coefficients:
            Estimate Std. Error t value Pr(>|t|)
(Intercept)  37.8846     2.0738   18.27  < 2e-16 ***
cyl          -2.8758     0.3224   -8.92 6.11e-10 ***
---
Signif. codes:  0 '***' 0.001 '**' 0.01 '*' 0.05 '.' 0.1 ' ' 1

Residual standard error: 3.206 on 30 degrees of freedom
Multiple R-squared:  0.7262, Adjusted R-squared:  0.7171
F-statistic: 79.56 on 1 and 30 DF,  p-value: 6.113e-10
```

Just as in earlier sections, you've received an intercept and a slope estimate; the latter is highly statistically significant, indicating that there is evidence against the true value of the slope being zero. Your fitted regression line is

$$\hat{y} = \hat{\beta}_0 + \hat{\beta}_1 x = 37.88 - 2.88x$$

where $\hat{y}$ is the average mileage and $x$ is numeric—the number of cylinders. For each single additional cylinder, the model says your mileage will decrease by 2.88 MPG, on average.

It's important to recognize the fact that you've fitted a continuous line to what is effectively categorical data. The right panel of Figure 20-6, created with the following lines, highlights this fact:

```
R> plot(mtcars$mpg~mtcars$cyl,xlab="Cylinders",ylab="MPG")
R> abline(carfit,lwd=2)
```

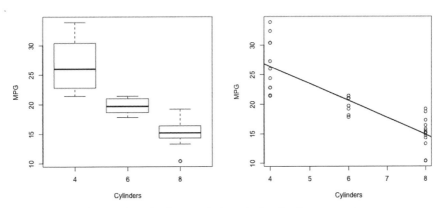

Figure 20-6: Left: Boxplots of mileage split by cylinders for the mtcars data set. Right: Scatterplot of the same data with fitted regression line (treating cyl as numeric-continuous) superimposed.

Some researchers fit categorical or discrete predictors as continuous variables purposefully. First, it allows interpolation; for example, you could use this model to evaluate the average MPG for a 5-cylinder car. Second, it means there are fewer parameters that require estimation; in other words, instead of $k - 1$ intercepts for a categorical variable with $k$ groups, you need only one parameter for the slope. Finally, it can be a convenient way to control for so-called nuisance variables; this will become clearer in Chapter 21. On the other hand, it means that you no longer get group-specific information. It can be misleading to proceed in this way if any differences in the mean response according to the predictor category of an observation are not well represented linearly—detection of significant effects can be lost altogether.

At the very least, it's important to recognize this distinction when fitting models. If you had only just now recognized that R had fitted the cyl variable as continuous and wanted to actually fit the model with cyl as categorical, you'd have to explicitly convert it into a factor vector beforehand or in the actual call to lm.

```
R> carfit <- lm(mpg~factor(cyl),data=mtcars)
R> summary(carfit)

Call:
lm(formula = mpg ~ factor(cyl), data = mtcars)

Residuals:
    Min      1Q  Median      3Q     Max
-5.2636 -1.8357  0.0286  1.3893  7.2364

Coefficients:
             Estimate Std. Error t value Pr(>|t|)
(Intercept)   26.6636     0.9718  27.437  < 2e-16 ***
factor(cyl)6  -6.9208     1.5583  -4.441 0.000119 ***
factor(cyl)8 -11.5636     1.2986  -8.905 8.57e-10 ***
---
Signif. codes:  0 '***' 0.001 '**' 0.01 '*' 0.05 '.' 0.1 ' ' 1

Residual standard error: 3.223 on 29 degrees of freedom
Multiple R-squared:  0.7325, Adjusted R-squared:  0.714
F-statistic:  39.7 on 2 and 29 DF,  p-value: 4.979e-09
```

Here, by wrapping cyl in a call to factor when specifying the formula for lm, you can see you've obtained regression coefficient estimates for the levels of cyl corresponding to 6- and 8-cylinder cars (with the reference level automatically set to 4-cylinder cars).

## 20.5.5 Equivalence with One-Way ANOVA

There's one final observation to make about regression models with a single nominal categorical predictor. Think about the fact that these models describe a mean response value for the *k* different groups. Does this remind you of anything? In this particular setting, you're actually doing the same thing as in one-way ANOVA (Section 19.1): comparing more than two means and determining whether there is statistical evidence that at least one mean is different from the others. You need to be able to make the same key assumptions of independence and normality for both techniques.

In fact, simple linear regression with a single categorical predictor, implemented using least-squares estimation, is just another way to perform one-way ANOVA. Or, perhaps more concisely, ANOVA is a special case of least-squares regression. The outcome of a one-way ANOVA test is a single *p*-value quantifying a level of statistical evidence against the null hypothesis that states that group means are equal. When you have one categorical predictor in a regression, it's exactly that *p*-value that's reported at the end of the summary of an lm object—something I've referred to a couple of times now as the "overall" or "global" significance test (for example, in Section 20.3.3).

Look back to the final result of that global significance test for the student height modeled by smoking status example—you had a *p*-value of 0.2147. This came from an *F* test statistic of 1.504 with $df_1 = 3$ and $df_2 = 205$. Now, suppose you were just handed the data and asked to perform a one-way ANOVA of height on smoking. Using the aov function as introduced in Section 19.1, you'd call something like this:

```
R> summary(aov(Height~Smoke,data=survey))
             Df Sum Sq Mean Sq F value Pr(>F)
Smoke         3    434  144.78   1.504  0.215
Residuals   205  19736   96.27
28 observations deleted due to missingness
```

Those same values are returned here; you can also find the square root of the MSE:

```
R> sqrt(96.27)
[1] 9.811728
```

This is in fact the "residual standard error" given in the lm summary. The two conclusions you'd draw about the impact of smoking status on height (one for the lm output, the other for the ANOVA test) are of course also the same.

The global test that lm provides isn't just there for the benefit of confirming ANOVA results. As a generalization of ANOVA, least-squares regression models provide more than just coefficient-specific tests. That global test is formally referred to as the *omnibus* F-*test*, and while it is indeed equivalent to one-way ANOVA in the "single categorical predictor" setting, it's also a useful overall, stand-alone test of the statistical contribution of several predictors to the outcome value. You'll explore this further in Section 21.3.5 after you've begun modeling your response variable using multiple explanatory variables.

## Exercise 20.2

Continue using the survey data frame from the package MASS for the next few exercises.

a. The survey data set has a variable named Exer, a factor with $k = 3$ levels describing the amount of physical exercise time each student gets: none, some, or frequent. Obtain a count of the number of students in each category and produce side-by-side boxplots of student height split by exercise.

b. Assuming independence of the observations and normality as usual, fit a linear regression model with height as the response variable and exercise as the explanatory variable (dummy coding). What's the default reference level of the predictor? Produce a model summary.

c. Draw a conclusion based on the fitted model from (b)—does it appear that exercise frequency has any impact on mean height? What is the nature of the estimated effect?

d. Predict the mean heights of one individual in each of the three exercise categories, accompanied by 95 percent prediction intervals.

e. Do you arrive at the same result and interpretation for the height-by-exercise model if you construct an ANOVA table using aov?

f. Is there any change to the outcome of (e) if you alter the model so that the reference level of the exercise variable is "none"? Would you expect there to be?

Now, turn back to the ready-to-use mtcars data set. One of the variables in this data frame is qsec, described as the time in seconds it takes to race a quarter mile; another is gear, the number of forward gears (cars in this data set have either 3, 4, or 5 gears).

g. Using the vectors straight from the data frame, fit a simple linear regression model with qsec as the response variable and gear as the explanatory variable and interpret the model summary.

h. Explicitly convert gear to a factor vector and refit the model. Compare the model summary with that from (g). What do you find?

i. Explain, with the aid of a relevant plot in the same style as the right image of Figure 20-6, why you think there is a difference between the two models (g) and (h).

## Important Code in This Chapter

| Function/operator | Brief description | First occurrence |
|---|---|---|
| lm | Fit linear model | Section 20.2.3, p. 455 |
| coef | Get estimated coefficients | Section 20.2.4, p. 457 |
| summary | Summarize linear model | Section 20.3.1, p. 458 |
| confint | Get CIs for estimated coefficients | Section 20.3.2, p. 460 |
| predict | Predict from linear model | Section 20.4.2, p. 463 |
| relevel | Change factor reference level | Section 20.5.3, p. 477 |

# 21

## MULTIPLE LINEAR REGRESSION

Multiple linear regression is a straight-forward generalization of the single-predictor models discussed in the previous chapter. It allows you to model your continuous response variable in terms of more than one predictor so you can measure the joint effect of several explanatory variables on the response variable. In this chapter, you'll see how to model your response variable in this way, and you'll use R to fit the model using least-squares. You'll also explore other key statistical aspects of linear modeling in the R environment, such as transforming variables and including interactive effects.

Multiple linear regression represents an important part of the practice of statistics. It lets you control or adjust for multiple sources of influence on the value of the response, rather than just measuring the effect of one explanatory variable (in most situations, there is more than one contributor to the outcome measurements). At the heart of this class of methods is the intention to uncover potentially causal relationships between your response variable and the (joint) effect of any explanatory variables. In reality, causality itself is extremely difficult to establish, but you can strengthen any evidence of causality by using a well-designed study supported by sound data collection and by fitting models that might realistically gauge the relationships present in your data.

## 21.1 Terminology

Before you look at the theory behind multiple regression models, it's important to have a clear understanding of some terminology associated with variables.

- A *lurking variable* influences the response, another predictor, or both, but goes unmeasured (or is not included) in a predictive model. For example, say a researcher establishes a link between the volume of trash thrown out by a household and whether the household owns a trampoline. The potential lurking variable here would be the number of children in the household—this variable is more likely to be positively associated with an increase in trash and chances of owning a trampoline. An interpretation that suggests owning a trampoline is a cause of increased waste would be erroneous.

- The presence of a lurking variable can lead to spurious conclusions about causal relationships between the response and the other predictors, or it can mask a true cause-and-effect association; this kind of error is referred to as *confounding*. To put it another way, you can think of confounding as the entanglement of the effects of one or more predictors on the response.

- A *nuisance* or *extraneous variable* is a predictor of secondary or no interest that has the potential to confound relationships between other variables and so affect your estimates of the other regression coefficients. Extraneous variables are included in the modeling as a matter of necessity, but the specific nature of their influence on the response is not the primary interest of the analysis.

These definitions will become clearer once you begin fitting and interpreting the regression models in Section 21.3. The main message I want to emphasize here, once more, is that correlation does not imply causation. If a fitted model finds a statistically significant association between a predictor (or predictors) and a response, it's important to consider the possibility that lurking variables are contributing to the results and to attempt to control any confounding before you draw conclusions. Multiple regression models allow you to do this.

## 21.2 Theory

Before you start using R to fit regression models, you'll examine the technical definitions of a linear regression model with multiple predictors. Here, you'll look at how the models work in a mathematical sense and get a glimpse of the calculations that happen "behind the scenes" when estimating the model parameters in R.

### 21.2.1   Extending the Simple Model to a Multiple Model

Rather than having just one predictor, you want to determine the value of a continuous response variable $Y$ given the values of $p > 1$ independent explanatory variables $X_1, X_2, \ldots, X_p$. The overarching model is defined as

$$Y = \beta_0 + \beta_1 X_1 + \beta_2 X_2 + \ldots + \beta_p X_p + \epsilon, \qquad (21.1)$$

where $\beta_0, \ldots, \beta_p$ are the regression coefficients and, as before, you assume independent, normally distributed residuals $\epsilon \sim \mathrm{N}(0, \sigma)$ around the mean.

In practice, you have $n$ data records; each record provides values for each of the predictors $X_j; j = \{1, \ldots, p\}$. The model to be fitted is given in terms of the mean response, conditional upon a particular realization of the set of explanatory variables

$$\hat{y} = \mathbb{E}[Y | X_1 = x_1, X_2 = x_2, \ldots, X_p = x_p] = \hat{\beta}_0 + \hat{\beta}_1 x_1 + \hat{\beta}_2 x_2 + \ldots + \hat{\beta}_p x_p,$$

where the $\hat{\beta}_j$s represent estimates of the regression coefficients.

In simple linear regression, where you have only one predictor variable, recall that the goal is to find the "line of best fit." The idea of least-squares estimation for linear models with multiple independent predictors follows much the same motivation. Now, however, in an abstract sense you can think of the relationship between response and predictors as a multidimensional plane or surface. You want to find the surface that best fits your multivariate data in terms of minimizing the overall squared distance between itself and the raw response data.

More formally, for your $n$ data records, the $\hat{\beta}_j$s are found as the values that minimize the sum

$$\sum_{i=1}^{n} \{y_i - (\beta_0 + \beta_1 x_{1,i} + \beta_2 x_{2,i} + \ldots + \hat{\beta}_p x_{p,i})\}^2, \qquad (21.2)$$

where $x_{j,i}$ is the observed value of individual $i$ for explanatory variable $X_j$ and $y_i$ is their response value.

### 21.2.2   Estimating in Matrix Form

The computations involved in minimizing this squared distance (21.2) are made much easier by a *matrix representation* of the data. When dealing with $n$ multivariate observations, you can write Equation (21.1) as follows,

$$Y = X \cdot \beta + \epsilon,$$

where $Y$ and $\epsilon$ denote $n \times 1$ column matrices such that

$$Y = \begin{bmatrix} y_1 \\ y_2 \\ \vdots \\ y_n \end{bmatrix} \quad \text{and} \quad \epsilon = \begin{bmatrix} \epsilon_1 \\ \epsilon_2 \\ \vdots \\ \epsilon_n \end{bmatrix}.$$

Here, $y_i$ and $\epsilon_i$ refer to the response observation and random error term for the $i$th individual. The quantity $\beta$ is a $(p + 1) \times 1$ column matrix of the regression coefficients, and then the observed predictor data for all individuals and explanatory variables are stored in an $n \times (p + 1)$ matrix $X$, called the *design matrix*:

$$\beta = \begin{bmatrix} \beta_0 \\ \beta_1 \\ \vdots \\ \beta_p \end{bmatrix} \quad \text{and} \quad X = \begin{bmatrix} 1 & x_{1,1} & \cdots & x_{p,1} \\ 1 & x_{1,2} & \cdots & x_{p,2} \\ \vdots & \vdots & \ddots & \vdots \\ 1 & x_{1,n} & \cdots & x_{p,n} \end{bmatrix}$$

The minimization of (21.2) providing the estimated regression coefficient values is then found with the following calculation:

$$\hat{\beta} = \begin{bmatrix} \hat{\beta}_0 \\ \hat{\beta}_1 \\ \vdots \\ \hat{\beta}_p \end{bmatrix} = (X^\top \cdot X)^{-1} \cdot X^\top \cdot Y \tag{21.3}$$

It's important to note the following:

- The symbol $\cdot$ represents matrix multiplication, the superscript $^\top$ represents the transpose, and $^{-1}$ represents the inverse when applied to matrices (as per Section 3.3).

- Extending the size of $\beta$ and $X$ (note the leading column of 1s in $X$) to create structures of size $p + 1$ (as opposed to just the number of predictors $p$) allows for the estimation of the overall intercept $\beta_0$.

- As well as (21.3), the design matrix plays a crucial role in the estimation of other quantities, such as the standard errors of the coefficients.

### 21.2.3   A Basic Example

You can manually estimate the $\beta_j$ ($j = 0, 1, \ldots, p$) in R using the functions covered in Chapter 3: %*% (matrix multiplication), t (matrix transposition), and solve (matrix inversion). As a quick demonstration, let's say you have two predictor variables: $X_1$ as continuous and $X_2$ as binary. Your target regression equation is therefore $\hat{y} = \hat{\beta}_0 + \hat{\beta}_1 x_1 + \hat{\beta}_2 x_2$. Suppose you collect the following data, where the response data, data for $X_1$, and data for $X_2$, for $n = 8$ individuals, are given in the columns y, x1, and x2, respectively.

```
R> demo.data <- data.frame(y=c(1.55,0.42,1.29,0.73,0.76,-1.09,1.41,-0.32),
                           x1=c(1.13,-0.73,0.12,0.52,-0.54,-1.15,0.20,-1.09),
                           x2=c(1,0,1,1,0,1,0,1))
R> demo.data
       y    x1 x2
1   1.55  1.13  1
2   0.42 -0.73  0
3   1.29  0.12  1
```

```
4   0.73  0.52   1
5   0.76 -0.54   0
6  -1.09 -1.15   1
7   1.41  0.20   0
8  -0.32 -1.09   1
```

To get your point estimates in $\boldsymbol{\beta} = [\beta_0, \beta_1, \beta_2]^\top$ for the linear model, you first have to construct $\boldsymbol{X}$ and $\boldsymbol{Y}$ as required by (21.3).

```
R> Y <- matrix(demo.data$y)
R> Y
       [,1]
[1,]   1.55
[2,]   0.42
[3,]   1.29
[4,]   0.73
[5,]   0.76
[6,]  -1.09
[7,]   1.41
[8,]  -0.32
R> n <- nrow(demo.data)
R> X <- matrix(c(rep(1,n),demo.data$x1,demo.data$x2),nrow=n,ncol=3)
R> X
     [,1]  [,2] [,3]
[1,]    1  1.13    1
[2,]    1 -0.73    0
[3,]    1  0.12    1
[4,]    1  0.52    1
[5,]    1 -0.54    0
[6,]    1 -1.15    1
[7,]    1  0.20    0
[8,]    1 -1.09    1
```

Now all you have to do is execute the line corresponding to (21.3).

```
R> BETA.HAT <- solve(t(X)%*%X)%*%t(X)%*%Y
R> BETA.HAT
           [,1]
[1,]   1.2254572
[2,]   1.0153004
[3,]  -0.6980189
```

You've just used least-squares to fit your model based on the observed data in demo.data, which results in the estimates $\hat{\beta}_0 = 1.225$, $\hat{\beta}_1 = 1.015$, and $\hat{\beta}_2 = -0.698$.

## 21.3 Implementing in R and Interpreting

Ever helpful, R automatically builds the matrices and carries out all the necessary calculations when you instruct it to fit a multiple linear regression model. As in simple regression models, you use lm and just include any additional predictors when you specify the formula in the first argument. So that you can focus on the R syntax and on interpretation, I'll focus only on *main effects* for the moment, and then you'll explore more complex relationships later in the chapter.

When it comes to output and interpretation, working with multiple explanatory variables follows the same rules as you've seen in Chapter 20. Any numeric-continuous variables (or a categorical variable being treated as such) have a slope coefficient that provides a "per-unit-change" quantity. Any $k$-group categorical variables (factors, formally unordered) are dummy coded and provide $k - 1$ intercepts.

### 21.3.1 Additional Predictors

Let's first confirm the manual matrix calculations from a moment ago. Using the demo.data object, fit the multiple linear model and examine the coefficients from that object as follows:

```
R> demo.fit <- lm(y~x1+x2,data=demo.data)
R> coef(demo.fit)
(Intercept)          x1          x2
  1.2254572   1.0153004  -0.6980189
```

You'll see that you obtain exactly the point estimates stored earlier in BETA.HAT.

With the response variable on the left as usual, you specify the multiple predictors on the right side of the ~ symbol; altogether this represents the formula argument. To fit a model with several main effects, use + to separate any variables you want to include. In fact, you've already seen this notation in Section 19.2.2, when investigating two-way ANOVA.

To study the interpretation of the parameter estimates of a multiple linear regression model, let's return to the survey data set in the MASS package. In Chapter 20, you explored several simple linear regression models based on a response variable of student height, as well as stand-alone predictors of handspan (continuous) and sex (categorical, $k = 2$). You found that handspan was highly statistically significant, with the estimated coefficient suggesting an average increase of about 3.12 cm for each 1 cm increase in handspan. When you looked at the same $t$-test using sex as the explanatory variable, the model also suggested evidence against the null hypothesis, with "being male" adding around 13.14 cm to the mean height when compared to the mean for females (the category used as the reference level).

What those models can't tell you is the *joint effect* of sex and handspan on predicting height. If you include both predictors in a multiple linear model, you can (to some extent) reduce any confounding that might otherwise occur in the isolated fits of the effect of either single predictor on height.

```
R> survmult <- lm(Height~Wr.Hnd+Sex,data=survey)
R> summary(survmult)
Call:
lm(formula = Height ~ Wr.Hnd + Sex, data = survey)

Residuals:
    Min      1Q   Median      3Q     Max
-17.7479  -4.1830   0.7749   4.6665  21.9253

Coefficients:
            Estimate Std. Error t value Pr(>|t|)
(Intercept) 137.6870     5.7131  24.100  < 2e-16 ***
Wr.Hnd        1.5944     0.3229   4.937 1.64e-06 ***
SexMale       9.4898     1.2287   7.724 5.00e-13 ***
---
Signif. codes:  0 '***' 0.001 '**' 0.01 '*' 0.05 '.' 0.1 ' ' 1

Residual standard error: 6.987 on 204 degrees of freedom
  (30 observations deleted due to missingness)
Multiple R-squared:  0.5062,  Adjusted R-squared:  0.5014
F-statistic: 104.6 on 2 and 204 DF,  p-value: < 2.2e-16
```

The coefficient for handspan is now only about 1.59, almost half of its corresponding value (3.12 cm) in the stand-alone simple linear regression for height. Despite this, it's still highly statistically significant in the presence of sex. The coefficient for sex has also reduced in magnitude when compared with its simple linear model and is also still significant in the presence of handspan. You'll interpret these new figures in a moment.

As for the rest of the output, the Residual standard error still provides you with an estimate of the standard error of the random noise term $\epsilon$, and you're also provided with an R-squared value. When associated with more than one predictor, the latter is formally referred to as the coefficient of *multiple* determination. The calculation of this coefficient, as in the single predictor setting, comes from the correlations between the variables in the model. I'll leave the theoretical intricacies to more advanced texts, but it's important to note that R-squared still represents the proportion of variability in the response that's explained by the regression; in this example, it sits at around 0.51.

You can continue to add explanatory variables in the same way if you need to do so. In Section 20.5.2, you examined smoking frequency as a stand-alone categorical predictor for height and found that this explanatory variable provided no statistical evidence of an impact on the mean response. But could the smoking variable contribute in a statistically significant way if you control for handspan and sex?

```
R> survmult2 <- lm(Height~Wr.Hnd+Sex+Smoke,data=survey)
R> summary(survmult2)

Call:
lm(formula = Height ~ Wr.Hnd + Sex + Smoke, data = survey)

Residuals:
     Min       1Q   Median       3Q      Max
-17.4869  -4.7617   0.7604   4.3691  22.1237

Coefficients:
            Estimate Std. Error t value Pr(>|t|)
(Intercept) 137.4056     6.5444  20.996  < 2e-16 ***
Wr.Hnd        1.6042     0.3301   4.860 2.36e-06 ***
SexMale       9.3979     1.2452   7.547 1.51e-12 ***
SmokeNever   -0.0442     2.3135  -0.019    0.985
SmokeOccas    1.5267     2.8694   0.532    0.595
SmokeRegul    0.9211     2.9290   0.314    0.753
---
Signif. codes:  0 '***' 0.001 '**' 0.01 '*' 0.05 '.' 0.1 ' ' 1

Residual standard error: 7.023 on 201 degrees of freedom
  (30 observations deleted due to missingness)
Multiple R-squared:  0.5085, Adjusted R-squared:  0.4962
F-statistic: 41.59 on 5 and 201 DF,  p-value: < 2.2e-16
```

Since it's a categorical variable with $k > 2$ levels, Smoke is dummy coded (with heavy smokers as the default reference level), giving you three extra intercepts for the three nonreference levels of the variable; the fourth is incorporated into the overall intercept.

In the summary of the latest fit, you can see that while handspan and sex continue to yield very small $p$-values, smoking frequency suggests no such evidence against the hypotheses of zero coefficients. The smoking variable has had little effect on the values of the other coefficients compared with the previous model in survmult, and the R-squared coefficient of multiple determination has barely increased.

One question you might now ask is, if smoking frequency doesn't benefit your ability to predict mean height in any substantial way, should you remove that variable from the model altogether? This is the primary goal of *model selection*: to find the "best" model for predicting the outcome, without fitting one that is unnecessarily complex (by including more explanatory variables than is required). You'll look at some common ways researchers attempt to achieve this in Section 22.2.

## 21.3.2 Interpreting Marginal Effects

In multiple regression, the estimation of each predictor takes into account the effect of all other predictors present in the model. A coefficient for a specific predictor Z should therefore be interpreted as the change in the mean response for a one-unit increase in Z, while holding all other predictors constant.

As you've determined that smoking frequency still appears to have no discernible impact on mean height when taking sex and handspan into consideration, return your focus to survmult, the model that includes only the explanatory variables of sex and handspan. Note the following:

- For students of the same sex (that is, focusing on either just males or just females), a 1 cm increase in handspan leads to an estimated increase of 1.5944 cm in mean height.

- For students of similar handspan, males on average will be 9.4898 cm taller than females.

- The difference in the values of the two estimated predictor coefficients when compared with their respective simple linear model fits, plus the fact that both continue to indicate evidence against the null hypothesis of "being zero" in the multivariate fit, suggests that confounding (in terms of the effect of both handspan and sex on the response variable of height) is present in the single-predictor models.

The final point highlights the general usefulness of multiple regression. It shows that, in this example, if you use only single predictor models, the determination of the "true" impact that each explanatory variable has in predicting the mean response is misleading since some of the change in height is determined by sex, but some is also attributed to handspan. It's worth noting that the coefficient of determination (refer to Section 20.3.3) for the survmult model is noticeably higher than the same quantity in either of the single-variate models, so you're actually accounting for more of the variation in the response by using multiple regression.

The fitted model itself can be thought of as

$$\text{"Mean height"} = 137.687 + 1.594 \times \text{"handspan"} + 9.49 \times \text{"sex"} \quad (21.4)$$

where "handspan" is the writing handspan supplied in centimeters and "sex" is supplied as either 1 (if male) or 0 (if female).

**NOTE**    *The baseline (overall) intercept of around 137.687 cm represents the mean height of a female with a handspan of 0 cm—again, this is clearly not directly interpretable in the context of the application. For this kind of situation, some researchers center the offending continuous predictor (or predictors) on zero by subtracting the sample mean of all the observations on that predictor from each observation prior to fitting the model. The* *centered predictor data are then used in place of the original (untranslated) data. The resulting fitted model allows you to use the mean value of the untranslated predictor (in this case handspan) rather than a zero value in order to directly interpret the intercept estimate $\hat{\beta}_0$.*

### 21.3.3   Visualizing the Multiple Linear Model

As shown here, "being male" simply changes the overall intercept by around 9.49 cm:

```
R> survcoefs <- coef(survmult)
R> survcoefs
(Intercept)      Wr.Hnd      SexMale
 137.686951    1.594446     9.489814
R> as.numeric(survcoefs[1]+survcoefs[3])
[1] 147.1768
```

Because of this, you could also write (21.4) as two equations. Here's the equation for female students:

$$\text{"Mean height"} = 137.687 + 1.594 \times \text{"handspan"}$$

Here's the equation for male students:

$$\text{"Mean height"} = (137.687 + 9.4898) + 1.594 \times \text{"handspan"}$$
$$= 147.177 + 1.594 \times \text{"handspan"}$$

This is handy because it allows you to visualize the multivariate model in much the same way as you can the simple linear models. This code produces Figure 21-1:

```
R> plot(survey$Height~survey$Wr.Hnd,
        col=c("gray","black")[as.numeric(survey$Sex)],
        pch=16,xlab="Writing handspan",ylab="Height")
R> abline(a=survcoefs[1],b=survcoefs[2],col="gray",lwd=2)
R> abline(a=survcoefs[1]+survcoefs[3],b=survcoefs[2],col="black",lwd=2)
R> legend("topleft",legend=levels(survey$Sex),col=c("gray","black"),pch=16)
```

First, a scatterplot of the height and handspan observations, split by sex, is drawn. Then, abline adds the line corresponding to females and adds a second one corresponding to males, based on those two equations.

Although this plot might look like two separate simple linear model fits, one for each level of sex, it's important to recognize that isn't the case. You're effectively looking at a representation of a multivariate model on a two-dimensional canvas, where the statistics that determine the fit of the two visible lines have been estimated "jointly," in other words, when considering both predictors.

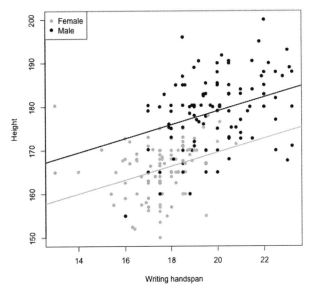

Figure 21-1: Visualizing the observed data and fitted multiple linear model of student height modeled by handspan and sex

### 21.3.4 Finding Confidence Intervals

As in Chapter 20, you can easily find confidence intervals for any of the regression parameters in multiple regression models with confint. Using survmult2, the object of the fitted model for student height including the smoking frequency predictor, the output of a call to confint looks like this:

```
R> confint(survmult2)
                  2.5 %      97.5 %
(Intercept) 124.5010442 150.310074
Wr.Hnd        0.9534078   2.255053
SexMale       6.9426040  11.853129
SmokeNever   -4.6061148   4.517705
SmokeOccas   -4.1312384   7.184710
SmokeRegul   -4.8543683   6.696525
```

Note that the Wr.Hnd and SexMale variables were shown to be statistically significant at the 5 percent level in the earlier model summary and that their 95 percent confidence levels do not include the null value of zero. On the other hand, all the coefficients for the dummy variables associated with the smoking frequency predictor are all nonsignificant, and their confidence intervals clearly include zero. This reflects the fact that the smoking variable isn't, as a whole, considered statistically significant in this particular model.

## 21.3.5 Omnibus F-Test

First encountered in Section 20.5.2 in the context of multilevel predictors, you can think of the omnibus $F$-test more generally for multiple regression models as a test with the following hypotheses:

$$H_0 : \beta_1 = \beta_2 = \ldots = \beta_p = 0$$

$$H_A : \text{At least one of the } \beta_j \neq 0 \ (\text{for } j = 1, \ldots, p) \qquad (21.5)$$

The test is effectively comparing the amount of error attributed to the "null" model (in other words, one with an intercept only) with the amount of error attributed to the predictors when all the predictors are present. In other words, the more the predictors are able to model the response, the more error they explain, giving you a more extreme $F$ statistic and therefore a smaller $p$-value. The single result makes the test especially useful when you have many explanatory variables. The test works the same regardless of the mix of predictors you have in a given model: one or more might be continuous, discrete, binary, and/or categorical with $k > 2$ levels. When multiple regression models are fitted, the amount of output alone can take time to digest and interpret, and care must be taken to avoid Type I errors (incorrect rejection of a true null hypothesis—refer to Section 18.5).

The $F$-test helps boil all that down, allowing you to conclude either of the following:

1. Evidence against $H_0$ if the associated $p$-value is smaller than your chosen significance level $\alpha$, which suggests that your regression—your combination of the explanatory variables—does a significantly better job of predicting the response than if you removed *all* those predictors.

2. No evidence against $H_0$ if the associated $p$-value is larger than $\alpha$, which suggests that using the predictors has no tangible benefit over having an intercept alone.

The downside is that the test doesn't tell you which of the predictors (or which subset thereof) is having a beneficial impact on the fit of the model, nor does it tell you anything about their coefficients or respective standard errors.

You can compute the $F$-test statistic using the coefficient of determination, $R^2$, from the fitted regression model. Let $p$ be the number of regression parameters requiring estimation, excluding the intercept $\beta_0$. Then,

$$\mathcal{F} = \frac{R^2(n - p - 1)}{(1 - R^2)p}, \qquad (21.6)$$

where $n$ is the number of observations used in fitting the model (after records with missing values have been deleted). Then, under $H_0$ in (21.5), $\mathcal{F}$ follows an $F$ distribution (see Section 16.2.5 and also Section 19.1.2) with $df_1 = p$, $df_2 = n - p - 1$ degrees of freedom. The $p$-value associated with (21.6) is yielded as the upper-tail area of that $F$ distribution.

As a quick exercise to confirm this, turn your attention back to the fitted multiple regression model survmult2 in Section 21.3.1, which is the model for student height by handspan, sex, and smoking status from survey. You can extract the coefficient of multiple determination from the summary report (using the technique noted in Section 20.3.4).

```
R> R2 <- summary(survmult2)$r.squared
R> R2
[1] 0.508469
```

This matches the multiple R-squared value from Section 21.3.1. Then, you can get $n$ as the original size of the data set in survey minus any missing values (reported as 30 in the earlier summary output).

```
R> n <- nrow(survey)-30
R> n
[1] 207
```

You get $p$ as the number of estimated regression parameters (minus 1 for the intercept).

```
R> p <- length(coef(survmult2))-1
R> p
[1] 5
```

You can then confirm the value of $n - p - 1$, which matches the summary output (201 degrees of freedom):

```
R> n-p-1
[1] 201
```

Finally, you find the test statistic $\mathcal{F}$ as dictated by (21.6), and you can use the pf function as follows to obtain the corresponding $p$-value for the test:

```
R> Fstat <- (R2*(n-p-1))/((1-R2)*p)
R> Fstat
[1] 41.58529
R> 1-pf(Fstat,df1=p,df2=n-p-1)
[1] 0
```

You can see that the omnibus $F$-test for this example gives a $p$-value that's so small, it's effectively zero. These calculations match the relevant results reported in the output of summary(survmult2) completely.

Looking back at the student height multiple regression fit based on handspan, sex, and smoking in survmult2 in Section 21.3.1, it's little surprise that with two of the predictors yielding small $p$-values, the omnibus $F$-test suggests strong evidence against $H_0$ based on (21.5). This highlights the "umbrella" nature of the omnibus test: although the smoking frequency variable itself doesn't appear to contribute anything statistically important, the *F*-test for that model still suggests survmult2 should be preferred over a "no-predictor" model, because both handspan and sex are important.

### 21.3.6   Predicting from a Multiple Linear Model

Prediction (or *forecasting*) for multiple regression follows the same rules as for simple regression. It's important to remember that point predictions found for a particular *covariate profile*—the collection of predictor values for a given individual—are associated with the mean (or *expected value*) of the response; that confidence intervals provide measures for mean responses; and that prediction intervals provide measures for raw observations. You also have to consider the issue of interpolation (predictions based on *x* values that fall within the range of the originally observed covariate data) versus extrapolation (prediction from *x* values that fall outside the range of said data). Other than that, the R syntax for predict is identical to that used in Section 20.4.

As an example, using the model fitted on student height as a linear function of handspan and sex (in survmult), you can estimate the mean height of a male student with a writing handspan of 16.5 cm, together with a confidence interval.

```
R> predict(survmult,newdata=data.frame(Wr.Hnd=16.5,Sex="Male"),
        interval="confidence",level=0.95)
       fit      lwr      upr
1 173.4851 170.9419 176.0283
```

The result indicates that you have an expected value of about 173.48 cm and that you can be 95 percent confident the true value lies somewhere between 170.94 and 176.03 (rounded to 2 d.p.). In the same way, the mean height of a female with a handspan of 13 cm is estimated at 158.42 cm, with a 99 percent prediction interval of 139.76 to 177.07.

```
R> predict(survmult,newdata=data.frame(Wr.Hnd=13,Sex="Female"),
        interval="prediction",level=0.99)
       fit      lwr      upr
1 158.4147 139.7611 177.0684
```

There are in fact two female students in the data set with writing handspans of 13 cm, as you can see in Figure 21-1. Using your knowledge of subsetting data frames, you can inspect these two records and select the three variables of interest.

```
R> survey[survey$Sex=="Female" & survey$Wr.Hnd==13,c("Sex","Wr.Hnd","Height")]
        Sex Wr.Hnd Height
45   Female     13 180.34
152  Female     13 165.00
```

Now, the second female's height falls well inside the prediction interval, but the first female's height is significantly higher than the upper limit. It's important to realize that, technically, nothing has gone wrong here in terms of the model fitting and interpretation—it's still possible that an observation can fall outside a prediction interval, even a wide 99 percent interval, though it's perhaps improbable. There could be any number of reasons for this occurring. First, the model could be inadequate. For example, you might be excluding important predictors in the fitted model and therefore have less predictive power. Second, although the prediction is within the range of the observed data, it has occurred at one extreme end of the range, where it's less reliable because your data are relatively sparse. Third, the observation itself may be tainted in some way—perhaps the individual recorded her handspan incorrectly, in which case her invalid observation should be removed prior to model fitting. It's with this critical eye that a good statistician will appraise data and models; this is a skill that I'll emphasize further as this chapter unfolds.

### Exercise 21.1

In the MASS package, you'll find the data frame cats, which provides data on sex, body weight (in kilograms), and heart weight (in grams) for 144 household cats (see Venables and Ripley, 2002, for further details); you can read the documentation with a call to ?cats. Load the MASS package with a call to library("MASS"), and access the object directly by entering cats at the console prompt.

a.  Plot heart weight on the vertical axis and body weight on the horizontal axis, using different colors or point characters to distinguish between male and female cats. Annotate your plot with a legend and appropriate axis labels.

b.  Fit a least-squares multiple linear regression model using heart weight as the response variable and the other two variables as predictors, and view a model summary.
    i.   Write down the equation for the fitted model and interpret the estimated regression coefficients for body weight and sex. Are both statistically significant? What does this say about the relationship between the response and predictors?
    ii.  Report and interpret the coefficient of determination and the outcome of the omnibus $F$-test.

c.  Tilman's cat, Sigma, is a 3.4 kg female. Use your model to estimate her mean heart weight and provide a 95 percent prediction interval.

d. Use predict to superimpose continuous lines based on the fitted linear model on your plot from (a), one for male cats and one for female. What do you notice? Does this reflect the statistical significance (or lack thereof) of the parameter estimates?

The boot package (Davison and Hinkley, 1997; Canty and Ripley, 2015) is another library of R code that's included with the standard installation but isn't automatically loaded. Load boot with a call to library("boot"). You'll find a data frame called nuclear, which contains data on the construction of nuclear power plants in the United States in the late 1960s (Cox and Snell, 1981).

e. Access the documentation by entering ?nuclear at the prompt and examine the details of the variables. (Note there is a mistake for date, which provides the date that the construction permits were issued—it should read "measured in years since January 1 **1900** to the nearest month.") Use pairs to produce a quick scatterplot matrix of the data.

f. One of the original objectives was to predict the cost of further construction of these power plants. Create a fit and summary of a linear regression model that aims to model cost by t1 and t2, two variables that describe different elapsed times associated with the application for and issue of various permits. Take note of the estimated regression coefficients and their significance in the fitted model.

g. Refit the model, but this time also include an effect for the date the construction permit was issued. Contrast the output for this new model against the previous one. What do you notice, and what does this information suggest about the relationships in the data with respect to these predictors?

h. Fit a third model for power plant cost, using the predictors for "date of permit issue," "power plant capacity," and the binary variable describing whether the plant was sited in the north-eastern United States. Write down the fitted model equation and provide 95 percent confidence intervals for each estimated coefficient.

The following table gives an excerpt of a historical data set compiled between 1961 and 1973. It concerns the annual murder rate in Detroit, Michigan; the data were originally presented and analyzed by Fisher (1976) and are reproduced here from Harraway (1995). In the data set you'll find the number of murders, police officers, and gun licenses issued per 100,000 population, as well as the overall unemployment rate as a percentage of the overall population.

| Murders | Police | Unemployment | Guns |
|---------|--------|--------------|---------|
| 8.60 | 260.35 | 11.0 | 178.15 |
| 8.90 | 269.80 | 7.0 | 156.41 |
| 8.52 | 272.04 | 5.2 | 198.02 |
| 8.89 | 272.96 | 4.3 | 222.10 |
| 13.07 | 272.51 | 3.5 | 301.92 |
| 14.57 | 261.34 | 3.2 | 391.22 |
| 21.36 | 268.89 | 4.1 | 665.56 |
| 28.03 | 295.99 | 3.9 | 1131.21 |
| 31.49 | 319.87 | 3.6 | 837.60 |
| 37.39 | 341.43 | 7.1 | 794.90 |
| 46.26 | 356.59 | 8.4 | 817.74 |
| 47.24 | 376.69 | 7.7 | 583.17 |
| 52.33 | 390.19 | 6.3 | 709.59 |

i. Create your own data frame in your R workspace and produce a scatterplot matrix. Which of the variables appears to be most strongly related to the murder rate?

j. Fit a multiple linear regression model using the number of murders as the response and all other variables as predictors. Write down the model equation and interpret the coefficients. Is it reasonable to state that all relationships between the response and the predictors are causal?

k. Identify the amount of variation in the response attributed to the joint effect of the three explanatory variables. Then refit the model excluding the predictor associated with the largest (in other words, "most nonsignificant") $p$-value. Compare the new coefficient of determination with that of the previous model. Is there much difference?

l. Use your model from (k) to predict the mean number of murders per 100,000 residents, with 300 police officers and 500 issued gun licenses. Compare this to the mean response if there were no gun licenses issued and provide 99 percent confidence intervals for both predictions.

# 21.4 Transforming Numeric Variables

Sometimes, the linear function as strictly defined by the standard regression equation, (21.1), can be inadequate when it comes to capturing relationships between a response and selected covariates. You might, for example, observe curvature in a scatterplot between two numeric variables to which a perfectly straight line isn't necessarily best suited. To a certain extent, the requirement that your data exhibit such linear behavior in order for a linear regression model to be appropriate can be relaxed by simply transforming (typically in a nonlinear fashion) certain variables before any estimation or model fitting takes place.

*Numeric transformation* refers to the application of a mathematical function to your numeric observations in order to rescale them. Finding the square root of a number and converting a temperature from Fahrenheit to Celsius are both examples of a numeric transformation. In the context of regression, transformation is generally applied only to continuous variables and can be done in any number of ways. In this section, you'll limit your attention to examples using the two most common approaches: *polynomial* and *logarithmic* transformations. However, note that the appropriateness of the methods used to transform variables, and any modeling benefits that might occur, can only really be considered on a case-by-case basis.

Transformation in general doesn't represent a universal solution to solving problems of nonlinearity in the trends in your data, but it can at least improve how faithfully a linear model is able to represent those trends.

### 21.4.1   Polynomial

Following on from a comment made earlier, let's say you observe a curved relationship in your data such that a straight line isn't a sensible choice for modeling it. In an effort to fit your data more closely, a polynomial or *power* transformation can be applied to a specific predictor variable in your regression model. This is a straightforward technique that, by allowing *polynomial curvature* in the relationships, allows changes in that predictor to influence the response in more complex ways than otherwise possible. You achieve this by including additional terms in the model definition that represent the impact of progressively higher powers of the variable of interest on the response.

To clarify the concept of polynomial curvature, consider the following sequence between $-4$ and $4$, as well as the simple vectors computed from it:

```
R> x <- seq(-4,4,length=50)
R> y <- x
R> y2 <- x + x^2
R> y3 <- x + x^2 + x^3
```

Here, you're taking the original value of x and calculating specific functionals of it. The vector y, as a copy of x, is clearly linear (in technical terms, this is a "polynomial of order 1"). You assign y2 to take on an additionally squared valued of x, providing *quadratic* behavior—a polynomial of order 2. Lastly, the vector y3 represents the results of a *cubic* function of the values of x, with the inclusion of x raised to the power of 3—a polynomial of order 3.

The following three lines of code produce, separately, the plots from left to right in Figure 21-2.

```
R> plot(x,y,type="l")
R> plot(x,y2,type="l")
R> plot(x,y3,type="l")
```

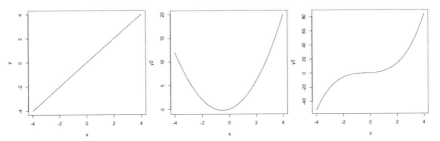

Figure 21-2: Illustrating linear (left), quadratic (middle), and cubic functions (right) of x

Perhaps a bit more generally, let's say you have data for a continuous predictor, $X$, that you want to use to model your response, $Y$. Following estimation in the usual way, linearly, the simple model is $\hat{y} = \hat{\beta}_0 + \hat{\beta}_1 x$; a quadratic trend in $X$ can be modeled via the multiple regression $\hat{y} = \hat{\beta}_0 + \hat{\beta}_1 x + \hat{\beta}_2 x^2$; a cubic relationship can be captured by $\hat{y} = \hat{\beta}_0 + \hat{\beta}_1 x + \hat{\beta}_2 x^2 + \hat{\beta}_3 x^3$; and so on. From the plots in Figure 21-2, a good way to interpret the effects of including these extra terms is in the complexity of the curves that can be captured. At order 1, the linear relationship allows no curvature. At order 2, a quadratic function of any given variable allows one "bend." At order 3, the model can cope with two bends in the relationship, and this continues if you keep adding terms corresponding to increasing powers of the covariate. The regression coefficients associated with these terms (all implied to be 1 in the code that produced the previous plots) are able to control the specific appearance (in other words, the *strength* and *direction*) of the curvature.

### Fitting a Polynomial Transformation

Return your attention to the built-in mtcars data set. Consider the disp variable, which describes engine displacement volume in cubic inches, against a response variable of miles per gallon. If you examine a plot of the data in Figure 21-3, you can see that there does appear to be a slight yet noticeable curve in the relationship between displacement and mileage.

```
R> plot(mtcars$disp,mtcars$mpg,xlab="Displacement (cu. in.)",ylab="MPG")
```

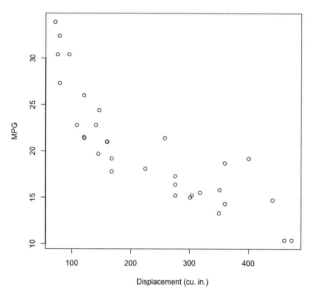

*Figure 21-3: Scatterplot of miles per gallon and engine displacement, for the* mtcars *data*

Is the straight line that a simple linear regression model would provide really the best way to represent this relationship? To investigate this, start by fitting that basic linear setup.

```
R> car.order1 <- lm(mpg~disp,data=mtcars)
R> summary(car.order1)

Call:
lm(formula = mpg ~ disp, data = mtcars)

Residuals:
    Min     1Q  Median     3Q     Max
-4.8922 -2.2022 -0.9631  1.6272  7.2305

Coefficients:
             Estimate Std. Error t value Pr(>|t|)
(Intercept) 29.599855   1.229720  24.070  < 2e-16 ***
disp        -0.041215   0.004712  -8.747 9.38e-10 ***
---
Signif. codes:  0 '***' 0.001 '**' 0.01 '*' 0.05 '.' 0.1 ' ' 1

Residual standard error: 3.251 on 30 degrees of freedom
Multiple R-squared:  0.7183,     Adjusted R-squared:  0.709
F-statistic: 76.51 on 1 and 30 DF,  p-value: 9.38e-10
```

This clearly indicates statistical evidence of a negative linear impact of displacement on mileage—for each additional cubic inch of displacement, the mean response decreases by about 0.041 miles per gallon.

Now, try to capture the apparent curve in the data by adding a quadratic term in disp to the model. You can do this in two ways. First, you could create a new vector in the workspace by simply squaring the mtcars$disp vector and then supplying the result to the formula in lm. Second, you could specify disp^2 directly as an additive term in the formula. If you do it this way, it's essential to wrap that particular expression in a call to I as follows:

```
R> car.order2 <- lm(mpg~disp+I(disp^2),data=mtcars)
R> summary(car.order2)

Call:
lm(formula = mpg ~ disp + I(disp^2), data = mtcars)

Residuals:
    Min      1Q  Median      3Q     Max
-3.9112 -1.5269 -0.3124  1.3489  5.3946

Coefficients:
              Estimate Std. Error t value Pr(>|t|)
(Intercept)  3.583e+01  2.209e+00  16.221 4.39e-16 ***
disp        -1.053e-01  2.028e-02  -5.192 1.49e-05 ***
I(disp^2)    1.255e-04  3.891e-05   3.226   0.0031 **
---
Signif. codes:  0 '***' 0.001 '**' 0.01 '*' 0.05 '.' 0.1 ' ' 1

Residual standard error: 2.837 on 29 degrees of freedom
Multiple R-squared:  0.7927,     Adjusted R-squared:  0.7784
F-statistic: 55.46 on 2 and 29 DF,  p-value: 1.229e-10
```

Use of the I function around a given term in the formula is necessary when said term requires an arithmetic calculation—in this case, disp^2—before the model itself is actually fitted.

Turning to the fitted multiple regression model itself, you can see that the contribution of the squared component is statistically significant—the output corresponding to I(disp^2) shows a $p$-value of 0.0031. This implies that even if a linear trend is taken into account, the model that includes a quadratic component (which introduces a curve) is a better-fitting model. This conclusion is supported by a noticeably higher coefficient of determination compared to the first fit (0.7927 against 0.7183). You can see the fit of this quadratic curve in Figure 21-4 (code for which follows shortly).

Here you might reasonably wonder whether you can improve the ability of the model to capture the relationship further by adding yet another higher-order term in the covariate of interest. To that end:

```
R> car.order3 <- lm(mpg~disp+I(disp^2)+I(disp^3),data=mtcars)
R> summary(car.order3)

Call:
lm(formula = mpg ~ disp + I(disp^2) + I(disp^3), data = mtcars)

Residuals:
    Min      1Q  Median      3Q     Max
-3.0896 -1.5653 -0.3619  1.4368  4.7617

Coefficients:
              Estimate Std. Error t value Pr(>|t|)
(Intercept)  5.070e+01  3.809e+00  13.310 1.25e-13 ***
disp        -3.372e-01  5.526e-02  -6.102 1.39e-06 ***
I(disp^2)    1.109e-03  2.265e-04   4.897 3.68e-05 ***
I(disp^3)   -1.217e-06  2.776e-07  -4.382  0.00015 ***
---
Signif. codes:  0 '***' 0.001 '**' 0.01 '*' 0.05 '.' 0.1 ' ' 1

Residual standard error: 2.224 on 28 degrees of freedom
Multiple R-squared:  0.8771,     Adjusted R-squared:  0.8639
F-statistic: 66.58 on 3 and 28 DF,  p-value: 7.347e-13
```

The output shows that a cubic component also offers a statistically significant contribution. However, if you were to continue adding higher-order terms, you'd find that fitting a polynomial of order 4 to these data isn't able to improve the fit at all, with several coefficients being rendered nonsignificant (the order 4 fit isn't shown).

So, letting $\hat{y}$ be miles per gallon and $x$ be displacement in cubic inches, and expanding the e-notation from the previous output, the fitted multiple regression model is

$$\hat{y} = 50.7 - 0.3372x + 0.0011x^2 - 0.000001x^3,$$

which is precisely what the order 3 line in the left panel of Figure 21-4 reflects.

## Plotting the Polynomial Fit

To address the plot itself, you visualize the data and the first (simple linear) model in car.order1 in the usual way. To begin Figure 21-4, execute the following code:

```
R> plot(mtcars$disp,mtcars$mpg,xlab="Displacement (cu. in.)",ylab="MPG")
R> abline(car.order1)
```

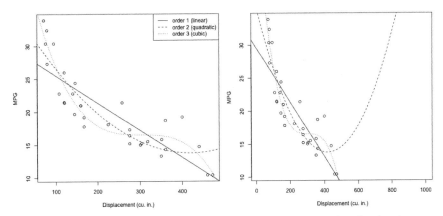

Figure 21-4: Three different models, polynomials of orders 1, 2, and 3, fitted to the "mileage per displacement" relationship from the `mtcars` data set. Left: Visible plot limits constrained to the data. Right: Visible plot limits widened considerably to illustrate unreliability in extrapolation.

It's a little more difficult to add the line corresponding to either of the polynomial-termed models since `abline` is equipped to handle only straight-line trends. One way to do this is to make use of `predict` for each value in a sequence that represents the desired values of the explanatory variable. (I favor this approach because it also allows you to simultaneously calculate confidence and prediction bands if you want.) To add the line for the order 2 model only, first create the required sequence over the observed range of `disp`.

```
R> disp.seq <- seq(min(mtcars$disp)-50,max(mtcars$disp)+50,length=30)
```

Here, the sequence has been widened a little by minus and plus 50 to predict a small amount on either side of the scope of the original covariate data, so the curve meets the edges of the graph. Then you make the prediction itself and superimpose the fitted line.

```
R> car.order2.pred <- predict(car.order2,newdata=data.frame(disp=disp.seq))
R> lines(disp.seq,car.order2.pred,lty=2)
```

You use the same technique, followed by the final addition of the legend, for the order 3 polynomial.

```
R> car.order3.pred <- predict(car.order3,newdata=data.frame(disp=disp.seq))
R> lines(disp.seq,car.order3.pred,lty=3)
R> legend("topright",lty=1:3,
          legend=c("order 1 (linear)","order 2 (quadratic)","order 3 (cubic)"))
```

The result of all this is on the left panel of Figure 21-4. Even though you've used raw data from only one covariate, `disp`, the example illustrated here is considered multiple regression because more than one parameter (in addition to the universal intercept $\beta_0$) required estimation in the order 2 and 3 models.

The different types of trend lines fitted to the mileage and displacement data clearly show different interpretations of the relationship. Visually, you could reasonably argue that the simple linear fit is inadequate at modeling the relationship between response and predictor, but it's harder to come to a clear conclusion when choosing between the order 2 and order 3 versions. The order 2 fit captures the curve that tapers off as disp increases; the order 3 fit additionally allows for a bump (in technical terms a *saddle* or *inflection*), followed by a steeper downward trend in the same domain.

So, which model is "best"? In this case, the statistical significance of the parameters suggests that the order 3 model should be preferred. Having said that, there are other things to consider when choosing between different models, which you'll think about more carefully in Section 22.2.

### Pitfalls of Polynomials

One particular drawback associated with polynomial terms in linear regression models is the instability of the fitted trend when trying to perform any kind of extrapolation. The right plot in Figure 21-4 shows the same three fitted models (MPG by displacement), but this time with a much wider scale for displacement. As you can see, the validity of these models is questionable. Though the order 2 and 3 models fit MPG acceptably within the range of the observed data, if you move even slightly outside the maximum threshold of observed displacement values, the predictions of the mean mileage go wildly off course. The order 2 model in particular becomes completely nonsensical, suggesting a rapid improvement in MPG once the engine displacement rises over 500 cubic inches. You must keep this natural mathematical behavior of polynomial functions in mind if you're considering using higher-order terms in your regression models.

To create this plot, the same code that created the left plot can be used; you simply use xlim to widen the *x*-axis range and define the disp.seq object to a correspondingly wider sequence (in this case, I just set xlim=c(10,1000) with matching from and to limits in the creation of disp.seq).

**NOTE** *Models like this are still referred to as* linear *regression models, which might seem a bit confusing since the fitted trends for higher-order polynomials are clearly nonlinear. This is because* linear regression *refers to the fact that the function defining the mean response is linear in terms of the regression parameters $\beta_0$, $\beta_1$, ..., $\beta_p$. As such, any transformation applied to individual variables doesn't affect the linearity of the function with respect to the coefficients themselves.*

## 21.4.2   Logarithmic

In statistical modeling situations where you have positive numeric observations, it's common to perform a log transformation of the data to dramatically reduce the overall range of the data and bring extreme observations closer to a measure of centrality. In that sense, transforming to a logarithmic scale can help reduce the severity of heavily skewed data (see Section 15.2.4). In the context of regression modeling, log transformations can be used to capture trends where apparent curves "flatten off," without the same kind of instability outside the range of the observed data that you saw with some of the polynomials.

If you need to refresh your memory on logarithms, turn back to Section 2.1.2; it suffices here to note that the logarithm is the power to which you must raise a base value in order to obtain an $x$ value. For example, in $3^5 = 243$, the logarithm is 5 and 3 is the base, expressed as $\log_3 243 = 5$. Because of the ubiquity of the exponential function in common probability distributions, statisticians almost exclusively work with the natural log (logarithm to the base $e$). From here, assume all mentions of the log transformation refer to the natural log.

To briefly illustrate the typical behavior of the log transformation, take a look at Figure 21-5, achieved with the following:

```
R> plot(1:1000,log(1:1000),type="l",xlab="x",ylab="",ylim=c(-8,8))
R> lines(1:1000,-log(1:1000),lty=2)
R> legend("topleft",legend=c("log(x)","-log(x)"),lty=c(1,2))
```

This plots the log of the integers 1 to 1000 against the raw values, as well as plotting the negative log. You can see the way in which the log-transformed values taper off and flatten out as the raw values increase.

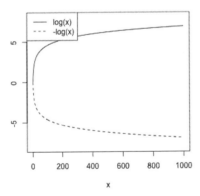

Figure 21-5: The log function applied to integers 1 to 1000

### Fitting the Log Transformation

As noted, one use of the log transformation in regression is to allow this kind of curvature in situations when a perfectly straight line doesn't suit the observed relationship. For an illustration, return to the mtcars examples and consider mileage as a function of both horsepower and transmission type (variables hp and am, respectively). Create a scatterplot of MPG against horsepower, with different colors distinguishing between automatic and manual cars.

```
R> plot(mtcars$hp,mtcars$mpg,pch=19,col=c("black","gray")[factor(mtcars$am)],
       xlab="Horsepower",ylab="MPG")
R> legend("topright",legend=c("auto","man"),col=c("black","gray"),pch=19)
```

The plotted points shown in Figure 21-6 suggest that curved trends in horsepower may be more appropriate than straight-line relationships. Note that you have to explicitly coerce the binary numeric mtcars$am vector to a factor here in order to use it as a selector for the vector of two colors. You'll add the lines in after fitting the linear model.

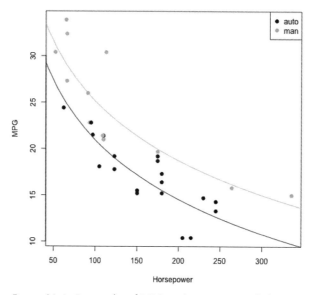

Figure 21-6: Scatterplot of MPG on horsepower, split by transmission type, with lines corresponding to a multiple linear regression using a log-scaled effect of horsepower superimposed

Let's do so using the log transformation of horsepower to try to capture the curved relationship. Since, in this example, you also want to account for the potential of transmission type to affect the response, this is included as an additional predictor variable as usual.

```
R> car.log <- lm(mpg~log(hp)+am,data=mtcars)
R> summary(car.log)
```

```
Call:
lm(formula = mpg ~ log(hp) + am, data = mtcars)
```

```
Residuals:
    Min     1Q  Median     3Q     Max
-3.9084 -1.7692 -0.1432  1.4032  6.3865

Coefficients:
             Estimate Std. Error t value Pr(>|t|)
(Intercept)  63.4842     5.2697  12.047 8.24e-13 ***
log(hp)      -9.2383     1.0439  -8.850 9.78e-10 ***
am            4.2025     0.9942   4.227 0.000215 ***
---
Signif. codes:  0 '***' 0.001 '**' 0.01 '*' 0.05 '.' 0.1 ' ' 1

Residual standard error: 2.592 on 29 degrees of freedom
Multiple R-squared:  0.827,      Adjusted R-squared:  0.8151
F-statistic: 69.31 on 2 and 29 DF,  p-value: 8.949e-12
```

The output indicates jointly statistically significant effects of both log-horsepower and transmission type on mileage. Keeping transmission constant, the mean MPG drops by around 9.24 for each additional unit of log-horsepower. Having a manual transmission increases the mean MPG by roughly 4.2 (estimated in this order owing to the coding of am—0 for automatic, 1 for manual; see ?mtcars). The coefficient of determination shows 82.7 percent of the variation in the response is explained by this regression, suggesting a satisfactory fit.

### Plotting the Log Transformation Fit

To visualize the fitted model, you first need to calculate the fitted values for all desired predictor values. The following code creates a sequence of horsepower values (minus and plus 20 horsepower) and performs the required prediction for both transmission types.

```
R> hp.seq <- seq(min(mtcars$hp)-20,max(mtcars$hp)+20,length=30)
R> n <- length(hp.seq)
R> car.log.pred <- predict(car.log,newdata=data.frame(hp=rep(hp.seq,2),
                                        am=rep(c(0,1),each=n)))
```

In the above code, since you want to plot predictions for both possible values of am, when using newdata you need to replicate hp.seq twice. Then, when you provide values for am to newdata, one series of hp.seq is paired with an appropriately replicated am value of 0, the other with 1. The result of this is a vector of predictions of length twice that of hp.seq, car.log.pred, with the first n elements corresponding to automatic cars and the latter n to manuals. Now you can add these lines to Figure 21-6 with the following:

```
R> lines(hp.seq,car.log.pred[1:n])
R> lines(hp.seq,car.log.pred[(n+1):(2*n)],col="gray")
```

By examining the scatterplot, you can see that the fitted model appears to do a good job of estimating the joint relationship between horsepower/transmission and MPG. The statistical significance of transmission type in this model directly affects the difference between the two added lines. If am weren't significant, the lines would be closer together; in that case, the model would be suggesting that one curve would be sufficient to capture the relationship. As usual, extrapolation too far outside the range of the observed predictor data isn't a great idea, though it's less unstable for log-transformed trends than for polynomial functions.

### 21.4.3  Other Transformations

Transformation can involve more than one variable of the data set and isn't limited to just predictor variables either. In their original investigation into the mtcars data, Henderson and Velleman (1981) also noted the presence of the same curved relationships you've uncovered between the response and variables such as horsepower and displacement. They argued that it's preferable to use gallons per mile (GPM) instead of MPG as the response variable to improve linearity. This would involve modeling a transformation of MPG, namely, that GPM = 1/MPG.

The authors also commented on the limited influence that both horsepower and displacement have on GPM if the weight of the car is included in a fitted model, because of the relatively high correlations present among these three predictors (known as *multicollinearity*). To address this, the authors created a new predictor variable calculated as horsepower divided by weight. This measures, in their words, how "overpowered" a car is—and they proceeded to use that new predictor instead of horsepower or displacement alone. This is just some of the experimentation that took place in the search for an appropriate way to model these data.

To this end, however you choose to model your own data, the objective of transforming numeric variables should always be to fit a valid model that represents the data and the relationships more realistically and accurately. When reaching for this goal, there's plenty of freedom in how you can transform numeric observations in applications of regression methods. For a further discussion on transformations in linear regression, Chapter 7 of Faraway (2005) provides an informative introduction.

### Exercise 21.2

The following table presents data collected in one of Galileo's famous "ball" experiments, in which he rolled a ball down a ramp of different heights and measured how far it traveled from the base of the ramp. For more on this and other interesting examples, look at "Teaching Statistics with Data of Historic Significance" by Dickey and Arnold (1995).

| Initial height | Distance |
|---|---|
| 1000 | 573 |
| 800 | 534 |
| 600 | 495 |
| 450 | 451 |
| 300 | 395 |
| 200 | 337 |
| 100 | 253 |

a.  Create a data frame in R based on this table and plot the data points with distance on the *y*-axis.

b.  Galileo believed there was a quadratic relationship between initial height and the distance traveled.

    i.  Fit an order 2 polynomial in height, with distance as the response.

    ii.  Fit a cubic (order 3) and a quartic (order 4) model for these data. What do they tell you about the nature of the relationship?

c.  Based on your models from (b), choose the one that you think best represents the data and plot the fitted line on the raw data. Add 90 percent confidence bands for mean distance traveled to the plot.

The contributed R package `faraway` contains a large number of data sets that accompany a linear regression textbook by Faraway (2005). Install the package and then call `library("faraway")` to load it. One of the data sets is `trees`, which provides data on the dimensions of felled trees of a certain type (see, for example, Atkinson, 1985).

d.  Access the data object at the prompt and plot volume against girth (the latter along the *x*-axis).

e.  Fit two models with `Volume` as the response: one quadratic model in `Girth` and the other based on log transformations of both `Volume` and `Girth`. Write down the model equations for each and comment on the similarity (or difference) of the fits in terms of the coefficient of determination and the omnibus *F*-test.

f.  Use predict to add lines to the plot from (d) for each of the two models from (e). Use different line types; add a corresponding legend. Also include 95 percent prediction intervals, with line types matching those of the fitted values (note that for the model that involves log transformation of the response and the predictor, any returned values from predict will themselves be on the log scale; you have to back-transform these to the original scale using exp before the lines for that model can be superimposed). Comment on the respective fits and their estimated prediction intervals.

Lastly, turn your attention back to the mtcars data frame.

g.  Fit and summarize a multiple linear regression model to determine mean MPG from horsepower, weight, and displacement.

h.  In the spirit of Henderson and Velleman (1981), use I to refit the model in (g) in terms of GPM = 1/MPG. Which model explains a greater amount of variation in the response?

## 21.5  Interactive Terms

So far, you've looked only at the joint main effects of how predictors affect the outcome variable (and one-to-one transformations thereof). Now you'll look at interactions between covariates. An *interactive effect* between predictors is an additional change to the response that occurs at particular combinations of the predictors. In other words, an interactive effect is present if, for a given covariate profile, the values of the predictors are such that they produce an effect that augments the stand-alone main effects associated with those predictors.

### 21.5.1  Concept and Motivation

Diagrams such as those found in Figure 21-7 are often used to help explain the concept of interactive effects. These diagrams show your mean response value, $\hat{y}$, on the vertical axis, as usual, and a predictor value for the variable $x_1$ on the horizontal axis. They also show a binary categorical variable $x_2$, which can be either zero or one. These hypothetical variables are labeled as such in the images.

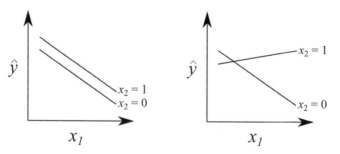

Figure 21-7: Concept of an interactive effect between two predictors $x_1$ and $x_2$, on the mean response value $\hat{y}$. Left: Only main effects of $x_1$ and $x_2$ influence $\hat{y}$. Right: An interaction between $x_1$ and $x_2$ is needed in addition to their main effects in order to model $\hat{y}$.

The left diagram shows the limit of the models you've considered so far in this chapter—that both $x_1$ and $x_2$ affect $\hat{y}$ independently of each other. The right diagram, however, clearly shows that the effect of $x_1$ on $\hat{y}$ changes completely depending on the value of $x_2$. On the left, only main effects of $x_1$ and $x_2$ are needed to determine $\hat{y}$; on the right, main effects *and* an interactive effect between $x_1$ and $x_2$ are present.

**NOTE** *When estimating regression models, you always have to accompany interactions with the main effects of the relevant predictors, for reasons of interpretability. Since interactions are themselves best understood as an augmentation of the main effects, it makes no sense to remove the latter and leave in the former.*

For a good example of an interaction, think about pharmacology. Interactive effects between medicines are relatively common, which is why health care professionals often ask about other medicines you might be taking. Consider statins—drugs commonly used to reduce cholesterol. Users of statins are told to avoid grapefruit juice because it contains natural chemical compounds that inhibit the efficacy of the enzyme responsible for the correct metabolization of the drug. If an individual is taking statins and not consuming grapefruit, you would expect a negative relationship between cholesterol level and statin use (think about "statin use" either as a continuous or as a categorical dosage variable)—as statin use increases or is affirmative, the cholesterol level decreases. On the other hand, for an individual on statins who *is* consuming grapefruit, the nature of the relationship between cholesterol level and statin use could easily be different—weakened negative, neutral, or even positive. If so, since the effect of the statins on cholesterol changes according to the value of another variable—whether or not grapefruit is consumed—this would be considered an interaction between those two predictors.

Interactions can occur between categorical variables, numeric variables, or both. It's most common to find *two-way* interactions—interactions between exactly two predictors—which is what you'll focus on in Sections 21.5.2 to 21.5.4. Three-way and higher-order interactive effects are technically possible but less common, partly because they are difficult to interpret in a real-world context. You'll consider an example of these in Section 21.5.5.

### 21.5.2 One Categorical, One Continuous

Generally, a two-way interaction between a categorical and a continuous predictor should be understood as effecting a change in the slope of the continuous predictor with respect to the nonreference levels of the categorical predictor. In the presence of a term for the continuous variable, a categorical variable with $k$ levels will have $k - 1$ main effect terms, so there will be a further $k - 1$ interactive terms between all the alternative levels of the categorical variable and the continuous variable.

The different slopes for $x_1$ by category of $x_2$ for $\hat{y}$ can be seen clearly on the right of Figure 21-7. In such a situation, in addition to the main effects for $x_1$ and $x_2$, there would be one interactive term in the fitted model corresponding to $x_2 = 1$. This defines the additive term needed to change the slope in $x_1$ for $x_2 = 0$ to the new slope in $x_1$ for $x_2 = 1$.

For an example, let's access a new data set. In Exercise 21.2, you looked at the faraway package (Faraway, 2005) to access the trees data. In this package, you'll also find the diabetes object—a cardiovascular disease data set detailing characteristics of 403 African Americans (originally investigated and reported in Schorling et al., 1997; Willems et al., 1997). Install faraway if you haven't already and load it with library("faraway"). Restrict your attention to the total cholesterol level (chol—continuous), age of the individual (age—continuous), and body frame type (frame—categorical with $k = 3$ levels: "small" as the reference level, "medium", and "large"). You can see the data in Figure 21-8, which will be created momentarily.

You'll look at modeling total cholesterol by age and body frame. It seems logical to expect that cholesterol is related to both age and body type, so it makes sense to also consider the possibility that the effect of age on cholesterol is different for individuals of different body frames. To investigate, let's fit the multiple linear regression and include a two-way interaction between the two variables. In the call to lm, you specify the main effects first, using + as usual, and then specify an interactive effect of two predictors by using a colon (:) between them.

```
R> dia.fit <- lm(chol~age+frame+age:frame,data=diabetes)
R> summary(dia.fit)

Call:
lm(formula = chol ~ age + frame + age:frame, data = diabetes)

Residuals:
    Min     1Q  Median     3Q     Max
-131.90  -26.24   -5.33  22.17  226.11

Coefficients:
                 Estimate Std. Error t value Pr(>|t|)
(Intercept)      155.9636    12.0697  12.922  < 2e-16 ***
age                0.9852     0.2687   3.667  0.00028 ***
framemedium       28.6051    15.5503   1.840  0.06661 .
framelarge        44.9474    18.9842   2.368  0.01840 *
age:framemedium   -0.3514     0.3370  -1.043  0.29768
age:framelarge    -0.8511     0.3779  -2.252  0.02490 *
---
Signif. codes:  0 '***' 0.001 '**' 0.01 '*' 0.05 '.' 0.1 ' ' 1

Residual standard error: 42.34 on 384 degrees of freedom
  (13 observations deleted due to missingness)
```

Multiple R-squared: 0.07891,     Adjusted R-squared: 0.06692
F-statistic: 6.58 on 5 and 384 DF, p-value: 6.849e-06

---

Inspecting the estimated model parameters in the output, you can see a main effect coefficient for age, main effect coefficients for the two levels of frame (that aren't the reference level), and two further terms for the interactive effect of age with those same nonreference levels.

**NOTE** *There's actually a shortcut to doing this in R—the cross-factor notation. The same model shown previously could have been fitted by using chol~age*frame in lm; the symbol * between two variables in a formula should be interpreted as "include an intercept, all main effects, and the interaction." I'll use this shortcut from now on.*

The output shows the significance of age and some evidence to support the presence of a main effect of frame. There's also slight indication of significance of the interaction, though it's weak. Assessing significance in this case, where one predictor is categorical with $k > 2$ levels, follows the same rule as noted in the discussion of multilevel variables in Section 20.5.2—if at least one of the coefficients is significant, the entire effect should be deemed significant.

The general equation for the fitted model can be written down directly from the output.

$$\text{"Mean total cholesterol"} = 155.9636 + 0.9852 \times \text{"age"}$$
$$+ 28.6051 \times \text{"medium frame"}$$
$$+ 44.9474 \times \text{"large frame"}$$
$$- 0.3514 \times \text{"age : medium frame"}$$
$$- 0.8511 \times \text{"age : large frame"} \qquad (21.7)$$

I've used a colon (:) to denote the interactive terms to mirror the R output.

For the reference level of the categorical predictor, body type "small," the fitted model can be written down straight from the output.

$$\text{"Mean total cholesterol"} = 155.9636 + 0.9852 \times \text{"age"}$$

For a model with the main effects only, changing body type to "medium" or "large" would affect only the intercept—you know from Section 20.5 that the relevant effect is simply added to the outcome. The presence of the interaction, however, means that *in addition* to the change in the intercept, the main effect slope of age must now also be changed according to the relevant interactive term. For an individual with a "medium" frame, the model is

$$\text{"Mean total cholesterol"} = 155.9636 + 0.9852 \times \text{"age"} + 28.6051$$
$$- 0.3514 \times \text{"age"}$$
$$= 184.5687 + (0.9852 - 0.3514) \times \text{"age"}$$
$$= 184.5687 + 0.6338 \times \text{"age"}$$

and for an individual with a "large" frame, the model is

$$\text{"Mean total cholesterol"} = 155.9636 + 0.9852 \times \text{"age"} + 44.9474$$
$$- 0.8511 \times \text{"age"}$$
$$= 200.911 + (0.9852 - 0.8511) \times \text{"age"}$$
$$= 200.911 + 0.1341 \times \text{"age"}$$

You can easily calculate these in R by accessing the coefficients of the fitted model object:

```
R> dia.coef <- coef(dia.fit)
R> dia.coef
     (Intercept)            age      framemedium       framelarge
     155.9635868       0.9852028       28.6051035       44.9474105
age:framemedium   age:framelarge
      -0.3513906       -0.8510549
```

Next, let's sum the relevant components of this vector. Once you have the sums, you'll be able to plot the fitted model.

```
R> dia.small <- c(dia.coef[1],dia.coef[2])
R> dia.small
(Intercept)          age
155.9635868    0.9852028
R> dia.medium <- c(dia.coef[1]+dia.coef[3],dia.coef[2]+dia.coef[5])
R> dia.medium
(Intercept)          age
184.5686904    0.6338122
R> dia.large <- c(dia.coef[1]+dia.coef[4],dia.coef[2]+dia.coef[6])
R> dia.large
(Intercept)          age
200.9109973    0.1341479
```

The three lines are stored as numeric vectors of length 2, with the intercept first and the slope second. This is the form required by the optional coef argument of abline, which allows you to superimpose these straight lines on a plot of the raw data. The following code produces Figure 21-8.

```
R> cols <- c("black","darkgray","lightgray")
R> plot(diabetes$chol~diabetes$age,col=cols[diabetes$frame],
        cex=0.5,xlab="age",ylab="cholesterol")
R> abline(coef=dia.small,lwd=2)
R> abline(coef=dia.medium,lwd=2,col="darkgray")
R> abline(coef=dia.large,lwd=2,col="lightgray")
R> legend("topright",legend=c("small frame","medium frame","large frame"),
          lty=1,lwd=2,col=cols)
```

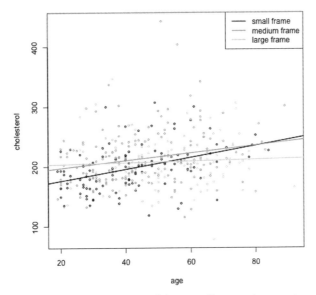

Figure 21-8: Fitted linear model, main effects, and interaction
for mean total cholesterol by age and body frame

If you examine the fitted model in Figure 21-8, it's clear that inclusion
of an interaction between age and body frame has allowed more flexibility
in the way mean total cholesterol relates to the two predictors. The non-
parallel nature of the three plotted lines reflects the concept illustrated in
Figure 21-7.

I walked through this to illustrate how the concept works, but in practice
you don't need to go through all of these steps to find the point estimates
(and any associated confidence intervals). You can predict from a fitted linear model with interactions in the same way as for main-effect-only models
through the use of predict.

### 21.5.3  Two Categorical

You met the concept of interactions between two categorical explanatory
variables in the introduction to two-way ANOVA in Section 19.2. There, you
uncovered evidence of an interactive effect of wool type and tension on the
mean number of warp breaks in lengths of yarn (based on the ready-to-use
warpbreaks data frame). You then visualized the interaction with an interaction plot (Figure 19-2 on page 447), not unlike the diagrams in Figure 21-7.

Let's implement the same model as the last warpbreaks example in Section 19.2.2 in an explicit linear regression format.

```
R> warp.fit <- lm(breaks~wool*tension,data=warpbreaks)
R> summary(warp.fit)

Call:
lm(formula = breaks ~ wool * tension, data = warpbreaks)
```

```
Residuals:
     Min      1Q   Median      3Q      Max
-19.5556  -6.8889  -0.6667  7.1944  25.4444

Coefficients:
               Estimate Std. Error t value Pr(>|t|)
(Intercept)      44.556      3.647  12.218 2.43e-16 ***
woolB           -16.333      5.157  -3.167 0.002677 **
tensionM        -20.556      5.157  -3.986 0.000228 ***
tensionH        -20.000      5.157  -3.878 0.000320 ***
woolB:tensionM   21.111      7.294   2.895 0.005698 **
woolB:tensionH   10.556      7.294   1.447 0.154327
---
Signif. codes:  0 '***' 0.001 '**' 0.01 '*' 0.05 '.' 0.1 ' ' 1

Residual standard error: 10.94 on 48 degrees of freedom
Multiple R-squared:  0.3778,     Adjusted R-squared:  0.3129
F-statistic: 5.828 on 5 and 48 DF,  p-value: 0.0002772
```

Here I've used the cross-factor symbol *, rather than wool + tension + wool:tension. When both predictors in a two-way interaction are categorical, there will be a term for each nonreference level of the first predictor combined with all nonreference levels of the second predictor. In this example, wool is binary with only $k = 2$ levels and tension has $k = 3$; therefore, the only interaction terms present are the "medium" (M) and "high" (H) tension levels ("low", L, is the reference level) with wool type B (A is the reference level). Therefore, altogether in the fitted model, there are terms for B, M, H, B:M, and B:H.

These results provide the same conclusion as the ANOVA analysis—there is indeed statistical evidence of an interactive effect between wool type and tension on mean breaks, on top of the contributing main effects of those predictors.

The general fitted model can be understood as

$$\text{"Mean warp breaks"} = 44.556 - 16.333 \times \text{"wool type B"}$$
$$- 20.556 \times \text{"medium tension"}$$
$$- 20.000 \times \text{"high tension"}$$
$$+ 21.111 \times \text{"wool type B : medium tension"}$$
$$+ 10.556 \times \text{"wool type B : high tension"}$$

The additional interaction terms work the same way as the main effects—when only categorical predictors are involved, the model can be seen as a series of additive terms to the overall intercept. Exactly which ones you use in any given prediction depends on the covariate profile of a given individual.

Let's have a quick series of examples: for wool A at low tension, the mean number of warp breaks is predicted as simply the overall intercept; for wool A at high tension, you have the overall intercept and the main effect term for high tension; for wool B at low tension, you have the overall intercept and the main effect for wool type B only; and for wool B at medium tension, you have the overall intercept, the main effect for wool type B, the main effect for medium tension, *and* the interactive term for wool B with medium tension.

You can use predict to estimate the mean warp breaks for these four scenarios; they're accompanied here with 90 percent confidence intervals:

```
R> nd <- data.frame(wool=c("A","A","B","B"),tension=c("L","H","L","M"))
R> predict(warp.fit,newdata=nd,interval="confidence",level=0.9)
       fit      lwr      upr
1 44.55556 38.43912 50.67199
2 24.55556 18.43912 30.67199
3 28.22222 22.10579 34.33866
4 28.77778 22.66134 34.89421
```

### 21.5.4   Two Continuous

Finally, you'll look at the situation when the two predictors are continuous. In this case, an interaction term operates as a modifier on the continuous plane that's fitted using the main effects only. In a similar way to an interaction between a continuous and a categorical predictor, an interaction between two continuous explanatory variables allows the slope associated with one variable to be affected, but this time, that modification is made in a continuous way (that is, according to the value of the other continuous variable).

Returning to the mtcars data frame, consider MPG once more as a function of horsepower and weight. The fitted model, shown next, includes the interaction in addition to the main effects of the two continuous predictors. As you can see, there is a single estimated interactive term, and it is deemed significantly different from zero.

```
R> car.fit <- lm(mpg~hp*wt,data=mtcars)
R> summary(car.fit)

Call:
lm(formula = mpg ~ hp * wt, data = mtcars)

Residuals:
    Min      1Q  Median      3Q     Max
-3.0632 -1.6491 -0.7362  1.4211  4.5513

Coefficients:
            Estimate Std. Error t value Pr(>|t|)
```

```
(Intercept) 49.80842     3.60516  13.816 5.01e-14 ***
hp          -0.12010     0.02470  -4.863 4.04e-05 ***
wt          -8.21662     1.26971  -6.471 5.20e-07 ***
hp:wt        0.02785     0.00742   3.753 0.000811 ***
---
Signif. codes:  0 '***' 0.001 '**' 0.01 '*' 0.05 '.' 0.1 ' ' 1

Residual standard error: 2.153 on 28 degrees of freedom
Multiple R-squared:  0.8848,       Adjusted R-squared:  0.8724
F-statistic: 71.66 on 3 and 28 DF,  p-value: 2.981e-13
```

The model is written as

$$\text{“Mean MPG”} = 49.80842 - 0.12010 \times \text{“horsepower”}$$
$$- 8.21662 \times \text{“weight”}$$
$$+ 0.02785 \times \text{“horsepower : weight”}$$
$$= 49.80842 - 0.12010 \times \text{“horsepower”}$$
$$- 8.21662 \times \text{“weight”}$$
$$+ 0.02785 \times \text{“horsepower”} \times \text{“weight”}$$

The second version of the model equation provided here reveals for the first time an interaction expressed as the *product* of the values of the two predictors, which is exactly how the fitted model is used to predict the response. (Technically, this is the same as when at least one of the predictors is categorical—but the dummy coding simply results in zeros and ones for the respective terms, so multiplication just amounts to the presence or absence of a given term, as you've seen.)

You can interpret an interaction between two continuous predictors by considering the sign (+ or −) of the coefficient. Negativity suggests that as the values of the predictors increase, the response is reduced after computing the result of the main effects. Positivity, as is the case here, suggests that as the values of the predictors increase, the effect is an additional increase, an amplification, on the mean response.

Contextually, the negative main effects of hp and wt indicate that mileage is naturally reduced for heavier, more powerful cars. However, positivity of the interactive effect suggests that this impact on the response is "softened" as horsepower or weight is increased. To put it another way, the negative relationship imparted by the main effects is rendered "less extreme" as the values of the predictors get bigger and bigger.

Figure 21-9 contrasts the main-effects-only version of the model (obtained using lm with the formula mpg~hp+wt; not explicitly fitted in this section) with the interaction version of the model fitted just above as the object car.fit.

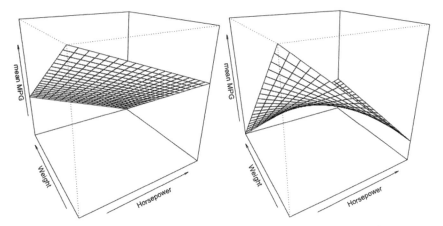

Figure 21-9: Response surfaces for mean MPG by horsepower and weight, for a main-effects-only model (left), and one that includes the two-way interaction between the continuous predictors (right)

The plotted *response surfaces* show the mean MPG on the vertical z-axis and the two predictor variables on the horizontal axes as marked. You can interpret the predicted mean MPG, based on a given horsepower and weight value, as a point on the surface. Note that both surfaces decrease in MPG (vertically along the z-axis) as you move to larger values of either predictor along the respective horizontal axes.

I'll show how these plots are created in Chapter 25. For now, they serve simply to highlight the aforementioned "softening" impact of the interaction in car.fit. On the left, the main-effects-only model shows a flat plane decreasing according to the negative linear slopes in each predictor. On the right, however, the presence of the positive interactive term flattens this plane out, meaning the rate of decrease is slowed as the values of the predictor variables increase.

### 21.5.5   Higher-Order Interactions

As mentioned, two-way interactions are the most common kind of interactions you'll encounter in applications of regression methods. This is because for three-way or higher-order terms, you need a lot more data for a reliable estimation of interactive effects, and there are a number of interpretative complexities to overcome. Three-way interactions are far rarer than two-way effects, and four-way and above are rarer still.

In Exercise 21.1, you used the nuclear data set found in the boot package (provided with the standard R installation), which includes data on the constructions of nuclear power plants in the United States. In the exercises, you focused mainly on date and time predictors related to construction permits to model the mean cost of construction for the nuclear power plants. For the sake of this example, assume you don't have the data on these predictors. Can the cost of construction be adequately modeled using only the variables that describe characteristics of the plant itself?

Load the boot package and access the ?nuclear help page to find details on the variables: cap (continuous variable describing the capacity of the plant); cum.n (treated as continuous, describing the number of similar constructions the engineers had previously worked on); ne (binary, describing whether the plant was in the northeastern United States); and ct (binary, describing whether the plant had a cooling tower).

The following model is fitted with the final construction cost of the plant as the response; a main effect for capacity; and main effects of, and all two-way interactions and the three-way interaction among, cum.n, ne, and ct:

```
R> nuc.fit <- lm(cost~cap+cum.n*ne*ct,data=nuclear)
R> summary(nuc.fit)

Call:
lm(formula = cost ~ cap + cum.n * ne * ct, data = nuclear)

Residuals:
    Min      1Q  Median      3Q     Max
-162.475  -50.368  -8.833  43.370  213.131

Coefficients:
              Estimate Std. Error t value Pr(>|t|)
(Intercept)  138.0336    99.9599   1.381 0.180585
cap            0.5085     0.1127   4.513 0.000157 ***
cum.n        -24.2433     6.7874  -3.572 0.001618 **
ne          -260.1036   164.7650  -1.579 0.128076
ct          -187.4904    76.6316  -2.447 0.022480 *
cum.n:ne      44.0196    12.2880   3.582 0.001577 **
cum.n:ct      35.1687     8.0660   4.360 0.000229 ***
ne:ct        524.1194   200.9567   2.608 0.015721 *
cum.n:ne:ct  -64.4444    18.0213  -3.576 0.001601 **
---
Signif. codes:  0 '***' 0.001 '**' 0.01 '*' 0.05 '.' 0.1 ' ' 1

Residual standard error: 107.3 on 23 degrees of freedom
Multiple R-squared:  0.705,      Adjusted R-squared:  0.6024
F-statistic: 6.872 on 8 and 23 DF,  p-value: 0.0001264
```

In this code, you specify the higher-order interactions by extending the number of variables connected with a * (using * instead of : since you want to include all the lower-order effects of those three predictors as well).

In the estimated results, the main effect for cap is positive, showing that an increased power capacity is tied to an increased construction cost. All other main effects are negative, which at face value seems to imply that a reduced construction cost is associated with more experienced engineers, plants constructed in the Northeast, and plants with a cooling tower. However, this isn't an accurate statement since you haven't yet considered the interactive terms in those predictors. All estimated two-way interactive effects are positive—having more experienced engineers means a higher construction cost in the Northeast regardless of whether there's a cooling tower, and having more experienced engineers also means higher costs for plants *with* a cooling tower, regardless of region.

Cost is also dramatically increased for plants in the Northeast with a cooling tower, regardless of the experience of the engineer. All that being said, the negative three-way interaction suggests that the increased cost associated with more experienced engineers working in the Northeast *and* on a plant with a cooling tower is lessened somewhat after the main effects and two-way interactive effects are calculated.

At the least, this example highlights the complexities associated with interpreting model coefficients for higher-order interactions. It's also possible that statistically significant high-order interactions crop up due to lurking variables that have gone unaccounted for, that is, that the significant interactions are a spurious manifestation of patterns in the data that simpler terms involving those missing predictors could explain just as well (if not better). In part, this motivates the importance of adequate *model selection*, which is up next in the discussion.

## Exercise 21.3

Return your attention to the cats data frame in package MASS. In the first few problems in Exercise 21.1, you fitted the main-effect-only model to predict the heart weights of domestic cats by total body weight and sex.

a. Fit the model again, and this time include an interaction between the two predictors. Inspect the model summary. What do you notice in terms of the parameter estimates and their significance when compared to the earlier main-effect-only version?

b. Produce a scatterplot of heart weight on body weight, using different point characters or colors to distinguish the observations according to sex. Use abline to add two lines denoting the fitted model. How does this plot differ from the one in Exercise 21.1 (d)?

c. Predict the heart weight of Tilman's cat using the new model (remember that Sigma is a 3.4 kg female) accompanied by a 95 percent prediction interval. Compare it to the main-effects-only model from the earlier exercise.

In Exercise 21.2, you accessed the trees data frame in the contributed faraway package. After loading the package, access the ?trees help file; you'll find the volume and girth measurements you used earlier, as well as data on the height of each tree.

d.  Without using any transformations of the data, fit and inspect a main-effects-only model for predicting volume from girth and height. Then, fit and inspect a second version of this model including an interaction.

e.  Repeat (d), but this time use the log transformation of all variables. What do you notice about the significance of the interaction between the untransformed and transformed models? What does this suggest about the relationships in the data?

Turn back to the mtcars data set and remind yourself of the variables in the help file ?mtcars.

f.  Fit a linear model for mpg based on a two-way interaction between hp and factor(cyl) and their main effects, as well as a main effect for wt. Produce a summary of the fit.

g.  Interpret the estimated coefficients for the interaction between horsepower and the (categorical) number of cylinders.

h.  Suppose you're keen on purchasing a 1970s performance car. Your mother advises you to purchase a "practical and economical" car that's capable of an average MPG value of at least 25. You see three vehicles advertised: car 1 is a four-cylinder, 100 horsepower car that weighs 2100 lbs; car 2 is an eight-cylinder, 210 horsepower car that weighs 3900 lbs; and car 3 is a six-cylinder, 200 horsepower car that weighs 2900 lbs.

    i.   Use your model to predict the mean MPG for each of the three cars; provide 95 percent confidence intervals. Based on your point estimates only, which car would you propose to your mother?

    ii.  You still want the most gas-guzzling car you can own with your mother's blessing, so you decide to be sneaky and base your decision on what the confidence intervals tell you instead. Does this change your choice of vehicle?

## Important Code in This Chapter

| Function/operator | Brief description | First occurrence |
| --- | --- | --- |
| I | Include arithmetic term | Section 21.4.1, p. 505 |
| : | Interaction term | Section 21.5.2, p. 516 |
| * | Cross-factor operator | Section 21.5.3, p. 519 |

# 22

# LINEAR MODEL SELECTION AND DIAGNOSTICS

You've now spent a fair amount of time on many aspects of linear regression models. In this chapter, I'll cover how formal R tools and techniques can be used to investigate two other, and no less important, aspects of regression: choosing an appropriate model for your analysis and assessing the validity of the assumptions you've made.

## 22.1   Goodness-of-Fit vs. Complexity

The overarching goal of fitting any statistical model is to faithfully represent the data and the relationships held within them. In general, fitting statistical models boils down to a balancing act between two things: goodness-of-fit and complexity. *Goodness-of-fit* refers to the goal of obtaining a model that best represents the relationships between the response and the predictor (or predictors). *Complexity* describes how complicated a model is; this is always tied to the number of terms in the model that require estimation—the inclusion of more predictors and additional functions (such as polynomial transformations and interactions) leads to a more complex model.

## 22.1.1 Principle of Parsimony

Statisticians refer to the balancing act between goodness-of-fit and complexity as the *principle of parsimony*, where the goal of the associated *model selection* is to find a model that's as simple as possible (in other words, with relatively low complexity), without sacrificing too much goodness-of-fit. We'd say that a model that satisfies this notion is a *parsimonious* fit. You'll often hear of researchers talking about choosing the "best" model—they're actually referring to the idea of parsimony.

So, how do you decide where to draw the line on such a balance? Naturally, statistical significance plays a role here—and model selection often simply comes down to assessing the significance of the effect of predictors or functions of predictors on the response. In an effort to impart some amount of objectivity to such a process, you can use systematic selection algorithms, such as those you'll learn about in Section 22.2, to decide between multiple explanatory variables and any associated functions.

## 22.1.2 General Guidelines

Performing any kind of model selection or comparing several models against one another involves decision making regarding the inclusion of available predictor variables. On this topic, there are several guidelines you should always follow.

- First, it's important to remember that you can't remove individual levels of a categorical predictor in a given model; this makes no sense. In other words, if one of the nonreference levels is statistically significant but all others are nonsignificant, you should treat the categorical variable, as a whole, as making a statistically significant contribution to the determination of the mean response. You should only really consider *entire* removal of that categorical predictor if *all* nonreference coefficients are associated with a lack of evidence (against being zero). This also holds for interactive terms involving categorical predictors.

- If an interaction is present in the fitted model, all lower-order interactions and main effects of the relevant predictors must remain in the model. This was touched upon in Section 21.5.1, when I discussed interpretation of interactive effects as augmentations of lower-order effects. As an example, you should only really consider removing the main effect of a predictor if there are no interaction terms present in the fitted model involving that predictor (even if that main effect has a high *p*-value).

- In models where you've used a polynomial transformation of a certain explanatory variable (refer to Section 21.4.1), keep all lower-order polynomial terms in the model if the highest is deemed significant. A model containing an order 3 polynomial transformation in a predictor, for example, must also include the order 1 and order 2 transformations of that variable. This is because of the mathematical behavior of polynomial functions—only by explicitly separating out the linear, quadratic, and cubic (and so on) effects as distinct terms can you avoid confounding said effects with one another.

## 22.2   Model Selection Algorithms

The job of a model selection algorithm is to sift through your available explanatory variables in some systematic fashion in order to establish which are best able to jointly describe the response, as opposed to fitting models by examining specific combinations of predictors in isolation, as you've done so far.

Model selection algorithms can be controversial. There are several different methods, and no single approach is universally appropriate for every regression model. Different selection algorithms can result in different final models, as you'll see. In many cases, researchers will have additional information or knowledge about the problem that influences the decision—for example, that certain predictors must always be included or that it makes no sense to ever include them. This must be considered at the same time as other complications, such as the possibility of interactions or unobserved lurking variables influencing significant relationships and the need to ensure any fitted model is statistically valid (which you'll look at in Section 22.3).

It's helpful to keep in mind this famous quote from celebrated statistician George Box (1919–2013): "All models are wrong, but some are useful."

Any fitted model you produce can never be assumed to be the truth, but a model that's fitted and checked carefully and thoroughly can reveal interesting features of the data and so have the potential to reveal associations and relationships by providing quantitative estimates thereof.

### 22.2.1   Nested Comparisons: The Partial F-Test

The *partial F-test* is probably the most direct way to compare several different models. It looks at two or more *nested* models, where the smaller, less complex model is a reduced version of the bigger, more complex model. Formally, let's say you've fitted two linear regressions models as follows:

$$\hat{y}_{\mathrm{redu}} = \hat{\beta}_0 + \hat{\beta}_1 x_1 + \hat{\beta}_2 x_2 + \ldots + \hat{\beta}_p x_p$$

$$\hat{y}_{\mathrm{full}} = \hat{\beta}_0 + \hat{\beta}_1 x_1 + \hat{\beta}_2 x_2 + \ldots + \hat{\beta}_p x_p + \ldots + \hat{\beta}_q x_q$$

Here, the reduced model, predicting $\hat{y}_{\mathrm{redu}}$, has $p$ predictors, plus one intercept. The full model, predicting $\hat{y}_{\mathrm{full}}$, has $q$ predictor terms. The notation implies that $q > p$ and that, along with the standard inclusion of an intercept $\hat{\beta}_0$, the full model involves all $p$ predictors of the reduced model defined by $\hat{y}_{\mathrm{redu}}$, as well as $q - p$ additional terms. This emphasizes the fact that the model for $\hat{y}_{\mathrm{redu}}$ is nested within $\hat{y}_{\mathrm{full}}$.

It's important to note that increasing the number of predictors in a regression model will always improve $R^2$ and any other measures of goodness-of-fit. The real question, however, is whether that improvement in goodness-of-fit is large enough to make the additional complexity involved with including any additional predictor terms "worth it." This is precisely the question that the partial $F$-test tries to answer in the context of nested regression models. Its goal is to test whether including those extra $q - p$ terms, which produce the full model rather than the reduced model, provide a statistically significant improvement in goodness-of-fit. The partial $F$-test addresses these hypotheses:

$$H_0 : \beta_{p+1} = \beta_{p+2} = \ldots = \beta_q = 0$$
$$H_A : \text{At least one of the } \beta_j \neq 0 \text{ (for } j = p, \ldots, q) \quad (22.1)$$

The calculation of the test statistic to address these hypotheses follows the same ideas behind the omnibus $F$-test automatically produced by R when summarizing a fitted linear model object (detailed in Section 21.3.5). Denote the coefficient of determination for the reduced and full models with $R^2_{\text{redu}}$ and $R^2_{\text{full}}$, respectively. If $n$ refers to the sample size of the data used to fit both models, the test statistic given by

$$\mathcal{F} = \frac{(R^2_{\text{full}} - R^2_{\text{redu}})(n - q - 1)}{(1 - R^2_{\text{full}})(q - p)} \quad (22.2)$$

follows an $F$ distribution with $\text{df}_1 = q - p$, $\text{df}_2 = n - q$ degrees of freedom under the assumption of $H_0$ in (22.1). The $p$-value is found as the upper-tail area from $\mathcal{F}$ as usual; the smaller it is, the greater the evidence against the null hypothesis, which states that one or more of the additional parameters has no impact on the response variable.

Take the model objects survmult and survmult2 from Section 21.3.1 as an example. The survmult model aims to predict mean student height from writing handspan and sex based on the survey data frame from the MASS package; survmult2 adds smoking status to these predictors. If you need to, return to Section 21.3.1 to refit these two models. Printing the objects to the console screen previews the two fits and makes it easy to confirm that the smaller model is indeed nested within the larger model in terms of its explanatory variables:

```
R> survmult

Call:
lm(formula = Height ~ Wr.Hnd + Sex, data = survey)

Coefficients:
(Intercept)       Wr.Hnd       SexMale
    137.687        1.594         9.490

R> survmult2

Call:
```

```
lm(formula = Height ~ Wr.Hnd + Sex + Smoke, data = survey)

Coefficients:
(Intercept)      Wr.Hnd     SexMale   SmokeNever   SmokeOccas   SmokeRegul
   137.4056      1.6042      9.3979      -0.0442       1.5267       0.9211
```

Once you've fitted your nested models, R can carry out partial $F$-tests using the anova function (partial $F$-tests fall within the suite of analysis of variance methodologies). To determine whether adding Smoke as a predictor provides any statistically significant improvement in fit, simply start with the reduced model and supply the model objects as arguments.

```
R> anova(survmult,survmult2)
Analysis of Variance Table

Model 1: Height ~ Wr.Hnd + Sex
Model 2: Height ~ Wr.Hnd + Sex + Smoke
  Res.Df    RSS Df Sum of Sq      F Pr(>F)
1    204 9959.2
2    201 9914.3  3    44.876 0.3033  0.823
```

The output provides the quantities associated with calculation of $R^2_{\mathrm{redu}}$ and $R^2_{\mathrm{full}}$ and the test statistic $\mathcal{F}$ from (22.2), given in the resulting table as F, which is of the most interest. Using the values of $p$ and $q$ from printing survmult and survmult2, you should be able to confirm, for example, the values of $\mathrm{df}_1$ and $\mathrm{df}_2$ appearing in the second row of the table in the columns Df and Res.Df, respectively.

The result of this particular test, obtained from a test statistic of $\mathcal{F} = 0.3033$ associated with $\mathrm{df}_1 = 3$, $\mathrm{df}_2 = 201$, is a high $p$-value of 0.823, suggesting no evidence against $H_0$. This means that adding Smoke to the reduced model, which includes only the explanatory variables Wr.Hnd and Sex, offers no tangible improvement in fit when it comes to modeling student height. That conclusion isn't surprising, given the nonsignificant $p$-values of all non-reference levels of Smoke, seen previously in Section 21.3.1.

This is how partial $F$-tests are used for model selection—in the current example, the reduced model would be the more parsimonious fit and preferred over the full model.

You can conduct comparisons among several nested models in a given call to anova, which can be useful for investigating things such as the inclusion of interactive terms or including polynomial transformations of predictors since there's a natural hierarchy that requires you to retain any lower-order terms.

For an example, let's use the diabetes data frame in the faraway package from Section 21.5.2, with the model fitted to predict cholesterol level (chol) against age (age) and body frame (frame) and the interaction between those two predictors. Before using the partial $F$-tests to compare nested variants, you need to ensure you're using the same records for each model and that there aren't missing values for any of the predictors that are then going to be unavailable to the "fuller" models (so the sample size is the same for each comparison). To do this, you just need to first define a version of diabetes that removes records with missing values in the predictors you're using.

Load the faraway package and use logical subsetting to identify and delete any individuals with a missing value for age *or* for frame. Define this new version of the diabetes object:

```
R> diab <- diabetes[-which(is.na(diabetes$age) | is.na(diabetes$frame)),]
```

Now, fit the following four models using your new diab object:

```
R> dia.model1 <- lm(chol~1,data=diab)
R> dia.model2 <- lm(chol~age,data=diab)
R> dia.model3 <- lm(chol~age+frame,data=diab)
R> dia.model4 <- lm(chol~age*frame,data=diab)
```

The first model is just an intercept, the second adds age as a predictor, the third has age and frame, and the fourth includes the interaction. Nesting is evident, and you can now compare the significance of the improvements in goodness-of-fit as you increase the complexity of the model at each step.

```
R> anova(dia.model1,dia.model2,dia.model3,dia.model4)
Analysis of Variance Table

Model 1: chol ~ 1
Model 2: chol ~ age
Model 3: chol ~ age + frame
Model 4: chol ~ age * frame
  Res.Df    RSS Df Sum of Sq       F    Pr(>F)
1    389 747265
2    388 712078  1     35187 19.6306 1.227e-05 ***
3    386 697527  2     14551  4.0589   0.01801 *
4    384 688295  2      9233  2.5755   0.07743 .
---
Signif. codes:  0 '***' 0.001 '**' 0.01 '*' 0.05 '.' 0.1 ' ' 1
```

If you hadn't deleted the records containing missing values in those predictors, you would've received an error telling you that the data sets for the four models were not equal sizes.

The results themselves suggest that including age provides a significant improvement to modeling chol; including a main effect for frame provides a further mild improvement; and there's very weak evidence, if any, that including an interactive effect is beneficial to goodness-of-fit. From this, you might prefer to use dia.mod3, the main-effects-only model, as the most parsimonious representation of mean cholesterol out of these four models.

## 22.2.2 Forward Selection

Partial *F*-tests are a natural way to investigate nested models but can be difficult to manage if you have many different models to fit when, for example, you have many predictor variables.

This is where *forward selection* (also referred to as *forward elimination*) comes in. The idea is to start with an intercept-only model and then perform a series of independent tests to determine which of your predictor variables significantly improves the goodness-of-fit. Then you update your model object by adding that term and execute the series of tests again for all remaining terms to determine which of those would further improve the fit. The process repeats until there aren't any more terms that improve the fit in a statistically significant way. The ready-to-use R functions add1 and update perform the series of tests and update your fitted regression model.

You'll use the nuclear data frame in the boot library from Exercise 21.1 on page 499 and Section 21.5.5 as an example. The goal is to choose the most informative model for prediction of construction cost. Load boot and access the help file ?nuclear to remind yourself of the variable definitions. First fit the model for construction cost with an overall intercept term only.

```
R> nuc.0 <- lm(cost~1,data=nuclear)
R> summary(nuc.0)

Call:
lm(formula = cost ~ 1, data = nuclear)

Residuals:
    Min      1Q  Median      3Q     Max
-254.05 -151.24  -13.46  150.40  419.68

Coefficients:
            Estimate Std. Error t value Pr(>|t|)
(Intercept)   461.56      30.07   15.35 4.95e-16 ***
---
Signif. codes:  0 '***' 0.001 '**' 0.01 '*' 0.05 '.' 0.1 ' ' 1

Residual standard error: 170.1 on 31 degrees of freedom
```

You know from earlier exploits that this particular model is rather inadequate for the reliable prediction of cost. So, consider the following line of code to start the forward selection (I've suppressed the output, which I'll show separately and discuss in a moment):

```
R> add1(nuc.0,scope=.~.+date+t1+t2+cap+pr+ne+ct+bw+cum.n+pt,test="F")
```

The first argument to add1 is always the model you're aiming to update. The second argument, scope, is critical—you must supply a formula object defining the "fullest," most complex model you'd consider fitting. For this you would typically use the .~. notation, in which the dots refer to the definition of the model in the first argument. Specifically, the dots stand for "what is already there." In other words, through scope you're telling add1 that the fullest model you'd consider has cost as the response, an intercept, and main effects of all other predictors in the nuclear data frame (I'll restrict the full model to main effects only for ease of demonstration). You don't need to supply the data frame as an argument since those data are contained within the model object in the first argument. Lastly, you tell add1 the test to perform. There are a handful of variants available (see ?add1), but here you'll stick with test="F" for partial *F*-tests.

Now, focus on the output that's provided directly after the execution of add1.

```
Single term additions

Model:
cost ~ 1
        Df Sum of Sq    RSS    AIC F value     Pr(>F)
<none>               897172 329.72
date     1    334335 562837 316.80 17.8205 0.0002071 ***
t1       1    186984 710189 324.24  7.8986 0.0086296 **
t2       1        27 897145 331.72  0.0009 0.9760597
cap      1    199673 697499 323.66  8.5881 0.0064137 **
pr       1      9037 888136 331.40  0.3052 0.5847053
ne       1    128641 768531 326.77  5.0216 0.0325885 *
ct       1     43042 854130 330.15  1.5118 0.2284221
bw       1     16205 880967 331.14  0.5519 0.4633402
cum.n    1     67938 829234 329.20  2.4579 0.1274266
pt       1    305334 591839 318.41 15.4772 0.0004575 ***
---
Signif. codes:  0 '***' 0.001 '**' 0.01 '*' 0.05 '.' 0.1 ' ' 1
```

The output comprises a series of rows, starting with <none> (doing nothing to the current model). You receive the Sum of Sq and RSS values, directly related to calculating the test statistic. The differences in degrees of freedom are also reported. Another measure of parsimony, AIC, is also provided (you'll look at that in more detail in Section 22.2.4).

Most relevant are the test outcomes; with test="F", each row corresponds to an independent partial *F*-test comparing the model in the first argument, as $\hat{y}_{\text{redu}}$, with the model that results from having added that row term only as $\hat{y}_{\text{full}}$. Usually, therefore, you would update your model by adding only the term with the largest (and "most significant") improvement.

Here, you should be able to see that adding date as a predictor offers the largest significant improvement to modeling cost. So, let's update nuc.0 to include that term with the code.

```
R> nuc.1 <- update(nuc.0,formula=.~.+date)
R> summary(nuc.1)
```

```
Call:
lm(formula = cost ~ date, data = nuclear)

Residuals:
    Min      1Q  Median      3Q     Max
-176.00 -105.27  -25.24   58.63  359.46

Coefficients:
            Estimate Std. Error t value Pr(>|t|)
(Intercept) -6553.57    1661.96  -3.943 0.000446 ***
date          102.29      24.23   4.221 0.000207 ***
---
Signif. codes:  0 '***' 0.001 '**' 0.01 '*' 0.05 '.' 0.1 ' ' 1

Residual standard error: 137 on 30 degrees of freedom
Multiple R-squared:  0.3727,      Adjusted R-squared:  0.3517
F-statistic: 17.82 on 1 and 30 DF,  p-value: 0.0002071
```

In update you provide the model you want to update as the first argument, and the second argument, formula, tells update how to update the model. Again using the .~. notation, the instruction is to update nuc.0 by adding date as a predictor, resulting in a fitted model object of the same class of the first argument. Call a summary of the new model, nuc.1, to see this.

So, let's keep going! Call add1 again, but now pass nuc.1 as your first argument.

```
R> add1(nuc.1,scope=.~.+date+t1+t2+cap+pr+ne+ct+bw+cum.n+pt,test="F")
Single term additions

Model:
cost ~ date
       Df Sum of Sq    RSS    AIC F value    Pr(>F)
<none>              562837 316.80
t1      1     15322 547515 317.92  0.8115 0.3750843
t2      1     68161 494676 314.67  3.9959 0.0550606 .
cap     1    189732 373105 305.64 14.7471 0.0006163 ***
pr      1      4027 558810 318.57  0.2090 0.6509638
ne      1     92256 470581 313.07  5.6854 0.0238671 *
ct      1     54794 508043 315.52  3.1277 0.0874906 .
bw      1      1240 561597 318.73  0.0640 0.8020147
cum.n   1      4658 558179 318.53  0.2420 0.6264574
pt      1     90587 472250 313.18  5.5628 0.0252997 *
---
Signif. codes:  0 '***' 0.001 '**' 0.01 '*' 0.05 '.' 0.1 ' ' 1
```

Note that there's now no row for adding in date; it's already there in nuc.1. It seems the next most informative addition would be cap. Update nuc.1 to that effect.

```
R> nuc.2 <- update(nuc.1,formula=.~.+cap)
```

Now keep going, testing, and updating. By calling add1 on nuc.2 (output not shown here), you'll find that the next most significant addition is pt (by a small margin). Update to a new object named nuc.3, which includes the following term:

```
R> nuc.3 <- update(nuc.2,formula=.~.+pt)
```

Then test again, using add1 on nuc.3. You'll find weak evidence to additionally include a main effect for ne, so update with that inclusion to create nuc.4.

```
R> nuc.4 <- update(nuc.3,formula=.~.+ne)
```

At this point, you may be reasonably certain there won't be any more useful additions, but check with one final call to add1 on the latest fit to be thorough.

```
R> add1(nuc.4,scope=.~.+date+t1+t2+cap+pr+ne+ct+bw+cum.n+pt,test="F")
Single term additions

Model:
cost ~ date + cap + pt + ne
        Df Sum of Sq    RSS    AIC F value Pr(>F)
<none>               222617 293.12
t1       1     107.0 222510 295.10  0.0125 0.9118
t2       1   19229.9 203387 292.23  2.4583 0.1290
pr       1    5230.8 217386 294.36  0.6256 0.4361
ct       1   15764.7 206852 292.77  1.9815 0.1711
bw       1     448.0 222169 295.06  0.0524 0.8207
cum.n    1   13819.9 208797 293.07  1.7209 0.2010
```

Indeed it appears that none of the remaining covariates, if included in the model, would yield a statistically significant improvement in goodness-of-fit, so your final model will stay at nuc.4.

```
R> summary(nuc.4)

Call:
lm(formula = cost ~ date + cap + pt + ne, data = nuclear)

Residuals:
      Min      1Q   Median      3Q      Max
```

```
 -157.894  -38.424   -2.493   35.363  267.445

Coefficients:
               Estimate Std. Error t value Pr(>|t|)
(Intercept) -4.756e+03  1.286e+03  -3.699 0.000975 ***
date         7.102e+01  1.867e+01   3.804 0.000741 ***
cap          4.198e-01  8.616e-02   4.873 4.29e-05 ***
pt          -1.289e+02  4.950e+01  -2.605 0.014761 *
ne           9.940e+01  3.864e+01   2.573 0.015908 *
---
Signif. codes:  0 '***' 0.001 '**' 0.01 '*' 0.05 '.' 0.1 ' ' 1

Residual standard error: 90.8 on 27 degrees of freedom
Multiple R-squared:  0.7519,        Adjusted R-squared:  0.7151
F-statistic: 20.45 on 4 and 27 DF,  p-value: 7.507e-08
```

This method may seem a little cumbersome, and it's sometimes difficult to decide on the fullest model to be used as the scope, but it's a remarkably good way to stay involved at every stage of the selection process so you can consider each addition carefully. Note, however, that there's an element of subjectivity; it's possible to arrive at different final models by choosing one addition over another, such as if you'd added pt instead of date (they had similar levels of significance in the very first call to add1).

### 22.2.3   Backward Selection

After learning forward selection, understanding *backward selection* (or *elimination*) isn't much of a stretch. As you might have guessed, where forward selection starts from a reduced model and works its way up to a final model by adding terms, backward selection starts with your fullest model and systematically drops terms. The R functions for this process are drop1 to inspect the partial *F*-tests and update.

The choice of forward versus backward model selection is usually made on a case-by-case basis. If your fullest model isn't known or is difficult to define and fit, then forward selection is typically preferred. On the other hand, if you do have a natural and easily fitted fullest model, then backward selection can be more convenient to implement. Sometimes, researchers will perform both to see whether the final model they arrive at is different (a perfectly possible occurrence).

Revisit the nuclear example. First, define the fullest model as that which predicts cost by main effects of all available covariates (as you did in your use of scope in the forward selections).

```
R> nuc.0 <- lm(cost~date+t1+t2+cap+pr+ne+ct+bw+cum.n+pt,data=nuclear)
R> summary(nuc.0)

Call:
lm(formula = cost ~ date + t1 + t2 + cap + pr + ne + ct + bw +
```

```
       cum.n + pt, data = nuclear)

Residuals:
    Min      1Q   Median      3Q     Max
-128.608  -46.736   -2.668  39.782  180.365

Coefficients:
              Estimate Std. Error t value Pr(>|t|)
(Intercept) -8.135e+03  2.788e+03  -2.918 0.008222 **
date         1.155e+02  4.226e+01   2.733 0.012470 *
t1           5.928e+00  1.089e+01   0.545 0.591803
t2           4.571e+00  2.243e+00   2.038 0.054390 .
cap          4.217e-01  8.844e-02   4.768 0.000104 ***
pr          -8.112e+01  4.077e+01  -1.990 0.059794 .
ne           1.375e+02  3.869e+01   3.553 0.001883 **
ct           4.327e+01  3.431e+01   1.261 0.221008
bw          -8.238e+00  5.188e+01  -0.159 0.875354
cum.n       -6.989e+00  3.822e+00  -1.829 0.081698 .
pt          -1.925e+01  6.367e+01  -0.302 0.765401
---
Signif. codes:  0 '***' 0.001 '**' 0.01 '*' 0.05 '.' 0.1 ' ' 1

Residual standard error: 82.83 on 21 degrees of freedom
Multiple R-squared:  0.8394,       Adjusted R-squared:  0.763
F-statistic: 10.98 on 10 and 21 DF,  p-value: 2.844e-06
```

There are clearly several predictors that appear not to contribute significantly to the response, and these same results are evident the first time you use drop1 to examine the impact on goodness-of-fit that would occur from dropping each variable.

```
R> drop1(nuc.0,test="F")
Single term deletions

Model:
cost ~ date + t1 + t2 + cap + pr + ne + ct + bw + cum.n + pt
       Df Sum of Sq    RSS    AIC F value    Pr(>F)
<none>              144065 291.19
date    1     51230 195295 298.93  7.4677 0.0124702 *
t1      1      2034 146099 289.64  0.2965 0.5918028
t2      1     28481 172546 294.97  4.1517 0.0543902 .
cap     1    155943 300008 312.67 22.7314 0.0001039 ***
pr      1     27161 171226 294.72  3.9592 0.0597943 .
ne      1     86581 230646 304.25 12.6207 0.0018835 **
ct      1     10915 154980 291.53  1.5911 0.2210075
bw      1       173 144238 289.23  0.0252 0.8753538
cum.n   1     22939 167004 293.92  3.3438 0.0816977 .
```

```
pt       1        627 144692 289.33  0.0914 0.7654015
---
Signif. codes:  0 '***' 0.001 '**' 0.01 '*' 0.05 '.' 0.1 ' ' 1
```

One handy feature of drop1 is that its scope argument is optional. If you don't include scope, it defaults to the intercept-only model as the "most-reduced" model, which is usually a reasonable choice.

Before diving right into the deletion process, remind yourself of the interpretation of what you're doing. Just as adding any term will always improve the goodness-of-fit in forward selection, deleting any term in backward selection will always worsen the goodness-of-fit. The real question is the perceived significance of these changes in fit quality. In the same way as earlier, where you wanted to add only those terms that offer a *statistically significant improvement* in goodness-of-fit, when dropping terms, you only want to remove those that *do not* result in a statistically significant *detriment* to goodness-of-fit. As such, backward selection is the complete reverse of forward selection in the way it's carried out.

So, from the output of drop1, you want to choose the term to remove from the model that has the least significant effect of reducing the goodness of the fit. In other words, you're looking for the term with the largest, nonsignificant *p*-value for its partial *F*-test—because dropping a term with a significantly small *p*-value would significantly worsen the predictive capability of the regression model.

In the current example, it seems the predictor bw has the single least significant effect on reducing the goodness-of-fit, so let's start the update by removing that term from nuc.0.

```
R> nuc.1 <- update(nuc.0,.~.-bw)
```

Use of update in this selection algorithm is the same as before; now, though, you use a - to signify the deletion of a term following the standard "what's already there" .~. notation.

The process is then repeated using the latest model nuc.1:

```
R> drop1(nuc.1,test="F")
Single term deletions

Model:
cost ~ date + t1 + t2 + cap + pr + ne + ct + cum.n + pt
       Df Sum of Sq    RSS    AIC F value    Pr(>F)
<none>              144238 289.23
date    1     55942 200180 297.72  8.5326  0.007913 **
t1      1      3124 147362 287.92  0.4765  0.497245
t2      1     30717 174955 293.41  4.6852  0.041546 *
cap     1    159976 304214 311.11 24.4005 6.098e-05 ***
pr      1     27140 171377 292.75  4.1395  0.054122 .
ne      1     86408 230646 302.25 13.1795  0.001479 **
ct      1     11815 156053 289.75  1.8021  0.193153
```

```
cum.n   1      24048 168286 292.17  3.6680  0.068557 .
pt      1        930 145168 287.44  0.1419  0.710039
---
Signif. codes:  0 '***' 0.001 '**' 0.01 '*' 0.05 '.' 0.1 ' ' 1
```

It would seem that pt is the next most sensible main effect to drop. Do so and name the resulting object nuc.2.

```
R> nuc.2 <- update(nuc.1,.~.-pt)
```

Now keep going, rechecking with a call to drop1 (not shown), and you'll find that the predictor t1 reveals itself as another viable deletion. Update your model with that predictor deleted; name the model object nuc.3.

```
R> nuc.3 <- update(nuc.2,.~.-t1)
```

Recheck the new nuc.3 with drop1. You should now find the effect of ct remains nonsignificant, so delete that and update again, giving you a new nuc.4.

```
R> nuc.4 <- update(nuc.3,.~.-ct)
```

Perform yet another check with drop1, this time on nuc.4. At this point, you might hesitate in removing any more predictors, with significance at varying strengths being associated with the effect of their deletion. Note, however, that for at least three of the remaining predictors, t2, pr, and cum.n, the statistical significance should probably be considered borderline at best—all of their $p$-values lie between the conventional cutoff levels of $\alpha = 0.01$ and $\alpha = 0.05$. This again emphasizes the active role a researcher must play in model selection algorithms such as forward or backward elimination; whether you should delete any more variables from here is a difficult question to answer and is left up to your judgment.

Let's remain with nuc.4 as the final model. Summarizing, you're able to see the estimated regression parameters and the usual post-fit statistics.

```
R> summary(nuc.4)

Call:
lm(formula = cost ~ date + t2 + cap + pr + ne + cum.n, data = nuclear)

Residuals:
     Min      1Q   Median      3Q     Max
-152.851  -53.929   -8.827   53.382  155.581

Coefficients:
             Estimate Std. Error t value Pr(>|t|)
(Intercept) -9.702e+03  1.294e+03  -7.495 7.55e-08 ***
```

```
date         1.396e+02  1.843e+01   7.574 6.27e-08 ***
t2           4.905e+00  1.827e+00   2.685 0.012685 *
cap          4.137e-01  8.425e-02   4.911 4.70e-05 ***
pr          -8.851e+01  3.479e+01  -2.544 0.017499 *
ne           1.502e+02  3.400e+01   4.419 0.000168 ***
cum.n       -7.919e+00  2.871e+00  -2.758 0.010703 *
---
Signif. codes:  0 '***' 0.001 '**' 0.01 '*' 0.05 '.' 0.1 ' ' 1

Residual standard error: 80.8 on 25 degrees of freedom
Multiple R-squared:  0.8181,      Adjusted R-squared:  0.7744
F-statistic: 18.74 on 6 and 25 DF,  p-value: 3.796e-08
```

Immediately, you can see that your final model from forward selection in Section 22.2.2 is different from the final model selected here, despite the fullest model being the same for both. How has that occurred?

The answer, simply put, is that the predictors present in a model affect each other. Remember that the estimated coefficients of present predictors easily change in value as you control for different variables. As the number of predictor terms increases, these relationships become more and more complex, so both the order and direction of the selection algorithm have the potential to lead you on different paths through the selection process and arrive at different final destinations, which is exactly what's happened here.

As a perfect example of this, consider the main effect of pt in the nuclear data. In forward selection, pt was added because it offered the "most significant" improvement to the model cost~date+cap. In backward selection, pt was removed early, since it offered the least reduction in goodness-of-fit if taken from the model cost~date+t1+t2+cap+pr+ne+ct+cum.n+pt. What this means is that for the latter model, the contribution that pt might make in terms of predicting the outcome is already explained by the other present predictor terms. In the smaller model, that effect had not yet been explained, and so pt was an attractive addition.

All this serves to highlight the fickle nature of most selection algorithms, in spite of the systematic way they're implemented. It's important to acknowledge that a final model fit will probably vary between approaches and that you should view these selection methods more as helpful guidelines for finding the most parsimonious model and not as providing a universal, definitive solution.

### 22.2.4  Stepwise AIC Selection

The application of a series of partial *F*-tests is the most common *test-based* model selection method, but it's not the only tool a researcher has at their disposal. You can also locate parsimony by adopting a *criterion-based* approach. One of the most famous criterion measures is known as *Akaike's*

*Information Criterion (AIC)*. You'll have noticed this value as one of the columns in the output of `add1` and `drop1`.

For a given linear model, AIC is calculated as follows:

$$\text{AIC} = -2 \times \mathcal{L} + 2 \times (p + 2) \tag{22.3}$$

Here, $\mathcal{L}$ is a measure of goodness-of-fit named the *log-likelihood*, and $p$ is the number of regression parameters in the model, excluding the overall intercept. The value of $\mathcal{L}$ is a direct outcome of the estimation procedure used to fit the model, though its exact calculation is beyond the scope of this text. The thing to know is that it takes on larger values for better-fitting models.

Equation (22.3) produces a value that rewards goodness-of-fit with the $-2 \times \mathcal{L}$ but simultaneously penalizes complexity with the $2 \times (p + 2)$. The negative sign associated with $\mathcal{L}$ coupled with the positive sign of the $p + 2$ means that smaller values of AIC refer to more parsimonious models.

To find the AIC for a fitted linear model, you use the `AIC` or `extractAIC` functions on the object resulting from `lm`; take a look at the help files of these functions to see the technical differences between the two. The value of $\mathcal{L}$ (and therefore also the AIC) has no direct interpretation and is useful only when you compare it against the AIC of another model. You can base model selection on the AIC by identifying the fit with the lowest AIC value. This is the reason it's directly reported in the output of `add1` and `drop1`—you could decide on which term to add or drop based on the change that results in a shift to the smallest AIC, instead of focusing exclusively on the *significance* of the change via the *F*-test.

Let's go even further and combine the ideas of forward and backward selection. *Stepwise* model selection allows the option to either delete a present term *or* add a missing term and is typically implemented with respect to AIC. That is, a term is chosen to be added or deleted based on the one move out of all possible moves that yields the single biggest reduction in AIC. This affords you more flexibility in exploring candidate models on your way to the final model fit—determined as the model from which no addition or deletion would reduce the AIC value further.

It's possible to implement stepwise AIC selection yourself using either `add1` or `drop1` at each stage, but fortunately R provides the built-in `step` function to do it for you. Take the `mtcars` data from the `MASS` package from the past couple of chapters. Let's finally try to obtain a model for mean mileage that offers the opportunity to include every predictor that's available.

First, take a look at the documentation in `?mtcars` and a scatterplot matrix of the data again to remind yourself of the variables and their format in the R data frame object. Then define the starting model (often called the *null* model) as the intercept-only model.

```
R> car.null <- lm(mpg~1,data=mtcars)
```

Your starting model can be anything you like, provided it falls within the domain of the models described by your scope argument to be supplied to step. In this example, define scope as the fullest model to be considered—set this to be the overly complex model with a four-way interaction among wt, hp, cyl, and disp (and all relevant lower-order interactions and main effects, via the cross-factor operator), as well as main effects for am, gear, drat, vs, qsec, and carb. The two multilevel categorical variables, cyl and gear, are explicitly converted to factors to avoid them being treated as numeric (refer to Section 20.5.4).

The potential for interactions in the final model will serve to highlight an especially important (and convenient) feature of add1, drop1, and step. These functions all respect the hierarchy imposed by interactions and main effects. That is, for add1 (and step), an interactive term will not be provided as an option for addition unless all relevant lower-order effects are already present in the current fitted model; similarly, for drop1 (and step), an interactive term or main effect will not be provided as an option for deletion unless all relevant higher-order effects are already gone from the current fitted model.

The step function itself returns a fitted model object and by default provides a comprehensive report of each stage of selection. Let's call it now; for print reasons, some of the output has been snipped out, so you're encouraged to bring this up on your own machine.

```
R> car.step <- step(car.null,scope=.~.+wt*hp*factor(cyl)*disp+am
                                      +factor(gear)+drat+vs+qsec+carb)
Start:  AIC=115.94
mpg ~ 1
```

|               | Df | Sum of Sq | RSS     | AIC     |
|---------------|----|-----------|---------|---------|
| + wt          | 1  | 847.73    | 278.32  | 73.217  |
| + disp        | 1  | 808.89    | 317.16  | 77.397  |
| + factor(cyl) | 2  | 824.78    | 301.26  | 77.752  |
| + hp          | 1  | 678.37    | 447.67  | 88.427  |
| + drat        | 1  | 522.48    | 603.57  | 97.988  |
| + vs          | 1  | 496.53    | 629.52  | 99.335  |
| + factor(gear)| 2  | 483.24    | 642.80  | 102.003 |
| + am          | 1  | 405.15    | 720.90  | 103.672 |
| + carb        | 1  | 341.78    | 784.27  | 106.369 |
| + qsec        | 1  | 197.39    | 928.66  | 111.776 |
| <none>        |    |           | 1126.05 | 115.943 |

```
Step:  AIC=73.22
mpg ~ wt
```

|               | Df | Sum of Sq | RSS     | AIC     |
|---------------|----|-----------|---------|---------|
| + factor(cyl) | 2  | 95.26     | 183.06  | 63.810  |
| + hp          | 1  | 83.27     | 195.05  | 63.840  |
| + qsec        | 1  | 82.86     | 195.46  | 63.908  |
| + vs          | 1  | 54.23     | 224.09  | 68.283  |
| + carb        | 1  | 44.60     | 233.72  | 69.628  |

```
+ disp           1       31.64   246.68   71.356
+ factor(gear)   2       40.37   237.95   72.202
<none>                           278.32   73.217
+ drat           1        9.08   269.24   74.156
+ am             1        0.00   278.32   75.217
- wt             1      847.73  1126.05  115.943

Step:  AIC=63.81
mpg ~ wt + factor(cyl)

                 Df Sum of Sq    RSS    AIC
+ hp             1     22.281 160.78 61.657
+ wt:factor(cyl) 2     27.170 155.89 62.669
<none>                        183.06 63.810
+ qsec           1     10.949 172.11 63.837
+ carb           1      9.244 173.81 64.152
+ vs             1      1.842 181.22 65.487
+ disp           1      0.110 182.95 65.791
+ am             1      0.090 182.97 65.794
+ drat           1      0.073 182.99 65.798
+ factor(gear)   2      6.682 176.38 66.620
- factor(cyl)    2     95.263 278.32 73.217
- wt             1    118.204 301.26 77.752

Step:  AIC=61.66
mpg ~ wt + factor(cyl) + hp

--snip--

Step:  AIC=55.9
mpg ~ wt + factor(cyl) + hp + wt:hp

--snip--

Step:  AIC=52.8
mpg ~ wt + hp + wt:hp

--snip--

Step:  AIC=52.57
mpg ~ wt + hp + qsec + wt:hp

                 Df Sum of Sq    RSS    AIC
<none>                        121.04 52.573
- qsec           1      8.720 129.76 52.799
+ factor(gear)   2      9.482 111.56 53.962
+ am             1      1.939 119.10 54.056
```

| | | | | |
|---|---|---|---|---|
| + carb | 1 | 0.080 | 120.96 | 54.551 |
| + drat | 1 | 0.012 | 121.03 | 54.570 |
| + vs | 1 | 0.010 | 121.03 | 54.570 |
| + disp | 1 | 0.008 | 121.03 | 54.571 |
| + factor(cyl) | 2 | 0.164 | 120.88 | 56.529 |
| - wt:hp | 1 | 65.018 | 186.06 | 64.331 |

Each block of output displays the current model fit, its AIC value, and a table showing the possible moves (either adding +, deleting -, or doing nothing <none>). The AIC value that would result from each move alone is listed, and these potential single moves are ranked from smallest to largest AIC value.

As the algorithm proceeds, you see the <none> row creeping its way up the table. For example, in the first table, the value of the AIC for the intercept-only model is 115.94. The biggest reduction in AIC would result from adding a main effect for wt; that move is made, and the effect of subsequent moves on the AIC is reassessed. Also note that the addition of the two-way interaction term between wt and factor(cyl) is considered only at the third step, after the main effects of those predictors have been added. That particular two-way interaction never ends up being included, though, because the main effect of hp is preferable at that third step, and subsequent interactions involving hp then offer a much better reduction in the AIC value in the fourth step. In fact, at the fifth step, actually *deleting* the main effect for factor(cyl) is deemed to reduce the AIC most, and so the tables for the sixth and seventh steps no longer include that wt:factor(cyl) term as an option. The sixth step suggests that adding the main effect for qsec offers a minor reduction in the AIC, so this is done. The seventh table signals the end of the algorithm because doing nothing offers the lowest AIC value and doing anything else would increase the AIC (shown through <none> taking pole position in that last table).

The final model is stored as the object car.step; by summarizing it, you'll note that almost 90 percent of the variation in the response is explained by weight, horsepower, and their interaction, as well as the slightly curious main effect of qsec (which itself is not deemed statistically significant).

```
R> summary(car.step)

Call:
lm(formula = mpg ~ wt + hp + qsec + wt:hp, data = mtcars)

Residuals:
    Min      1Q  Median      3Q     Max
-3.8243 -1.3980  0.0303  1.1582  4.3650

Coefficients:
            Estimate Std. Error t value Pr(>|t|)
(Intercept) 40.310410   7.677887   5.250 1.56e-05 ***
```

```
wt             -8.681516    1.292525   -6.717 3.28e-07 ***
hp             -0.106181    0.026263   -4.043 0.000395 ***
qsec            0.503163    0.360768    1.395 0.174476
wt:hp           0.027791    0.007298    3.808 0.000733 ***
---
Signif. codes:  0 '***' 0.001 '**' 0.01 '*' 0.05 '.' 0.1 ' ' 1

Residual standard error: 2.117 on 27 degrees of freedom
Multiple R-squared:  0.8925,      Adjusted R-squared:  0.8766
F-statistic: 56.05 on 4 and 27 DF,  p-value: 1.094e-12
```

From this, it seems it would be worthwhile to investigate this predictor further, to establish validity of the fitted model (see Section 22.3), and perhaps to try transformations of the data (such as modeling GPM instead of MPG; see Section 21.4.3) to see whether this effect persists in subsequent runs of the stepwise AIC algorithm. The presence of qsec in the final model illustrates the fact that the selection of the model wasn't based solely on the significance of predictor contribution but on a criterion-based measure aiming for its own definition of parsimony.

The AIC is sometimes criticized for a tendency to err on the side of more complexity and higher *p*-values. To balance this, you can increase the penalizing effect of extra predictors by increasing the multiplicative contribution of the $(p + 2)$ on the right of Equation (22.3); though the standard multiplicative factor of 2 is used in the majority of cases (in step you can use the optional argument k to change this). That being said, criterion-based measures are incredibly useful when you have models that aren't nested (ruling out the partial *F*-test) and you want to compare them for quick identification of the one that, potentially, provides the most parsimonious representation of the data.

---

## Exercise 22.1

In Sections 22.2.2 and 22.2.3, you used forward and backward selection approaches to find a model for predicting the cost of the construction of nuclear power plants (based on the nuclear data frame in the boot package).

a.  Using the same fullest model (in other words, main effects of all present predictors only), use stepwise AIC selection to find a suitable model for the data.

b.  Does the final model found in (a) match either of the models resulting from the earlier use of forward and backward selection? How does it differ?

Exercise 21.2 on page 512 detailed Galileo's ball data. Enter these as a data frame in your current R workspace if you haven't already.

c. Fit five linear models to these data with distance as the response—an intercept-only model and four separate polynomial models of increasing order 1 to 4 in height.

d. Construct a table of partial $F$-tests to identify your favored model for distance traveled. Is your selection consistent with Exercise 21.2 (b) and (c)?

You first encountered the diabetes data frame in the contributed faraway package in Section 21.5.2, where you modeled the mean total cholesterol. Load the package and inspect the documentation in ?diabetes to refresh your memory of the data set.

e. There are some missing values in diabetes that might interfere with model selection algorithms. Define a new version of the diabetes data frame that deletes all rows with a missing value in any of the following variables: chol, age, gender, height, weight, frame, waist, hip, location. Hint: Use na.omit or your knowledge of record extraction or deletion for a data frame. You can create the required vector of row numbers to be extracted or deleted using which and is.na, or you can try using the complete.cases function to obtain a logical flag vector—inspect its help file for details.

f. Use your data frame from (e) to fit two linear models with chol as the response. The null model object, named dia.null, should be an intercept-only model. The full model object, named dia.full, should be the overly complex model with a four-way interaction (and all lower-order terms) among age, gender, weight, and frame; a three-way interaction (and all lower-order terms) among waist, height, and hip; and a main effect for location.

g. Starting from dia.null and using the same terms as in dia.full for scope, implement stepwise selection by AIC to choose a model for mean total cholesterol and then summarize.

h. Use forward selection based on partial $F$-tests with a conventional significance level of $\alpha = 0.05$ to choose a model, again starting from dia.null. Is the result here the same as the model arrived at in (g)?

i. Stepwise selection doesn't have to start from the simplest model. Repeat (g), but this time, set dia.full to be the starting model (you don't need to supply anything to scope if you're starting from the most complex model). What is the final model selected via AIC if you start from dia.full? Is it different than the final model from (g)? Why is this or is this not the case, do you think?

Revisit the ubiquitous mtcars data frame from the MASS package.

j. In Section 22.2.4, you used stepwise AIC selection to model mean MPG. The selected model included a main effect for qsec. Rerun the same AIC selection process, but this time, do it in terms of GPM=1/MPG. Does this change the complexity of the final model?

### 22.2.5 Other Selection Algorithms

Any model selection algorithm will always aim to quantitatively define *parsimony* and suggest a model that optimizes that definition in light of the available data. There are alternatives to AIC, such as the *corrected AIC ($AIC_c$)* or the *Bayesian Information Criterion (BIC)*, both of which impose heavier penalties on complexity than the default AIC in (22.3).

Sometimes it's tempting to simply monitor $R^2$, the coefficient of determination, for a series of models. However, as mentioned in Section 22.2.1, this on its own is inadequate for choosing between models since it doesn't penalize complexity and will generally always increase as you continue to add predictors, whether they have a statistically significant impact or not. The *adjusted $R^2$ statistic*, denoted $\bar{R}^2$ and reported as Adjusted R-squared in summary, is a simple transformation of the original $R^2$ that does incorporate a penalty for complexity relative to the sample size $n$; calculated as

$$\bar{R}^2 = 1 - \frac{(1 - R^2)(n - 1)}{n - p - 1},$$

where $p$ is the number of predictor terms (excluding the intercept). The algorithms based on tests and criteria are always preferable (since interpretation of $\bar{R}^2$ can be difficult), but monitoring $\bar{R}^2$ can be useful as a quick check between nested models—a higher value points to a preferred model.

For further reading, Chapter 8 of Faraway (2005) provides some excellent commentary on the guideline-only nature of both test- and criterion-based model selection procedures. Regardless of which approach you employ, always remember that any final model reached by using these algorithms should still be subject to scrutiny.

## 22.3 Residual Diagnostics

In previous chapters, you examined the practical aspects of multiple linear regression models, such as fitting and interpreting, dummy coding, transforming, and so on, but you haven't yet looked at methods that are essential for determining the *validity* of your model. The final part of this chapter will introduce you to *model diagnostics*, the primary goal of which is to ensure that your regression model is valid and accurately represents the relationships in your data. For this, I'll return focus to the theoretical assumptions underpinning the multiple linear regression model that were noted early in Section 21.2.1.

As a refresher, in general when fitting these models, remember to keep these four things in mind:

**Errors**   The error term $\epsilon$, which defines the departure of any observation from the fitted mean-outcome model, is assumed to be normally distributed with a mean of zero and a constant variance denoted with $\sigma^2$. The error associated with a given observation is also assumed to be independent of the error of any other observation. If a fitted model suggests violation of any of these assumptions, you'll need to investigate further (usually involving refitting a variation of the model).

**Linearity**   It is critical to be able to assume that the mean response as a function is linear in terms of the regression parameters $\beta_0, \beta_1, \ldots, \beta_p$. Though transformations of individual variables and the presence of interactions can relax the specific nature of the estimated trends somewhat, any diagnostic suggestion that a relationship is nonlinear (and hence not being captured by the fitted model at hand) must be investigated.

**Extreme or unusual observations**   Always inspect extreme data points or data points that strongly influence the fitted model—for example, points that have been recorded incorrectly should be removed from the analysis.

**Collinearity**   Predictors highly correlated with one another can adversely affect an entire model, meaning it can be easy to misinterpret the effects of any included predictors. This should be avoided in any regression.

You investigate the first three after fitting the model using diagnostic tools. Any violation of these assumptions diminishes the reliability of your model, sometimes severely. Collinearity and/or extreme observations can be discovered by basic statistical explorations (for example, viewing scatterplot matrices) of the raw data pre-fit, but any consequential effects are appraised post-fit.

There are some statistical tests you can perform to diagnose a statistical model, but commonly a diagnostic inspection boils down to interpretation of the results of graphical tools designed to target specific assumptions. Interpreting these plots can be quite difficult and only really becomes easier with experience. Here, I'll provide an overview of these tools in R and describe some common things to look for. For a more detailed discussion, look to dedicated texts on regression methods such as Chatterjee et al. (2000), Faraway (2005), or Montgomery et al. (2012).

### 22.3.1   *Inspecting and Interpreting Residuals*

If you look back at the plot in Figure 20-2 on page 456, you'll see a good demonstration of the importance of interpreting the results given by $\hat{y}$ as a mean response value. Under the assumed model, any deviation of the raw observations from the fitted line is deemed to be the result of the (normally distributed) errors defined by the $\epsilon$ term in Equation (20.1) on page 453.

Of course, in practice, you don't have the true error values because you don't know the true model of your data. For the $i$th response observation $y_i$ and its fitted value from the model, $\hat{y}_i$, you would typically assess diagnostic plots using the estimated residuals $e_i = y_i - \hat{y}_i$. A call to summary even encourages a post-fit analysis of the residuals by providing you with a five-number summary of the $e_i$ above the table of estimated coefficients. This allows you to take a look at their values and do a preliminary numeric assessment of the symmetry of their distribution (as required by the assumption of normality—see Section 22.3.2).

As well as a diagnostic inspection of the raw residuals $e_i$, some diagnostic checks can also be done using their *standardized* (or *Studentized*) values. The standardized residuals rescale the raw residuals $e_i$ to ensure they all have the same variance, which is important if you need to directly compare them to one another. Formally, this is achieved with the calculation $e_i/(\hat{\sigma}\{1 - h_{ii}\}^{0.5})$, where $\hat{\sigma}$ is the estimate of the residual standard error and $h_{ii}$ is the *leverage* of the $i$th observation (you'll learn about leverage in Section 22.3.4).

Arguably the most common graphical tool used for a post-fit analysis of the residuals is a simple scatterplot of the "observed-minus-fitted" raw residuals on the vertical axis against their corresponding fitted-model values from the regression. If the assumptions concerning $\epsilon$ are valid, then the $e_i$ should appear randomly scattered around zero (since the errors aren't assumed to be related in any way to the value of the response). Any systematic pattern in the plot suggests the residuals don't agree with the error assumptions—this could be because of nonlinear relationships in your data or the presence of dependent observations (in other words, your data points are correlated and therefore not independent of one another). The plot can also be used to detect *heteroscedasticity*—a nonconstant variance in the residuals—commonly seen as a "fanning out" of the residuals around 0 as the fitted values increase.

Note once more that it's important these theoretical assumptions are valid because they affect the validity of the estimates of the regression coefficients and the reliability of their standard errors (and so statistical significance)—in other words, the correctness of your interpretation of their impact on the response.

To give you a better idea of all this, consider the three images in Figure 22-1. These provide residuals versus fitted values plots for three hypothetical scenarios.

The plot on the left is, more or less, what you're looking for—the residuals appear randomly scattered around zero, and their spread around zero appears constant (*homoscedasticity*). In the middle plot, however, you can see systematic behavior in the residuals. Though the variability still seems to remain constant throughout the range of fitted values, the apparent trend suggests the current model isn't explaining some of the relationship between response and predictor (or predictors). On the right, the residuals seem to be scattered randomly about zero again. However, the variability they exhibit isn't constant. Among other things, this kind of heteroscedasticity will affect the reliability of your confidence and prediction intervals.

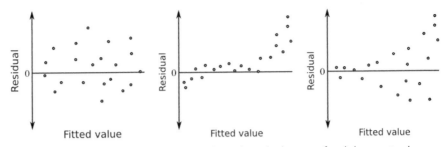

Figure 22-1: Three impressions of a hypothetical residuals versus fitted diagnostic plot from a linear regression: random (left), systematic (middle), and heteroscedastic (right)

It's important to know that even if your graphical diagnostics don't provide the well-behaved plot like the hypothetical example on the left of Figure 22-1, this is not a reason to immediately give up on the analysis. These plots can form an integral part in finding an appropriate model for your data. You can often reduce nonlinearity by including additional predictors or interactions, changing the treatment of a categorical variable, or performing nonlinear transformations of certain continuous predictors. Heteroscedasticity, especially the kind in Figure 22-1 where the variability is higher for higher fitted values, is common in some fields of research. A first step to remedy this problem often involves a simple log transformation of the response followed by a reinspection of the diagnostics.

It's time for an example. In Section 22.2.4, you used stepwise AIC selection to choose a model for MPG for the mtcars data, creating the object car.step. Let's now diagnose that same fit to see whether there are any problems with the assumptions of the model.

When you apply the plot function directly to an lm object, it can conveniently produce six types of diagnostic plot of the fit. By default, four of these plots are produced in succession. Follow the signal to the user in the console, Hit <Return> to see next plot, to progress through them. In the examples that follow, however, you'll select each plot individually using the optional which argument (specified by the integers 1 through 6; see ?plot.lm for the documentation). The residuals versus fitted plot is given with which=1; the following line produces the plot on the left in Figure 22-2:

```
R> plot(car.step,which=1)
```

As you can see, R adds a smoothed line to help the user interpret any trend, though this shouldn't be used exclusively in any judgment. By default, the three most extreme points from zero are annotated (according to the rownames attribute of the data frame used in the call that fitted the model). The model formula itself is specified below the horizontal axis label.

From this, you see that the residuals versus fitted plot for the car.step offers little, if any, cause for concern. There isn't much of a discernible trend, and you can take further comfort in the fact that the errors ($e_i$) appear homoscedastic in their distribution.

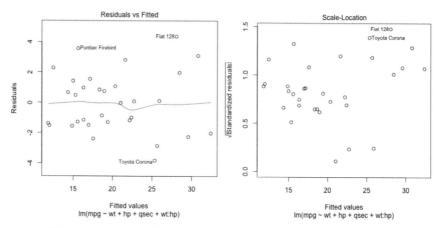

Figure 22-2: Residuals versus fitted and scale-location diagnostic plots for the car.step model

The *scale-location* plot is similar to the residuals versus fitted plot, though instead of the raw $e_i$ on the vertical axis, the scale-location plot provides $|e_i/(\hat{\sigma}\{1 - h_{ii}\}^{0.5})|^{0.5}$, that is, the square root of the absolute value (denoted by $| \cdot |$; this renders all negative values positive) of the standardized residuals. These are plotted against the respective fitted values on the horizontal axis. By restricting attention to the magnitude of each residual in this way, the scale-location plot is used to reveal trends in the size of the departure of each data point from its fitted value, as the fitted values increase. This means such a plot can, for example, be more useful than the raw residuals versus fitted plot in detecting things such as heteroscedasticity. Just as with the original residuals versus fitted plot, you're looking for a plot with no discernible pattern as an indication that no error assumptions have been violated.

The right plot of Figure 22-2 shows the scale-location plot for car.step, selected with which=3. This plot also demonstrates the ability to remove the default smoothed trend line with the add.smooth argument and to control how many extreme points are labeled using the id.n argument.

```
R> plot(car.step,which=3,add.smooth=FALSE,id.n=2)
```

As with the original residuals versus fitted plot, there doesn't seem to be much to be concerned about in the scale-location plot for this mtcars model.

Return to Galileo's ball-rolling data first laid out in Exercise 21.2 on page 512. In their use in the following example, the response variable "distance traveled" is given as column d, and the explanatory variable "height" is column h, in the data frame gal. I'll re-create some of the exercise to give you a couple of straightforward examples of cause for concern in residual diagnostic plots. Execute the following code to define the data frame of the seven observations and fit two regression models—the first linear in height and the second quadratic (refer to Section 21.4.1 for details on polynomial transformations).

```
R> gal <- data.frame(d=c(573,534,495,451,395,337,253),
                      h=c(1,0.8,0.6,0.45,0.3,0.2,0.1))
R> gal.mod1 <- lm(d~h,data=gal)
R> gal.mod2 <- lm(d~h+I(h^2),data=gal)
```

Now, take a look at the three images in Figure 22-3, created with the following code:

```
R> plot(gal$d~gal$h,xlab="Height",ylab="Distance")
R> abline(gal.mod1)
R> plot(gal.mod1,which=1,id.n=0)
R> plot(gal.mod2,which=1,id.n=0)
```

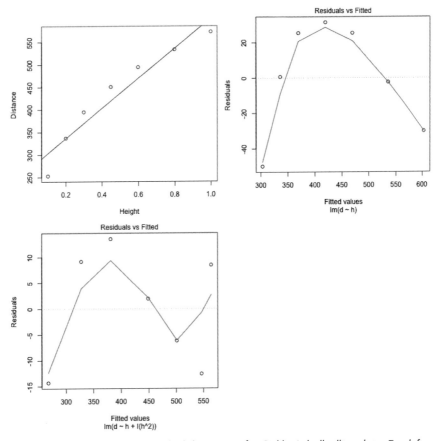

*Figure 22-3: Demonstrating residual diagnostics for Galileo's ball-rolling data. Top left: The raw data with a simple linear trend corresponding to* gal.mod1 *superimposed. Top right: Residuals versus fitted for the linear-trend-only model. Bottom: Residuals versus fitted for the quadratic model* gal.mod2.

The top-left plot shows the data and provides the straight line of the simple linear model. Although this clearly captures the increasing trend, this plot suggests some curvature is also present. The diagnostic residuals versus fitted plot (top-right) shows that the linear-trend-only model is inadequate—the systematic pattern throws up a red flag concerning the assumptions surrounding the linear model errors. The bottom image shows the residuals versus fitted plot based on the quadratic version of the model in gal.mod2. Including a quadratic term in "height" removes this prominent curve in the residuals. However, these latest $e_i$ values still seem to exhibit systematic behavior in a wavelike form, perhaps suggesting you try a cubic model, which is difficult with such a small sample size.

### 22.3.2   Assessing Normality

To assess the assumption that the error is normally distributed, you can use a normal QQ plot, as first discussed in Section 16.2.2. You select which=2 when calling plot on an lm object to produce a normal quantile-quantile plot of the (standardized) residuals. Return to the car.step model object and enter the following line to produce Figure 22-4.

```
R> plot(car.step,which=2)
```

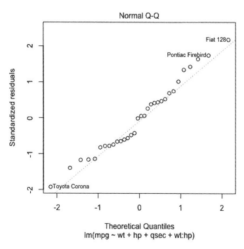

Figure 22-4: Normal QQ plot of the residuals from the car.step model

You interpret the QQ plot of the residuals in the same way as in Section 16.2.2. The gray diagonal line represents the true normal quantiles, and the plotted points are the corresponding numeric quantiles of the estimated regression errors. Normally distributed data should lie close to the straight line.

For the `car.step` regression model, the points generally seem to follow the path laid out by the theoretical normal quantiles. There is some deviation, which is to be expected, but no apparent major departure from normality.

There are also other ways to test for normality, such as the famous Shapiro-Wilk hypothesis test. The null hypothesis for the Shapiro-Wilk test is that the data are normally distributed, so a small *p*-value would suggest non-normality of your data (see Royston, 1982, for technical details). To execute the procedure, use the `shapiro.test` function in R. By first extracting the standardized residuals of your fitted model with `rstandard`, you'll see that this test applied to `car.step` offers up a large *p*-value.

```
R> shapiro.test(rstandard(car.step))

        Shapiro-Wilk normality test

data:  rstandard(car.step)
W = 0.97105, p-value = 0.5288
```

In other words, there's no evidence (according to this test) that the residuals of `car.step` aren't normal.

Being able to assume normality of the error term supports the methodology used to produce reliable estimates of the regression coefficients. As long as your data are *approximately* normal, though, you shouldn't be too concerned with mild indications of non-normality. Some transformations of your data, and an increase in your sample size, can reduce concerns about more severe indications of non-normal residuals.

### 22.3.3  Illustrating Outliers, Leverage, and Influence

It's always important to investigate any individual observations that appear unusual or extreme compared to the bulk of your observations. In general, an exploratory analysis of your data, perhaps involving summary statistics or scatterplot matrices, is a good idea since it can help you identify any such values—they have the potential to adversely affect your model fits. Before going further, it's important to clarify some frequently used terms.

**Outlier**   This is a general term used to describe an unusual observation in the context of the data, as you saw in Section 13.2.6. In linear regression, an outlier usually has a large residual but is identified as an outlier only if it doesn't conform to the trend of the fitted model. An outlier can, but doesn't always, significantly alter the trends described by the fitted model.

**Leverage**   This term refers to the extremity of the values of the present predictors. A high-leverage point is an observation with predictor values extreme enough to potentially significantly affect the slopes or trends in the fitted model. An outlier can have a high or low leverage.

**Influence** An observation with high leverage that *does* affect the estimated trends is deemed influential. In other words, influence is judged only when the response value is taken into account alongside the corresponding predictor values.

These definitions have some overlap, so a given observation can be described using a combination of these terms. Let's look at some hypothetical examples. Create the following two vectors of ten supposed responses (y) and explanatory (x) values:

```
R> x <- c(1.1,1.3,2.3,1.6,1.2,0.1,1.8,1.9,0.2,0.75)
R> y <- c(6.7,7.9,9.8,9.3,8.2,2.9,6.6,11.1,4.7,3)
```

These will form the bulk of the data. Now, consider the following six objects, p1x to p3y, which will be used to store the predictor and response values for three additional observation points:

```
p1x <- 1.2
p1y <- 14
p2x <- 5
p2y <- 19
p3x <- 5
p3y <- 5
```

That is, point 1 is $(1.2, 14)$; point 2 is $(5, 19)$; and point 3 is $(5, 5)$.

Next, the following four uses of lm provide four simple linear model fits. The first is the regression of y on x only. The next three additionally include points 1, 2, and 3, separately, as an 11th observation.

```
R> mod.0 <- lm(y~x)
R> mod.1 <- lm(c(y,p1y)~c(x,p1x))
R> mod.2 <- lm(c(y,p2y)~c(x,p2x))
R> mod.3 <- lm(c(y,p3y)~c(x,p3x))
```

Now, you can use these objects to visually clarify the definitions of *outlier*, *leverage*, and *influence*, as shown in Figure 22-5. Enter the following code to initialize the scatterplot with set axis limits of x and y:

```
R> plot(x,y,xlim=c(0,5),ylim=c(0,20))
```

Then use points, abline, and text to build the top-left plot of Figure 22-5, as follows:

```
R> points(p1x,p1y,pch=15,cex=1.5)
R> abline(mod.0)
R> abline(mod.1,lty=2)
R> text(2,1,labels="Outlier, low leverage, low influence",cex=1.4)
```

Create the middle and right plots by replacing `p1x`, `p1y`, and `mod.1` with those corresponding to points 2 and 3 and altering the `labels` argument in text.

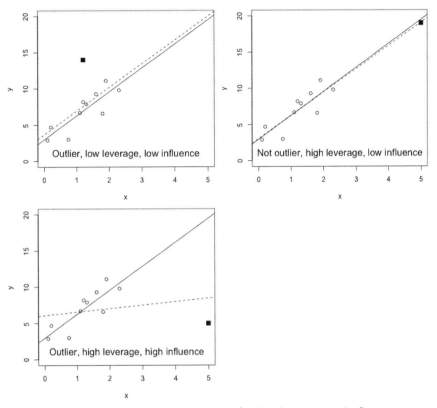

*Figure 22-5: Three examples of the definitions of outlier, leverage, and influence in a linear regression context. In each plot, the solid line represents the model fitted to the original observations in x and y, and the dashed line represents the model fitted including the extra point, plotted with ■.*

In the top-left plot of Figure 22-5, the additional point is an example of an outlier since it sits away from the bulk of the data *and* doesn't conform to the trend suggested by the original observations. Despite this, it's considered to have low leverage only because of its predictor value of 1.2 (`p1x`), which isn't deemed unusual compared to that of the other values of x. In fact, its proximity to the overall mean of the x values indicates that the effect of including it, incorporated in `mod.1`, is mainly a modification to the original intercept. You could even classify this as a low influence point—the overall change to the fitted model seems minimal.

In the top-right plot, you can see an example of an observation that would *not* be considered an outlier. Although point 2 does sit apart from the 10 original observations, it conforms quite well to the model fitted to only x and y, which is important in the context of regression. That being said, point 2 is considered a high leverage point since it sits at an extreme predictor value compared to all other values of x (in other words, it has the *potential* to dramatically alter the fit should its response value be different). As it stands, it's a low influence point since the model fit itself is barely affected by its inclusion.

Lastly, the bottom plot shows a clear example of an outlier in a high-leverage position that also has a high influence—it sits away from the original 10 observations and isn't a clear part of the original trend; its extreme predictor value means high leverage; and its inclusion substantially alters the entire model by dragging down the slope and raising the intercept. These ideas remain the same in higher dimensions (that is, when you have several predictors) for multiple linear regression models.

### 22.3.4  Calculating Leverage

Leverage itself is calculated using the design matrix structure $X$ defined in Section 21.2.2. Specifically, if you have $n$ observations, then the leverage of the $i$th point ($i = 1, \ldots, n$) is denoted $h_{ii}$. These are the diagonal elements ($i$th row, $i$th column) of the $n \times n$ matrix $H$ such that

$$H = X(X^\top X)^{-1} X^\top \qquad (22.4)$$

In R, constructing the design matrix for the 10 illustrative predictor observations you defined in Section 22.3.3 as x is achieved in a straightforward fashion using knowledge of cbind (refer to Section 3.1.2). $H$ is subsequently calculated using the corresponding functions for matrix multiplication (%*%), matrix transposition (t), matrix inversion (solve), and diagonal element extraction (diag). Then you can plot the values $h_{ii}$ against the values of x themselves. The following code produces Figure 22-6:

```
R> X <- cbind(rep(1,10),x)
R> hii <- diag(X%*%solve(t(X)%*%X)%*%t(X))
R> hii
 [1] 0.1033629 0.1012107 0.3487221 0.1302663 0.1001345 0.3723971 0.1711595
 [8] 0.1980630 0.3261232 0.1485607
R> plot(hii~x,ylab="Leverage",main="",pch=4)
```

You would typically use the built-in R function hatvalues, named after the style of the matrix algebra in Equation (22.4), to obtain the leverages (rather than manually constructing the design matrix $X$ and doing the math yourself). Simply provide hatvalues with your fitted model object. You can confirm your earlier calculations by using the corresponding lm object fitted to the x and y data (mod.0 created earlier).

```
R> hatvalues(mod.0)
        1         2         3         4         5         6         7
0.1033629 0.1012107 0.3487221 0.1302663 0.1001345 0.3723971 0.1711595
        8         9        10
0.1980630 0.3261232 0.1485607
```

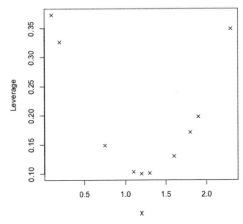

Figure 22-6: Plotting the leverage of the 10 illustrative
predictor observations in x

Looking at Figure 22-6, the appearance of the leverages plotted against the corresponding predictor values themselves makes sense—leverage gets progressively higher as you move away from the mean of the predictor data in either direction. This is essentially the pattern you'll see for any plot of the raw leverages.

### 22.3.5   Cook's Distance

Of course, leverage alone isn't enough to determine the overall influence of each observation on a fitted model. For that, the response value must also be taken into account.

Arguably the most well-known measure of influence is *Cook's distance*, which estimates the magnitude of the effect of deleting the $i$th value from the fitted model. Cook's distance for observation $i$ (denoted $D_i$) is given with the following equation:

$$D_i = \sum_{j=1}^{n} \frac{(\hat{y}_j - \hat{y}_j^{(-i)})^2}{(p+1)\hat{\sigma}^2}; \qquad i,j = 1,\ldots,n \qquad (22.5)$$

It turns out that this equation is a specific function of a point's leverage and residual. Here, the value $\hat{y}_j$ is the predicted mean response of observation $j$ for the model fitted with all $n$ observations, and $\hat{y}_j^{(-i)}$ represents the predicted mean response of observation $j$ for the model fitted *without* the $i$th observation. As usual, the term $p$ is the number of predictor regression parameters (excluding the intercept), and the value $\hat{\sigma}$ is the estimate of the residual standard error.

Put simply, the larger the value of $D_i$, the larger the influence the $i$th observation has on the fitted model, meaning outlying observations in high-leverage positions will correspond to higher values of $D_i$. The important question is, how big does $D_i$ have to be in order for point $i$ to be considered influential? In practice, this is difficult to universally answer, and there's no formal hypothesis test for it, but there are several rule-of-thumb cutoff values. One such rule states that if $D_i > 1$, the point should be considered influential; another, more sensitive rule suggests $D_i > 4/n$ (see, for example, Bollen and Jackman, 1990; Chatterjee et al., 2000). It's generally advised that you compare multiple Cook's distances for a given fitted model rather than analyzing one single value, and that any point corresponding to a comparatively large $D_i$ might need further inspection.

Continue with the objects created in Section 22.3.3, with the 10 illustrative observations in x and y, as well as the additional point defined in p1x and p1y. The linear regression model fitted to those data was stored as the object mod.1. It's a good exercise to write some code that calculates the Cook's distance measures following (22.5).

To that end, enter the following code in the R editor:

```
x1 <- c(x,p1x)
y1 <- c(y,p1y)
n <- length(x1)
param <- length(coef(mod.1))
yhat.full <- fitted(mod.1)
sigma <- summary(mod.1)$sigma
cooks <- rep(NA,n)
for(i in 1:n){
    temp.y <- y1[-i]
    temp.x <- x1[-i]
    temp.model <- lm(temp.y~temp.x)
    temp.fitted <- predict(temp.model,newdata=data.frame(temp.x=x1))
    cooks[i] <- sum((yhat.full-temp.fitted)^2)/(param*sigma^2)
}
```

First, create new objects x1 and y1 to hold all 11 observations. The objects n, param, and sigma extract the data set size, the total number of estimated regression parameters (two in this case), and the estimated residual standard error for the model originally fitted to all 11 data points. The latter two items, param and sigma, represent $(p + 1)$ and $\hat{\sigma}$ in Equation (22.5), respectively. The object yhat.full uses the fitted function on the object mod.1 to provide the fitted mean response values, representing the $\hat{y}_j$ values in (22.5).

To store the Cook's distances, a vector cooks of 11 positions is created (initialized to be filled with NAs) with rep. Now, to calculate each $D_i$ value, set a for loop (see Chapter 10) to scroll through each index from 1 to 11. The first step of the loop is to create two temporary vectors temp.x and temp.y to be x1 and y1 with the observation at index i removed. A new temporary linear model is fitted to temp.y based on temp.x; then predict finds the mean responses from temp.model for each of the 11 predictor values (in other words, including the one that was deleted). As such, the resulting vector temp.fitted represents the $\hat{y}_j^{(-i)}$ values in Equation (22.5). Finally, sum and the product of param with sigma^2 compute $D_i$, and the result is stored in cooks[i].

After highlighting and executing the code, the resulting Cook's distances are as follows:

```
R> cooks
 [1] 2.425993e-03 4.060891e-07 1.027322e-01 1.844150e-03 2.923667e-03
 [6] 7.213229e-02 1.387284e-01 3.021075e-02 7.099904e-03 1.251882e-01
[11] 3.136855e-01
```

Unsurprisingly, the largest value of these is the last one, at around 0.314. This corresponds to the 11th observation in x1 and y1, which is the additional point originally defined in p1x and p1y. The value 0.314 is less than 1 and less than $4/11 = 0.364$, the cutoffs defined by the earlier rules of thumb. This ties in with the assessment of the top-left image in Figure 22-5—that the influence of the point p1x and p1y is minimal when compared to the influence of a point like p3x and p3y in the rightmost image.

Just as the hatvalues function computes the leverages for you, the built-in cooks.distance function does the same for the $D_i$. You can confirm the previous values in cooks, which are based on mod.1.

```
R> cooks.distance(mod.1)
           1            2            3            4            5            6
2.425993e-03 4.060891e-07 1.027322e-01 1.844150e-03 2.923667e-03 7.213229e-02
           7            8            9           10           11
1.387284e-01 3.021075e-02 7.099904e-03 1.251882e-01 3.136855e-01
```

R automatically calculates and provides Cook's distances as a diagnostic plot of a fitted linear model object when you select which=4 in the relevant usage of plot. The following code uses mod.1, mod.2, and mod.3 from earlier to produce the three images in Figure 22-7; these correspond to the three data sets in Figure 22-5.

```
R> plot(mod.1,which=4)
R> plot(mod.2,which=4)
R> plot(mod.3,which=4)
R> abline(h=c(1,4/n),lty=2)
```

The $D_i$ displayed on the top left of Figure 22-7 match the values you manually calculated earlier, stored in cooks. The influences of all the data points in the top-right plot remain relatively small, which reflects what you can see in the top-right plot of Figure 22-5, where the additional point (p2x, p2y) doesn't greatly affect the overall fit. In the bottom plot, abline superimposes two horizontal lines marking off the values 1 (highest line) and $4/11 = 0.364$, both of which are clearly breached by the 11th point (p3x, p3y), just as you'd expect given the corresponding bottom image in Figure 22-5.

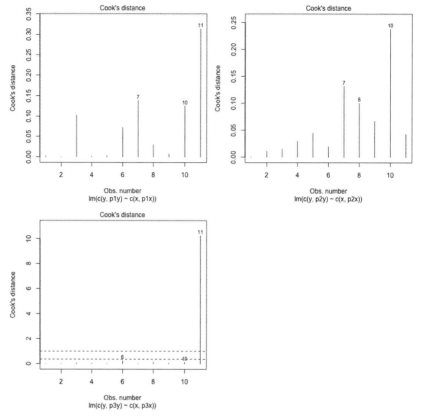

Figure 22-7: Three illustrative examples of the Cook's distance plots produced in R, based on mod.1 *(top left)*, mod.2 *(top right)*, and mod.3 *(bottom)*

Turn your attention back to the car.step model, where you modeled MPG using the mtcars data set, with the final fit achieved using stepwise AIC selection in Section 22.2.4. You've already looked at the residuals versus fitted values and QQ plot in Figures 22-2 and 22-4. Figure 22-8 provides a plot of Cook's distances for the model with these two lines:

```
R> plot(car.step,which=4)
R> abline(h=4/nrow(mtcars),lty=2)
```

The plot labels the three points with the highest $D_i$ by default; two of these breach the $4/n = 4/32 = 0.125$ mark. In light of the fitted model based on various effects of car weight (wt), horsepower (hp), and quarter-mile time (qsec), the Chrysler Imperial and Toyota Corolla are deemed to be in high-leverage positions with residuals large enough to be designated as highly influential. It should also be noted that the Fiat 128, though it doesn't quite breach the 0.125 line, is still rather influential and was in fact also one of the extreme labeled points in the residual plots (Figure 22-2) and the QQ plot (Figure 22-4).

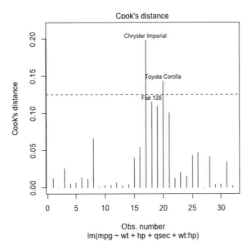

Figure 22-8: Cook's distances for the model in car.step; a dashed horizontal line marks off 4/n for the mtcars data frame

This could reasonably suggest investigating these highly influential observations further. Was everything recorded correctly? Has your model been selected carefully? Are there alternative options for the model, such as additional predictor terms or transformations? You could explore these options and continue to monitor a plot of the Cook's distances (along with the other diagnostics).

Whatever your result, the presence of influential points doesn't necessarily mean there is a serious problem with your model—this is more a tool to help you detect observations that are extreme in terms of their specific combination of the predictor values *and* that have a larger residual, suggesting their response value sits away from the trends predicted by the model itself. This is especially useful in multiple regression, when high dimensionality of the response-predictor data can make conventional visualization of the raw data in a single plot difficult.

### 22.3.6   Graphically Combining Residuals, Leverage, and Cook's Distance

The last two diagnostic plots from plot combine the standardized residual, the leverage, and the Cook's distance for the *i*th observation. These combination plots are especially useful in allowing you to see whether it is the high leverage or large residual, or both, that contributes more to a high influence observation.

Using the data models mod.1, mod.2, and mod.3, enter the following code with which=5 to produce the three images in the left column of Figure 22-9:

```
R> plot(mod.1,which=5,add.smooth=FALSE,cook.levels=c(4/11,0.5,1))
R> plot(mod.2,which=5,add.smooth=FALSE,cook.levels=c(4/11,0.5,1))
R> plot(mod.3,which=5,add.smooth=FALSE,cook.levels=c(4/11,0.5,1))
```

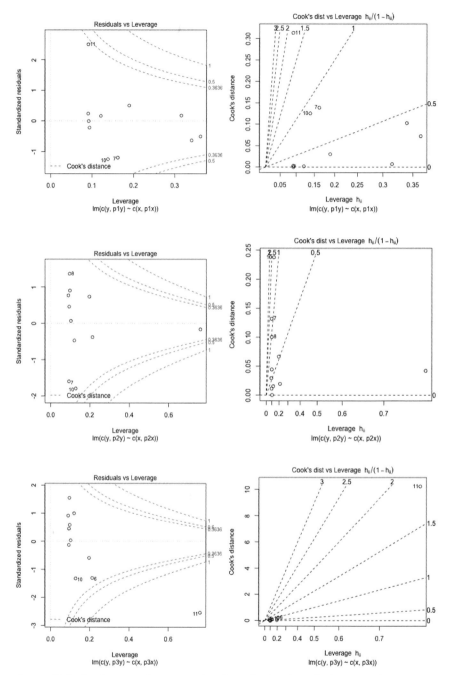

*Figure 22-9: Combination diagnostic plot of standardized residuals against leverage (left column) and Cook's distance against leverage (right column) for the three illustrative models* mod.1 *(top),* mod.2 *(middle), and* mod.3 *(bottom)*

These plots show leverage on the horizontal axis and the standardized residuals on the vertical axis for each observation. As a function of residual and leverage, the Cook's distances can be plotted as *contours* on each of these scatterplots. These contours delineate the spatial areas of the plots that correspond to high influence (in the right extreme corners).

The closer a point falls to the horizontal line at zero, the smaller its residual. A point that lies more left than right has a smaller leverage. If a point lies far enough from the horizontal line given its leverage (*x*-axis) position, it will breach the contours marking off certain values of $D_i$ (defaulting to just 0.5 and 1), indicating a high influence. Indeed, you can see by the narrowing of the contours as you move from left to right on these plots that classification as a high-influence point is easier if a given observation is in a high-leverage position, which makes perfect sense. In the previous calls to plot with which=5, the optional cook.levels argument is used to include a contour for the rule-of-thumb value of 4/11 for these three examples.

The plot for mod.1 shows that the added point (p1x, p1y) has a large residual, but it doesn't breach the 4/11 contour because it's in a low-leverage position. The plot for mod.2 shows that the added point (p2x,p2y) is in a high-leverage position but isn't influential because its residual is small. Lastly, the plot for mod.3 clearly identifies the added point (p3x, p3y) as highly influential—with a large residual and high leverage, it's in clear breach of the high-level contours. Looking back at all the previous plots of these three illustrative data sets, it's easy to note that these three plots clearly reflect the nature of each of the individually added extra observations.

The final diagnostic plot, using which=6, displays the same information as the which=5 combination diagnostic, but this time the vertical axis displays Cook's distance, and the horizontal axis displays a transformation of the leverage, namely, $h_{ii}/(1 - h_{ii})$. This transformation amplifies larger leverage points in terms of their horizontal position, an effect that, in part, indirectly displays itself as a "stretched-out" scale on the *x*-axis—useful if you're particularly interested in the extremity of the observations with respect to the collection of predictor variables.

As such, the contours now define standardized residuals as a function of the scaled leverage and Cook's distance. The following three lines produce the three images in the rightmost column of Figure 22-9:

```
R> plot(mod.1,which=6,add.smooth=FALSE)
R> plot(mod.2,which=6,add.smooth=FALSE)
R> plot(mod.3,which=6,add.smooth=FALSE)
```

Points positioned further to the right are high-leverage points; points positioned higher up are high-influence points. Looking down the right column of plots in Figure 22-9, you can see that the three additional points are found where you would expect them to be, according to their characteristics in mod.1, mod.2, and mod.3.

For a real-data example, return to the model stored in `car.step`. Enter the following code to produce the two combination diagnostic plots in Figure 22-10:

```
R> plot(car.step,which=5,cook.levels=c(4/nrow(mtcars),0.5,1))
R> plot(car.step,which=6,cook.levels=c(4/nrow(mtcars),0.5,1))
```

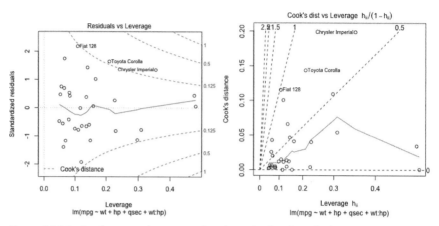

Figure 22-10: Combination diagnostic plot of standardized residuals against leverage (left) and Cook's distance against leverage (right) for the `car.step` model

The two images in Figure 22-10 show the Corolla and the Imperial as influential observations with $D_i$ values greater than the `4/nrow(mtcars)` rule-of-thumb cutoff. Interestingly, this plot reveals that the Imperial (which was shown to have the largest $D_i$ by far in Figure 22-8) actually has a smaller residual than both the Corolla and the Fiat 128. Its high influence is clearly because of its high-leverage position with respect to the predictor values of the variables present in `car.step`. The Fiat 128, on the other hand, has one of the largest residuals in the entire data set (which is why it was flagged in some of the earlier diagnostic plots) but just misses out on being labeled a high-influence observation because of its relatively low-leverage position (based purely on the rule-of-thumb cutoff).

Any linear regression model will have observations that influence the model more than others, and these plots aim to help you identify them. But deciding what to actually do with highly influential observations can be difficult and is application specific. Although it's not ideal to have a single observation exerting heavy influence over the final estimated model, it's also extremely unwise to remove these observations without careful thought since they might be pointing to other issues, such as deficiencies in your current fit or previously undetected trends.

In Section 22.2.2, you used the `nuclear` data frame in the `boot` package to illustrate forward selection, where a model was selected for `cost` as a function of main effects of `date`, `cap`, `pt`, and `ne`.

a.  Access the data frame; fit and summarize the model described earlier.

b.  Inspect the raw residuals versus fitted values and a normal QQ plot of the residuals and comment on your interpretations—do the assumptions underpinning the error component of the linear model appear satisfied in this case?

c.  Determine the rule-of-thumb cutoff for influential observations based on the Cook's distances. Produce a plot of the Cook's distances and add a horizontal line corresponding to the cutoff. Comment on your findings.

d.  Produce a combination diagnostic plot of the standardized residuals against leverage. Set the Cook's distance contours to include the cutoff value from (c) as well as the default contours. Interpret the plot—how are any individually influential points characterized?

e.  Based on (c) and (d), you should be able to identify the record in `nuclear` exerting the largest influence on the fitted model. For the sake of argument, let's assume the observation was recorded incorrectly. Refit the model from (a), this time omitting the offending row from the data frame. Summarize the model—which coefficients have changed the most? Produce the diagnostic plots from (b) for the new model and compare them to the ones from earlier.

Load the `faraway` package and access the `diabetes` data frame. In Exercise 22.1 (g), you used stepwise AIC selection to choose a model for `chol`.

f.  Using `diabetes`, fit the multiple linear model identified in the earlier exercise, that is, with main effects and a two-way interaction between `age` and `frame` and a main effect for `waist`. By summarizing the fit, determine the number of records that contained missing values in `diabetes` that were deleted from the estimation.

g.  Produce the raw residuals versus fitted and QQ diagnostic plots for the model in (f). Comment on the validity of the error assumptions.

h. Investigate influential points. Make use of the familiar rule-of-thumb cutoff (note you'll need to subtract the number of missing values from the total size of the data frame to get the effective sample size for your model). In the combination plot of the standardized residuals against leverage, use one, three, and five times the cutoff as the Cook's distance contours.

Recall the discussion of reading in web-based files in Section 8.2.3. There, you called in a data frame containing data on the prices of 308 diamonds (in Singapore dollars), as well as weight (in carats—continuous), color (categorical—six levels from D, the least yellow and the reference level, to I, the most yellow), clarity (categorical—five levels with IF, essentially flawless and the reference level, VVS1, VVS2, VS1, and VS2, with the last being the least clear), and certification (categorical—three levels for different diamond certification bodies with levels GIA as the reference, HRD and IGI). Seek out the freely available article by Chu (2001) for more information on these data. With an Internet connection, run the following lines, which will read in the data as the object diamonds and name each variable column appropriately.

```
R> dia.url <- "http://www.amstat.org/publications/jse/v9n2/4cdata.txt"
R> diamonds <- read.table(dia.url)
R> names(diamonds) <- c("Carat","Color","Clarity","Cert","Price")
```

i. Using either base R graphics or ggplot2, to get a feel for the data, produce a scatterplot of the price on the y-axis and carat weight on the x-axis. Experiment with using plotting color to split the points according to the following:
   - Diamond clarity
   - Diamond color
   - Diamond certification

j. Fit a multiple linear model with Price as the response and main effects for the other variables as the predictors. Summarize the model and produce the three diagnostic plots that tell you about the assumptions surrounding the error term. Comment on the plots—are you satisfied that this is an appropriate model for the diamond prices? Why or why not?

k. Repeat (j) but use the log transformation of Price. Again, inspect and comment on the validity of the error assumptions.

l. Repeat (k), but in modeling the log-price, this time include an additional quadratic term for Carat (refer to Section 21.4.1 for details on polynomial transformations). How do the residual diagnostics look now?

# 22.4 Collinearity

This final aspect of fitting regression models isn't technically classified as a diagnostic check but still has substantial potential to adversely affect the validity of any conclusions you draw from a fitted model and occurs frequently enough to warrant discussion here. *Collinearity* (also referred to as *multicollinearity*) is when two or more of the explanatory variables are highly correlated with each other.

## 22.4.1 Potential Warning Signs

High correlation between two predictors implies there will be some level of redundancy in terms of the information they contain when it comes to the response variable. It's a problem since it can destabilize the ability to reliably fit a model and, as noted earlier, therefore be detrimental to any subsequent model-based inference.

The following items serve as potential warnings of collinearity when you're inspecting a model summary:

- The omnibus $F$-test (Section 21.3.5) result is statistically significant, but none of the individual $t$-test results for the regression parameters are significant.

- The sign of a given coefficient estimate contradicts what you would reasonably expect to see, for example, drinking more wine resulting in a lower blood alcohol level.

- Parameter estimates are associated with unusually high standard errors or vary wildly when the model is fitted to different random record subsets of the data.

As the last point notes, collinearity tends to have more of a detrimental effect on the standard errors of the coefficients (and associated outcomes such as confidence intervals, significance tests, and prediction intervals) than it does on point predictions per se. In most cases, you can avoid collinearity simply by being careful. Be aware of the variables present and how the data have been collected. For example, ensure any given predictors you intend to include in the model don't just represent a rescaled value of another included predictor. It's also advisable to perform an exploratory analysis of your data, producing summary statistics and basic statistical plots. You can tabulate counts between categorical variables or look at estimated correlation coefficients between continuous variables, for example. In the latter case, as a rough guide, some statisticians suggest that a correlation of 0.8 or more could lead to potential problems.

## 22.4.2 Correlated Predictors: A Quick Example

Consider the survey data of the statistics students again, located in the MASS package. In most models you've looked at for these data, you've attempted to predict student height from certain explanatory variables, often including the handspan of the writing hand (Wr.Hnd). The help page ?survey shows that data have also been collected on the nonwriting handspan (NW.Hnd). It's reasonable to expect that these two variables will be highly correlated, which is precisely why I've avoided any use of NW.Hnd previously. Indeed, executing

```
R> cor(survey$Wr.Hnd,survey$NW.Hnd,use="complete.obs")
[1] 0.9483103
```

reveals a high correlation coefficient, suggesting a strong positive linear association between the writing and nonwriting handspans of the students. In other words, these two variables should represent much the same information in any given model.

Now, you know from previously fitted models that writing handspan has a significant and positive impact on predicting mean student height. The following code quickly confirms this via a simple linear regression.

```
R> summary(lm(Height~Wr.Hnd,data=survey))

Call:
lm(formula = Height ~ Wr.Hnd, data = survey)

Residuals:
    Min      1Q   Median      3Q     Max
-19.7276  -5.0706  -0.8269   4.9473  25.8704

Coefficients:
            Estimate Std. Error t value Pr(>|t|)
(Intercept) 113.9536     5.4416   20.94   <2e-16 ***
Wr.Hnd        3.1166     0.2888   10.79   <2e-16 ***
---
Signif. codes:  0 '***' 0.001 '**' 0.01 '*' 0.05 '.' 0.1 ' ' 1

Residual standard error: 7.909 on 206 degrees of freedom
  (29 observations deleted due to missingness)
Multiple R-squared:  0.3612,     Adjusted R-squared:  0.3581
F-statistic: 116.5 on 1 and 206 DF,  p-value: < 2.2e-16
```

The high positive correlation between Wr.Hnd and NW.Hnd suggests that using NW.Hnd instead should have a similar effect.

```
R> summary(lm(Height~NW.Hnd,data=survey))

Call:
lm(formula = Height ~ NW.Hnd, data = survey)

Residuals:
    Min      1Q   Median      3Q     Max
-21.8285  -5.1397  -0.2867   4.5611  25.5750
```

```
Coefficients:
            Estimate Std. Error t value Pr(>|t|)
(Intercept) 118.0324    5.2912   22.31   <2e-16 ***
NW.Hnd        2.9107    0.2818   10.33   <2e-16 ***
---
Signif. codes:  0 '***' 0.001 '**' 0.01 '*' 0.05 '.' 0.1 ' ' 1

Residual standard error: 8.032 on 206 degrees of freedom
  (29 observations deleted due to missingness)
Multiple R-squared:  0.3412,    Adjusted R-squared:  0.338
F-statistic: 106.7 on 1 and 206 DF,  p-value: < 2.2e-16
```

You can see from these results that this is certainly the case.

Look, however, at what happens if you try to model height based on both of these predictors at the same time:

```
R> summary(lm(Height~Wr.Hnd+NW.Hnd,data=survey))

Call:
lm(formula = Height ~ Wr.Hnd + NW.Hnd, data = survey)

Residuals:
    Min      1Q  Median      3Q     Max
-20.0144 -5.0533 -0.8558  4.7486 25.8380

Coefficients:
            Estimate Std. Error t value Pr(>|t|)
(Intercept) 113.9962    5.4545  20.900   <2e-16 ***
Wr.Hnd        2.7451    1.0728   2.559   0.0112 *
NW.Hnd        0.3707    1.0309   0.360   0.7195
---
Signif. codes:  0 '***' 0.001 '**' 0.01 '*' 0.05 '.' 0.1 ' ' 1

Residual standard error: 7.926 on 205 degrees of freedom
  (29 observations deleted due to missingness)
Multiple R-squared:  0.3616,    Adjusted R-squared:  0.3554
F-statistic: 58.06 on 2 and 205 DF,  p-value: < 2.2e-16
```

Since the effects of Wr.Hnd and NW.Hnd on Height are intermingled with one another, including both at the same time heavily masks any individual contribution to modeling the response. Statistical significance of the predictors is almost nonexistent; at the least, the effects are both associated with much, much higher $p$-values than in the individual single-predictor fits. That said, the omnibus $F$-test remains highly significant, giving an example of the first warning sign noted previously.

## Important Code in This Chapter

| Function/operator | Brief description | First occurrence |
|---|---|---|
| anova | Partial $F$-tests | Section 22.2.1, p. 531 |
| add1 | Review single-term additions | Section 22.2.2, p. 533 |
| update | Make changes to fitted model | Section 22.2.2, p. 534 |
| drop1 | Review single-term deletions | Section 22.2.3, p. 538 |
| step | Stepwise AIC model selection | Section 22.2.4, p. 543 |
| plot (used on lm object) | Model diagnostics | Section 22.3.1, p. 551 |
| rstandard | Extract standardized residuals | Section 22.3.2, p. 555 |
| shapiro.test | Shapiro-Wilk test of normality | Section 22.3.2, p. 555 |
| hatvalues | Calculate leverages | Section 22.3.4, p. 558 |
| cooks.distance | Calculate Cook's distances | Section 22.3.5, p. 561 |

# PART V

## ADVANCED GRAPHICS

# 23

## ADVANCED PLOT CUSTOMIZATION

Many users are first drawn to R because of its impressive graphical flexibility and the ease with which you can control and tailor the resulting visuals. In this chapter, you'll take a closer look at the base R graphics device, and at how you can fine-tune the plots you're already familiar with, to get the most use out of your visualizations. In the chapters that follow, you'll then expand your repertoire in both ggplot2 and traditional R graphics.

Much of this chapter will assume you're familiar with the content of Chapters 7 and 14. In general, I'll also assume you're using the standard base R application (for example, *R.app* on a Mac or *Rgui.exe* in Windows—see Appendix A), since the behavior and availability of some of the commands can vary if you're working with R in a different context.

Depending on your operating system, the default software drivers used to render graphical displays on your computer screen are also different. In the standard *R.app* application on a Mac, for example, you'll notice that producing a live plot will open a window with a banner title that looks something like *Quartz 2 [\*]*—the default graphics device driver for OS X is the Quartz window system. On a Windows machine, you'll see *R Graphics: Device 2 (ACTIVE)*. The numbering of any graphics devices always starts at 2; Device 1 is referred to as the *null device*, meaning there's nothing active currently.

**NOTE** *For a list of devices your R session has available, enter ?Devices at the prompt. You'll note that the list includes commands such as png and pdf, which are the so-called silent graphics devices that enable direct-to-file plotting, as detailed in Chapter 8. You can use a different device for any given plot if you want, though the default is almost always appropriate if you're plotting directly to your screen.*

# 23.1 Handling the Graphics Device

So far, your plotting has dealt with one image at a time. It's possible to have multiple graphics devices open, but only one will be deemed active at any given time (the banner titles highlight the currently active device with the *[\*]* or the *(ACTIVE)*). This is useful when you're working on several plots at once or want to view or alter one plot without closing any others.

## 23.1.1 *Manually Opening a New Device*

The typical base R commands you've met already (such as plot, hist, boxplot, and so on) will automatically open a device for plotting and draw the desired plot, if nothing is currently open. You can also open new device windows using dev.new; this newest window will immediately become active, and any subsequent plotting commands will affect that particular device.

As an example, first close any open graphics windows and then enter the following at the R prompt:

```
R> plot(quakes$long,quakes$lat)
```

This will generate a plot of the spatial locations of the occurrences of the 1000 seismic events in the ready-to-use quakes data frame. If the only device currently available is Device 1, the null device, any plotting command that refreshes a plotting window and produces a new image (such as plot here or more specialized commands such as hist or boxplot) will automatically open a new instance of the default graphics device before actually plotting the data. On my machine, I see *Quartz 2 [\*]* open and display the plot of the spatial coordinates.

Now, let's say you'd also like to see a histogram of the number of stations that detected each event. Execute the following to open a new plotting window:

```
R> dev.new()
```

This new window will be numbered 3 (it usually sits itself on top of the previously open window, so you may want to move it to one side with your mouse). Importantly, you'll see that this becomes the active device: on a Mac the *[\*]* is now on the Device 3 banner; in Windows, Device 3 will say *(ACTIVE)*, and Device 2 will now say *(inactive)*.

At this point, you can enter the usual command to bring up the desired histogram in Device 3:

```
R> hist(quakes$stations)
```

If you hadn't used dev.new, the histogram would've just overwritten the plot of the spatial locations in Device 2.

### 23.1.2   Switching Between Devices

To change something in Device 2 without closing Device 3, use dev.set followed by the device number you want to make active. The following code activates Device 2 and replots the locations of the seismic events so that the size of each point is proportional to the number of stations that detected the event. It also tidies up the axis labels.

```
R> dev.set(2)
quartz
     2
R> plot(quakes$long,quakes$lat,cex=0.02*quakes$stations,
        xlab="Longitude",ylab="Latitude")
```

Using dev.set always confirms the newly active device by printing to the console; the specific text will vary according to your operating system and type of device.

Switching back to Device 3, as a final tweak, add a vertical line marking off the mean number of detecting stations.

```
R> dev.set(3)
quartz
     3
R> abline(v=mean(quakes$stations),lty=2)
```

Figure 23-1 shows the two graphics devices after making these modifications.

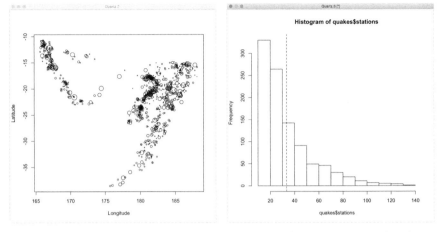

Figure 23-1: My two visible graphics devices, Device 2 (left) and Device 3 (right), showing the final results of producing and manipulating two plots of the quakes data

### 23.1.3   Closing a Device

To close a graphics device, either click the $X$ with your mouse as you would to close any window or use the dev.off function (you originally saw this command in Chapter 8 when closing a direct-to-file device). Calling dev.off() with no arguments simply closes the currently active device. Otherwise, you can specify the device number just as when using dev.set. To close the plot of the spatial locations, leaving the histogram as the active device, call dev.off with an argument of 2:

```
R> dev.off(2)
quartz
    3
```

Then repeat the call without an argument to close the remaining device:

```
R> dev.off()
null device
          1
```

Similar to dev.set, the printed output tells you what the newly active device is after you close one. When you close the last available actionable device, you're returned to the null device.

### 23.1.4   Multiple Plots in One Device

You can also control the number of individual plots in any one device. There are a few ways to do this; I'll describe the two easiest ways here.

## Setting the mfrow Parameter

Recall the par function is used to control various graphical parameters of traditional R plots. The mfrow argument instructs a new (or the currently active) device to "invisibly" divide itself into a grid of the specified dimensions, with each cell holding one plot. You pass the mfrow option a numeric integer vector of length 2 in the order of c(*rows,columns*); as you might guess, its default is c(1,1).

In your R session, make sure there are no plotting windows open. Now, say you want the two plots of the quakes data side by side in the same device. You would set mfrow as a 1 × 2 grid with the vector c(1,2)—one row of plots and two columns.

```
R> dev.new(width=8,height=4)
R> par(mfrow=c(1,2))
R> plot(quakes$long,quakes$lat,cex=0.02*quakes$stations,
        xlab="Longitude",ylab="Latitude")
R> hist(quakes$stations)
R> abline(v=mean(quakes$stations),lty=2)
```

The first line uses the optional arguments width and height to preset the dimensions of the new device, in inches, so it is twice as wide as it is high. Figure 23-2 shows exactly how the images are displayed, with the creation of a new plot filling the available cells as governed by the value of mfrow.

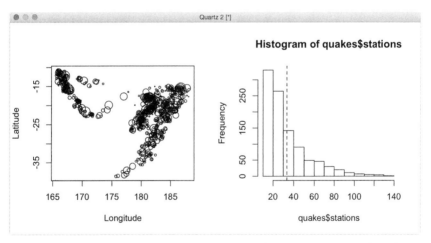

Figure 23-2: Using mfrow in par to generate a grid of plots in a single graphics device, showing the two plots of the quakes data

If you close any graphics devices and rerun this code without the lead-in call to dev.new, executing par(mfrow=c(1,2)) will automatically open a graphics device of the default square size of 7 × 7 inches. The two plots will still appear side by side but will be squashed. You can manually resize the device with your mouse to something more appropriate to the set value of mfrow, and then when you replot, the visualizations and their axes will be clearer. You'll find you use this trial-and-error approach quite often to produce multiple plots in a single device, especially if you don't want to be concerned with explicitly calling dev.new and setting width and height.

Note that any use of par in this way will affect *only* the currently active device. Subsequent calls to dev.new will open new devices with, for example, mfrow set back to be the default "one plot only" with c(1,1). In other words, if you want to tailor the options of any new graphics device (including the direct-to-file devices), you need to set the required values of par after opening the device but before executing any plotting commands.

### Defining a Particular Layout

You can refine the arrangements of plots in a single device using the layout function, which offers more ways to individualize the panels into which the plots will be drawn.

Return to the student survey data in survey in the MASS package. Suppose you want an array of three statistical plots—a scatterplot of height on writing handspan, side-by-side boxplots of height split by smoking status, and a barplot of the frequencies of exercise of the students. If you want the plots arranged in a square device (as opposed to a single row or column of three plots), using only mfrow in par may not work best. You could set a square grid with par(mfrow=c(2,2)), but you'd end up with a blank space for the cell with no image assigned to it.

When you use layout, you provide the dimensions in a matrix mat as the first argument; these govern an invisible rectangular grid, just like controlling the mfrow option. The difference now is that you can use numeric integer entries in mat to tell layout which plot number will go where. Examine the following object:

```
R> lay.mat <- matrix(c(1,3,2,3),2,2)
R> lay.mat
     [,1] [,2]
[1,]   1    2
[2,]   3    3
```

The dimensions of this matrix create a $2 \times 2$ grid of plotting cells, but the values inside lay.mat tell R that you want plot 1 to take the upper-left cell, plot 2 to take the upper-right cell, and plot 3 to stretch itself over the two bottom cells.

Calling layout as follows will either silently initialize the active device based on lay.mat or open a new one (if the null device is the only device currently available) and initialize it.

```
R> layout(mat=lay.mat)
```

If you're ever unsure of the result of your specification, you can use the layout.show function to see how plots will be placed. The following line produces the image on the left of Figure 23-3.

```
R> layout.show(n=max(lay.mat))
```

Then, once you've loaded the MASS package by calling library("MASS") so you can access survey, run the following lines to place the plots in the order the plotting commands are executed, matched to the integers in lay.mat.

```
R> plot(survey$Wr.Hnd,survey$Height,
        xlab="Writing handspan",ylab="Height")
R> boxplot(survey$Height~survey$Smoke,
        xlab="Smoking frequency",ylab="Height")
R> barplot(table(survey$Exer),horiz=TRUE,main="Exercise")
```

Note that if you've already closed down the plot arising from layout.show, then you'll need to reinitialize a new device with the same call to layout for these three plots to display as intended. The result should look like the right of Figure 23-3.

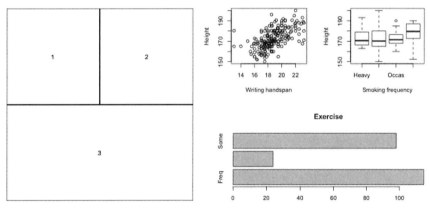

*Figure 23-3: Left: Using* layout.show *to visualize the planned plotting layout and order. Right: Demonstrating three plots of the* survey *data arranged according to* lay.mat *through* layout.

Probably the biggest benefit of layout is its ability to relax the rigidity of plotting cells when compared to using the mfrow par option, as you've just seen. Additional arguments to layout, namely, widths and heights, even allow you to preset the relative widths and heights of the cells as structured in the mat argument. See the documentation in ?layout for details; you'll find some other examples of its flexibility at the bottom of the file.

**NOTE**    *An unfortunate consequence of the two methods discussed here is the inability to edit a previous plot once you've finished it and moved on to the next. There is a* split.screen *function, which does allow you to set up several "screens" in a single device and switch between them. However, this method requires a lot of extra coding and in general doesn't behave well with regard to plotting regions and margins (see the next section) in R. Many users (myself included) prefer to remain with* layout, *even if it means going through a bit of trial and error.*

## 23.2   Plotting Regions and Margins

Although the main concern when plotting is of course the data set or model being visualized, it's additionally important to ensure the plot is annotated clearly and accurately to facilitate correct interpretation. To help do this, you need to know how to manipulate and draw in all visible areas of a given device, not just the area where your data lie.

For any single plot created using base R graphics, there are three regions that make up the image.

- The *plot region* is all you've dealt with so far. This is where your actual plot appears and where you'll usually be drawing your points, lines, text, and so on. The plot region uses the *user coordinate system*, which reflects the value and scale of the horizontal and vertical axes.

- The *figure region* is the area that contains the space for your axes, their labels, and any titles. These spaces are also referred to as the *figure margins*.

- The *outer region*, also referred to as the *outer margins*, is additional space around the figure region that is not included by default but can be specified if it's needed.

You can explicitly measure and set margin space in a few different ways. One typical way is in terms of *lines*—specifically, the number of lines of text that can fit on top of one another parallel to each edge. You specify these as vectors of length 4 in a particular order; each of the four elements corresponds to one of the four sides: c(*bottom, left, top, right*). The graphical parameters oma (outer margin) and mar (figure margin) are used to control these amounts; like mfrow, they are initialized through a call to par before you begin to draw any new plot.

### 23.2.1   Default Spacing

You can find your default figure margin settings with a call to par in R.

```
R> par()$oma
[1] 0 0 0 0
R> par()$mar
[1] 5.1 4.1 4.1 2.1
```

You can see here that oma=c(0, 0, 0, 0)—there is no outer margin set by default. The default figure margin space is mar=c(5.1, 4.1, 4.1, 2.1)—in other words, 5.1 lines of text on the bottom, 4.1 on the left and top, and 2.1 on the right.

To illustrate these regions, consider the image on the left of Figure 23-4, created in a fresh graphics device with the following:

```
R> plot(1:10)
R> box(which="figure",lty=2)
```

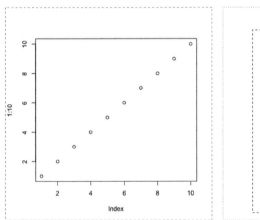 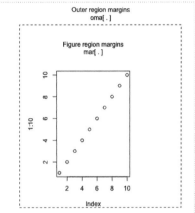

Figure 23-4: Illustrating graphical device regions as treated by traditional (base) R graphics; solid-line boxes show the plot region, dashed-line boxes show the figure region, and a dotted-line box shows the outer region area. Left: Default settings. Right: User specification, through par, of the outer and figure margin areas in "lines of text" via oma and mar, respectively.

If you use the box function with the optional argument which set to "figure", it shows you the figure region (the additional specification of lty=2 draws dashed lines).

If you're looking at this plot in your onscreen graphics device, you should note the dashed lines snug up against the window edges. Examining the default values in mar, you can see, relatively speaking, that they correctly correspond to the spacing on the four sides of the plot region (given with the default solid box). The widest figure margin, parallel to the bottom of the plot region, is 5.1 lines; the narrowest figure margin, parallel to the right of the plot region, is 2.1 lines.

### 23.2.2 Custom Spacing

Let's produce the same plot but with tailored outer margins so that the bottom, left, top, and right areas are one, four, three, and two lines, respectively, and the figure margins are four, five, six, and seven lines. The result of the following code is given on the right of Figure 23-4.

```
R> par(oma=c(1,4,3,2),mar=4:7)
R> plot(1:10)
R> box("figure",lty=2)
R> box("outer",lty=3)
```

Notice that the irregular margins have squashed the plot region in the default square device to accommodate the defined spacing around the edges. If you set graphical parameters that squash the plot region into nonexistence, R will throw an error stating figure margins too large.

Since you'd usually manipulate margin space to accommodate particular annotations of the plot, let's look at the mtext function, used specifically to produce text in the figure or outer margins. By default, the argument outer is FALSE, meaning the text will be written in the figure margin. Setting outer=TRUE positions the text in the outer region. If you've kept the most recent plot open, the following lines provide the additional margin annotation visible on the right of Figure 23-4:

```
R> mtext("Figure region margins\nmar[ . ]",line=2)
R> mtext("Outer region margins\noma[ . ]",line=0.5,outer=TRUE)
```

Here, you provide the text you want written in a character string as the first argument, and the argument line instructs how many lines of space away from the inside border the text should appear. There's also an optional argument side in mtext, which dictates where the text appears. It defaults to 3, setting the text at the top, but you can set side=1 to place the text on the bottom, use side=2 to set it on the left, and use side=4 to set it on the right. Look to ?mtext for details on even more arguments available for the margin text.

You might also like to investigate the ready-to-use function title, which is a specialized implementation of mtext often used if figure margin annotation for the four axes of a plot (beyond the basic capabilities of specifying things such as main, xlab, or ylab) is the primary concern.

### 23.2.3    Clipping

Controlling *clipping* allows you to draw in or add elements to the margin regions with reference to the user coordinates of the plot itself. For example, you might want to place a legend outside the plotting area, or you might want to draw an arrow that extends beyond the plot region to embellish a particular observation.

The graphical parameter xpd controls clipping in base R graphics. By default, xpd is set to FALSE, so all drawing is clipped to the available plot region only (with the exception of special margin-addition functions such as mtext). Setting xpd to TRUE allows you to draw things outside the formally defined plot region into the figure margins but not into any outer margins.

Setting xpd to NA will permit drawing in all three areas—plot region, figure margins, and the outer margins.

For example, take a look at the images in Figure 23-5, showing side-by-side boxplots of mileage split by number of cylinders, created with the following code:

```
R> dev.new()
R> par(oma=c(1,1,5,1),mar=c(2,4,5,4))
R> boxplot(mtcars$mpg~mtcars$cyl,xaxt="n",ylab="MPG")
R> box("figure",lty=2)
R> box("outer",lty=3)
R> arrows(x0=c(2,2.5,3),y0=c(44,37,27),x1=c(1.25,2.25,3),y1=c(31,22,20),
        xpd=FALSE)
R> text(x=c(2,2.5,3),y=c(45,38,28),c("V4 cars","V6 cars","V8 cars"),
        xpd=FALSE)
```

The particular result of this code is the top-left image in Figure 23-5. I've defined the device region itself with particular figure and outer margins for the purpose of illustration. The plotting of the horizontal axis is suppressed with xaxt="n" in the call to boxplot; calls to box add the boundaries of the figure and outer margins (dashed and dotted lines, respectively). Lastly, calls to arrows and text point to and annotate each boxplot; the label for V4 cars extends into the outer margin, the label for V6 cars extends into the figure region, and the label for V8 cars remains contained within the plot region.

Note that the graphical parameter xpd is specified only in the two "add-to-current-plot" functions arrows and text, explicitly set as the default FALSE. This means all plotting is restricted to the plot region.

If you run the code chunk again but now set xpd=TRUE in the calls to arrows and text, you'll get the image in the top right of Figure 23-5. This allows the label for the V6 car to be printed in the margin, instead of being chopped off. Finally, rerunning the code with xpd=NA produces the lower plot in Figure 23-5, where all drawing outside the plot region is permitted.

This effect is usually desirable when you need to annotate the main plot somehow, especially when there isn't enough space in the plot region to squeeze in any additions. Plots that I've created in earlier chapters, such as the bottom image of Figure 16-6 on page 349 (where the legend sits outside the main plot) and Figure 17-3 on page 380 (where I annotated the critical values), were created specifying xpd=TRUE in the relevant functions (legend, text, segments, and arrows).

As demonstrated, you'd typically set xpd in the specific commands (in other words, on a line-by-line basis), so only the results of that particular command will be produced with the given clipping rule. This offers a bit more control over what is and isn't visible outside the plot region. You can, however, set xpd alongside oma and mar in the initial call to par to make the value of xpd "universal" to that device.

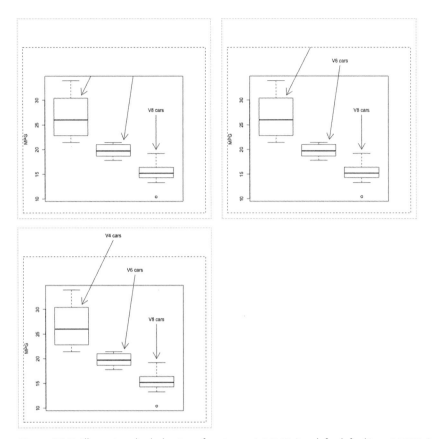

Figure 23-5: Illustrating the behavior of setting xpd=FALSE (top left, default), xpd=TRUE (top right), and xpd=NA (bottom) in relevant plotting commands to enable drawing in figure and outer margins with respect to the user coordinates of the plot region

## 23.3   Point-and-Click Coordinate Interaction

Your dealings with the graphics device don't need to be solely command based. Under typical circumstances, R can read mouse clicks you make inside the device.

### 23.3.1   Retrieving Coordinates Silently

The locator command allows you to find and return user coordinates. To see how it works, first execute a call to plot(1,1) to bring up a simple plot with a single point in the middle. To use locator, you simply execute the function (with no arguments for default behavior), which will "hang" the console, without returning you to the prompt. Then, on an active graphics device, your mouse cursor will change to a + symbol (you may need to first click your device once to bring it to the foreground of your computer desktop). With your cursor as the +, you can perform a series of (left) mouse clicks inside the device, and R will silently record the precise user coordinates. To stop this, simply right-click to terminate the command (other options to stop are system dependent and are mentioned in the help file ?locator), and once you do, the coordinates you identified in the device are returned as a list with components $x and $y. These are printed to the console screen unless you specifically assign the call to locator to an R object.

On my machine, I silently identified four points at arbitrary locations around the plotted point at $(1, 1)$, from top left clockwise around to the bottom left. The following is the output printed to my console:

```
R> plot(1,1)
R> locator()
$x
[1] 0.8275456 1.1737525 1.1440526 0.8201909

$y
[1] 1.1581795 1.1534442 0.9003221 0.8630254
```

This silent use of locator is useful if you need to, for example, identify approximate user coordinates in the plot region where you need to place future annotations.

### 23.3.2 Visualizing Selected Coordinates

You can also use locator to plot the points you select as either individual points or as lines. Running the following code produces Figure 23-6:

```
R> plot(1,1)
R> Rtist <- locator(type="o",pch=4,lty=2,lwd=3,col="red",xpd=TRUE)
R> Rtist
$x
 [1] 0.5013189 0.6267149 0.7384407 0.7172250 1.0386740 1.2765699
 [7] 1.4711542 1.2352573 1.2220592 0.8583484 1.0483300 1.0091491

$y
 [1] 0.6966016 0.9941945 0.9636752 1.2819852 1.2766579 1.4891270
 [7] 1.2439071 0.9630832 0.7625887 0.7541716 0.6394519 0.9618461
```

Drawing using locator requires you to specify the plot type, as covered in Chapter 7. Selecting type="o" (as opposed to the silent default, type="n") is what produces the overplotted points and lines in Figure 23-6. For just points, use type="p"; for just lines, use type="l". The graphical parameters controlling other relevant features, such as point/line type and color, can also be used, as you've seen in conventionally produced plots in Chapter 7. I also used xpd=TRUE, shown earlier, to allow the locator points and/or lines to protrude into the figure region margins. The call to locator is directly assigned to a new object Rtist, illustrating how you can use the clicked coordinates later if needed.

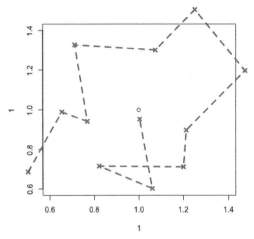

*Figure 23-6: Using* locator *to draw an arbitrary sequence of overplotted points and lines*

### 23.3.3   Ad Hoc Annotation

The locator function also allows you to place ad hoc annotations, such as legends, on your plot—remember, since locator returns valid R user coordinates, these results can directly form the positional argument of most standard annotation functions.

Return to the student survey data in the MASS package, first loading the package by calling library("MASS"). The following call produces the scatterplot used to illustrate a multiple linear model of mean student height as a function of handspan and sex in Section 21.3.3.

```
R> plot(survey$Height~survey$Wr.Hnd,pch=16,
       col=c("gray","black")[as.numeric(survey$Sex)],
       xlab="Writing handspan",ylab="Height")
```

For the plot in Section 21.3.3 (Figure 21-1 on page 495), I simply used the string "topleft" to position the legend. This time, call the following:

```
R> legend(locator(n=1),legend=levels(survey$Sex),pch=16,
       col=c("gray","black"))
```

An optional argument to locator, n, takes a positive integer for an upper limit on how many points you want to select; it defaults to 512. If you specify n=1, locator will automatically terminate after you left-click once in the device, so you don't need to manually exit the function with a right-click.

When the code is executed, the + cursor will appear on the graphics device, and you simply need to click once for the desired location of the legend. I chose to click the blank space above the cloud of points, producing the image in Figure 23-7.

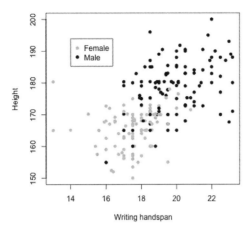

Figure 23-7: Ad hoc placement of a legend on a
scatterplot of the survey data

## Exercise 23.1

a.  In Section 20.5.4 (page 478), I gave you code showing a simple
    linear model fitted with a categorical predictor being treated as
    continuous (the mtcars data with mpg as the response and cyl as
    the explanatory variable). Reproduce the side-by-side boxplots
    and the scatterplot (with fitted line) from Figure 20-6, but this
    time, use mfrow to present the two plots as a vertical column in
    one device.

b.  Create the appropriate layout matrices to reproduce the follow-
    ing three plots (as they appear in a square device):

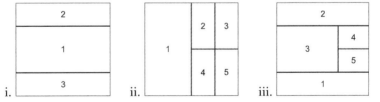

c.  By opening a new device of dimension 9 × 4.5 inches, set the
    following layout:

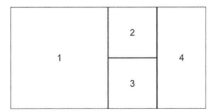

Then, produce the following combined set of plots exactly:

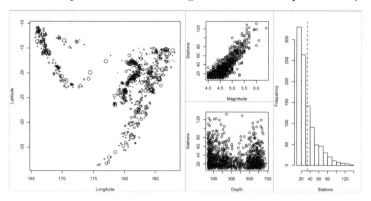

To achieve this, note the following:

–  After you open the device and setting the layout, the plot margins should be reset to four lines, four lines, two lines, and one line of space on the bottom, left, top, and right, respectively.

–  After each plot, add a gray box corresponding to the figure region to achieve the visible partitions.

–  Plots 1 and 4 are the same as the two plots shown in Figures 23-1 and 23-2.

–  Plots 2 and 3 are scatterplots showing the number of detecting stations on the *y*-axis, with magnitude and depth on the *x*-axis, respectively.

–  Do not place main titles on any plots, and ensure the axis titles are neat (that is, compared to their defaults).

d.  Write a little R function named `interactive.arrow`. The purpose of this function is to superimpose an arrow upon any base R plot using two clicks of your mouse. The details are as follows:

–  The crux of your function will be the use of `locator` to read exactly two mouse clicks. You may assume a suitable active graphics device is already open whenever the function is called. The first click should mean the beginning of the arrow, and the second click should mean the tip of the arrow (where it's pointing to).

–  In the function, the coordinates returned by `locator` should be passed to `arrows` to do the actual drawing.

–  The function should take an ellipsis as its first argument, intended to hold additional arguments to be passed directly to `arrows`.

- The function should take an optional logical argument label, which defaults to NA but should be intended to have an optional character string. If label is not NA, then locator should be invoked once more (separately, after drawing the arrow) to select exactly one coordinate. That point will be passed to text so that the user can additionally place the character string given to label as an annotation (intended to be for the interactively placed arrow). The call to text should consistently allow completely relaxed clipping (in other words, any text added in this fashion will still be visible in the figure region and outer margins, if there are any).

Take another look at the rightmost plot of Figure 14-6 on page 298, a stand-alone boxplot of the magnitude data from the quakes data frame. Arrows and labels were superimposed externally pointing out the various statistics summarized by a boxplot. Create the same boxplot and use interactive.arrow to annotate the same features to your own satisfaction (you'll likely have to use the ellipsis to relax the clipping associated with each arrow). My result is given here:

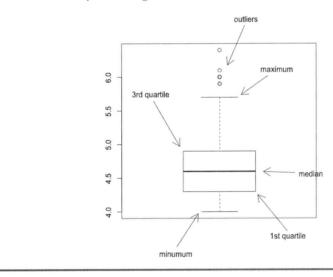

## 23.4  Customizing Traditional R Plots

Now that you're familiar with the way R places and handles plots in the graphics device, it's time to focus on common features of plots. So far, you've largely left the default settings in place.

### 23.4.1  Graphical Parameters for Style and Suppression

If you want finer control over an R plot, you'll typically want to begin with a "clean slate." To do this, you need to be aware of the default settings of certain graphical parameters when calling a plotting function and of how to suppress things such as boxes and axes. This is where you'll start.

For an example image, let's plot MPG against horsepower (from the ready-to-use mtcars data set) and set each plotted point to be sized proportionally to the weight of each car. For convenience, create the following objects:

```
R> hp <- mtcars$hp
R> mpg <- mtcars$mpg
R> wtcex <- mtcars$wt/mean(mtcars$wt)
```

The last object is the car weight vector scaled by its sample mean. This creates a vector where cars less than the average weight have a value < 1 and cars more than the average weight have a value > 1, making it ideal for the cex parameter to scale the size of the plotted points accordingly (refer to Chapter 7).

Let's start by focusing on some more graphical parameters usually used in the first instance of a call to plot, paving the way for using the box and axis commands. Executing the following line gives you the default appearance of the plot and its box, axes, and labeling; this is shown as the leftmost image in Figure 23-8.

```
R> plot(hp,mpg,cex=wtcex)
```

There are two axis "styles," controlled by the graphical parameters xaxs and yaxs. Their sole purpose is to decide whether to impose the small amount of additional horizontal and vertical buffer space that's present at the ends of each axis to prevent points being chopped off at the end of the plotting region. The default, xaxs="r" and yaxis="r", is to include that space. The alternative, setting one or both of these to "i", instructs the plot region to be *strictly* defined by the upper and lower limits of the data (or by those optionally supplied to xlim and/or ylim), that is, with *no* additional padding space.

For example, the following line produces the middle plot in Figure 23-8.

```
R> plot(hp,mpg,cex=wtcex,xaxs="i",yaxs="i")
```

This plot is almost the same as the default, but note now that there's no padding space at the end of the axes; the most extreme data points sit right on the axes. Generally, the default axis style "r" is fine, but on occasions where you need finer control over the axes' scales and the corresponding plot region, that additional buffer space can be problematic. On those occasions, you'll often see xlim/ylim being used in conjunction with xaxs="i"/yaxs="i".

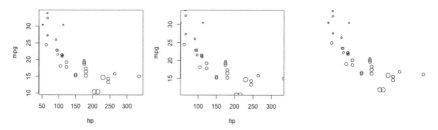

Figure 23-8: Plotting MPG against horsepower for the mtcars data; point size proportional to car weight, using a call to plot only. Left: Default appearance. Middle: Setting xaxs="i" and yaxs="i" to prevent the buffer spacing on the limits of the axes. Right: Using xaxt, yaxt, xlab, ylab, and bty to suppress all box, axis, and label drawing (alternatively achieved by setting axes=FALSE and ann=FALSE).

If you want total control over the specific appearance of any boxes, axes, and their labels, you'll want to start a plot with none of these and add them as per your design. The rightmost plot in Figure 23-8 is the result of suppressing the default drawing of these by a call to either

```
R> plot(hp,mpg,cex=wtcex,xaxt="n",yaxt="n",bty="n",xlab="",ylab="")
```

or

```
R> plot(hp,mpg,cex=wtcex,axes=FALSE,ann=FALSE)
```

You can achieve this either by setting the parameters xaxt, yaxt, and bty to "n" and setting the default axis labels xlab and ylab to the empty string "", or by simply setting both axes and ann to FALSE (the former suppressing all axes and the box, the latter suppressing any annotation). Although the first way might seem overcomplicated, it affords you greater flexibility in suppressing each aspect of a given plot (as opposed to the "total" suppression enforced by the second approach).

### 23.4.2   Customizing Boxes

When you're starting with a suppressed-box or suppressed-axis plot, to add a box specific to the current plot region in the active graphics device, you use box and specify its type with bty. For example, if you start with a plot like the one on the right of Figure 23-8 (just run the most recent line of code to get this), then additionally calling the following line provides you with the image given on the left of Figure 23-9.

```
R> box(bty="u")
```

The bty argument is supplied a single character: "o" (default), "l", "7", "c", "u", "]", or "n". The help file entry for bty in ?par tells you that based on one of these values, the resulting box boundaries will follow the appearance of the corresponding uppercase letter, with the exception of "n" (which, as you saw a moment ago, will suppress the box).

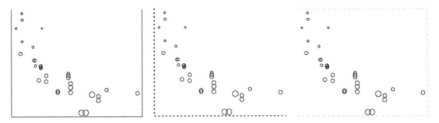

*Figure 23-9: Various box configurations added to the* `mtcars` *scatterplot*

You can use other relevant parameters that you've met already, such as `lty`, `lwd`, and `col`, to further control the appearance of a box. Replot the data as on the right of Figure 23-8 and then call the following to produce the image in the middle of Figure 23-9:

```
R> box(bty="l",lty=3,lwd=2)
```

The final example on the right of Figure 23-9 is created with this line:

```
R> box(bty="]",lty=2,col="gray")
```

### 23.4.3   Customizing Axes

Once you have the box the way you want it, you can focus on the axes. The axis function allows you to control the addition and appearance of an axis on any of the four sides of the plot region in greater detail. The first argument it takes is side, provided with a single integer: 1 (bottom), 2 (left), 3 (top), or 4 (right). These numbers are consistent with the positions of the relevant margin-spacing values when you're setting graphical parameter vectors like `mar`.

The first thing you might want to change on an axis is where the tick marks are drawn. By default, R uses the built-in function pretty to find a "neat" sequence of values for the scale of each axis, but you can set your own by passing the at argument to axis. The following lines create the plot on the left of Figure 23-10.

```
R> hpseq <- seq(min(hp),max(hp),length=10)
R> plot(hp,mpg,cex=wtcex,xaxt="n",bty="n",ann=FALSE)
R> axis(side=1,at=hpseq)
R> axis(side=3,at=round(hpseq))
```

First, an evenly spaced sequence of 10 values spanning the range of hp is stored as hpseq. The initial call to plot suppresses the *x*-axis, the box, and any default axis labels; however, the *y*-axis is permitted to appear as per its default. Then axis is instructed to draw the *x*-axis (side=1), with tick marks at hpseq. To provide a comparison to that, an axis is also drawn along the top (side=3), but this time the tick marks are drawn at hpseq after it's rounded to the nearest integer.

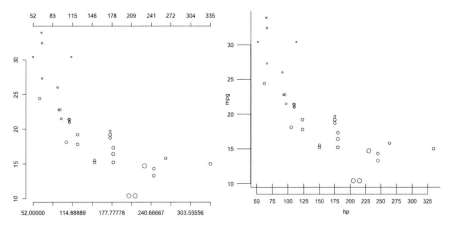

Figure 23-10: Customizing axes of the mtcars scatterplot

As you can see on the left of the figure, the custom *x*-axis I've created on the bottom shows 10 tick marks at the sequence of values supplied to at. R may suppress some of the labels so they don't overlap one another, which is what has occurred here. Since these "decimal" values mightn't be aesthetically pleasing, the axis that's been drawn along the top has tick marks drawn at the nearest integers of hpseq, achieved using round in the final call to axis shown earlier. Although, strictly speaking, this now means that the tick marks are no longer exactly evenly spaced, the rounded values mean shorter default axis labels that can all be displayed in the current device.

You can see from these difficulties that tick mark locations are generally best left to R, unless you have specific axis values that you know you want marked out—you'll see an example of this in Section 23.6. For now, let's look at some other tweaks you can make to your axes. In particular, the tcl (length of the ticks), las (orientation of the labels), and mgp (axis spacing) parameters are arguably among those more frequently used. The following code creates the plot on the right of Figure 23-10.

```
R> hpseq2 <- seq(50,325,by=25)
R> plot(hp,mpg,cex=wtcex,axes=FALSE)
R> box(bty="l")
R> axis(side=2,tcl=-2,las=1,mgp=c(3,2.5,0))
R> axis(side=1,at=hpseq2,tcl=1.5,mgp=c(3,1.5,1))
```

After a new sequence—hpseq2—is defined as all integers that lie within the recorded range of the data and are evenly spaced by 25 units, the plot is initialized. The box and axes are suppressed, but the default variable titles (mpg and hp) along the axes remain.

Now, an L-shaped box and the *y*-axis (side=2) are added. In the latter, the tcl parameter governs the length of each tick mark in "parallel lines of text" (recall this is a standard unit measurement for margin spacing in an R plot); it defaults to -0.5. When the value is negative, it draws the tick marks away from the plot region; when it's positive, the tick marks are drawn inward. For this side=2 axis, tcl=-2, meaning that the ticks will point outward from the plot but be four times the length they usually are (two whole lines of text as opposed to half a line).

The las parameter controls the way the labels for each tick mark are oriented; setting it to 1 instructs R to produce all tick labels *horizontally*, regardless of axis side. The default, las=0, writes all labels *parallel* to the corresponding axis; the alternative las=2 means labels are always *perpendicular* to the corresponding axes; and using las=3 orients all labels to be read *vertically*, regardless of axis.

Next up, the mgp parameter controls three further aspects of axis spacing and, as such, is supplied a vector of length 3 as per the following definition: c(*axis title,axis labels,axis line*). Once more, these arguments are expressed in "lines of text." The default value of mgp is c(3,1,0)—meaning that in every axis you've seen so far, the title has sat three lines of text away from the plot region, the tick mark labels one line of text away, and the axis line itself zero lines of text away from the plot region (so it's flush with any drawn plot region box). When used in axis, only the second and third elements of mgp are relevant. In the vertical axis in the plot on the right of Figure 23-10, the only alteration from the default was to set the second element (spacing of the axis labels) to 2.5—pushing the axis labels out to the left, further away from the plot region. The tick marks themselves are considerably lengthened by tcl, so this is required to avoid the axis tick mark labels going through those ticks. Try replotting the image and that axis, but without specifying mgp, and you'll see that unappealing result.

Moving to the addition of the *x*-axis (side=1), you can see tick marks at hpseq2 being placed via at. This time, a positive value has been supplied to tcl, instructing the axis to have *inward*-facing tick marks of 1.5 lines of text in length. In mgp, note that the third element of the vector is now set to 1, meaning you want the axis line itself to sit one line of text away from the plot region. Looking at the right of Figure 23-10, you can see that the entire axis has been moved downward, away from the plot region. To account for this in terms of the spacing of the tick mark labeling, the second element of mgp has been increased a little from its default, to be 1.5.

## 23.5   Specialized Text and Label Notation

Now you'll investigate some immediately accessible tools for controlling fonts and displaying special notation, such as Greek symbols and mathematical expressions.

### 23.5.1   Font

The displayed font is controlled by two graphical parameters: family for the specific font family and font, an integer selector for controlling bold and italic typeface.

Available fonts depend on both your operating system and the graphics device you're using. That said, there are three generic families—"sans" (the default), "serif", and "mono"—that are always available. These are paired with the four possible values of font—1 (normal text, default), 2 (bold), 3 (italic), and 4 (bold and italic). You can set these two graphical parameters universally for a device using par, but like the use of xpd, it's just as common (if not more so) to set family and font in the relevant annotation functions.

Figure 23-11 shows you some variants alongside the corresponding values of family and font. To create it, start with an empty plot region with preset *x*- and *y*-limits, created with the following:

```
R> par(mar=c(3,3,3,3))
R> plot(1,1,type="n",xlim=c(-1,1),ylim=c(0,7),xaxt="n",yaxt="n",ann=FALSE)
```

Then, the image with six possible variants is completed by executing the following lines:

```
R> text(0,6,label="sans text (default)\nfamily=\"sans\", font=1")
R> text(0,5,label="serif text\nfamily=\"serif\", font=1",
        family="serif",font=1)
R> text(0,4,label="mono text\nfamily=\"mono\", font=1",
        family="mono",font=1)
R> text(0,3,label="mono text (bold, italic)\nfamily=\"mono\", font=4",
        family="mono",font=4)
R> text(0,2,label="sans text (italic)\nfamily=\"sans\", font=3",
        family="sans",font=3)
R> text(0,1,label="serif text (bold)\nfamily=\"serif\", font=2",
        family="serif",font=2)
R> mtext("some",line=1,at=-0.5,cex=2,family="sans")
R> mtext("different",line=1,at=0,cex=2,family="serif")
R> mtext("fonts",line=1,at=0.5,cex=2,family="mono")
```

Here, text is used to place the content at predetermined coordinates, and mtext is used to add to the top figure margin.

Figure 23-11: Displaying font styles through use of the family and font graphical parameters

## 23.5.2   Greek Symbols

For statistically or mathematically technical plots, annotation may occasionally require Greek symbols or mathematical markup. You can display these using the expression function, which, among other things, is capable of invoking the plotmath mode of R (Murrell and Ihaka, 2000; Murrell, 2011). Use of expression returns a special object that has a class of the same name and can subsequently be passed to any argument in a plotting function that requires the character string to be displayed.

Focusing for the moment on Greek symbols, consider Figure 23-12, which is produced with the following code:

```
R> par(mar=c(3,3,3,3))
R> plot(1,1,type="n",xlim=c(-1,1),ylim=c(0.5,4.5),xaxt="n",yaxt="n",
        ann=FALSE)
R> text(0,4,label=expression(alpha),cex=1.5)
R> text(0,3,label=expression(paste("sigma: ",sigma,"    Sigma: ",Sigma)),
        family="mono",cex=1.5)
R> text(0,2,label=expression(paste(beta," ",gamma," ",Phi)),cex=1.5)
R> text(0,1,label=expression(paste(Gamma,"(",tau,") = 24 when ",tau," = 5")),
        family="serif",cex=1.5)
R> title(main=expression(paste("Gr",epsilon,epsilon,"k")),cex.main=2)
```

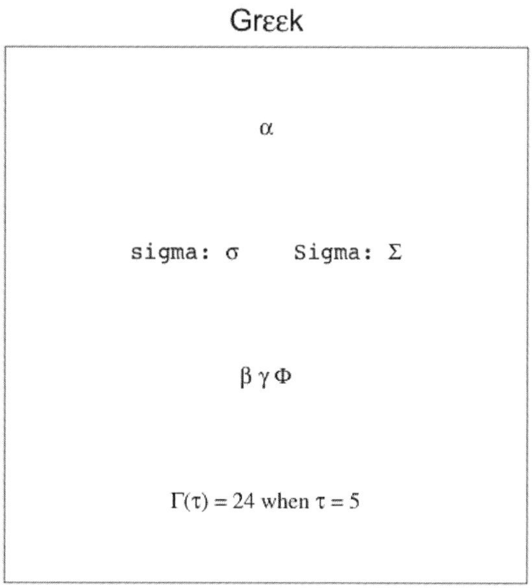

Greek

α

sigma: σ     Sigma: Σ

β γ Φ

Γ(τ) = 24 when τ = 5

*Figure 23-12: Displaying Greek symbols using* expression

If you just want a single special character by itself, then something like expression(alpha) is all you need to produce $\alpha$ in the plot, as in the first call to text shown in the code chunk. Note that the specification of the special characters is done *without* quotes around the name of the desired symbol. More commonly, however, you'll want a character to appear alongside other components, such as regular text or in an equation. For that, you need to use paste inside the call to expression, separating the components with commas. These are shown in the remaining three calls to text.

You can use cex to control size, though use of family and font affects only quoted regular text, not symbols, as the final call to text demonstrates.

The title function, which allows you to add axis and main titles, is then used to add the title "Greek" by supplying the corresponding expression to main. I use cex.main=2 in the same call to double its size (the slightly different tag cex.main is required there to distinguish between the size of the main title and any axis titles, controlled via cex.lab).

### 23.5.3 *Mathematical Expressions*

Formatting entire mathematical expressions to appear in R plots is a bit more complicated and is reminiscent of using markup languages like LaTeX. Because of this, I won't give a full exposition of the syntax required here, but I'll provide some examples of the kinds of things that are possible, as shown in Figure 23-13. To create the image, I first defined four expression objects as follows:

```
R> expr1 <- expression(c^2==a[1]^2+b[1]^2)
R> expr2 <- expression(paste(pi^{x[i]},(1-pi)^(n-x[i])))
R> expr3 <- expression(paste("Sample mean:    ",
                        italic(n)^{-1},
                        sum(italic(x)[italic(i)],
                            italic(i)==1,
                            italic(n))
                        ==frac(italic(x)[1]+...+italic(x)[italic(n)],
                            italic(n))))
R> expr4 <- expression(paste("f(x","|",alpha,",",beta,
                        ")"==frac(x^{alpha-1}~(1-x)^{beta-1},
                            B(alpha,beta))))
```

And then I used them in the following code:

```
R> par(mar=c(3,3,3,3))
R> plot(1,1,type="n",xlim=c(-1,1),ylim=c(0.5,4.5),xaxt="n",yaxt="n",
        ann=FALSE)
R> text(0,4:1,labels=c(expr1,expr2,expr3,expr4),cex=1.5)
R> title(main="Math",cex.main=2)
```

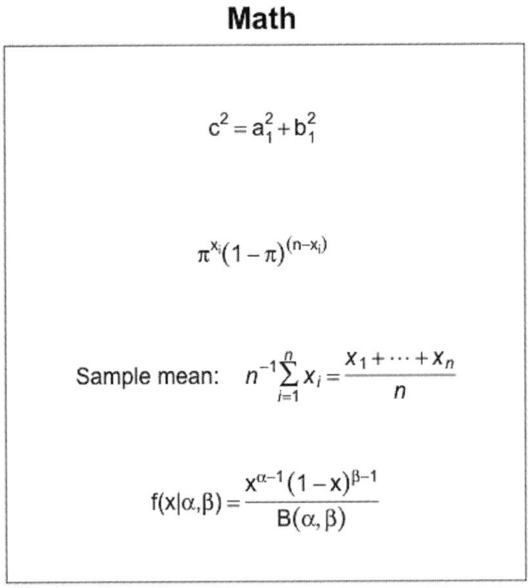

**Math**

$$c^2 = a_1^2 + b_1^2$$

$$\pi^{x_i}(1-\pi)^{(n-x_i)}$$

Sample mean: $\quad n^{-1}\sum_{i=1}^{n} x_i = \dfrac{x_1 + \cdots + x_n}{n}$

$$f(x|\alpha,\beta) = \frac{x^{\alpha-1}(1-x)^{\beta-1}}{B(\alpha,\beta)}$$

*Figure 23-13: Some examples of typesetting mathematical expressions in R plots*

All Greek and mathematical markup is contained within a call to expression. It's necessary to use paste if you require separate components (separated by commas), some of which may or may not be regular text (in other words, in quotes) to produce the final result. Here are some key notes:

- Superscripts are given by ^ and subscripts by [ ]; for example, c^2 provides $c^2$ in expr1, and the a[1]^2 component provides $a_1^2$.

- You can group components with parentheses ( ), which are visible (for example, the (1-pi)^(n-x[i]) component of expr2), or with braces { }, which aren't (for example, the pi^{x[i]} component).

- Italicized alphabetic variables are drawn with italic(); for example, italic(n) produces $n$ in expr3.

- Constructs for common arithmetic operators already exist, such as sum( , , ) and frac( , ); for example, in expr3, calling sum(italic(x)[italic(i)],italic(i)==1,italic(n)) produces a result that looks like this:

$$\sum_{i=1}^{n} x_i$$

  and frac(italic(x)[1]+...+italic(x)[italic(n)],italic(n)) produces an expression like this:

$$\frac{x_1 + \ldots + x_n}{n}$$

- There are additional markup tools for proper formatting of expressions, such as combining regular text in quotes directly next to mathematical markup and creating spaces between components without needing to insert quotes. The need for these depends where the markup contents are exactly (in other words, as a stand-alone component of the call to paste or as a component of an operator tool like frac). See, for example, the ")"==frac( , ) part of expr4, and the space-separation of the two components x^{alpha-1}~(1-x)^{beta-1} with a ~ (these sit on the numerator of the fraction).

There's an extensive amount of functionality built in to R for this type of string formatting in graphical displays that I haven't covered here. If you're interested in seeing more, see the help file accessed by entering ?plotmath at the prompt as a first step. There's also an extremely useful demonstration, which you can view in R by entering demo(plotmath), that shows off much of what's possible, alongside the relevant syntax for expression.

## 23.6 A Fully Annotated Scatterplot

To provide a final example that covers most of the concepts you've considered so far, let's create a detailed plot of the MPG by horsepower data that was used in Sections 23.4.1 to 23.4.3. The images in Figure 23-14 show the final result as the largest plot on the bottom, with three smaller interim plots to illustrate the various stages of production appearing along the top.

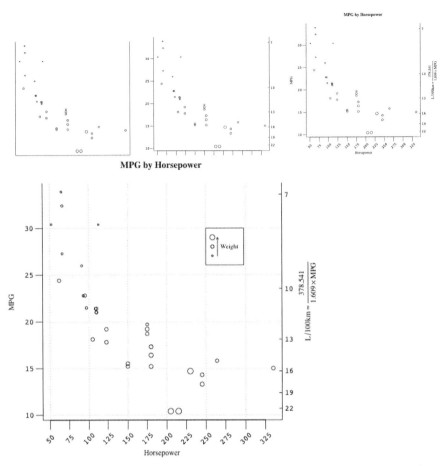

Figure 23-14: A detailed version of the `mtcars` scatterplot of MPG by horsepower, with point size proportional to weight

First, ensure you have the objects `mpg`, `hp`, `wtcex`, and `hpseq2` (defined in Sections 23.4.1 and 23.4.3) ready in your workspace, since you'll use them to ease the length of the code. Here they are again:

```
R> hp <- mtcars$hp
R> mpg <- mtcars$mpg
R> wtcex <- mtcars$wt/mean(mtcars$wt)
R> hpseq2 <- seq(50,325,by=25)
```

The plot, with a slightly wider right margin than default and a U-shaped box, is started with the following code:

```
R> dev.new()
R> par(mar=c(5,4,4,4))
R> plot(hp,mpg,cex=wtcex,axes=FALSE,ann=FALSE)
R> box(bty="u")
```

This provides the top-left image of Figure 23-14. I've used dev.new to explicitly open a new graphics device on my machine, which defaults to $7 \times 7$ inches. You can use width and height arguments supplied to dev.new to alter this on your machine if you want.

Now add some axes:

```
R> axis(2,las=1,tcl=-0.8,family="mono")
R> axis(1,at=hpseq2,labels=FALSE,tcl=-1)
```

These two lines add the left vertical axis for MPG; the tick marks are lengthened a little using tcl, their labels are made horizontal through las, and a "mono" font is requested. For the horizontal axis (horsepower), longer outward tick marks are drawn at the values in hpseq2, but their labels are suppressed by setting labels=FALSE. You'll populate those in a moment.

Rather than using MPG as a measure of fuel efficiency, many countries use "liters per one hundred kilometers" (L/100km). So, for their benefit, let's say you want to provide a second vertical axis on the right of the plot that provides L/100km. To do this, you need the conversion formula. Based on a US gallon, an approximate conversion between the two is given by the following:

$$\text{MPG} = \frac{378.541}{1.609 \times (\text{L/100km})}$$

It turns out that this function is *involutory*. That is, to convert from MPG back to L/100km, simply swap those two variables in the equation.

A little experimentation with the conversion formula, based on the limits of the observed MPG data, has given me a sensible collection of L/100km values at which to mark the right axis.

```
R> L100 <- seq(22,7,by=-3)
R> L100
[1] 22 19 16 13 10  7
```

Note that these are in decreasing order for convenience, since you can see that once you convert these to MPG, the results are increasing:

```
R> MPG.L100 <- (100/L100*3.78541)/1.609
R> MPG.L100
[1] 10.69385 12.38236 14.70405 18.09729 23.52648 33.60925
```

This makes sense—a smaller number for L/100km means a more fuel-efficient car.

Why do you need the MPG version of these numbers? Well, remember that the plot itself is on the scale of MPG, so to instruct R to mark off the appropriate tick marks on the right side, you need the L/100km values in MPG "coordinates."

With that done, one final call to axis leaves you with the top-middle plot in Figure 23-14.

```
R> axis(4,at=MPG.L100,labels=L100,las=1,tcl=0.3,mgp=c(3,0.3,0),family="mono")
```

Of particular note here is that you've used at to specify the tick marks on the MPG scale, at the values in MPG.L100, but since they correspond to the L/100km sequence in L100, it's the latter vector that you supply to labels to actually label said tick marks.

Next, it's time to annotate the axes with some titles and provide the labels for the tick marks on the horizontal axis. Before doing that, construct an expression for the MPG-to-L/100km conversion to clarify the right vertical axis.

```
R> express.L100 <- expression(paste(L/100,"km"%~~%frac(378.541,1.609%*%MPG)))
```

In express.L100, the %~~% provides an "approximately equal to" sign ($\approx$), and %*% provides an explicit multiplication symbol ($\times$).

Then, by running the following lines, you get the top-right image in Figure 23-14.

```
R> title(main="MPG by Horsepower",xlab="Horsepower",ylab="MPG",
        family="serif")
R> mtext(express.L100,side=4,line=3,family="serif")
R> text(hpseq2,rep(7.5,length(hpseq2)),labels=hpseq2,srt=45,
        xpd=TRUE,family="mono")
```

The first line provides a main title and *x*- and *y*-axis titles in a "serif" style. Then mtext places a "serif" version of the arithmetic expression just created in an appropriate position line=3 on the right axis (side=4). The third line places the "mono"-style tick mark labels in hpseq2 along the *x*-axis at the appropriate user coordinates in the same vector, with a vertical position of 7.5, after a little trial and error. Since you're using text to draw in the figure margin, you must set xpd to TRUE. Special to text is the optional srt graphical parameter, which allows you to rotate the labels. Here, they've been rotated 45 degrees.

You're now ready to put the final touches on the plot. So far, the scaled sizes of the points according to car weight have been ignored. It'd be helpful to provide at least minimal information about that and other features that can aid in interpretation of the relationship (especially since there are two vertical axes), like an overlaid grid.

Superimposition of a grid on such a plot is straightforward.

```
R> grid(col="darkgray")
```

You can specify the number of cells along the horizontal and vertical axes using the optional arguments nx and ny, respectively; if not, R will draw the grid lines at what would be the default *x*- and *y*-axis tick marks (as I've let it do here). Other aesthetics can be altered in the usual way, using arguments such as col (color) and lty (line type).

The fun part is now trying to work out how to reference the sizes of the plotted points by the weights of the cars. There are a number of ways you might achieve this. For this last example, I'll manhandle the legend function in an effort to produce the graphic. The following three lines provide the end result.

```
R> legend(250,30,legend=rep("          ",3),pch=rep(1,3),pt.cex=c(1.5,1,0.5))
R> arrows(265,27,265,29,length=0.05)
R> text(locator(1),labels="Weight",cex=0.8,family="serif")
```

The legend is placed at user coordinates (250, 30), and three points of the default pch type 1 are included—one large, one standard, and one small—using pt.cex set to 1.5, 1, and 0.5, respectively. Instead of writing text labels for each of these three points by using the legend argument, I simply assign them to be empty strings made up of 10 spaces. What this does is widen the box around the legend, creating a space for what will instead accompany the three points—a small-headed arrow pointing upward and the word *Weight*. Finding suitable user coordinates for the arrow to fit inside the artificially empty legend box took a little trial and error, and I place the "Weight" text by invoking the interactive locator function as you saw used in Section 23.3.

Playing with R functionality to produce intricate plots like this is an excellent way to start learning how to handle the traditional graphical abilities of the language. It's not uncommon to use trial and error and little cheats to reach an end result, though of course that kind of thing comes at the expense of the robustness of your code. For example, resizing the graphics device even moderately and attempting to reproduce the mtcars scatterplot shown earlier will likely result in a displeasing misalignment of the arrow in the legend. If you want to learn more, the authoritative reference on graphics in R is arguably Murrell (2011), which is a good text to consult once you're familiar with the fundamentals discussed here and want a comprehensive guide on all things visual in R.

## Exercise 23.2

For the following tasks, you'll work with the diamond-pricing data as analyzed by Chu (2001). You'll need an Internet connection for this. Read the data in and name the columns as you've done previously with the following:

```
R> dia.url <- "http://www.amstat.org/publications/jse/v9n2/4cdata.txt"
R> diamonds <- read.table(dia.url)
R> names(diamonds) <- c("Carat","Color","Clarity","Cert","Price")
```

a. Open a new graphics device of 6 × 6 inches. Initialize the margin spacing to be zero, four, two, and zero lines on the bottom, left, top, and right of the plot region, respectively. Then, complete the following:

  i. Produce side-by-side boxplots of the diamond prices in Singapore dollars (SGD$) split by certification. Suppress all axes and the surrounding box—note that the boxplot command requires you to set frame=FALSE for suppressing the box (as opposed to bty="n" in plot). Use the same command to provide an appropriate title.

  ii. Next, insert a vertical axis. The axis should have tick marks ranging from SGD$0 to SGD$18000, progressing in steps of SGD$2000. However, the axis should be clipped to the plotting region. The axis tick marks should point inward and be one line in length. The axis labels should sit only half a line away from the axis and should be horizontally readable.

  iii. Finally, use locator in conjunction with text to add an appropriate title sitting at the top of the y-axis; note that clipping will need to be relaxed. Use the same approach to add text, sitting inside each boxplot, denoting the corresponding certification (GIA, HRD, or IGI).

  My version of the plot looks like this:

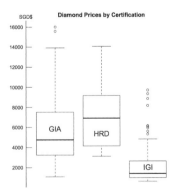

b.  Now, open a new graphics device of 8 × 7 inches. Set the figure margins to be two, five, three, and five lines on the bottom, left, top, and right, respectively. Also allow one line of outer margin space on each side other than the bottom, which should get two lines of outer margin.

    i.  Produce a scatterplot of diamond price on the vertical axis and carat weight on the horizontal axis. Use the colors red, green, and blue to distinguish the points according to certification. Suppress all axes, boxes, labels, and titles in the initial plot, but then add a U-shaped box.

    ii.  Add the horizontal axis. Use axis to place tick marks at an evenly spaced sequence of carat values between 0.2 and 1.1, in steps of 0.1. Use a bold, italic, sans-style font for the labels and adjust the labels to be only half a line from the axis. Then add smaller, outward-facing tick marks between the existing ones. To do this, make a second call to the axis function and place the ticks at a sequence of values from 0.15 to 1.05 at steps of 0.1. Set these secondary tick marks to have a length of one-quarter of a line and suppress the axis labels.

    iii.  Add the vertical axes. On the left, ticks should appear at SGD\$1000–17000. Labels should be horizontally readable and in the same font style as the horizontal axis. On the right, axis ticks should be made in the equivalent of US dollars (USD\$) at the sequence USD\$1000–11000 in steps of USD\$1000 and should be labeled as such. To do this, use the conversion USD\$ = 1.37 × SGD\$. Label orientation and font should match the other axes.

    iv.  Fit a linear model of price on a quadratic polynomial of carat weight for the data. Provide a prediction of the model for a sequence of carat values spanning the range of the observed values; include estimation of a 95 percent prediction interval. Use this information to superimpose a gray solid line for the fitted values and gray dashed lines for the prediction interval upon the scatterplot.

    v.  Set up expression objects for labeling the approximate US dollar conversion and the regression equation. Name the conversion expr1; it should look something like USD\$ ≈ 1.37 × SGD\$. The regression equation should look similar to Price = $\beta_0 + \beta_1$Carat + $\beta_2$Carat$^2$; name it expr2.

    vi.  Use mtext to add an appropriate main title and titles for all three individual axes. You may need to experiment a little with line depth for each one, as well as whether to write in the outer margin or the figure margin, depending on your own spacing preference. The rightmost axis title should make use of expr1.

vii. Either via trial and error to find appropriate coordinates or by using the interactive.arrow function from Exercise 23.1, place an arrow pointing to the fitted polynomial regression line and label it with expr2.

viii. Finally, use a call to locator to place a legend in any appropriate location, referencing the color of the points according to the appropriate certification.

My version of the plot looks like this:

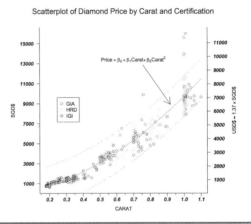

## Important Code in This Chapter

| Function/operator | Brief description | First occurrence |
|---|---|---|
| dev.new | Open new graphics device | Section 23.1.1, p. 576 |
| dev.set | Change active device | Section 23.1.2, p. 577 |
| dev.off | Close device | Section 23.1.3, p. 578 |
| par | Set graphical parameters | Section 23.1.4, p. 579 |
| layout | Open new graphics device | Section 23.1.4, p. 580 |
| box | Add box to plot | Section 23.2.1, p. 583 |
| mtext | Write text in margins | Section 23.2.2, p. 584 |
| locator | Interactive coordinates | Section 23.3.1, p. 587 |
| axis | Add axis to plot | Section 23.4.3, p. 594 |
| expression | Render Greek/math in plot | Section 23.5.2, p. 598 |
| title | Add main/axis titles | Section 23.5.2, p. 598 |
| italic | Italicize text | Section 23.5.3, p. 600 |
| grid | Add grid to plot | Section 23.6, p. 605 |

# 24

## GOING FURTHER WITH THE GRAMMAR OF GRAPHICS

You covered the basics of the ggplot2 package—which offers an alternative to traditional R graphics—in Section 7.4 and throughout Chapter 14. In this chapter, you'll look at a couple of this package's more popular and useful features, as well as its relatively young cousin, ggvis, which provides an interactive, browser-based experience.

## 24.1   ggplot or qplot?

So far when creating the relatively simple ggplot2 graphics, you've used the qplot function to initialize the visual object. In fact, the more general ggplot command is the core function of ggplot2. There are several key differences between these two initialization functions:

- qplot is a shortcut version of ggplot; it's used if you just want a quick look at your data or if you're working directly in the R console.

- qplot is designed to be reminiscent of the base R plot function—you pass it *x*- and *y*-coordinate vectors and then tell it what to do. By contrast, ggplot prefers its data argument as a data frame object, and you tell it what to do by explicitly adding geom layers.

- A call to qplot alone can produce a graphic. When using ggplot, layers have to be added before anything becomes visible.

- To access the full power and flexibility of ggplot2 graphics, ggplot is the recommended function; this comes at the cost of providing a little more explicit instruction than qplot requires.

All in all, you can create most plots using either qplot or ggplot. Many users make the decision based on the form their data are in (in other words, a data frame or as separate vectors in the global environment) and whether they want polished visuals (for example, for publication purposes) or just a quick look at the data while working directly in the console.

As a quick example of the difference in syntax, flip back to the code on page 297 used to create the histogram on the right of Figure 14-5. You could argue that the numerous modifications made to that particular plot warrant a more compartmentalized approach than qplot offers. Load ggplot2 with a call to library("ggplot2") and create the following three objects:

```
R> gg.static <- ggplot(data=mtcars,mapping=aes(x=hp)) +
                 ggtitle("Horsepower") + labs(x="HP")
R> mtcars.mm <- data.frame(mm=c(mean(mtcars$hp),median(mtcars$hp)),
                    stats=factor(c("mean","median")))
R> gg.lines <- geom_vline(mapping=aes(xintercept=mm,linetype=stats),
                 show.legend=TRUE,data=mtcars.mm)
```

The first object, gg.static, represents the part of the plot that will stay the same throughout, say, if you wanted to experiment with adding other features later. Note that the call to ggplot differs from qplot in that the first argument is the entire data frame of interest, allowing access to all data columns within the frame for any subsequent geoms or annotations. You then add the ggtitle and labs functions to set the main title and the horizontal axis title. The second object, mtcars.mm, stores the horsepower mean and median as a "dummy" data frame. The mean and median lines are then superimposed on the histogram by the third object, gg.lines, which is a single call to the geom_vline function with the same content used in the earlier code, albeit in a slightly modified form to stay true to the initial use of ggplot.

Nothing is displayed until you make a call that prints the ggplot2 object (as noted in Section 7.4). The following call reproduces the image on the right of Figure 14-5:

```
R> gg.static + geom_histogram(color="black",fill="white",
                    breaks=seq(0,400,25),closed="right") + gg.lines +
              scale_linetype_manual(values=c(2,3)) + labs(linetype="")
```

The pieces are put together in much the same way as for the creation of Figure 14-5: the addition of the geom_histogram layer to gg.static invokes the plot, and the addition of gg.lines with changes to the default line types made with scale_linetype_manual marks off the mean and median. If you wanted to produce the histogram without these lines, you would simply print the gg.static object plus geom_histogram.

As you get more experienced with ggplot2, you'll find yourself leaning toward either ggplot or qplot, depending on the application. The help file in ?ggplot provides a good description of the typical ways ggplot is used and how it stacks up against qplot. For further information, refer to *ggplot2: Elegant Graphics for Data Analysis* by Wickham (2009). I'll use ggplot for the rest of the plots in this chapter to provide some examples of the syntax of the ggplot command to compare with the earlier uses of qplot.

# 24.2   Smoothing and Shading

Data visualization using the ggplot2 package is particularly powerful when you want to split features of the plot by one or more categorical variables. This is especially apparent when you're enhancing your plot with features that are more difficult to achieve using base R commands.

## 24.2.1   Adding LOESS Trends

When you're looking at raw data, it's sometimes difficult to get an overall impression of trends without fitting a parametric model (for example, via linear regression), which means making assumptions about the nature of these trends. This is where *nonparametric smoothing* comes in—you can use certain methods to determine how your data appear to behave without fitting a specific model. These methods are a flexible aid for interpreting overall trends, whatever their form, but the trade-off is that you're not provided with any specific numeric details of the relationships between response and predictors (since you're not estimating any coefficients such as slopes or intercepts) and you lose any reliable ability to extrapolate.

*Locally weighted scatterplot smoothing (LOESS or LOWESS)* is a nonparametric smoothing technique that produces the smoothed trend by using regression methods on localized subsets of the data, step-by-step over the entire range of the explanatory variable.

**NOTE**   *For theoretical details, Chapter 6 of* Applied Nonparametric Regression *(Härdle, 1990), as well as Chapters 2 and 3 of* Introduction to Nonparametric Regression *(Takezawa, 2006), provide clear discussions of LOESS smoothers.*

For an illustration, load the MASS package and return your attention to the survey data frame. First, create a new data frame object with any missing values deleted to avoid default warning messages:

```
R> surv <- na.omit(survey[,c("Sex","Wr.Hnd","Height")])
```

Then, after loading ggplot2, execute the following to produce the image on the left of Figure 24-1.

```
R> ggplot(surv,aes(x=Wr.Hnd,y=Height)) +
       geom_point(aes(col=Sex,shape=Sex)) + geom_smooth(method="loess")
```

The call to ggplot initializes the object and sets the default mapping of handspan on the *x*-axis and height on the *y*-axis. The addition of geom_point adds the points, using color and point type to differentiate between males and females. The addition of geom_smooth superimposes the LOESS smoother. By default, a 95 percent confidence interval for the estimated trend is marked off by a transparent gray-shaded area.

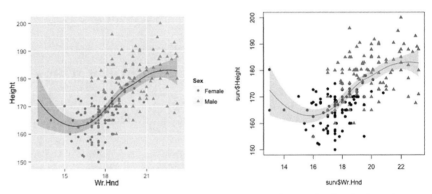

*Figure 24-1: Showcasing* ggplot2 *(left) and base R graphics (right) for display of nonparametrically estimated trends via LOESS*

Now I'll demonstrate how to produce a similar result using base R graphics. Although there are base R functions, such as scatter.smooth, that can produce a scatterplot with the smoothed trend relatively quickly, to be able to do things such as shade in the confidence interval region, it's helpful to be able to build the plot up piece by piece. Compare the relative ease of the ggplot2 approach with the following base R code, which produces the image on the right of Figure 24-1:

```
R> plot(surv$Wr.Hnd,surv$Height,col=surv$Sex,pch=c(16,17)[surv$Sex])
R> smoother <- loess(Height~Wr.Hnd,data=surv)
R> handseq <- seq(min(surv$Wr.Hnd),max(surv$Wr.Hnd),length=100)
R> sm <- predict(smoother,newdata=data.frame(Wr.Hnd=handseq),se=TRUE)
R> lines(handseq,sm$fit)
R> polygon(x=c(handseq,rev(handseq)),
           y=c(sm$fit+2*sm$se,rev(sm$fit-2*sm$se)),
           col=adjustcolor("gray",alpha.f=0.5),border=NA)
```

The first line plots the raw data, and the second line uses the built-in loess function to provide the smoothed trend—the syntax is identical to that of lm. Just as with linear models fitted by lm, for drawing to begin, you need to set up a fine sequence of values of the *x*-axis variable at which to obtain the point estimates and their standard errors; this is achieved with seq in the third line, followed by predict in the fourth line with the se argument set to TRUE. This results in the object sm, a list with components $fit and $se as usual.

The smoothed trend is then drawn, using sm$fit in a call to lines. Finally, a rough 95 percent confidence interval is calculated for each of the predicted values as the sm$fit elements plus and minus twice the corresponding standard errors in sm$se. This is done directly in the call to polygon, which draws the gray band based on the vertices formed by the confidence interval (therein, the rev command is used to reverse the entries in the given handseq vector). You need to instruct the gray-filled shape to be transparent with a call to the ready-to-use adjustcolor command (the argument alpha.f takes a value from 0, which is fully transparent, to 1, which is fully opaque); setting alpha.f=0.5 sets 50 percent opacity of the specified "gray".

All that, and a legend hasn't been put in yet! This example certainly exposes the extra effort the base R version of the image requires, not just in terms of the length of the script but also for the whole process of thinking about its construction (for example, putting together the vertices of a polygon for the confidence region appropriately and remembering to adjust the opacity of the filled shape to prevent any preplotted content being covered up). This becomes even more apparent the moment you become a little more ambitious with such features. Suppose you wanted to superimpose smoothers for each sex separately; this would require separate estimation of the LOESS functions and a rethink of the plotting strategy. However, this addition is simple in ggplot2 terms, simply requiring a change in the aesthetic mapping of the relevant geom. The following code produces Figure 24-2:

```
R> ggplot(surv,aes(x=Wr.Hnd,y=Height,col=Sex,shape=Sex)) +
       geom_point() + geom_smooth(method="loess")
```

All that's happened is that the aesthetic mapping for color and point type (col=Sex and shape=Sex, respectively) has shifted so that instead of being specific to the plotted points only, it's part of the default mapping declared in the initialization call to ggplot. Any layer added afterward (that doesn't re-assign the mapping) will follow this default, as is the case for both geom_point and geom_smooth.

**NOTE**  *The implementation of LOESS and other trend smoothers depends on you specifying the amount of smoothing you want; this is controlled by the proportion of the data to use as each localized weighted subset, for each step/location in the estimation procedure. A larger proportion leads to a smoother, less variable trend estimate than a smaller proportion. This value, referred to as the* span, *can be set by the optional argument* span *in either* loess *or* geom_smooth. *For quick data exploration, however, the default value of 0.75 is usually adequate. You can try experimenting with this on the example plots in this section to see the effects on the respective trends.*

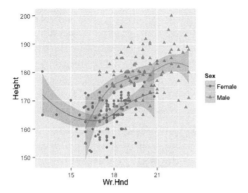

Figure 24-2: Illustrating separate LOESS smoothers for categorical subsets of data, a result of a simple change in aesthetic mapping, with `ggplot2` functionality

## 24.2.2 Constructing Smooth Density Estimates

The idea of smoothing isn't limited to scatterplot trends. *Kernel density estimation (KDE)* is a method for producing a smooth estimate of a probability density function, based on observed data. Briefly, KDE involves assigning a scaled probability function (the *kernel*) to each observation in a data set and summing them all to give an impression of the distribution of the data set as a whole. It's basically a sophisticated version of a histogram. For theoretical details, the text by Wand and Jones (1995) is a good reference.

To illustrate this method, consider the built-in `airquality` data frame; enter `?airquality` at the prompt to open the documentation, which tells you it contains a number of measurements taken of the air in New York over several months. A basic plot of the kernel estimate of the density of the temperature measurements is provided with the following line and shown on the left of Figure 24-3:

```
R> ggplot(data=airquality,aes(x=Temp)) + geom_density()
```

Such a plot is relatively easy to create with base R graphics as well, using the built-in `density` command to implement KDE for a given data vector. However, `ggplot2` lets you dress up the plot using aesthetic mappings with relative ease—a big draw for fans of `ggplot2`. For example, suppose you want to visualize the density estimates for temperature separately according to the month of observation. First, execute the following code:

```
R> air <- airquality
R> air$Month <- factor(air$Month,
                  labels=c("May","June","July","August","September"))
```

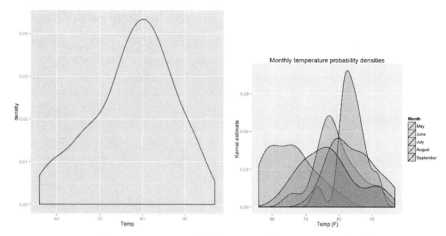

Figure 24-3: Visualizing temperature distributions in the airquality data frame via KDE, using ggplot2 functionality

This has created a copy of the airquality data frame in your workspace and recoded the originally numeric Month vector as a factor vector (as required for ggplot2 mappings), labeling the entries appropriately. Then, using air, the following code produces the right-hand plot in Figure 24-3:

```
R> ggplot(data=air,aes(x=Temp,fill=Month)) + geom_density(alpha=0.4) +
        ggtitle("Monthly temperature probability densities") +
        labs(x="Temp (F)",y="Kernel estimate")
```

The different densities are clearly identified by different color fills, set using fill=Month in aes in the plot initialization mapped out by ggplot. You additionally supply alpha=0.4 to geom_density to set 40 percent opacity so you can see all five curves clearly. The remaining calls to ggtitle and labs simply tidy up the main and axis titles. Features of the distributions of these measurements are as you might expect—temperatures for July, the hottest month, are centered over a far higher range of values than, say, those for May.

**NOTE**  *Just like LOESS techniques, the precise appearance of kernel-estimated probability density function is dependent on the amount of smoothing employed. Like the binwidth in the construction of a histogram, the quantity of interest in KDE is referred to as the* bandwidth *or* smoothing parameter—*a larger bandwidth imposes greater smoothing over the range of the data. By default, the bandwidth is automatically chosen using a data-driven technique in these examples. This default level of smoothing is generally acceptable for simple exploration of your data.*

## 24.3 Multiple Plots and Variable-Mapped Facets

In Section 23.1.4, you saw different ways in which several traditional R plots can be viewed or laid out in a single graphics device. The same methods, such as setting mfrow in a call to par or compartmentalizing the device using layout, can't be used for ggplot2 graphics. There are other functions, though, that allow independent ggplot2 plots to populate a single device. True to form, ggplot2 also offers a convenient way to consider multiple-plot graphics using facets, where the images are all drawn in one go.

### 24.3.1 Independent Plots

First, let's say you have several ggplot2 plots that you've created independently of one another and that you'd like to arrange as a single image. A quick way to do this is to use the grid.arrange function provided in the contributed gridExtra package (Auguie, 2012). Install the package by running install.packages("gridExtra") at the prompt (you'll need an Internet connection).

To illustrate the use of grid.arrange, continue with the air object—the copy of airquality you created in Section 24.2 with the factor Month column. Now, consider the following three ggplot2 objects, which I'll explain further in a moment:

```
R> gg1 <- ggplot(air,aes(x=1:nrow(air),y=Temp)) +
            geom_line(aes(col=Month)) +
            geom_point(aes(col=Month,size=Wind)) +
            geom_smooth(method="loess",col="black") +
            labs(x="Time (days)",y="Temperature (F)")
R> gg2 <- ggplot(air,aes(x=Solar.R,fill=Month)) +
            geom_density(alpha=0.4) +
            labs(x=expression(paste("Solar radiation (",ring(A),")")),
                 y="Kernel estimate")
R> gg3 <- ggplot(air,aes(x=Wind,y=Temp,color=Month)) +
            geom_point(aes(size=Ozone)) +
            geom_smooth(method="lm",level=0.9,fullrange=FALSE,alpha=0.2) +
            labs(x="Wind speed (MPH)",y="Temperature (F)")
```

Execute library("gridExtra") to load the required package. To view gg1, gg2, and gg3 in one window, simply call the following, which produces Figure 24-4:

```
R> grid.arrange(gg1,gg2,gg3)
```

Note that you'll likely see some warning messages telling you there are missing values in the air data frame and recommending to resize the window containing the plots.

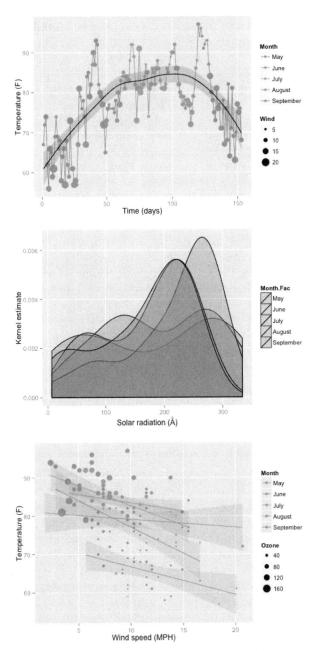

*Figure 24-4: A series of three* ggplot2 *graphics of the* airquality *data, plotted in the same device window via* grid.arrange *in the* gridExtra *package. Top: A time series of the temperatures by day, distinguishing between months and wind speed, with an overall LOESS trend with a 95 percent CI. Middle: Kernel density estimates of the distributions of solar radiation by month. Bottom: A scatterplot of temperature by wind speed, using color to delineate month and point size to reference ozone level. Separately fitted simple linear models of temperature on wind speed, split by month, along with 90 percent CIs, are superimposed. See Plate I in the color insert.*

As you can see, `grid.arrange` is easy to use—you simply create your `ggplot2` images first and store them as objects, then provide them directly to the arrangement function. `grid.arrange` decides how to produce the final layout based on the number of objects you give it (in this case, it's a column of the three plots). You can control the order of the plots by changing the order in which the objects are supplied. There are more optional arguments, which you can read about in the documentation file `?grid.arrange`.

The plots `gg1`, `gg2`, and `gg3` also provide an opportunity to discuss even more `ggplot2` capabilities. Since there's a lot going on, especially in `gg1` and `gg3`, I'll discuss the code for each object separately.

`gg1`  The first plot is of the daily temperature. In setting the default aesthetics in `ggplot`, I create a sequence of integers matching the number of rows in `air`, to be paired with the relevant `Temp` element. Then `geom_line` and `geom_point` add the interconnecting lines and the raw observations themselves, to be added to the default aesthetic mapping. The interconnecting lines are set to change color according to `Month`. The raw observations also change color according to `Month`, and the point sizes change to be proportional to the wind speed readings. I include an overall LOESS smoother with its default color changed to `"black"`. I remain with the default mapping here—I don't want separate smoothed trends for each month. The final addition of `labs` merely clarifies the axis titles as you've already seen it used.

`gg2`  The second plot is a variant of the plot on the right of Figure 24-3. This time, it shows the estimated densities of the solar radiation readings (in angstroms). The opacity is set, as you saw earlier, using `alpha` in `geom_density`. It's also worth noting that I used `expression` in `labs` to approximate the angstrom unit symbol, Å, using `ring(A)`.

`gg3`  The last is a scatterplot of temperature by wind speed, where you can see a negative relationship. The color, again to be assigned to each month, is also set as the default aesthetic mapping in `ggplot`. In the call to `geom_point`, the aesthetic enhancement is instructed to plot point size as proportional to the ozone reading (this is to ensure correct formatting of the corresponding legend, in light of the next addition). Here you can see a different kind of use for `geom_smooth`. In setting `method="lm"`, the line (or lines) I want superimposed correspond to simple linear model fits according to the x and y aesthetic mappings as predictor and response, respectively. Additionally including the factor `Month` in the default mapping ensures separate simple linear models are fitted for the temperature on wind speed data for each month and colored appropriately (it's important to note that the plotted lines do not reflect a multiple linear model that includes all the variables used in the plot). Light, transparent 90 percent CIs are included with each regression (`level=0.9` and `alpha=0.2`), and setting `fullrange=FALSE` restricts each regression line only to the width of the observed data for each month.

## 24.3.2  Facets Mapped to a Categorical Variable

If you want independently created ggplot2 graphics to appear in the same window, grid.arrange is arguably the best way to deal with them. However, the ggplot2 package offers a flexible alternative to quickly view multiple plots. Often, when exploring a data set, you'll want to create several plots of the same variables based on the levels of one or more important categorical variables. This behavior, referred to as *faceting*, is familiar territory for ggplot2 using either the facet_wrap or the facet_grid command.

Let's focus on the simplest case where you have one categorical variable. Remaining with the air data frame object, the following line creates a ggplot2 object of the density plots of the New York temperatures shown on the right of Figure 24-3:

```
R> ggp <- ggplot(data=air,aes(x=Temp,fill=Month)) + geom_density(alpha=0.4) +
          ggtitle("Monthly temperature probability densities") +
          labs(x="Temp (F)",y="Kernel estimate")
```

Rather than view all density estimates together, you can create a plot of each one separately, displaying them in the same device, with the following three uses of facet_wrap; the results are at the top left, top right, and bottom of Figure 24-5.

```
R> ggp + facet_wrap(~Month)
R> ggp + facet_wrap(~Month,scales="free")
R> ggp + facet_wrap(~Month,nrow=1)
```

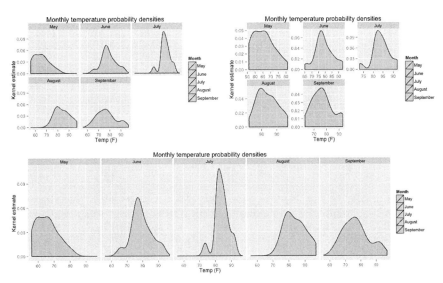

*Figure 24-5: Three examples of using* facet_wrap *to display kernel density estimates of the temperature data, split by month*

The `facet_wrap` function automates the layout of the multiple plots; a formula specifies the faceting variable. In all the previous plots, this is set to ~Month, to be read as "by Month." The code for the first visualization, given at the top left, offers no additional arguments, and without arguments the *x*- and *y*-axes of each facet are fixed, so you can compare the plots on the same scale. If you don't want that, you can instruct the axes to be "free," which means each plot is produced on scales specific to its own contents. You can see this in the second plot, at the top right in Figure 24-5, the line of code for which specifies `scales="free"`. You can also opt to free only the horizontal or vertical axis with `scales="free_x"` or `scales="free_y"`, respectively. Finally, note that facet placement can be tailored by using the `nrow` and `ncol` arguments. In the third plot, setting `nrow=1` instructs R to place the plots in one row only, giving the horizontal arrangement on the bottom of Figure 24-5. For further details on placement, you can find the documentation for this command at `?facet_wrap`.

The alternative to `facet_wrap`, `facet_grid`, does much the same thing but isn't able to wrap the plots if you're faceting by only one categorical variable. The formula *var1* ~ *var2* is interpreted as "facet by *var1* as rows and by *var2* as columns." If you are indeed interested in faceting by only one grouping variable in either columns or rows, then simply replace either *var1* or *var2* with a dot (`.`). The third image in Figure 24-5, for example, can just as easily be achieved via `facet_grid` as follows:

```
R> ggp + facet_grid(.~Month)
```

The next example, however, shows `facet_grid` in action with two grouping variables. Turn your attention again to the `diabetes` data frame in the `faraway` package. After loading the package, the following code creates the object with `diab` as the data frame of interest and with missing-value rows deleted, producing Figure 24-6:

```
R> diab <- na.omit(diabetes[,c("chol","weight","gender","frame","age",
                                "height","location")])
R> ggplot(diab,aes(x=age,y=chol)) +
       geom_point(aes(shape=location,size=weight,col=height)) +
       facet_grid(gender~frame) + geom_smooth(method="lm") +
       labs(y="cholesterol")
```

The initial call to `ggplot` tells R to use `diab` and plot total cholesterol against age. Then an addition of `geom_point` sets the shape, size, and color of each plotted point to change according to the county location, the weight, and the height of the individuals, respectively (as you've already seen for point size based on a continuous variable, point color is also automatically changed to vary on a continuum if the correspondingly mapped aesthetic variable isn't a factor).

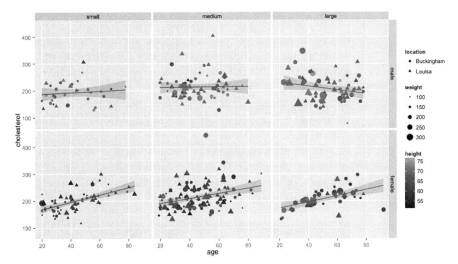

*Figure 24-6: Illustrating two-way faceting in* `ggplot2`, *using the* `diabetes` *data frame from the* `faraway` *package. Plots of cholesterol level against age, along with simple linear model fits, are faceted according to sex (rows) and body frame type (columns). Points are colored and sized according to weight and height, respectively, and two different point types differentiate the two county locations in Virginia of the study participants.*

So far, those commands have still defined only a single scatterplot. With the addition of the call to `facet_grid`, the formula `gender~frame` separates the plots into a different scatterplot for males/females (as rows) and for each of the three body frame types: small/medium/large (as columns). You set simple linear model fits to accompany each plot based on the default aesthetic mapping (cholesterol on age) with a call to `geom_smooth`, and a final call to `labs` clarifies the vertical axis title.

The plots themselves reveal, generally, some of the trends you'll have picked up on an earlier analysis of these data (Section 21.5.2). Increasing age tends to be associated with an increase in mean cholesterol, though that relationship seems, at least visually, less prominent for males. The overall smaller size of the points in the small frame column on the left makes sense—those with a smaller frame will typically weigh less than those with larger frames. There is also a tendency for the plots on the bottom row (females) to be of a darker shade than those along the top row—indicating that on average females are typically shorter than males. Any differences between the participants from the two counties, however, is difficult to discern—there doesn't seem to be a systematic departure of the pattern of the Buckingham symbols (•) from that of the Louisa symbols (▲). (Remember, though, if you're trying to understand the complex, potentially interactive relationships in your multivariate data, fitting an appropriate statistical model is preferable over plots alone.)

These elegant plots serve to further highlight the relative ease with which the `ggplot2` package can produce complex graphics—typically involving partitioning the observations by one or more factors—both in terms of an individual image or an arrangement of images. Although similar plots are of course still possible using base R methods, that approach demands a somewhat finer or lower-level handling of the details of your data subsets, as well as any varying aesthetic features. This doesn't mean base R graphics are redundant or should be ignored (you'll see some nice new plots achieved with traditional commands in Chapter 25)—it's just that you can create certain graphical displays with less coding effort (and usually a prettier end result) by utilizing Wickham's well-received implementation of the grammar of graphics.

---

### Exercise 24.1

Load the `MASS` package and inspect the help file for the `UScereal` data. This data frame provides nutritional and other information concerning breakfast cereals for sale in the United States in the early 1990s.

a. Create a copy of the data frame; name it `cereal`. To ease plotting, collapse the `mfr` column (manufacturer) of `cereal` to be a factor with only three levels, with the corresponding labels `"General Mills"`, `"Kelloggs"`, and `"Other"`. Also, convert the `shelf` variable (shelf number from floor) to a factor.

b. Using `cereal`, construct and store two `ggplot` objects.
   i. A scatterplot of calories on protein. Points should be colored according to shelf position and shaped according to manufacturer. Include simple linear regression lines for calories on protein, split according to shelf position. Ensure tidy axis and legend titles.
   ii. A set of kernel estimates of calories, with filled color differentiating shelf positions. Use 50 percent opaque fills, and again ensure tidy axis and legend titles.

c. Arrange the two plots in (b) on a single device.

d. Produce a faceted graphic of calories on protein, with each panel corresponding to a manufacturer as defined in your `cereal` object. A LOESS smoother with a 90 percent span should be superimposed upon each scatterplot. In addition, the points should be colored according to sugar content, sized according to sodium content, and shaped according to shelf position.

Load the `car` package (downloading and installing it first if you haven't already) and consider the `Salaries` object—a data frame detailing the salaries (in US dollars) of 397 academics working in the United States during the 2008–2009 school year (Fox and Weisberg, 2011). An inspection of the help file ?Salaries informs you of the present variables, which, in addition to the salary figure, include each academic's rank, sex, and research discipline (as factors) as well as the number of years of service.

e. Produce a ggplot object, named gg1, of a scatterplot of salary on the vertical axis against years of service on the horizontal axis. Color should be used to distinguish between males and females, along with sex-specific LOESS trends, and ensure axis and legend titles are understandable. View your plot.

f. Create the following three additional plot objects, again ensuring tidy axis and legend titles. Name the following gg2, gg3, and gg4, respectively:

    i. Side-by-side boxplots of salary, split by rank. Each boxplot should be further split up according to sex (this can be done simply in the default aesthetic mapping—try assigning the sex variable to either col or fill).

    ii. Side-by-side boxplots of salary, split by discipline, with each discipline split further by sex using color or fill.

    iii. Kernel density estimates of salary, using 30 percent opaque fills to distinguish rank.

g. Display your four plot objects (gg1, gg2, gg3, and gg4) from (e) and (f) in a single device.

h. Finally, plot the following:

    i. A series of kernel density estimates of salary using 70 percent opaque fills to distinguish between males and females, faceted by academic rank.

    ii. Scatterplots of salary on years of service, using color to distinguish between males and females, faceted by discipline as rows and by academic rank as columns. Each scatterplot should have a sex-specific simple linear regression line with confidence band superimposed and have free horizontal scales.

## 24.4 Interactive Tools in ggvis

To wrap up this chapter, I'll touch on a relatively new addition to the "gg" family, ggvis, by Chang and Wickham (2015). The package enables you to design flexible statistical plots that the end user can interact with. The results are provided as web graphics. You'll see the image pop up as a new tab in your default web browser (if you're using the RStudio IDE—see Appendix B—the ggvis graphics are embedded within the Viewer pane).

As a cautionary note, be aware that ggvis is, at the time of writing, still under development by its authors. New functionality is being added and bugs addressed. If you're interested in the functionality, visit the ggvis website at *http://ggvis.rstudio.com/*. The site contains a beginner-friendly tutorial and recipe book of things that are currently possible. Here, I'll just give you an overview of ggvis.

Install the ggvis package along with its dependencies and then load it with a call to library("ggvis"). Also make sure you have access to the student survey data, survey, by loading the MASS package. Create the following object to be used in the upcoming examples:

```
R> surv <- na.omit(survey[,c("Sex","Wr.Hnd","Height","Smoke","Exer")])
```

The common way to begin a ggvis graphic is to declare the data frame of interest, followed by a call to ggvis that defines the variables to be used and then to pile on the layers. When you use the variables from the data frame, they must be prefaced by a ~, which explicitly tells R that you're referring to a column of that data frame and not another object of the same name somewhere else. To add functions in the object definition, you don't use + as in ggplot2, but %>% (called a *pipe*). The equivalents of the geom_ functions in ggplot2 are prefaced by layer_ in ggvis.

Let's start with a simple static plot. The topmost image of Figure 24-7, a histogram of the height measurements, can be obtained with the following execution:

```
R> surv %>% ggvis(x=~Height) %>% layer_histograms()
Guessing width = 2 # range / 25
```

The surv data frame is declared, and then you pipe to ggvis, which instructs the ~Height variable to be mapped to the *x*-axis. Last, a pipe to layer_histograms produces the graphic, which assigns a default binwidth based on the range of the *x*-mapped data.

So what? You've already created lot of histograms. But wouldn't it be great if you could play with the value of the binwidth without needing to create static plot after static plot? The input_ collection of commands in ggvis allows you to instruct the resulting graphic to take interactive input. Consider the following code; Figure 24-7 shows my result.

```
R> surv %>% ggvis(x=~Height) %>%
    layer_histograms(width=input_slider(1,15,label="Binwidth:"),fill:="gray")
Showing dynamic visualisation. Press Escape/Ctrl + C to stop.
```

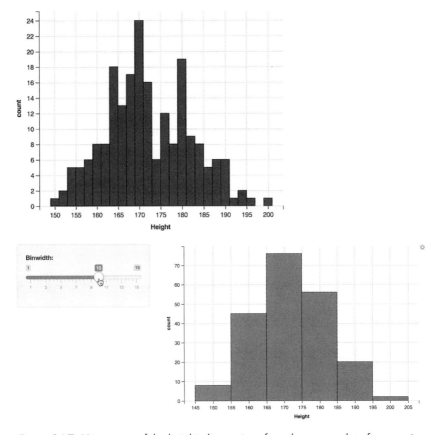

*Figure 24-7: Histograms of the height observations from the* survey *data frame, using* ggvis. *Top: Default static plot. Bottom: The result of incorporating a slider button tied to the binwidth—the user can interact with, and immediately see, the effect of altering the bins.*

Here, the width argument that controls the feature of interest is instructed to take the result of input_slider, which sets up an interactive slider button. The range of the slider values for width is set as 1 to 15 (inclusive), and the optional argument label provides a title for the interactive gadget. Last, using fill in layer_histogram sets the color of the bars. Note the particular assignment fill:="gray" uses :=, not just =. The = alone is used in ggvis for mapping variables, that is, when the feature of interest is to be passed a variable subject to change, essentially like an aesthetic mapping in ggplot2. The combination of := should be interpreted as a set constant, that is, when you simply intend to universally fix a certain feature.

Once the code successfully executes, you can experiment with sliding the button for smaller and larger binwidths. It's interesting to gauge just how much your interpretation of the distribution changes along with it. As the text printed out beneath the executed commands in the console tells you, you have to exit the interactive plot to use R again. Pressing ESC will terminate the interactivity and return control to the user at the prompt.

Other kinds of interactive abilities include input_select (for a drop-down menu), input_radiobuttons (radio button options), and input_checkbox (for checkboxes). You can even set up interactive text or numeric input boxes with input_numeric. See the relevant help files or the ggvis website for further details.

As another example based on the surv data frame, let's try a scatterplot. Starting again with a simple static plot, run the following:

```
R> surv %>% ggvis(x=~Wr.Hnd,y=~Height,size:=200,opacity:=0.3) %>%
     layer_points()
```

I won't show this result here, but you can see from the call to ggvis that you'll be plotting height against handspan and that you're universally enlarging the points as well as setting a universal level of 30 percent opacity. The last pipe to layer_points produces the image. As with the static plot of the histogram, since there's no interactivity, you don't need to "exit" the plot—you're returned control at the console prompt immediately.

For a more interesting graphic, try this:

```
R> filler <- input_radiobuttons(c("Sex"="Sex","Smoking status"="Smoke",
                                   "Exercise frequency"="Exer"),map=as.name,
                                 label="Color points by...")
R> sizer <- input_slider(10,300,label="Point size:")
R> opacityer <- input_slider(0.1,1,label="Opacity:")
R> surv %>% ggvis(x=~Wr.Hnd,y=~Height,fill=filler,
                  size:=sizer,opacity:=opacityer) %>%
     layer_points() %>% add_axis("x",title="Handspan") %>%
     add_legend("fill",title="")
Showing dynamic visualisation. Press Escape/Ctrl + C to stop.
```

First, three objects are created for the interactive bits. A set of radio buttons specifies the color of the plotted points according to one of three possible categorical variables (Sex, Smoke, or Exer), and two slider buttons control the point size and opacity. Note that when you intend to use variables from the data frame as ingredients for interactive behavior, you need to supply their exact names as a vector of character strings and set the optional map=as.name; this is done when defining the filler object. In the subsequent call to ggvis, you pass filler to fill, using =. The two slider buttons in the objects sizer and opacityer are passed to the relevant arguments with := since they don't depend on variables in the data frame. The call to layer_points generates the plot, and additional pipes to add_axis and add_legend simply tidy up the x-axis and legend titles from their defaults.

The top of Figure 24-8 shows a screenshot of the result, where I've selected the point color to vary according to the exercise frequency variable, reduced the point size, and chosen a moderate-to-high level of opacity.

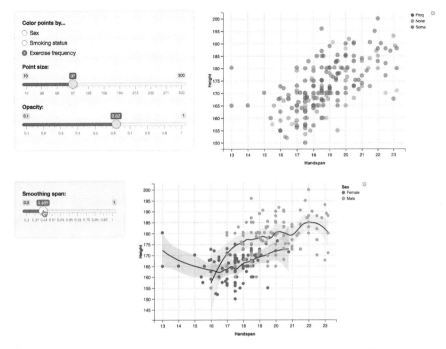

Figure 24-8: Two examples of ggvis scatterplots of height on handspan of the student survey data. Top: Color (fill) changed using radio buttons based on sex, smoking status, or exercise frequency and slider buttons for point size and opacity. Bottom: Splitting points by sex using color and superimposing sex-specific LOESS smoothers and corresponding CIs; their smoothing span is controllable via a slider button.

Finally, let's produce the same scatterplot but settle on using sex to color the points. You can add separate LOESS smoothers to males and females and dynamically control the degree of smoothing using a slider button. This last example, a screenshot of which appears on the bottom of Figure 24-8, is the result of executing the following code:

```
R> surv %>% ggvis(x=~Wr.Hnd,y=~Height,fill=~Sex) %>% group_by(Sex) %>%
     layer_smooths(span=input_slider(0.3,1,value=0.75,
                                     label="Smoothing span:"),
                   se=TRUE) %>% layer_points() %>%
     add_axis("x",title="Handspan")
Showing dynamic visualisation. Press Escape/Ctrl + C to stop.
```

LOESS smoothers are added with the layer_smooths command, whose span argument, the target parameter of interest, is assigned an input_slider. Its range of possible values is set as usual, and the optional value argument (also applicable in other input_ functions) sets the starting value when the plot is first initialized. Additionally, the se=TRUE argument to layer_smooths ensures 95 percent CIs accompany the smoothed trends. Note that the pipe to layer_smooths is preempted by a pipe through group_by(Sex). Without it, the smoothers would simply be applied to the x and y data as a whole (also note, at the time of writing, you don't preface the variable name with ~ in group_by).

As such, ggvis shows great potential for a more dynamic experience for visual data exploration. These tools are especially useful in activities such as presentations or website designs, where you can offer your audience an interactive appreciation of your data in a grammar of graphics style. If you're interested in using these tools, I strongly encourage you to keep abreast of developments at the ggvis website.

---

### Exercise 24.2

Ensure the car and ggvis packages are loaded. Revisit the Salaries data frame you looked at in Exercise 24.1; inspect the help file ?Salaries to remind yourself of the present variables.

a. Produce an interactive scatterplot of salary on the vertical axis and the years of service on the horizontal axis. Employ radio buttons to color points according to either academic rank, research discipline, or sex. Use pipes to add_legend and add_axis to omit a legend title and to tidy up the axis titles, respectively.

b. A pipe to layer_densities (which you've not yet met) is used to produce kernel density estimates, similar to those appearing in Figure 24-5.

    i. Use ggvis to create a static plot of kernel density estimates of salary distributions, split up according to academic rank. To do this, assign the salary variable to x and the rank variable to fill, followed by a pipe to group_by to explicitly instruct grouping by the rank variable. Lastly, piping to layer_densities (just use all default argument values in this instance) will generate the graphic. Your result should resemble the gg4 object from Exercise 24.1.

    ii. Just like the width argument to layer_histograms is used to control the appearance of a histogram, the adjust argument in layer_densities is used to control the degree of smoothness of the kernel estimates. Reproduce the rank-specific kernel estimates from the previous plot, but this time, the graphic should be interactive—implement a slider button with a range of 0.2 to 2 and a label of "Smoothness" to control the smoothing adjustment. At your discretion, either suppress or clarify the axis and legend titles of the result.

Ensure you have the MASS package loaded, once more gaining access to the UScereal data frame. If you haven't already done so, inspect the help file ?UScereal and re-create the cereal object exactly as specified in Exercise 24.1 (a). Then do the following:

c.  Set up an object for radio buttons to choose among the manufacturer, the shelf, and the vitamins variables. Make sure the labels for each radio button are clear, and set up an appropriate title label for what will form the collection of options to color the points. Name the object filler.

d.  Borrowing the sizer and opacityer objects created in Section 24.4 and using the object you just created in (c) to control fill, create an interactive scatterplot of calories on protein. Tidy up the axis titles and suppress the legend title for the point color fill. The result should essentially be the same, in terms of functionality, as the graphic appearing as the topmost screenshot in Figure 24-8.

e.  Create a new object for the same radio buttons as specified in (c) that will control the shape of the points (in other words, the characters used to plot points). Modify the title label accordingly. Name this object shaper.

f.  Finally, re-create the interactive scatterplot of calories on protein exactly as in (d), but this time additionally assigning shaper from (e) to the shape modifier in your call to ggvis. To prevent the legends for the two sets of radio buttons from overlapping each other, you need to add the following pipes to your code:

```
add_legend("shape",title="",
           properties=legend_props(legend=list(y=100)))
```

and

```
set_options(duration=0)
```

The first simply moves the legend for the shape modifier vertically downward, and the second eliminates the slight "animation delay" that occurs by default when switching between options in the interactive graphic. Once more, use additional calls to add_axis and add_legend to clarify or suppress axis and legend titles.

## Important Code in This Chapter

| Function/operator | Brief description | First occurrence |
|---|---|---|
| ggplot | Initialize ggplot2 plot | Section 24.1, p. 610 |
| geom_smooth | Trend line geom | Section 24.2.1, p. 612 |
| loess | Calculate LOESS (base R) | Section 24.2.1, p. 612 |
| rev | Reverse vector elements | Section 24.2.1, p. 612 |
| adjustcolor | Alter color opacity (base R) | Section 24.2.1, p. 612 |
| geom_density | Kernel density geom | Section 24.2.2, p. 614 |
| ggtitle | Add ggplot2 title | Section 24.2.2, p. 615 |
| grid.arrange | Multiple ggplot2 plots | Section 24.3.1, p. 616 |
| facet_wrap | One-factor faceting | Section 24.3.2, p. 619 |
| facet_grid | Two-factor faceting | Section 24.3.2, p. 620 |
| ggvis | Initialize ggvis plot | Section 24.4, p. 624 |
| %>% | Pipe to ggvis layer | Section 24.4, p. 624 |
| layer_histograms | ggvis histogram layer | Section 24.4, p. 624 |
| input_slider | Interactive slider | Section 24.4, p. 624 |
| := | Constant ggvis assignment | Section 24.4, p. 624 |
| layer_points | ggvis points layer | Section 24.4, p. 626 |
| input_radiobuttons | Interactive buttons | Section 24.4, p. 626 |
| add_legend | Add/alter ggvis legend | Section 24.4, p. 626 |
| layer_smooths | ggvis trend line layer | Section 24.4, p. 627 |
| add_axis | Add/alter ggvis axis | Section 24.4, p. 627 |

# 25

## DEFINING COLORS AND PLOTTING IN HIGHER DIMENSIONS

Now that you've mastered some fundamental visualization skills, you can go beyond the standard *x*- and *y*-axes by, for example, coloring points according to some additional value or variable or adding a *z*-axis for constructing a 3D plot. Higher-dimensional plots like this allow you to visually explore your data or models using more variables than would be possible otherwise.

In this chapter, you'll get into more detail when it comes to handling colors and color palettes in R, and then you'll look at four new plots: 3D scatterplots, contour plots, pixel image plots, and perspective plots.

## 25.1 Representing and Using Color

Color plays a key role in many plots. As you've already seen, color can be used purely for aesthetic enhancement, or it can be a critical aid to interpreting your data/models by distinguishing between values and variables. Before learning about some more complicated data and model visualization tools, it's useful to understand a little about how R formally represents and handles colors. In this section, you'll examine common ways to create and represent specific colors and how to define and use a cohesive collection of colors; the latter is referred to as a *palette*.

### 25.1.1   Red-Green-Blue Hexadecimal Color Codes

When specifying colors in plots, your instruction to R so far has been given either in the form of an integer value from 1 to 8 or as a character string (see the relevant comments in Section 7.2.3). For programming purposes, you need a more objective representation of these colors.

One of the most common methods of color specification is to specify different *saturations* or *intensities* of three primaries—red, green, and blue (RGB)—which are then mixed to form the resulting target color. Each primary component of the standard RGB system is assigned an integer from 0 to 255 (inclusive). Such mixtures are therefore able to form a total of $256^3 = 16,777,216$ possible colors.

You always express these values in (R, G, B) order; the result is commonly referred to as a *triplet*. For example, $(0,0,0)$ represents pure black, $(255,255,255)$ represents pure white, and $(0,255,0)$ is full green.

The col argument lets you select one of eight colors when you supply it an integer from 1 to 8. You can find these eight colors with the following call:

```
R> palette()
[1] "black"  "red"    "green3" "blue"   "cyan"   "magenta"
[7] "yellow" "gray"
```

These are but a small subset of the 650+ named colors that you can list by entering colors() at the R prompt. All of these named colors can also be expressed in the standard RGB format. To find the RGB values for a color, supply the desired color names as a vector of character strings to the built-in col2rgb function. Here's an example:

```
R> col2rgb(c("black","green3","pink"))
      [,1] [,2] [,3]
red      0    0  255
green    0  205  192
blue     0    0  203
```

The result is a matrix of RGB values, with each column representing one of your specified colors. This is what R actually means, in an RGB sense, when you ask it to plot these colors using the corresponding character string.

These RGB triplets are frequently expressed as *hexadecimals*, a numeric coding system often used in computing. In R, a hexadecimal, or *hex code*, is a character string with a # followed by six alphanumeric characters: valid characters are the letters *A* through *F* and the digits 0 through 9. The first pair of characters represents the red component, and the second and third pairs represent green and blue, respectively. If you have or create one or more RGB triplets, you can turn them into hex codes for R to use in any subsequent plotting through the rgb function. This command takes a matrix of RGB values, though note that it expects each (R, G, B) color to be a row of that matrix (as opposed to the columns provided from a call to, say, col2rgb).

You'll also need to tell rgb that your maximum color value, as per the standard RGB format, is 255 (since by default it scales this and uses 1). The following code performs a matrix transpose (refer to Section 3.3) on the result of the previous call to col2rgb, putting my three colors as RGB triplets in the required form as rows, and specifies the maxColorValue accordingly:

```
R> rgb(t(col2rgb(c("black","green3","pink"))),maxColorValue=255)
[1] "#000000" "#00CD00" "#FFC0CB"
```

The output tells you the hexadecimal codes for the RGB values R refers to with the names "black", "green3", and "pink", respectively.

I won't go into the specifics of converting a standard RGB triplet to a hexadecimal here because it's beyond the scope of this book, but it's important to know that R represents any colors you create using RGB triplets as hex codes, so you should be able to at least recognize a hexadecimal when you're working with colors and color palettes in plotting.

For an even more colorful exploration, let's write a modest little function to plot points in individual colors and label them appropriately with RGB triplets and corresponding hex codes. Consider the following in the editor:

```
pcol <- function(cols){
        n <- length(cols)
        dev.new(width=7,height=7)
        par(mar=rep(1,4))
        plot(1:5,1:5,type="n",xaxt="n",yaxt="n",ann=FALSE)
        for(i in 1:n){
                pt <- locator(1)
                rgbval <- col2rgb(cols[i])
                points(pt,cex=4,pch=19,col=cols[i])
                text(pt$x+1,pt$y,family="mono",
                    label=paste("\"",cols[i],"\"","\nR: ",rgbval[1],
                        " G: ",rgbval[2]," B: ",rgbval[3],
                        "\nhex: ",rgb(t(rgbval),maxColorValue=255),
                        sep=""))
        }
}
```

The function pcol takes one argument, cols, intended to be a character vector of color names recognized by R. When you execute pcol, it opens a new graphics device and equalizes the figure margin settings to be one line on each side. A plot is begun, fully suppressed except for a box. This is so you can use locator (see Section 23.3) to place points in the plot region, implemented in a for loop, one after the other. Each point represents one of the cols, and after its coordinates are returned from locator, points puts down a large dot of the color at hand, with text providing annotation to the right of each point achieved using paste (refer to Section 4.2). This annotation includes the R color name, the RGB triplet, and the hex code on top of one another; the latter two are found using col2rgb and rgb exactly as demonstrated earlier.

The following code sets up the device, first storing 14 valid R color names (chosen randomly) in the character vector mycols. After exhausting these with mouse clicks in different areas of the plot region, the execution is complete.

```
R> mycols <- c("black","blue","royalblue2","pink","magenta","purple",
            "violet","coral","lightgray","seagreen4","red","red2",
            "yellow","lemonchiffon3")
R> pcol(mycols)
```

When I execute pcol as shown here, I click through the 14 points, producing rough columns on my graphics device. Figure 25-1 shows the result.

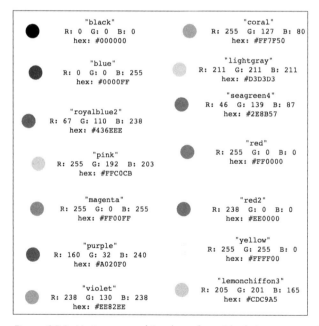

Figure 25-1: Various named R colors alongside their corresponding RGB triplets and hex codes. See Plate II in the color insert.

In essence, you can obtain any color you want (in other words, far more than the named ones that are built into R) by specifying an RGB value and obtaining its hex code. These hexadecimals can be supplied as is to any of the traditional R graphics functions where you specify color (commonly to a col argument). You'll see this as the chapter progresses. Naturally, you can also assign a hex code or a vector of hex codes (for example, if you're creating your own custom colors) to a new object in your R workspace so you can use it in subsequent plotting.

## 25.1.2   Built-in Palettes

Being able to implement your own RGB colors is most useful when you need many colors, the collection of which is referred to as a *palette*. You'll typically need a palette when color is used to describe something on a continuum, like the various shades of blue used for the height measurements in Figure 24-6 on page 621.

There are a number of color palettes built into the base R installation. These are defined by the functions rainbow, heat.colors, terrain.colors, topo.colors, cm.colors, gray.colors, and gray. With the exception of gray, you directly specify the number of colors you want, and they'll be returned as a character vector of hex codes representing an equally spaced sequence over the entire color range of that particular palette.

It's easiest to see this in action with a visualization. The following code generates exactly 600 colors from each palette:

```
R> N <- 600
R> rbow <- rainbow(N)
R> heat <- heat.colors(N)
R> terr <- terrain.colors(N)
R> topo <- topo.colors(N)
R> cm <- cm.colors(N)
R> gry1 <- gray.colors(N)
R> gry2 <- gray(level=seq(0,1,length=N))
```

Note that instead of a single integer, gray expects a numeric vector of values between 0 (total black) and 1 (total white) to provide a grayscale. Its counterpart function, gray.colors, works the same as the other built-in palettes but defaults to a slightly narrower visual range between the extremes of black and white. These can be reset using the optional arguments start and end, which you'll see shortly.

The next code chunk uses skills from Chapter 23 to initialize a new plot and uses vector repetition to place 600 points for each palette in a single call to points, coloring them appropriately as per the vectors of hex codes.

```
R> dev.new(width=8,height=3)
R> par(mar=c(1,8,1,1))
R> plot(1,1,xlim=c(1,N),ylim=c(0.5,7.5),type="n",xaxt="n",yaxt="n",ann=FALSE)
R> points(rep(1:N,7),rep(7:1,each=N),pch=19,cex=3,
          col=c(rbow,heat,terr,topo,cm,gry1,gry2))
R> axis(2,at=7:1,labels=c("rainbow","heat.colors","terrain.colors",
                          "topo.colors","cm.colors","gray.colors","gray"),
         family="mono",las=1)
```

Figure 25-2 shows the result.

Figure 25-2: Showcasing the color ranges of the built-in palettes, with default limits used in gray.colors. See Plate III in the color insert.

For more information, access the help files ?gray.colors and ?gray for the respective grayscale palettes, with the others all appearing under ?rainbow.

### 25.1.3 Custom Palettes

You're not restricted to the ready-to-use color designs. The function colorRampPalette allows you to create your own palettes; you supply two or more desired key colors to an argument of the same name, and it creates a palette that transitions between them. The result of a call to colorRampPalette is itself a function—one that behaves exactly like the built-in palette functions noted earlier.

Let's say you'd like to be able to generate colors on a scale between purple and yellow. You specify the key colors to be interpolated, in the desired order, as a character vector of names from the collection that R recognizes. The following line creates this palette function:

```
R> puryel.colors <- colorRampPalette(colors=c("purple","yellow"))
```

Let's create another one, this time picking one that will show up a little clearer in case a color plot that ends up using it is printed in grayscale (in which case sticking to monochromatic palettes is a good idea).

```
R> blues <- colorRampPalette(colors=c("navyblue","lightblue"))
```

Here are a couple more, using more than two colors this time:

```
R> fours <- colorRampPalette(colors=c("black","hotpink","seagreen4","tomato"))
R> patriot.colors <- colorRampPalette(colors=c("red","white","blue"))
```

Having created a handful of custom palette functions, you can now generate any number of colors from each range just like before (done here using the previously stored N value of 600 each). After doing so, you can adapt the earlier plotting code to get the image in Figure 25-3.

```
R> py <- puryel.colors(N)
R> bls <- blues(N)
R> frs <- fours(N)
R> pat <- patriot.colors(N)
R> dev.new(width=8,height=2)
R> par(mar=c(1,8,1,1))
R> plot(1,1,xlim=c(1,N),ylim=c(0.5,4.5),type="n",xaxt="n",yaxt="n",ann=FALSE)
R> points(rep(1:N,4),rep(4:1,each=N),pch=19,cex=3,col=c(py,bls,frs,pat))
R> axis(2,at=4:1,labels=c("peryel.colors","blues","fours","patriot.colors"),
          family="mono",las=1)
```

Figure 25-3: Some examples of custom color palettes created using colorRampPalette. See Plate IV in the color insert.

## 25.1.4 Using Color Palettes to Index a Continuum

You've now seen a few times how color can be used to identify groups based on a categorical variable (the data corresponding to a certain level are simply given a distinct color from the others), which is pretty easy to do. However, assigning colors appropriately to values on a continuum requires a little more thought. There are two methods for this: through categorization or through normalization of your continuous values. Let's look first at the former approach.

### Via Categorization

One way to color values according to a continuous variable is to turn it into the familiar problem of coloring points of a categorical variable. You can do this by binning your continuous values into a fixed number of $k$ categories, generating $k$ colors from your palette, and matching each observation to the appropriate color based on the bin it falls into.

In Section 20.1, you plotted height against writing handspan for the survey data from the MASS package. This time, let's use color to additionally inform the nonwriting handspan variable. Load the package and execute the following line:

```
R> surv <- na.omit(survey[,c("Wr.Hnd","NW.Hnd","Height")])
```

This creates the data frame object surv, which is made up of only the three required columns. Any rows with missing values are removed via a call to na.omit (refer to Section 6.1.3).

Now, the first thing to do is decide on your color palette.

```
R> NW.pal <- colorRampPalette(colors=c("red4","yellow2"))
```

This will generate colors that go from a dark, dull red at the lower end of the scale to a slightly faded yellow at the higher end (similar to the built-in heat.colors palette; see Figure 25-2/Plate III). Next, you need to decide how many bins, $k$, you're going to construct for the continuous values. This determines how many distinct colors to generate from NW.pal. For these data, set $k = 5$.

```
R> k <- 5
R> ryc <- NW.pal(k)
R> ryc
[1] "#8B0000" "#A33B00" "#BC7700" "#D5B200" "#EEEE00"
```

Your five NW.pal colors, as hex codes, are available. Next, you need to actually bin the continuous values, which you can do using cut. First you need to set $k + 1$ break points for the bins (refer to Section 4.3.3 for a refresher), using seq.

```
R> NW.breaks <- seq(min(surv$NW.Hnd),max(surv$NW.Hnd),length=k+1)
R> NW.breaks
[1] 12.5 14.7 16.9 19.1 21.3 23.5
```

The six equally spaced values span the range of the students' nonwriting handspans, delineating your five intended bins. Then cut factorizes the nonwriting handspans with respect to those bins. You can use as.numeric to specifically return the indexes for extracting the appropriate color for each observation from your five ordered hex codes in ryc (full output is suppressed here for reasons of print).

```
R> NW.fac <- cut(surv$NW.Hnd,breaks=NW.breaks,include.lowest=TRUE)
R> as.numeric(NW.fac)
  [1] 3 4 3 4 3 3 3 4 3 3 2 4 3 3 4 4 4 5 4 3 4 4 5 4 3 3 4 4 2 3 5
 [32] 3 2 3 4 1 3 5 5 3 3 5 4 3 4 5 3 2 3 4 5 3 4 3 3 4 3 3 3 4 2 3
 [63] 2 3 3 3 3 4 3 5 3 3 3   --snip--
R> NW.cols <- ryc[as.numeric(NW.fac)]
R> NW.cols
  [1] "#BC7700" "#D5B200" "#BC7700" "#D5B200" "#BC7700" "#BC7700"
  [7] "#BC7700" "#D5B200" "#BC7700" "#BC7700"   --snip--
```

You're ready to plot; the result of the following is given on the left of Figure 25-4:

```
R> plot(surv$Wr.Hnd,surv$Height,col=NW.cols,pch=19)
```

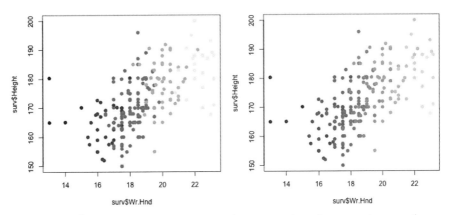

Figure 25-4: *Illustrating two ways to assign color to points based on a continuous value: via categorization (left) and via normalization (right)*

## Via Normalization

Using categorization to index a continuum with color is a little unsophisticated. There are plenty of ways you can bin your observations, for example, so your plot might look very different from the same plot designed by someone else. In a computational sense, it's more accurate (not to mention elegant) to leave your continuous data as is.

Recall the built-in gray palette mentioned in Section 25.1.2. This function behaved a little differently from the others. Instead of simply asking for a number of colors from the specified palette, you're required to provide a numeric vector of values to tell R, on a continuous scale from 0 through 1, how "far along" the palette to go. This type of behavior suits the current task perfectly, since your raw data are also on a continuous scale. To implement it, you need two things: a way to create a palette that will behave like gray and a *normalized* version of your continuous values that fall within the acceptable standardized range of 0 to 1 inclusive.

The colorRamp function allows you to create your palette and is used in the same way as colorRampPalette, but the result is a color palette function that expects a numeric vector as stated. You'll see that in a moment. To transform a collection of $n$ original values $\{x_1, \ldots, x_n\}$ to, say, $\{z_1, \ldots, z_n\}$, where $0 \le z_i \le 1; i = 1, \ldots, n$, you can employ the following equation:

$$z_i = \frac{x_i - \min x_i}{\max x_i - \min x_i} \tag{25.1}$$

Let's make that an R function by writing the following in the R editor:

```
normalize <- function(datavec){
    lo <- min(datavec,na.rm=TRUE)
    up <- max(datavec,na.rm=TRUE)
    datanorm <- (datavec-lo)/(up-lo)
    return(datanorm)
}
```

Based on a vector datavec as its only argument, normalize implements Equation (25.1), using the optional na.rm argument to ensure any missing values in datavec don't contaminate the calculation of the minimum and maximum values (see Section 13.2.1).

Import normalize and enter the following, which shows the original non-writing handspan values (from the object surv you created earlier) and their corresponding normalized values (output snipped for brevity):

```
R> surv$NW.Hnd
 [1] 18.0 20.5 18.9 20.0 17.7 17.7 17.3 19.5 18.5 17.2 16.0 20.2
[13] 17.0 18.0 19.2 20.5 20.9 22.0 20.7    --snip--
R> normalize(surv$NW.Hnd)
 [1] 0.50000000 0.72727273 0.58181818 0.68181818 0.47272727
 [6] 0.47272727 0.43636364 0.63636364    --snip--
```

Now, you need to create a new version of the color palette with colorRamp.

```
R> NW.pal2 <- colorRamp(colors=c("red4","yellow2"))
```

Generate the corresponding colors for each observation based on the normalized data.

```
R> ryc2 <- NW.pal2(normalize(surv$NW.Hnd))
```

If you actually look at the returned object in ryc2, you'll note it's a matrix of RGB triplets corresponding to each normalized value you supplied to your colorRamp function NW.pal2 (noninteger values end up being coerced to integers). These need to be converted to hex codes before you can use them in plotting. Using rgb just as you saw in Section 25.1.1, you get the vector you need (snipped for print).

```
R> NW.cols2 <- rgb(ryc2,maxColorValue=255)
R> NW.cols2
 [1] "#BC7700" "#D3AD00" "#C48A00" "#CEA200" "#B97000" "#B97000"
 [7] "#B66700" "#CA9700" "#C18100" "#B56500"    --snip--
```

Note the difference between the hex codes you obtain here in NW.cols2 and those in NW.cols. Here, you get a hex code for each unique value, but for the categorized NW.cols, you have only one hex code for each bin (so just $k = 5$ colors).

This line produces the image on the right of Figure 25-4.

```
R> plot(surv$Wr.Hnd,surv$Height,col=NW.cols2,pch=19)
```

In terms of this relatively simple example, the visual difference between the two approaches is minimal, though looking closely you can indeed pick out the smoother color transition in the normalized version. As you increase $k$ when using the categorization technique, the visual result will become closer to that of the normalization approach. That said, the normalization approach should generally be preferred, since it more closely fits the continuous nature of the values you're trying to visualize, and it works more effectively for values with a skewed distribution or when you're working with a complex color palette.

### 25.1.5   Including a Color Legend

Now that you can use color to significant effect in your plots, you need a legend to reference the color scale. It's possible to create a legend using base R tools alone, but it can be simpler to use contributed functionality in R instead.

One useful function for this is the colorlegend command. This is found in the shape package (Soetaert, 2014), so first download and install shape from CRAN. The following code then loads the package, reproduces the most recent plot (based on the surv object created earlier and shown on the right of Figure 25-4) with some tidier axis titles, and draws the color strip legend:

```
R> library("shape")
R> plot(surv$Wr.Hnd,surv$Height,col=NW.cols2,pch=19,
        xlab="Writing handspan (cm)",ylab="Height (cm)")
R> colorlegend(NW.pal(200),zlim=range(surv$NW.Hnd),zval=seq(13,23,by=2),
               posx=c(0.3,0.33),posy=c(0.5,0.9),main="Nonwriting handspan")
```

This result is given on the left of Figure 25-5.

The colorlegend functions assumes that you already have a plot present in an active graphics device, so you need to have one created first. The first thing you supply to colorlegend is the color span of the values you want to reference. This is easiest with a color palette function like those listed in the help file ?rainbow or created using colorRampPalette—in other words, a function that takes an integer value telling it how many colors to generate. Doing so with a large number of colors gives you a smooth color strip, so I use NW.pal(200). Next, you provide colorlegend with the range of the values that will be referenced by the legend using zlim, in this case, the range of the nonwriting handspans range(surv$NW.Hnd). The zval argument takes in the values that you want to mark off on the legend. The values of a sequence between 13 and 23, in steps of 2, are marked off.

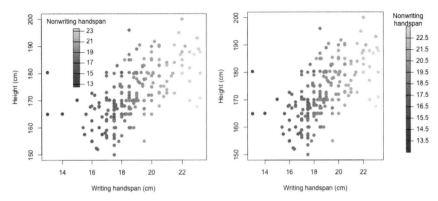

Figure 25-5: Two examples of implementing a color strip legend using the colorlegend function from the contributed shape package

The positioning and sizing of the color legend are done using the posx and posy arguments. Rather than taking user coordinates, each of these must be a vector of length 2 describing, in *relative device coordinates*, the horizontal (posx) and vertical (posy) lengths of the strip. In this example, posx=c(0.3,0.33) tells the function to draw the width of the legend from 30 percent of the left of the device to 33 percent so that the width is 3 percent of the overall device and positioned to the left of the center. Setting posy=c(0.5,0.9) says you want the length of the strip to span 40 percent of the device, from 50 percent of the way up from the bottom to 90 percent and up. Lastly, you can add the title to the legend by supplying a character string to main.

You'll probably need to experiment a bit in a trial-and-error fashion to get the positioning and sizing (and appropriate tick marks using zval) you want. The device-specific nature of posx and posy means that if you resize your device, you might well need to reevaluate the values of these arguments.

If you wanted the legend to appear outside the default plot region, you could easily use the xlim argument in the initial call to plot to widen the horizontal size of the plot, giving you extra space to draw a full-length legend. Alternatively, you could've changed the figure or outer margin spacing (refer to Section 23.2) to give you enough room to put the legend outside the plot region. This next chunk of code does just that by widening the right margin, replotting the scatterplot, and inserting a color legend into that extra space.

```
R> par(mar=c(5,4,4,6))
R> plot(surv$Wr.Hnd,surv$Height,col=NW.cols2,pch=19,
       xlab="Writing handspan (cm)",ylab="Height (cm)")
R> colorlegend(NW.pal(200),zlim=range(surv$NW.Hnd),zval=13.5:22.5,digit=1,
       posx=c(0.89,0.91),main="Nonwriting\nhandspan")
```

The result is given on the right of Figure 25-5. The legend is narrower than before, taking up only 2 percent of the device width on the right with posx=c(0.89,0.91). With no specification of posy, colorlegend has used the default of c(0.05,0.9), giving a color strip that spans almost the entire height of the device. The tick marks and labeling of the new legend are now placed in increments of 1 from 13.5 to 22.5; note that to display decimal places (in other words, *significant digits*), you need to increase the digit argument from its default, 0, to reveal them. Here, digit=1 prints the tick mark labels to one decimal place.

There are more properties that you can control with these legends, including labeling style and tick mark positioning; see the ?colorlegend help file for details. You may also like to investigate the similarly named function color.legend in the contributed plotrix package (Lemon, 2006) for a slightly different take on drawing color legends on existing R plots.

### 25.1.6  Opacity

Another useful skill is the ability to specify the opacity of any of the colors and color palettes discussed so far. All functions that provide the user with hex codes have an optional argument alpha, the valid range of which depends on the function (a quick check of the corresponding documentation will tell you). For example, the rgb function uses maxColorValue to set the upper bound on opacity, and palette functions like rainbow all use the normalized range from 0 through 1 (just like in the ggplot2 plots created throughout Chapter 24).

By default, R assumes full opacity when you're creating colors. However, hex codes change slightly when opacity is explicitly set using alpha. Rather than six characters after the #, eight will appear, with the last two containing the additional opacity information. Consider the following lines of code, which generate four different versions of red: default, zero opacity, 40 percent opacity ($0.4 \times 255 = 102$), and full opacity, respectively:

```
R> rgb(cbind(255,0,0),maxColorValue=255)
[1] "#FF0000"
R> rgb(cbind(255,0,0),maxColorValue=255,alpha=0)
[1] "#FF000000"
R> rgb(cbind(255,0,0),maxColorValue=255,alpha=102)
[1] "#FF000066"
R> rgb(cbind(255,0,0),maxColorValue=255,alpha=255)
[1] "#FF0000FF"
```

Note that the first and last colors are identical; it's just that the last hex code explicitly specifies full opacity.

You can always adjust the opacity of any color you've already got with the alpha.f argument (which takes values in the range 0 through 1) of the ready-to-use adjustcolor function. The following line takes the default red hex code created by the first line in the previous example and turns it into a 40 percent opaque version (the third line in the previous code):

```
R> adjustcolor(rgb(cbind(255,0,0),maxColorValue=255),alpha.f=0.4)
[1] "#FF000066"
```

You briefly came across this command already in Section 24.2.1, when creating a transparent gray confidence interval for a LOESS-smoothed trend using base R graphics. This approach is also applicable to hex codes generated after you've used a built-in or custom palette function to obtain a vector of colors.

You'll put opacity to the test using the built-in quakes data frame, which consists of data on 1,000 seismic events near Fiji. Let's re-create the plot in Figure 13-6 on page 284 showing "number of detecting stations" against "event magnitude" and dress it up using color to identify the continuous "depth" data. Since there are many overlapping observations, reducing opacity of the individual points would be a good idea for visualization. The code

```
R> keycols <- c("blue","red","yellow")
R> depth.pal <- colorRampPalette(keycols)
R> depth.pal2 <- colorRamp(keycols)
```

sets up a custom three-color palette both ways (in other words, as a function expecting an integer, depth.pal, and as a function expecting a value between 0 and 1, depth.pal2; refer to Sections 25.1.3 and 25.1.4). Then, the following line uses the normalization approach, with the normalize function defined in Section 25.1.4, to obtain the appropriate colors for the points to be plotted, according to the "depth" variable of the data set:

```
R> depth.cols <- rgb(depth.pal2(normalize(quakes$depth)),maxColorValue=255,
                     alpha=0.6*255)
```

The request for 60 percent opacity is made through alpha in the call to rgb. You can create the plot with the following call, which assigns the colors stored in depth.cols:

```
R> plot(quakes$mag,quakes$stations,pch=19,cex=2,col=depth.cols,
        xlab="Magnitude",ylab="No. of stations")
```

This plot affords another opportunity to showcase the colorlegend function from the shape package. Assuming you have shape already loaded in the current R session, the next line draws a corresponding color legend inside the plot region (on a default-size device):

```
R> colorlegend(adjustcolor(depth.pal(20),alpha.f=0.6),
               zlim=range(quakes$depth),zval=seq(100,600,100),
               posx=c(0.3,0.32),posy=c(0.5,0.9),left=TRUE,main="Depth")
```

Here you can see another demonstration of the use of adjustcolor, where the color sequence generated with the call to depth.pal(20) is then reduced to 60 percent opacity to match the plotted points. Again, posx and posy are used to position the legend, and the optional logical argument left is set to TRUE to make the tick marks and color legend labels appear on the left side of the strip. Figure 25-6 shows the final result.

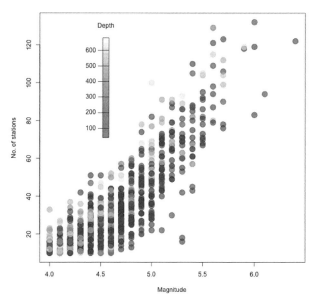

Figure 25-6: Altering the color opacity in a custom palette, used to index the continuous "event depth" observations in a plot of "number of stations" against "magnitude" for the quakes data set, and a corresponding color legend using colorlegend from the shape package

## 25.1.7    RGB Alternatives and Further Functionality

RGB triplets aren't the only way color can be represented in R. Other specifications include hue-saturation-value (HSV) and hue-chroma-luminance (HCL), available through the built-in hsv and hcl functions. These work in much the same way as rgb, where you specify the strength of influence of the three components and out pop corresponding character string hex codes that form valid R colors for any relevant plotting command. In fact, the HSV parameterization is what's used internally by the built-in palettes detailed in Section 25.1.2, such as rainbow and heat.colors.

Contributed functionality offers even more flexibility. The colorspace package (Ihaka et al., 2015), which translates between different color formats, is worth noting, as is RColorBrewer (Neuwirth, 2014), which is based directly on the well-received color schemes designed by Cynthia Brewer (see *http://colorbrewer2.org/*). RColorBrewer provides more options for creating palettes than are supplied by the built-in functionality colorRampPalette and colorRamp. That said, from an introductory perspective, you should find the use of RGB and the base R functionality as discussed here sufficient for most visual explorations of your data and models.

## Exercise 25.1

Ensure the car package is loaded. Revisit the Salaries data frame you looked at in Exercises 24.1 (page 622) and 24.2 (page 628) and take a look at the help file ?Salaries to remind yourself of the variables. Your task is to use color, point size, opacity, and point character type to reflect "years since Ph.D.," "sex," and "rank" in a scatterplot of "salary" against "years of service," by completing the following steps:

a.  Set up a custom color palette that goes from "black" to "red" to "yellow2". Create two versions of this palette—one that expects a number of colors and one that expects a vector of normalized values between 0 and 1.

b.  Create two vectors that will control point character and character expansion following the guidelines in (i) and (ii). Each of these can be achieved in a single line by vector subsetting/repetition based on a numeric coercion of the corresponding factor vector in the data frame.

    i.   Use the point characters 19, 17, and 15 to reference the three increasing academic ranks in that order.

    ii.  Use a character expansion of 1 for females and a character expansion of 1.5 for males.

c.  Use the normalize function defined in Section 25.1.4 to create a [0,1] normalized version of the range of values of the "years since Ph.D." variable. Then use the appropriate palette from (a) along with rgb to convert these to the required hex codes.

d.  Modify the vector of colors you just created in (c), adjusting opacity. Colors in the vector that correspond to females should be reduced to 90 percent opacity; colors that correspond to males should be reduced to 30 percent opacity.

e.  Now, start the plot; alter the default figure margins to be 5, 4, 4, and 6 lines wide on the bottom, left, top, and right, respectively. Plot salary on the y-axis against years of service on the x-axis. Set the corresponding point colors according to your vector from (d) and the point characters and character expansion according to your vectors from (b). Tidy up the x-axis and y-axis titles.

f.  Incorporate two separate legends following the guidelines in (i) and (ii). Both legends should be horizontal, and you should relax clipping to allow their placement in figure margins (refer to Section 23.2.3).

i. Place the first legend at the user coordinate given by x=-5 and y=265000. It should use the levels of the "rank" factor vector as the referencing text and pair these with the corresponding pch symbols as assigned. Include an appropriate title for the legend.

ii. The second legend should be placed next to the first, using an *x*-coordinate of 40 and the same *y*-coordinate value. This legend should show two points, both red and of type 19, but reference the two levels of sex by altering the character expansion and opacity of these to reference points as assigned.

g. Lastly, ensure the shape package is loaded and use the colorlegend function along with 50 colors generated from the appropriate palette from (a) to reference "years since Ph.D." You can leave the horizontal and vertical placements of the legend at their default values. The zlim range should simply be set to match the range of the observed data, and the tick mark values set via zval should be a sequence between 10 and 50, increasing in steps of 10. Include an appropriate title for the color legend.

After all this, my version of this plot is given here:

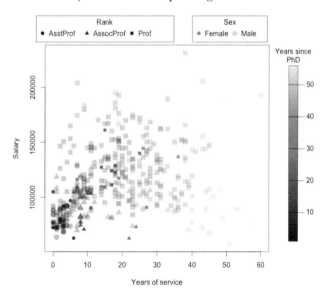

Your next task is a little different. The goal is to plot a standard normal probability density function but use color to shade in polygons underneath the curve to denote "distance from mean." To achieve this, complete the following:

h. Generate a vector of exactly 25 colors from the built-in palette terrain.colors and name it tcols. Then, using a reversed version of it obtained via tcols[25:1], append the two vectors together to form a new vector of length 50 containing the first 25 colors shading one way and then the same 25 shading the opposite way.

i. Next, create and store an evenly spaced sequence of exactly 51 values between −3 and 3 inclusive; name it vals. Use dnorm to calculate and store the corresponding 51 values of the standard normal density curve; name it normvals.

j. Draw the normal density curve by plotting the values in (i) as a line (recall type="l"). In the same call to plot, use knowledge from Chapter 23 to set both the x-axis and y-axis styles to be of type "i"; suppress both axis titles with empty strings; change the surrounding box to be an L shape; and suppress the drawing of the x-axis. Give the plot a suitable main title.

k. To shade the different colors underneath the curve, use a for loop, iterating through the integers 1 to 50. At each iteration, the loop should call polygon (refer to Section 15.2.3). Assuming your indexer is i, the vertices of each polygon should be formed by the vectors vals[rep(c(i,i+1),each=2)] and c(0,normvals[c(i,i+1)],0). Each polygon should suppress its border and be colored according to the relevant ith entry in your color vector of length 50 created in (h).

l. Lastly, ensure the shape package has been loaded and use your length 50 color vector to produce a color legend with default placement to reference "distance from mean." You can easily set the zlim and zval arguments in the call to colorlegend using vals. Include an appropriate title for the legend. For reference, my result is given here:

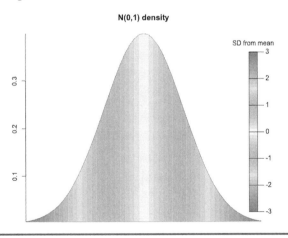

## 25.2 3D Scatterplots

This section will look at creating 3D scatterplots, which allow you to plot raw observations based on three continuous variables at once, as opposed to only two in a conventional 2D scatterplot. You'll then learn how to enhance your 3D scatterplot to represent more variables and make it easier to interpret. There are several ways to create three-variable scatterplots in R, but the go-to method is usually the `scatterplot3d` function in the contributed package of the same name (Ligges and Mächler, 2003).

### 25.2.1 Basic Syntax

The syntax of the `scatterplot3d` function is similar to the default `plot` function. In the latter, you supply a vector of *x*- and *y*-axis coordinates; in the former, you merely supply an additional third vector of values providing the *z*-axis coordinates. With that additional dimension, you can think of these three axes in terms of the *x*-axis increasing from left to right, the *y*-axis increasing from foreground to background, and the *z*-axis increasing from bottom to top.

Install and load the `scatterplot3d` package, and let's go straight into an example. Recall the famous iris flower data, which you first encountered in Section 14.4. This data set contains measurements on four continuous variables (petal length/width and sepal length/width) and one categorical variable (flower species); the iris data frame is immediately accessible from the R prompt, so there's no need to load anything. Enter the following so you have quick access to the measurement values that make up the data:

```
R> pwid <- iris$Petal.Width
R> plen <- iris$Petal.Length
R> swid <- iris$Sepal.Width
R> slen <- iris$Sepal.Length
```

The most basic 3D scatterplot of, say, petal length, petal width, and sepal width, is achieved with the following:

```
R> library("scatterplot3d")
R> scatterplot3d(x=pwid,y=plen,z=swid)
```

It's as simple as that—the result of this code is given on the left of Figure 25-7. Here you can observe a general positive relationship among all three plotted variables. There's also a clearly isolated cluster of observations in the foreground that have relatively large sepal widths but small petal measurements.

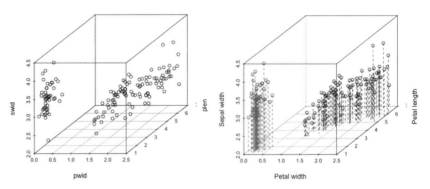

*Figure 25-7: Two 3D scatterplots of the famous* iris *data with petal width, petal length, and sepal width on the x-, y-, and z-axis, respectively. Left: Basic default appearance. Right: Tidying up titles and adding visual enhancements to emphasize 3D depth and legibility via color and vertical line marks.*

## 25.2.2 Visual Enhancements

It can be difficult to clearly perceive depth in the plotted cloud of points, even with the box and *x-y* plane grid lines that are drawn by default. For this reason, there are a couple of optional enhancements you can make to a scatterplot3d plot—coloring the points to help make the transition from foreground to background clearer and setting the type="h" argument to draw lines perpendicular to the *x-y* plane.

The right-hand plot in Figure 25-7 shows the plot with these enhancements and is the result of the following:

```
R> scatterplot3d(x=pwid,y=plen,z=swid,highlight.3d=TRUE,type="h",
                 lty.hplot=2,lty.hide=3,xlab="Petal width",
                 ylab="Petal length",zlab="Sepal width",
                 main="Iris Flower Measurements")
```

xlab, ylab, zlab, and main control the corresponding titles of the three axes and the plot itself.

The vertical lines make reading the values of the points much easier. By default, those lines in a type="h" plot are solid, but you can alter this with the lty.hplot argument (which behaves in the same way as the standard graphical parameter lty); setting lty.hplot=2 requests dashed lines. Similarly, you can alter the line type of the "nonvisible" sides of the box; setting lty.hide=3 instructs the plot to draw those lines as dotted.

Setting highlight.3d=TRUE emphasizes 3D depth by applying color transitioning from red to black based on the *y*-axis position of a point. This is useful, but there's an important consequence—it means you can no longer use color to represent a fourth variable with such a plot.

Along that line of thought, remember that the iris data has a fourth continuous variable, sepal length (stored as slen in Section 25.2.1), that you're not displaying in either of the plots in Figure 25-7. You're also not displaying the categorical variable of flower species, so let's fix that. First, set up a color band for the missing measurement variable, using your knowledge of having color palettes reference a continuous variable from Section 25.1.4.

```
R> keycols <- c("purple","yellow2","blue")
R> slen.pal <- colorRampPalette(keycols)
R> slen.pal2 <- colorRamp(keycols)
R> slen.cols <- rgb(slen.pal2(normalize(slen)),maxColorValue=255)
```

Note that for the last line to run, you'll need to have the normalize function defined in Section 25.1.4 available in your current session.

The following code produces the 3D scatterplot, which also uses the pch argument to distinguish among the three different species:

```
R> scatterplot3d(x=pwid,y=plen,z=swid,color=slen.cols,
                 pch=c(19,17,15)[as.numeric(iris$Species)],type="h",
                 lty.hplot=2,lty.hide=3,xlab="Petal width",
                 ylab="Petal length",zlab="Sepal width",
                 main="Iris Flower Measurements")
```

I've used the vector c(19,17,15), with the numeric coercion of the iris$Species vector passed to the square brackets, to pair pch character numbers as follows: 19 with *Iris setosa* (the first level of the factor), 17 with *Iris versicolor* (the second level), and 15 with *Iris virginica* (the third level), respectively (refer to Figure 7-5 on page 133 for the different types of point characters).

You can then insert a legend referencing species with a familiar call to legend.

```
R> legend("bottomright",legend=levels(iris$Species),pch=c(19,17,15))
```

And with a little experimentation, you can include a color strip legend too (making sure you've loaded the shape package so you have access to the colorlegend function as per Section 25.1.4).

```
R> colorlegend(slen.pal(200),zlim=range(slen),zval=5:7,digit=1,
               posx=c(0.1,0.13),posy=c(0.7,0.9),left=TRUE,
               main="Sepal length")
```

The final result of all this is the image in Figure 25-8.

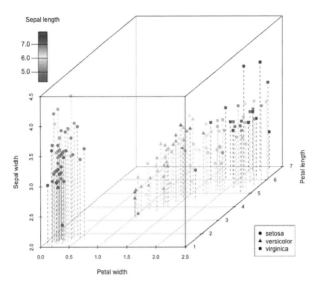

*Figure 25-8: A 3D scatterplot of the famous* `iris` *data, displaying all five present variables with the additional use of color (for sepal length) and point character (for species). See Plate V in the color insert.*

With the creative use of color and point type, you've now been able to display the five-dimensional data in a single 3D scatterplot. This reveals important information about the data. For instance, you can now identify *Iris setosa* as the clearly separate group of points in the foreground and see that while *Iris setosa* tend to have smaller petal widths and lengths and larger sepal widths than the other two species (especially *Iris versicolor*), the purple coloring at the lower end of the scale suggests they tend to have smaller sepal lengths.

## Exercise 25.2

Ensure the `scatterplot3d` library has been loaded in your current R session.

a.  Turn your attention back to the `diabetes` data frame found in the `faraway` package (you first looked at these data in Section 21.5.2). Your goal is to produce a `scatterplot3d` plot of weight, hip, and waist measurements as per the following guidelines:
    –  Hip, waist, and weight variables should correspond to the *x*-axis, *y*-axis, and *z*-axis, respectively; provide neat axis titles.
    –  Use built-in functionality to ensure the 3D depth is highlighted by color.

- Choose two different point characters to reflect gender.
- Place a simple legend referencing these two point characters and gender in the blank space in the upper-left area.

b. Create a 3D scatterplot of the built-in `airquality` data, which you first met in Section 24.2.2, according to the following guidelines:
- Create a copy of the data frame using `na.omit` to remove all rows that contain missing values and work with this copy.
- Plot wind speed and solar radiation against the *x*- and *y*-axes, respectively, using the *z*-axis to plot temperature.
- Apply vertical dotted lines reaching up from the *x-y* plane to each observation.
- The data in `airquality` are comprised of measurements taken over five months, from May to September. Each plotted point should take on the corresponding `pch` value from 1 to 5 respective to the order of these five months.
- With a vector of 50 colors generated from the built-in `topo.colors` palette, use the categorization approach to ensure each plotted point is colored according to its ozone value.
- Set a legend to reference the five point types according to month.
- Set a color legend (using functionality from the `shape` package) to reference the ozone value accordingly.
- Ensure the plot has neat axis, main, and legend titles.

## 25.3  Preparing a Surface for Plotting

In the rest of this chapter, you'll look at three types of 3D plots geared to visualize a *bivariate surface*. Such plots are required when you have two variables, based on which a function, estimate, or model has been defined, and you want to use the third available axis (in other words, the *z*-axis) to map out the resulting surface. You've seen examples of bivariate functions already, through the response surfaces for the `mtcars` data in Section 21.5.4 (where you looked at mean MPG as a function of car weight and horsepower) and through the study of diagnostic tools for linear regression models in Section 22.3.6 (where you saw how Cook's distance can be expressed as a function of residual and leverage).

Before you look at producing these plots, it's important to understand how they're created in R. The function/estimate/model of interest should be thought of as a plane or surface that can vary according to continuous, two-dimensional *x-y* coordinates. Plotting a completely continuous surface is technically impossible since that would require you to evaluate the function at an infinite number of coordinates. Therefore, evaluation of the surface is typically performed on a finite *grid* of evenly spaced coordinates along both the *x*- and *y*-axes. The result of the function at each unique pair of coordinates is stored in a corresponding position in an appropriately sized matrix (the size of which depends directly upon the resolution of the evaluation grid in the *x*- and *y*-axes), generically referred to as the *z-matrix*.

Since all the traditional R graphics commands that plot these bivariate functions operate in the same way—using this *z*-matrix—it's critical to understand how this matrix is constructed, arranged, and interpreted by those commands to ensure you're correctly drawing the outcome. In this section, you'll ready yourself for the specific plot types looked at in the remainder of this chapter by getting familiar with this construct in a hypothetical situation.

### 25.3.1    Constructing an Evaluation Grid

Say you have a bivariate function that results in a continuous surface that's defined between 1 and 6 on the *x*-axis and 1 and 4 on the *y*-axis. You can define evenly spaced sequences over each of these coordinate ranges using seq; for simplicity, let's just do so in straight-out integers.

```
R> xcoords <- 1:6
R> xcoords
[1] 1 2 3 4 5 6
R> ycoords <- 1:4
R> ycoords
[1] 1 2 3 4
```

What this implies is that you're planning to draw your surface based on evaluation of the bivariate function of interest upon the grid of *x-y* values defined by 24 unique positions.

When passed two vectors, the built-in expand.grid function explicitly generates all unique coordinate pairs by simply repeating each value in the second vector against the entire length of the first vector.

```
R> xycoords <- expand.grid(x=xcoords,y=ycoords)
```

The result is stored as a two-column data frame with 24 rows. If you look at xycoords object in the R console, you'll see x values from 1 to 6 all paired with a repeated y value of 1, then x from 1 to 6 paired with y as 2, and so on.

In practice, what you'd now do is use the evaluation grid coordinates in xycoords to calculate the result of your bivariate function. For this hypothetical example, let's just say that your bivariate function has resulted in the 24 letters *a* to *x*, corresponding to the order of the unique evaluation coordinates in xycoords. To make this even clearer, take a look at the following column-bind of the hypothetical function result with each evaluation coordinate (note that the ready-to-use `letters` object in R allows you to generate letters of the alphabet quickly):

```
R> z <- letters[1:24]
R> cbind(xycoords,z)
   x y z
1  1 1 a
2  2 1 b
3  3 1 c
4  4 1 d

--snip--

21 3 4 u
22 4 4 v
23 5 4 w
24 6 4 x
```

What this emphasizes is that each unique *x-y* evaluation coordinate, expressible via `expand.grid`, will have a *z* value associated with it. All together, these *z* values define the resulting surface.

### 25.3.2   Constructing the z-Matrix

The 3D plots used to visualize a bivariate function require the *z* values corresponding to the *x-y* evaluation grid in the form of an appropriately constructed matrix. The size of the *z*-matrix is determined directly by the resolution of the evaluation grid; the number of rows corresponds to the number of unique *x* grid values, and the number of columns corresponds to the number of unique *y* grid values.

You therefore need to take a little care turning your calculated *z* values into a matrix. When your vector of *z*-axis values corresponds to the evaluation grid arranged in the standard `expand.grid` fashion (in other words, where coordinates are stacked by increasing *x* values and repeated *y* values), be sure that your resulting *z*-matrix is filled in the default column-wise fashion (see Section 3.1.1), with the number of rows and columns being exactly representative of the number of values in each of the *x*- and *y*-value sequences, respectively (xcoords and ycoords shown earlier). In the current example, you know that the resulting *z*-matrix needs to be of size 6 × 4 because there are six *x* locations and four *y* locations.

The following is the correct matrix representation of the hypothetical "function result" vector z:

```
R> nx <- length(xcoords)
R> ny <- length(ycoords)
R> zmat <- matrix(z,nrow=nx,ncol=ny)
```

```
R> zmat
     [,1] [,2] [,3] [,4]
[1,] "a"  "g"  "m"  "s"
[2,] "b"  "h"  "n"  "t"
[3,] "c"  "i"  "o"  "u"
[4,] "d"  "j"  "p"  "v"
[5,] "e"  "k"  "q"  "w"
[6,] "f"  "l"  "r"  "x"
```

### 25.3.3   Conceptualizing the z-Matrix

The most important thing to be gained from this section is an idea of how the z-matrix in its current arrangement translates to *x-y* coordinate-based plotting. Comparing zmat to the earlier output, you can see that moving down a column of zmat translates to an increase in the *x*-coordinate value for a given *y*-coordinate value. In other words, when this hypothetical surface of letters is plotted, moving down a column of the matrix corresponds to moving horizontally from left to right on the corresponding plot, given a particular vertical *y* position.

Figure 25-9 provides a conceptual diagram of this illustrative surface, indexed by zmat as per the 24 unique coordinates defined via xcoords and ycoords. (The code to produce this is included in the R script files for this book, which can be found at *https://www.nostarch.com/bookofr/*.)

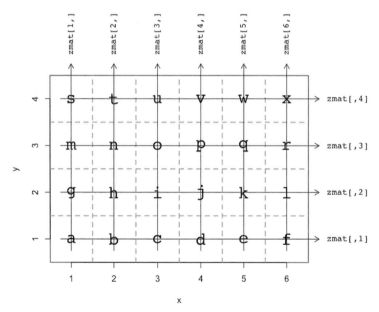

Figure 25-9: Conceptual diagram of a z-matrix for plotting bivariate functions, based on a 6 × 4 coordinate grid

As you now progress to plot some actual surfaces of interest, you should keep the concept of the z-matrix as illustrated in Figure 25-9 in mind. The 6 × 4 grid used in this hypothetical example is coarse. In practice, you'll usually use far finer grids in terms of the resolution of the *x*- and *y*-sequences to improve the visual appearance of the surface.

## 25.4   Contour Plots

One of the most common plots used to display a surface based on evaluation of a function over a grid of bivariate coordinates is the *contour plot*. Contour plots are perhaps most easily explained as a series of lines—the contours—drawn over the 2D evaluation grid, with each contour marking off a specific level of the surface of interest.

### 25.4.1   Drawing Contour Lines

Based on a given numeric *z*-matrix, the R function contour is what's used to produce the contours connecting *x-y* coordinates that share the same *z* value.

#### Example 1: Topographical Map

For an example, you'll use another ready-to-use data set—the volcano object. This data set is simply a matrix containing measurements of the elevation above sea level (in meters) of a dormant volcano over a rectangular area in the Auckland region of New Zealand; see the documentation in ?volcano for details. To view the topography, you need the volcano object (which is your *z*-matrix) and the relevant *x*- and *y*-coordinate sequences. In this case, just use integers corresponding to the size of the volcano matrix (row and column numbers can be obtained with a simple call to dim; see Section 3.1.3).

```
R> dim(volcano)
[1] 87 61
R> contour(x=1:nrow(volcano),y=1:ncol(volcano),z=volcano,asp=1)
```

The *x*- and *y*-sequences are provided to x and y, respectively, and the *z*-matrix to z. The optional argument asp=1, referring to the aspect ratio of the plot, forces a 1-to-1 unit treatment of the coordinate axes (this is relevant when the units have a physical size interpretation, like in plots of geographical regions—as is the case here).

Figure 25-10 shows the result of this example. By default, R automatically chooses the levels of z at which to draw the contours for an aesthetically pleasing result. Contours are also selectively labeled with their corresponding *z* value. Looking at the topography, you can see the highest peak is a rim on the left, marked by an oblong contour at 190 m, with a depression (at around 160 m) falling immediately to the right.

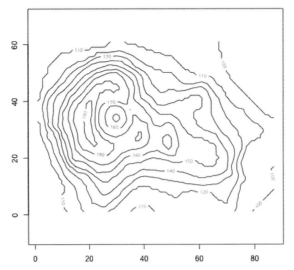

*Figure 25-10: Using* contour *to produce a topographic map of the* volcano *data*

Contours are able to show you not only the peaks and troughs in a surface like this but the "steepness" of any such features too. The closer together the contour lines lie, the more rapid the change in the overall level of the bivariate function.

### Example 2: Parametric Response Surface

As a different kind of example, consider the multiple linear model fitted to the mtcars data mentioned earlier—that is, of MPG modeled by horsepower, weight, and an interaction between the two predictors. As in Section 21.5.4, you can get the fitted model object with the following:

```
R> car.fit <- lm(mpg~hp*wt,data=mtcars)
R> car.fit

Call:
lm(formula = mpg ~ hp * wt, data = mtcars)

Coefficients:
(Intercept)           hp           wt        hp:wt
   49.80842     -0.12010     -8.21662      0.02785
```

The goal is to plot the response, mean mileage, as the previous function of horsepower and weight. To do this, you need to evaluate the mean MPG, according to the previous model, for a grid of horsepower and weight values. The following code does exactly that.

```
R> len <- 20
R> hp.seq <- seq(min(mtcars$hp),max(mtcars$hp),length=len)
```

```
R> wt.seq <- seq(min(mtcars$wt),max(mtcars$wt),length=len)
R> hp.wt <- expand.grid(hp=hp.seq,wt=wt.seq)
R> nrow(hp.wt)
[1] 400
R> hp.wt[1:5,]
          hp    wt
1   52.00000 1.513
2   66.89474 1.513
3   81.78947 1.513
4   96.68421 1.513
5 111.57895 1.513
```

First, this code sets up evenly spaced sequences (each of length 20, spanning the range of the observed data) in both hp and wt—these are your $x$- and $y$-sequences. This implies there will be $20 \times 20 = 400$ unique coordinates at which you'll be evaluating the fitted model; these coordinates are obtained using expand.grid as in Section 25.3.

Next, you can use predict to get the 400 corresponding mean MPG ($z$) values; since it's already a data frame in the required format, hp.wt can be passed directly to the newdata argument.

```
R> car.pred <- predict(car.fit,newdata=hp.wt)
```

Then, you simply need to arrange the resulting vector as the appropriate $20 \times 20$ $z$-matrix.

```
R> car.pred.mat <- matrix(car.pred,nrow=len,ncol=len)
```

Finally, you plot the result as contours, as shown in Figure 25-11.

```
R> contour(x=hp.seq,y=wt.seq,z=car.pred.mat,levels=32:8,lty=2,lwd=1.5,
           xaxs="i",yaxs="i",xlab="Horsepower",ylab="Weight",
           main="Mean MPG model")
```

In this call, you can see the use of the optional levels argument. Rather than let R automatically decide at which values of $z$ to show contours, you can supply a numeric vector to this argument with the specific levels at which to draw the lines. This numeric vector must be on the same scale as the resulting bivariate function of interest; here, I asked for contours at all integer levels from 32 through 8. I also employ the familiar arguments lty and lwd to control the appearance of the contour lines themselves, which are set here as dashed and slightly thicker than usual.

Furthermore, for contour plots in particular, you'll often want to deviate from the default axis limit style, because the small amount of additional "padding" space that's included in the default plot region (refer to Section 23.4.1) can be rather prominent—take another look at the volcano contour plot in Figure 25-10. As shown previously, setting xaxs and yaxs to "i" restricts all plotting to the exact limits imposed by x and y.

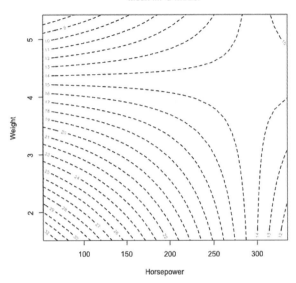

**Mean MPG model**

*Figure 25-11: Contour lines depicting the response surface based on a multiple linear model of MPG by horsepower and weight, from the* mtcars *data*

### Example 3: Nonparametric Bivariate Density Estimate (Earthquake Data)

Another useful role fulfilled by contour plots and the other plots in this chapter is to visualize bivariate density functions.

In Section 24.2.2, you looked at the idea of kernel density estimation (KDE) as a method by which to construct smooth estimates of the probability density functions of your data—essentially, sophisticated histograms. KDE extends to higher dimensions quite naturally so that you can also estimate the density of bivariate observations in the *x-y* plane. This again involves visualizing a *z*-matrix over a fixed grid of coordinates. For theoretical details on multivariate KDE, see Wand and Jones (1995).

Turn your attention back to the built-in quakes data frame and recall the plots of the spatial coordinates of the 1,000 seismic events (for example, Figure 13-1 on page 265 and Figure 23-1 on page 578). To estimate the probability density function of these points, you can use the kde2d function in the MASS package. Load MASS and execute the following line to produce the kernel estimate of the observed two-dimensional data:

```
R> quak.dens <- kde2d(x=quakes$long,y=quakes$lat,n=100)
```

You supply the bivariate data as the x and y arguments for the horizontal and vertical axes. The optional argument n is used to specify the number of evaluation coordinates (along each of the two axes) at which to actually return the estimated density surface. This defines the size of the matrix returned by a call to kde2d. Here, you've asked for KDE to be performed on a 100 × 100 evenly spaced grid over the range of the observed data.

The resulting object is simply a list with three members. The components accessed through $x and $y contain the evenly spaced evaluation grid coordinates in the corresponding axis directions, and $z provides you with the corresponding z-matrix. You can confirm by entering either quak.dens$x or quak.dens$y at the prompt that they are indeed increasing sequences spanning the ranges of the observed data. Entering the following confirms the size of the matrix of interest:

```
R> dim(quak.dens$z)
[1] 100 100
```

With that, you have all the ingredients you need to display contours of the KDE surface. The next line produces the default contour plot, given on the top left of Figure 25-12.

```
R> contour(quak.dens$x,quak.dens$y,quak.dens$z)
```

There are many more optional arguments available to contour for displaying your continuous surface. It can also be helpful to simultaneously view other data or raw observations (if they've been used in some way to create the surface, as is the case with bivariate KDE). The following code replots the quakes kernel estimate with unpadded axes, different contour levels to the defaults, and the raw observations; you can see the result on the top right of Figure 25-12:

```
R> contour(quak.dens$x,quak.dens$y,quak.dens$z,nlevels=50,drawlabels=FALSE,
           xaxs="i",yaxs="i",xlab="Longitude",ylab="Latitude")
R> points(quakes$long,quakes$lat,cex=0.7)
```

Rather than using levels to determine the exact levels at which to draw the contours (as you did with Example 2), you can use the nlevels argument to specify the *number* of levels to display, and the function will choose the specific values. This latest call to contour requested 50 levels to be drawn. You can suppress the automatic labeling of the displayed contours by setting drawlabels=FALSE, also done here, followed by a call to points to add the original observations to the image. Naturally, the smooth contours delineating the nonparametric density estimate reflect the heterogeneous spatial patterning of the data.

Changing the appearance of your plotted contours needn't be done universally; you can also alter the appearance of each individual contour level. This can be handy if, for example, you want to display contours at a handful of specific levels without the default labeling (to focus on the shape of the surface itself) but still want to be able to discern the values of those contours. You might also want to superimpose contours on an existing plot that already depicts other data or model-based results of interest. The third plot of the earthquake KDE surface, given on the bottom of Figure 25-12, shows how you can achieve both of these things.

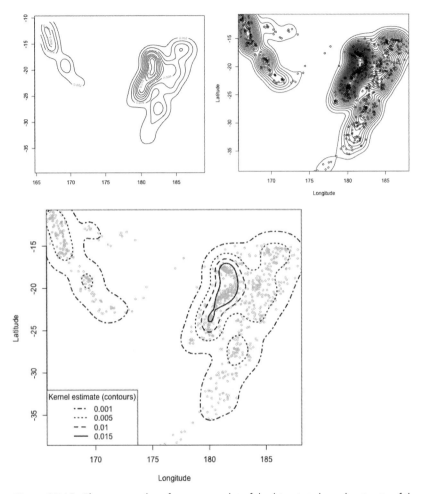

Figure 25-12: Three examples of a contour plot of the bivariate kernel estimate of the probability density function of the spatial earthquake locations given in the quakes data set

To start the plot, the spatial locations of the earthquake data are drawn as half-size gray dots using plot, the style of the axes are set using xaxs and yaxs to remove the artificial padding around the edges of the plot region, and axis titles are added.

```
R> plot(quakes$long,quakes$lat,cex=0.5,col="gray",xaxs="i",yaxs="i",
        xlab="Longitude",ylab="Latitude")
```

Then, before calling contour, store the desired levels at which to draw contours in a vector named quak.levs (again, choosing appropriate contour levels depends entirely on what kind of surface you're plotting; you need to be at least roughly aware of the values stored in the relevant z-matrix).

```
R> quak.levs <- c(0.001,0.005,0.01,0.015)
```

Now, remember that by default, contour refreshes the graphics device and starts a new plot, but you want to avoid that when adding contour lines to an existing plot. To do so, you need to explicitly specify add=TRUE. You then provide the four specified levels in quak.levs to levels and suppress labeling with drawlabels=FALSE. To control the appearance of contour lines at individual levels, you supply the sequence of integers 4:1 to lty, the first entry of which, 4, defines the line type of the contour at $z = 0.001$. The second entry, 3, specifies the line type of the $z = 0.005$ contour, and so on. Lastly, set all drawn contours to double-thickness with the single supplied value lwd=2. (You could supply a vector with four elements here too, if you want differing line thicknesses for the different contours. The same element-wise contour specification extends to other relevant aesthetics, such as color via col.)

```
R> contour(quak.dens$x,quak.dens$y,quak.dens$z,add=TRUE,levels=quak.levs,
           drawlabels=FALSE,lty=4:1,lwd=2)
```

As a final touch, since the automatic labeling was suppressed in contour, add a legend in the bottom-left corner of the plot region, referencing the values of the contours through the four different line types.

```
R> legend("bottomleft",legend=quak.levs,lty=4:1,lwd=2,
          title="Kernel estimate (contours)")
```

**NOTE**      *Many built-in and contributed base R plotting functions that by default initialize, refresh, or open a new plot include an add argument as shown here. This allows you to use the graphics produced by these functions as additions to an already existing graphic. Look in the relevant help file to see whether this is the case for a given command.*

### 25.4.2   *Color-Filled Contours*

For a straightforward variation on the contour plot, you can use color to fill the gaps between the different levels that are drawn. Combined with a color legend, this removes the need to label the contour lines and in certain cases can make it easier to visually interpret fluctuations in the plotted $z$-matrix surface.

The filled.contour function does this for you. You need to supply the increasing sequences of grid coordinates in both the $x$-axis and $y$-axis directions, as well as the corresponding $z$-matrix, to the arguments x, y, and z in the same way as in contour. The easiest way to specify the colors is to supply a color palette to the color.palette argument (which defaults to the built-in cm.colors palette; refer to Figure 25-2/Plate III), and R does the rest.

Let's use the mtcars response surface from Example 2 for a quick demonstration. If you don't already have them in your current workspace, use the code from Section 25.4.1 to obtain the relevant fitted multiple linear regression model, the evaluation grid coordinates, and the prediction thereof. With the objects hp.seq, wt.seq, and car.pred.mat defined as earlier, the following call produces Figure 25-13:

```
R> filled.contour(x=hp.seq,y=wt.seq,z=car.pred.mat,
                  color.palette=colorRampPalette(c("white","red4")),
                  xlab="Horsepower",ylab="Weight",
                  key.title=title(main="Mean MPG",cex.main=0.8))
```

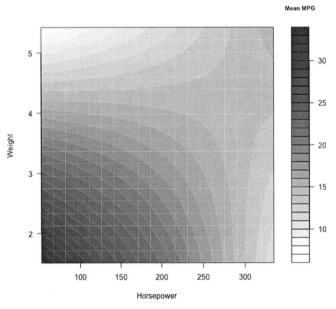

*Figure 25-13: Filled contour plot of the response surface for the fitted multiple linear model of the* mtcars *data*

Note in this plot that the default color palette hasn't been used. Instead, you've supplied a custom palette (produced as a direct result of an appropriate call to colorRampPalette; refer to Section 25.1.3) to the relevant argument, moving from white at the lower end to dark red at the upper end. Note also that although the *x*-axis and *y*-axis titles are provided as usual to xlab and ylab, you have to supply the title for the color legend in a particular way—inside a call to title to the key.title argument. This is because filled.contour actually produces two plots, one for the image itself and one for the color legend, and makes use of the layout command to place them next to one another.

This internal use of layout isn't directly a problem, but, as you saw in Section 23.1.4, it complicates matters somewhat if you want to annotate the filled contour plot after the fact (by, for example, adding points to an existing graphic) since the original user coordinate system is lost.

Turn your attention back to the two-dimensional kernel estimate of the spatial quakes data (use the code from Section 25.4.1 to re-create it if you haven't already got the quak.dens object in your workspace). The following code creates a filled contour plot of the density surface using the built-in topo.colors palette and modifies the number of drawn levels from the default of 20 to 30. In the same call, you can superimpose the points of the raw observations onto the image through special use of the optional plot.axes argument. Figure 25-14 shows the result.

```
R> filled.contour(x=quak.dens$x,y=quak.dens$y,z=quak.dens$z,
                  color.palette=topo.colors,nlevels=30,xlab="Longitude",
                  ylab="Latitude",key.title=title(main="KDE",cex.main=0.8),
                  plot.axes={axis(1);axis(2);
                             points(quakes$long,quakes$lat,cex=0.5,
                                    col=adjustcolor("black",alpha=0.3))})
```

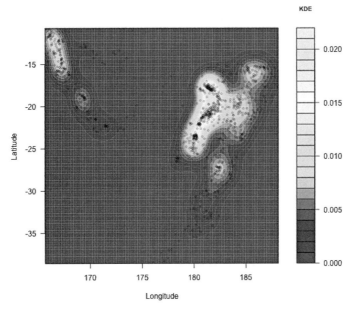

*Figure 25-14: Filled contour plot of the kernel estimate of the probability density function of the spatial quakes data, with raw observations superimposed. See Plate VI in the color insert.*

Take a look at the way in which plot.axes is used; it effectively takes a chunk of code. When plot.axes is invoked, you must explicitly tell it to mark the x- and y-axes if you want the labeled tick marks to remain. This is done with two calls to axis (refer to Section 23.4.3—axis(1) gives x, and axis(2) is used for y). You add the data points with a call to points; in this example, these are instructed to plot at half size, with 30 percent opacity imparted with adjustcolor. Since you're supplying multiple separate commands at once to the `plot.axes` argument, each command needs to be separated by a semicolon (;) inside braces ({ }).

Annotation of a filled contour plot in this fashion requires a little more forethought since you're required to manually add the axes via calls to `axis` and perform all subsequently desired plotting actions within the call to `filled.contour`. It won't work to, for example, produce a filled contour plot like the `quakes` KDE surface and then call `points` as a separate line of code. If you try it, you'll see the observed data points unable to align correctly with their original user coordinates as indicated on the axes.

## Exercise 25.3

Remember that you inspected various multiple linear regression models of the cost of nuclear power plant construction in Chapters 21 and 22. The goal now will be to visually assess the impact of including/excluding an interactive term between two continuous predictors using contours. Revisit the `nuclear` data set, available when you load the `boot` package, and bring up the help file to refresh your memory of the variables present.

a. Fit and summarize two linear models with construction cost as the response variable according to the following guidelines:
   i. The first should account for main effects of the two predictors concerning the date of issue of the construction permit and plant capacity.
   ii. The second, in addition to the two main effects, should include an interaction between permit issue date and capacity.

b. Set up appropriate $z$-matrices for plotting each of these response surfaces. Each one should be based on a $50 \times 50$ evaluation grid constructed using evenly spaced sequences in the capacity and date variables.

c. Specify `mfrow` in `par` so that you can display default contour plots for the two response surfaces from (a)(i) and (a)(ii) next to one another. Do they appear similar? Does this tie in with the statistical significance (or lack thereof) of the interaction term in (a)(ii)?

d. To directly compare the two surfaces, use your choice of built-in color palette to produce a filled contour plot of the main-effects-only model and superimpose the contour lines of the interactive model on it. Take note of the following:
   – This plot is achieved in a single call to `filled.contour`. Recall the special way you use `plot.axes` to draw additional features on an existing color-filled contour plot.

- The contour lines of the interactive model can be added with an appropriate call to contour. Recall the use of the optional argument add.
- The superimposed contours should be dashed lines of double thickness.
- The x- and y-axes should be included and given tidy titles.
- Add some brief text describing the filled contours versus the contour lines, with reference to the two versions of the construction cost model and with an additional call to text that makes use of a single mouse-clicked location from locator (see Section 23.3). Note that this call will need to fully relax clipping for the text to be visible in any of the margins.

My result is shown here.

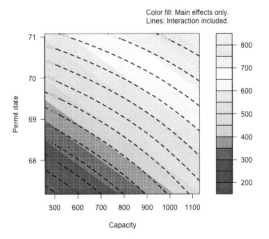

e. Another built-in data frame in R, faithful, contains observations of waiting times and durations of eruptions of the Old Faithful geyser in Yellowstone National Park, Wyoming. See the documentation in ?faithful for details. Plot the data with duration on the y-axis and waiting time on the x-axis.

f. Estimate the bivariate density of these data via KDE using a 100 × 100 evaluation grid and produce a default contour plot thereof.

g. Create a filled contour plot of the kernel estimate using a custom palette that ranges from "darkblue" to "hotpink"; include the raw data as half-size gray points. Label the axes and titles appropriately.

h. Replot the raw data as gray, 3/4-sized, type 2 point characters; set the style of the axes to restrict to exactly the ranges of the observed data; and ensure tidy axis titles and a main title. To this plot, add the contour lines of the density estimate at the specific levels obtained in a sequence from 0.002 to 0.014 in steps of 0.004. Suppress the labeling of the contours. The contour lines should be dark red and increase in line width thickness for higher levels of the density. Add a legend referencing the density level at each of these lines.

My plots for (g) and (h) are shown here.

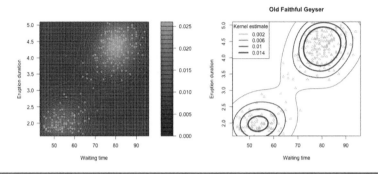

## 25.5 Pixel Images

A *pixel image* is arguably the most literal visual representation of a continuous surface approximated by a finite evaluation grid. Its appearance is similar to a filled contour plot, but an image plot gives you more direct control over the display of each entry of the relevant *z*-matrix.

### 25.5.1    One Grid Point = One Pixel

Consider each entry of your *z*-matrix as a little rectangle whose color depicts its relative value. These rectangles, or *pixels*, are exactly what's depicted as the cells formed by the dashed gray lines making up the conceptual diagram of the *z*-matrix in Figure 25-9 on page 656. This emphasizes the important fact that the fineness of your evaluation grid sequences (in both the *x*- and *y*-coordinate directions) directly defines the size of each pixel and therefore the smoothness of the resulting image. A smaller pixel means the *resolution* of the image is increased.

The built-in image function plots pixel images. Much as with contour, you supply your *x*- and *y*-axis evaluation grid coordinates as increasing sequences to the x and y arguments, with the corresponding *z*-matrix supplied to z. Going back to the volcano data set first looked at in Example 1 of Section 25.4.1, the following line produces Figure 25-15:

```
R> image(x=1:nrow(volcano),y=1:ncol(volcano),z=volcano,asp=1)
```

Note again that you use the optional argument asp=1 to enforce a one-to-one aspect ratio of the horizontal and vertical axes. This plot is comprised of exactly $87 \times 61 = 5307$ pixels; each one represents a particular entry in the volcano matrix. Visually, the reflection of this image in the contour plot of the same data in Figure 25-10 is clear.

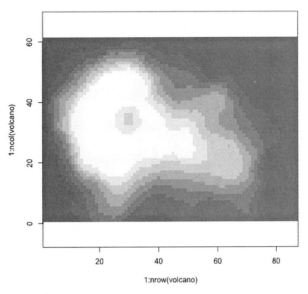

*Figure 25-15: Pixel image of the Auckland volcano topography*

The image command expects a vector of colors, usually supplied as hex codes from a palette, to be passed to its col argument. If this isn't specified, it defaults to heat.colors(12) using the built-in palette, as in the image plot of volcano. One immediate concern, however, is the lack of a color legend. Contributed tools such as the colorlegend function from the shape package (refer to Section 25.1.5) prove useful for these plots.

Return now to the mtcars response surface from Example 2 that fits the multiple linear regression model of MPG on horsepower and weight (and an interactive effect between the two predictors). The code for the necessary objects is reproduced here in a shortened form for convenience (refer to Section 25.4.1 for a fuller explanation of the operations):

```
R> car.fit <- lm(mpg~hp*wt,data=mtcars)
R> len <- 20
R> hp.seq <- seq(min(mtcars$hp),max(mtcars$hp),length=len)
R> wt.seq <- seq(min(mtcars$wt),max(mtcars$wt),length=len)
R> hp.wt <- expand.grid(hp=hp.seq,wt=wt.seq)
R> car.pred.mat <- matrix(predict(car.fit,newdata=hp.wt),nrow=len,ncol=len)
```

Just as earlier, you've set up a matrix of 400 elements in car.pred.mat, which is based on sequences of length 20 in both continuous predictors.

Now, make sure the shape package is loaded so you have access to the colorlegend function. The code that follows first sets up a custom palette of blue colors, sets new margin limits that widen the area on the rightmost axis, and then plots the predicted $20 \times 20$ response surface including a color legend; the result is given on the left of Figure 25-16.

```
R> blues <- colorRampPalette(c("cyan","navyblue"))
R> par(mar=c(5,4,4,5))
R> image(hp.seq,wt.seq,car.pred.mat,col=blues(10),
        xlab="Horsepower",ylab="Weight")
R> colorlegend(col=blues(10),zlim=range(car.pred.mat),zval=seq(10,30,5),
            main="Mean\nMPG")
```

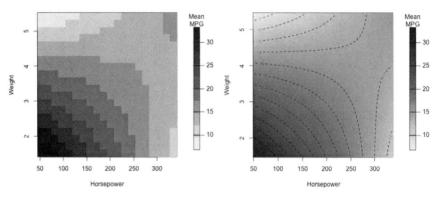

Figure 25-16: Two pixel images of the mtcars mean MPG response surface introduced in Example 2, with accompanying color legends. In terms of the evaluation grid in the horsepower and weight variables, the surface on the left has a resolution of $20^2$; the image on the right is based on a finer $50^2$ grid. Contours are superimposed upon the rightmost plot.

With a relatively coarse evaluation grid, the pixels making up the surface are prominent. You can easily increase the resolution of the parametric response surface by using finer sequences for the hp.seq and wt.seq evaluation grid. The code that follows does just that by increasing len to 50, overwriting the objects used previously:

```
R> car.fit <- lm(mpg~hp*wt,data=mtcars)
R> len <- 50
R> hp.seq <- seq(min(mtcars$hp),max(mtcars$hp),length=len)
R> wt.seq <- seq(min(mtcars$wt),max(mtcars$wt),length=len)
R> hp.wt <- expand.grid(hp=hp.seq,wt=wt.seq)
R> car.pred.mat <- matrix(predict(car.fit,newdata=hp.wt),nrow=len,ncol=len)
```

Then the right-hand image of Figure 25-16 is produced with the following code:

```
R> par(mar=c(5,4,4,5))
R> image(hp.seq,wt.seq,car.pred.mat,col=blues(100),
        xlab="Horsepower",ylab="Weight")
R> contour(hp.seq,wt.seq,car.pred.mat,add=TRUE,lty=2)
```

```
R> colorlegend(col=blues(100),zlim=range(car.pred.mat),zval=seq(10,30,5),
               main="Mean\nMPG")
```

The newly plotted surface consists of $50^2 = 2500$ pixels, as opposed to the previous image of merely $20^2 = 400$ pixels. The improvement in the picture is obvious. In plotting the new image, the number of colors used (from the custom blues palette) is increased to 100 to provide smoother color transitions. Note also the use of add in a call to contour to superimpose contour lines upon the image to provide further visual emphasis of the fluctuating surface over the evaluation grid. A legend is added with an appropriate call to colorlegend as a final touch.

### 25.5.2  Surface Truncation and Empty Pixels

Because of its one-to-one literal representation of the z-matrix, a pixel image is especially good when you want to plot a surface that fits irregularly over, or is smaller than, the standard rectangular evaluation grid spanning the x- and y-axes. To carefully demonstrate this kind of manipulation, let's turn to a new data set from the contributed spatstat package by Baddeley and Turner (2005). Install spatstat with a call to install.package("spatstat"). Note that spatstat has a number of dependencies; see Appendix A.2.3 if you have any trouble downloading and installing spatstat.

#### Example 4: Nonparametric Bivariate Density Estimate (Chorley-Ribble Data)

Once spatstat is installed and loaded in your current R session with a call to library("spatstat"), inspect the help file brought up by entering ?chorley at the prompt. This details the Chorely-Ribble cancer data—spatial locations of 1,036 cases of cancer of the larynx and lung collected in the late 1970s and early 1980s in a particular region of England (data first analyzed by Diggle, 1990). The chorley object is of a special class specific to spatstat (a "ppp" object—*planar point pattern*), but its components can be extracted just as if you're referencing members of a named list.

The coordinates of the observations can be retrieved as the components $x and $y. To view the spatial dispersion of the observations, the following line gives you the top-left image of Figure 25-17:

```
R> plot(chorley$x,chorley$y,xlab="Eastings (km)",ylab="Northings (km)")
```

Your goal is to display a kernel estimate of the two-dimensional probability density function of the cancer distribution, similar to what you did with the earthquake data in Example 3. You'll use the kde2d function for this—execute library("MASS") to gain access to it. Then, exactly as you used it for the spatial locations of quakes, the default KDE surface for the observed Chorley-Ribble data is given with the following:

```
R> chor.dens <- kde2d(x=chorley$x,y=chorley$y,n=256)
```

Note the specification of a fine $256 \times 256$ easting-northing evaluation grid.

To display the density estimate, use the built-in `rainbow` palette and use the optional start and end arguments to restrict the total range of the palette to begin at red at the lower end and end at magenta/pink at the upper end (these arguments were mentioned briefly in Section 25.1.2; refer to the help file ?rainbow for more details on the use of start and end). Prestore 200 colors from this palette with the following line:

```
R> rbow <- rainbow(200,start=0,end=5/6)
```

Then, the image is produced by calling this:

```
R> image(x=chor.dens$x,y=chor.dens$y,z=chor.dens$z,col=rbow)
```

Another component of `chorley`, named `$window`, contains the vertices of an irregular polygon. This polygon defines the geographical study region in which the observations themselves were made. The `$window` component also happens to be another special object class of spatstat, namely, `"owin"` for "observation window." Although it's possible to extract the specific vertices of the polygon and plot it manually with built-in functionality, the authors of spatstat have provided a standard `plot` method to use for this purpose.

After running the `image` command, calling the following code superimposes the border of the study region upon the pixel image:

```
R> plot(chorley$window,add=TRUE)
```

The final result is given on the top right of Figure 25-17.

You'll notice that the geographical region in which the data were collected is a little wider than the *x*- and *y*-ranges of the observations themselves, so the current plot hasn't been able to show the region in its entirety. The following code shows this numerically:

```
R> chor.WIN <- chorley$window
R> range(chorley$x)
[1] 346.6 364.1
R> WIN.xr <- chor.WIN$xrange
R> WIN.xr
[1] 343.45 366.45
R> range(chorley$y)
[1] 412.6 430.3
R> WIN.yr <- chor.WIN$yrange
R> WIN.yr
[1] 410.41 431.79
```

The *x*- and *y*-ranges of the study region can be obtained as the `$xrange` and `$yrange` components of the `$window` component (which is stored in the first line as the object chor.WIN). You can see that the overall study region is slightly larger when you compare its limits to the results of calling range on the raw data.

That's not the only problem, either. From the plot, you can also see that the KDE surface has been estimated and drawn in some areas that are actually *outside* the study region, so that will need to be fixed as well. (You'll look at that in a moment.)

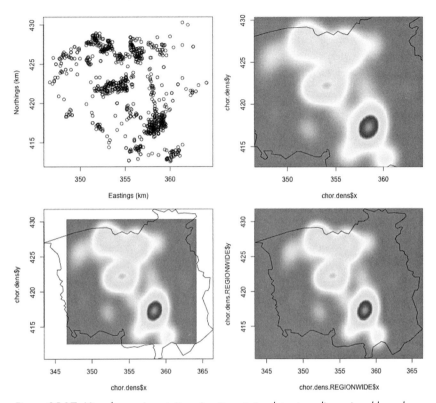

*Figure 25-17: Visual experimentations in attempts to plot a two-dimensional kernel estimate of the probability density function of the Chorley-Ribble cancer data as a pixel image. Top left: The raw data. Top right: The default* kde2d *result based on the data ranges with the study region superimposed. Bottom left: Expanding the* xlim *and* ylim *of the call to* image *when plotting the original density estimate. Bottom right: A revised density estimate, using the full x- and y-ranges of the study region to define the evaluation grid.*

So, first off, what can you do to ensure the entire geographical region is displayed? Well, you could of course use the ranges of the region as stored previously in the vectors WIN.xr and WIN.yr and supply them to the familiar optional xlim and ylim arguments when calling image.

```
R> image(chor.dens$x,chor.dens$y,chor.dens$z,col=rbow,
        xlim=WIN.xr,ylim=WIN.yr)
R> plot(chor.WIN,add=TRUE)
```

The result of these two lines is given on the bottom left of Figure 25-17. Unfortunately, the original density estimate is still defined in terms of the original *x*- and *y*-ranges of the raw data, which gives you a border of empty pixels; in addition, the aforementioned density areas still fall outside the observation window.

All this emphasizes the important fact that a *z*-matrix is specific to a predefined evaluation grid. The only way to get your density estimate to span the geographical study region for the Chorley-Ribble data is to revise your kernel estimate so that it's produced on an evaluation grid that spans the limits of the region. Fortunately, the kde2d function allows you to set optional *x-y* limits of the evaluation grid with the lims argument. This expects a numeric vector of length 4, with the *x*-axis lower and upper values followed by the *y*-axis lower and upper values, in that order. The following code reestimates the density using the study region limits and plots it. The result is given on the bottom right of Figure 25-17.

```
R> chor.dens.WIN <- kde2d(chorley$x,chorley$y,n=256,lims=c(WIN.xr,WIN.yr))
R> image(chor.dens.WIN$x,chor.dens.WIN$y,chor.dens.WIN$z,col=rbow)
R> plot(chor.WIN,add=TRUE)
```

With that, you've solved the problem of ensuring your surface spans the desired area. However, this definitely highlights the second problem—the data that were actually observed fall strictly within the defined polygon, but you can see plotted pixels outside the geographical region, which doesn't make sense. You can control precisely which pixels are plotted in any given pixel image by setting the relevant entries in your *z*-matrix to be NA if you don't want them drawn.

You'll need a mechanism that can decide whether a given cell entry in your *z*-matrix, namely, chor.dens.WIN$z, corresponds to a location inside or outside the polygon (the object chor.WIN). If it falls outside, you'll want to force that entry to be NA. In general, this type of decision making requires you to test each element of the matrix with respect to its coordinate value on the evaluation grid, possibly using your own R function. Fortunately, in this case, the inside.owin function of spatstat does exactly that, but the principle remains the same whenever you need control over precisely which pixels are plotted and which aren't.

Given one or more two-dimensional $(x, y)$ coordinates and an object of class "owin", the inside.owin function returns a corresponding logical vector with a TRUE for those coordinates inside the defined region and a FALSE for any other coordinate. As a quick demonstration, observe the following result:

```
R> inside.owin(x=c(355,345),y=c(420,415),w=chor.WIN)
[1]  TRUE FALSE
```

This confirms what you can see from Figure 25-17—that the coordinate $(355, 420)$ lies well within the polygon and that the coordinate $(345, 415)$ doesn't.

Now, you need to use the `inside.owin` function on every coordinate in the evaluation grid that *z*-matrix `chor.dens.WIN$z` sits on. First, create the full set of grid coordinates using `expand.grid`, in the same way as illustrated in Section 25.3.1.

```
R> chor.xy <- expand.grid(chor.dens.WIN$x,chor.dens.WIN$y)
R> nrow(chor.xy)
[1] 65536
```

Calling `nrow` on the resulting data frame of coordinates confirms you have exactly $256^2 = 65536$ grid points as defined in the `chor.dens.WIN` KDE object. The following call then takes the two columns of `chor.xy` and makes use of logical negation (using `!`) to produce a logical vector that flags grid coordinates that are located *outside* the defined geographical region.

```
R> chor.outside <- !inside.owin(x=chor.xy[,1],y=chor.xy[,2],w=chor.WIN)
```

The final step is now at hand.

```
R> chor.out.mat <- matrix(chor.outside,nrow=256,ncol=256)
R> chor.dens.WIN$z[chor.out.mat] <- NA
```

First, for clarity, recast the long `chor.outside` vector as a $256 \times 256$ matrix to emphasize that it corresponds exactly to the *z*-matrix of interest. Then this logical flag matrix is used to directly overwrite the "outside" entries in the *z*-matrix to be `NA`.

All that's left now is to plot the image with the newly manipulated *z*-matrix. Make sure you have the shape package loaded for the finishing touch of a color legend. The following code creates the KDE surface pixel image plot with pixel points restricted to the geographical region defined by `$window` only:

```
R> dev.new(width=7.5,height=7)
R> par(mar=c(5,4,4,7))
R> image(chor.dens.WIN$x,chor.dens.WIN$y,chor.dens.WIN$z,col=rbow,
        xlab="Eastings",ylab="Northings",bty="l",asp=1)
R> plot(chor.WIN,lwd=2,add=TRUE)
R> colorlegend(col=rbow,zlim=range(chor.dens.WIN$z,na.rm=TRUE),
            zval=seq(0,0.02,0.0025),main="KDE",digit=4,posx=c(0.85,0.87))
```

First you open a new graphics device and widen the right margin to incorporate the color legend. Next you invoke `image` to plot, specifically using an *L*-shaped box and a strict one-to-one *x-y* aspect ratio, and then you add the region polygon with slightly thicker lines. Finally you execute `colorlegend` to obtain an appropriately positioned legend referencing the color values (the specific positioning and tick marks of which were found after a little trial and error). You can see the final result in Figure 25-18.

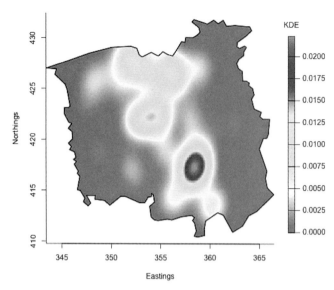

Figure 25-18: Final pixel image plot of the Chorley-Ribble KDE surface, restricted to the geographical study region of the originally collected data. See Plate VII in the color insert.

**NOTE** *In truncating the kernel estimate of the bivariate density estimate that was originally defined over the full rectangular evaluation grid, technically you no longer have a valid probability density function as a result (since the integral over the irregular region will no longer evaluate to a total probability of 1). A more mathematically sound approach requires a deeper knowledge of multivariate KDE and is beyond the scope of this text. Nevertheless, being able to truncate pixel plots like this is useful in any situation where you want to define your surface on a (possibly irregular) subset of an overall rectangular evaluation grid.*

## Exercise 25.4

Revisit the built-in airquality data set and take a look at the help file to refresh your memory of the variables present. Create a copy of the data frame: select the columns pertaining to daily temperature, wind speed, and ozone level and use na.omit to remove any records with missing values.

a.  From your explorations of these data in Chapter 24, there appears to be an association among daily temperature, wind speed, and ozone level. Fit a multiple linear regression model that aims to predict mean temperature based on the wind speed and ozone level, including an interactive effect. Summarize the resulting object.

b. Using the model from (a), construct a z-matrix of predicted mean daily temperature based on a $50 \times 50$ evaluation grid in both wind speed and ozone.

c. Create a pixel image of the response surface, superimposing the raw observations as per the following:
   - A graphics device should be initialized based on bottom, left, top, and right margin lines of 5, 4, 4, and 6, respectively.
   - 20 colors from the built-in topo.colors palette should be used to produce the image; include tidy axis titles.
   - Revisit the normalize function defined in Section 25.1.4 and use the built-in function gray to generate a vector of gray colors (refer to Section 25.1.2) based on the normalized raw temperature observations. Superimpose the raw observations based on wind speed and ozone onto the pixel image, using the gray color vector to indicate the corresponding temperature observations.
   - Two separate calls should then be made to colorlegend of the shape package. The first should appear in the space on the right margin, referencing the surface itself. The second should use the built-in gray.colors function, setting the optional arguments start=0 and end=1, to generate 10 shades of gray for use in the legend that references the raw temperature observations of the superimposed points. This legend should reside on top of the pixel image itself, in the upper-right quadrant where there are no raw observations.
   - Both legends should have appropriate titles, and you may need to experiment a little with the posx and posy arguments to find satisfactory placement.

My result of this plotting exercise appears here.

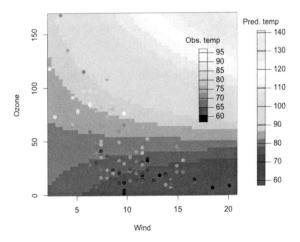

In Section 25.5.2, you used the chorley data set in creating a pixel image truncated to a subset of the overall rectangular evaluation grid. Ensure spatstat is loaded in your current R session and execute the following two lines:

```
R> fire <- split(clmfires)$intentional
R> firewin <- clmfires$window
```

This extracts the 1,786 locations of fires recorded as intentionally lit in a particular region of Spain. The spatial coordinates can be extracted as the $x and $y members of fire, and the geographical region itself is stored as a polygon in firewin (of the same class as the chorley$window object you looked at earlier). See the documentation obtained with ?clmfires for further details.

d.  Using the total *x*- and *y*-range of the study region, use kde2d from the MASS package to calculate a bivariate kernel estimate of the probability density function of the spatial dispersion of intentionally lit fires. The KDE surface should be calculated based on a 256 × 256 evaluation grid.

e.  Identify all points on the rectangular evaluation grid that fall outside the geographical region using expand.grid in conjunction with inside.owin. Set all corresponding pixels of the density surface to NA.

f.  Construct a pixel image of the truncated density, as per the following:
    -   The graphics device should have three lines of space on the bottom, left, and top of the plot region and should have seven lines on the right.
    -   In producing the image itself, you should use 50 colors generated from the built-in heat.colors palette. A one-to-one aspect ratio should be maintained, the axis titles should be suppressed, and the box type set to be an *L* shape.
    -   The geographical study region should be superimposed onto the image using a double-width line.
    -   Using colorlegend from shape, a color legend referencing the density with an appropriate title should be placed to the right of the image. You'll need to experiment with the posx argument for placement. Label the legend at a sequence from 5e-6 to 35e-6 in steps of 5e-6 (refer to Section 2.1.3 for an explanation of e-notation); also, ensure these labels are able to display up to six decimal places of precision.

For your reference, my result is given here.

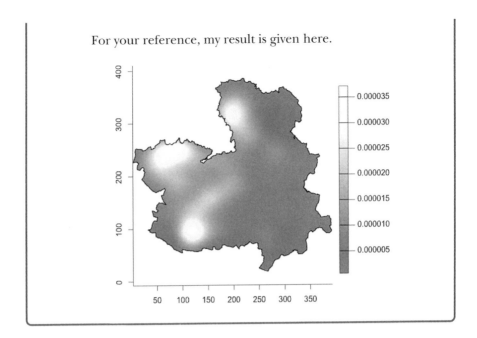

## 25.6 Perspective Plots

The last kind of plot you'll look at in this chapter is the *perspective plot*, some-times also referred to as a *wireframe*. Unlike contour plots and pixel images, where fluctuations in the surface are emphasized with line patterns and/or colors, a perspective plot uses a physical third dimension against which the *z* value is plotted.

### 25.6.1 Basic Plots and Angle Adjustment

Perspective plots are especially useful when you want to emphasize the fluctuating nature of the values populating your *z*-matrix. For example, in some applications you might want to get a good impression of the relative extremity of any present peaks and/or troughs in the plotted surface, which is harder to do in, for example, a pixel image or contour plot.

Recall the mtcars response surface plotted as contours and as pixel images in Sections 25.4.1 and 25.5.1. You created a $20 \times 20$ evaluation grid in the horsepower and weight variables, as well as a corresponding *z*-matrix of 400 giving the predicted mean MPG result:

```
R> car.fit <- lm(mpg~hp*wt,data=mtcars)
R> len <- 20
R> hp.seq <- seq(min(mtcars$hp),max(mtcars$hp),length=len)
R> wt.seq <- seq(min(mtcars$wt),max(mtcars$wt),length=len)
R> hp.wt <- expand.grid(hp=hp.seq,wt=wt.seq)
R> car.pred.mat <- matrix(predict(car.fit,newdata=hp.wt),nrow=len,ncol=len)
```

The built-in R function `persp` is used to create perspective plots. Its basic usage is the same as `contour`, `filled.contour`, and `image`. Your increasing sequences in the x- and y-axis directions, which define the evaluation grid, are passed to x and y, with your corresponding z-matrix passed to z. Bring up the default appearance for the 20 × 20 `mtcars` response surface with the following:

```
R> persp(x=hp.seq,y=wt.seq,z=car.pred.mat)
```

This appears in the top left of Figure 25-19.

Interpreting the perspective plot is straightforward. The default viewing angle shows the x-axis in the foreground, increasing from left to right, and the y-axis on the left side, increasing from the foreground to deeper in the background. In this way, the evaluation grid lies flat along the bottom in the 3D graphic, with the z-axis against which your surface is plotted increasing from the bottom vertically to the top.

The viewing angle is one of the most important aspects of such a plot. In `persp`, you can control it with the two optional arguments `theta`, which spins the plot around horizontally, and `phi`, which adjusts the vertical viewing position. Both are specified in degrees; `theta` defaults to 0, so you're looking directly at the x-axis spanning left to right in front of you, and `phi` defaults to 15 to give a slightly elevated viewing position so you can see the y-axis extending foreground to background. In general, you can think of the possible value of `theta` as anywhere from 0 to 360, representing a complete rotation all around the plot, and the possible value of `phi` as anywhere from 90 to -90, the range of which moves you from a bird's-eye view directly from the top looking down to a submarine view directly from the bottom looking up.

This second example demonstrates this behavior:

```
R> persp(x=hp.seq,y=wt.seq,z=car.pred.mat,theta=-30,phi=23,
        xlab="Horsepower",ylab="Weight",zlab="mean MPG")
```

In fact, it's this line of code that originally produced the rightmost image in Figure 21-9 on page 523 (when you were introduced to the concept of an interactive term between two continuous predictors in a multiple linear regression model). The graphic is reproduced here in the top right of Figure 25-19. The axis titles are tidied up using `xlab` and `ylab`, with `zlab` used to control the title for the third vertical axis. The use of `theta` and `phi` in this instance has elevated the viewing point slightly more than the default and rotated the plot so that the origin (in other words, the lower vertex denoting the lower limit of the x-y plane) is prominent in the foreground. It's worth noting that increasing `theta` from 0 rotates the plot in a clockwise-horizontal fashion, but you could also supply a negative value to that argument to rotate the plot in the other direction. Setting `theta=-30`, as shown here, has the same effect as setting `theta=330`.

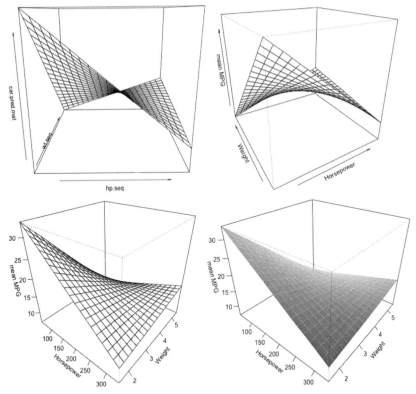

*Figure 25-19: Perspective plots of the 20 × 20* mtcars *response surface created using* persp. *Top left: Default appearance. Top right: Using* theta *and* phi *to adjust the viewing angle. Bottom left: Setting* ticktype="detailed" *to provide detailed axis labeling. Bottom right: Adding depth shading using* shade *and removing facet border lines with* border=NA.

By default, there are no tick marks or labels included, only directional arrows. You can remedy this by setting the optional ticktype argument to "detailed". You can find the result of the following in the bottom left of Figure 25-19, which also offers another viewing angle:

```
R> persp(x=hp.seq,y=wt.seq,z=car.pred.mat,theta=40,phi=30,ticktype="detailed",
        xlab="Horsepower",ylab="Weight",zlab="mean MPG")
```

The help file ?persp details a host of other arguments specific to controlling the presentation of any given perspective plot. As a few examples, you could shade the surface in grayscale to emphasize the 3D depth of the image, you could change the color or suppress the plotting of the grid lines making up the surface itself, or you could change the relative length of the z-axis. The final plot of the mtcars response surface illustrates such actions. The result of the following call is visible in the bottom right of Figure 25-19.

```
R> persp(x=hp.seq,y=wt.seq,z=car.pred.mat,theta=40,phi=30,ticktype="detailed",
        shade=0.6,border=NA,expand=0.8,
        xlab="Horsepower",ylab="Weight",zlab="mean MPG")
```

With the same viewing angle as the previous plot, this plot uses the
shade argument to shade the surface facets to produce a lighting-style effect,
enhancing the perceptive depth slightly. The calculations for the shading
rely on a non-negative numeric value; setting shade=0.6 provides a moderate-
strength effect. You might like to experiment with larger or smaller values.
If you're shading the surface in this way, it's usually best to suppress the grid
lines that by default make up the surface; you can set border=NA to achieve
this (the border argument can also be used to simply change the surface grid
color by supplying any valid R color to it). Finally, the expand argument is
used to adjust the size of the z-axis. Specifying expand=0.8 requests a vertical
axis that is 80 percent the size of the axes in the evaluation grid, producing
a slightly "squashed down" prism in which the surface is drawn. You could
also use values greater than 1 for expand, in which case the effect would be to
"stretch out" the plot along the vertical.

### 25.6.2   Coloring Facets

Like most traditional R plotting commands, you can use the optional col
argument to color the facets of a perspective surface. To color a perspective
surface with a constant color throughout, you would just provide col with a
single value.

If you're interested in col, however, it's often the case that you want
to color the surface according to the fluctuating z-values to highlight the
changing value of the bivariate function. To successfully do this for the
facets making up the surface, it's important to understand that these facets
aren't the same as the pixels that would make up a pixel image of the same
z-matrix. Where image pixels are directly represented by the entries of, say,
your $m \times n$-sized z-matrix, persp facets should be interpreted as the space
*between* the border lines drawn at those matrix entries, leaving you with
$(m - 1) \times (n - 1)$ facets. In other words, in a perspective plot, each z-matrix
entry lies at an intersection of the drawn lines—the z-matrix entries are *not*
situated in the middle of each facet.

To illustrate this, take another look at Figure 25-9 on page 656. When
you use image, R automatically calculates the pixel sizes based on your x- and
y-axis evaluation grid sequences and plots the surface based on the rectangles formed by the dashed gray lines, with the z-matrix entries a, b, c, and
so on, represented directly. When you use persp, however, the visible border lines are represented by the solid-line grid (of arrows), intersecting at
each entry, and so the facets of the resulting surface are formed by the space
between these lines, each one defined by four adjacent entries. Figure 25-20
shows a section of the hypothetical grid in Figure 25-9, where I've marked

off one pixel as interpreted by image and one facet as interpreted by persp. With that, you can see why, in Figure 25-9, there would be exactly $6 \times 4 = 24$ pixels in an image plot but $5 \times 3 = 15$ facets in a perspective plot.

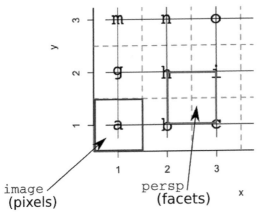

Figure 25-20: Illustrating the difference in treatment of the z-matrix in a pixel image and in a perspective plot. The highlighted box in the bottom-left corner represents an image pixel of the value a in the z-matrix; the highlighted box to the right represents a persp facet formed by the values b, h, i, and c. For coloring, the z-value of the highlighted facet will be calculated as the mean of those four entries, in other words, (b + h + i + c)/4.

The col argument needs to specify the $(m - 1) \times (n - 1)$ facet colors (assuming an $m \times n$ z-matrix passed to z). The typical way to find this in R if you're intending to color the facets according to the z-value is to first calculate each facet's z-value, which will be the average of the four adjacent z-matrix entries. Only thereafter can you deploy one of the color assignment approaches from Section 25.1.4.

Let's recast the pixel image of the Chorley-Ribble kernel density estimate (Example 4; Figure 25-18/Plate VII), complete with z-axis-specific coloring, as a perspective plot. First, make sure you have the packages spatstat and MASS already loaded. Then repeat the code from earlier to obtain the kernel estimate on the appropriate evaluation grid, truncated to the geographical study region.

```
R> chor.WIN <- chorley$window
R> chor.dens.WIN <- kde2d(chorley$x,chorley$y,n=256,
                    lims=c(chor.WIN$xrange,chor.WIN$yrange))
R> chor.xy <- expand.grid(chor.dens.WIN$x,chor.dens.WIN$y)
R> chor.out.mat <- matrix(!inside.owin(x=chor.xy[,1],y=chor.xy[,2],w=chor.WIN),
                    256,256)
R> chor.dens.WIN$z[chor.out.mat] <- NA
```

Next, you need to calculate all the facet *z* values; this can be done en masse with the following code:

```
R> zm <- chor.dens.WIN$z
R> nr <- nrow(zm)
R> nc <- ncol(zm)
R> zf <- (zm[-1,-1]+zm[-1,-nc]+zm[-nr,-1]+zm[-nr,-nc])/4
R> dim(zf)
[1] 255 255
```

The first three lines simply store the *z*-matrix as the object zm and its total rows and columns (both 256 in this case) as nr and nc, respectively, for compactness of the code.

The fourth line is where the relevant calculations happen, giving a matrix of the facet *z* values. It does this systematically, by element-wise summation of four versions of the original *z*-matrix: zm[-1,-1] (first row and first column omitted), zm[-1,-nc] (first row, last column omitted), zm[-nr,-1] (last row, first column omitted), and zm[-nr,-nc] (last row, last column omitted). When the four alternates are summed in this way and divided by 4 at the end, the result is a matrix zf, each element of which is the four-point average of each "rectangle" of four adjacent entries in the original *z*-matrix, exactly as noted in the discussion and caption of Figure 25-20. The final call to dim on zf confirms the size of the result. Since there are a total of 256 × 256 evaluation grid lines in the defined *z*-matrix, these encapsulate a total of 255 × 255 perspective facets.

The hard work is done, and all you need to do now is assign the colors from your palette to the calculated facet *z* values in zf. You can do this using either the categorization or normalization approach, as noted in Section 25.1.4; for simplicity, let's stick to categorization. Consider the following code:

```
R> rbow <- rainbow(200,start=0,end=5/6)
R> zf.breaks <- seq(min(zf,na.rm=TRUE),max(zf,na.rm=TRUE),length=201)
R> zf.colors <- cut(zf,breaks=zf.breaks,include.lowest=TRUE)
```

The first line is repeated from Section 25.5.2 to generate the same 200 colors from the built-in rainbow palette as were used in the pixel images. The second line sets up an evenly spaced sequence spanning the range of the calculated facet *z*-values to form the category break points that are required by the categorization approach. Note the use of na.rm=TRUE in the required calls to min and max to avoid all the NA entries present in zf (remember, the surface has been truncated to the irregular polygon representing the geographical study region). The sequence is one more in length than the number of generated colors—again, refer to Section 25.1.4 for this necessary feature of the categorization approach. Lastly, cut assigns each of the zf facet value entries an appropriate rank with respect to the 200 ordered bins. As you've learned, the zf.colors ranks are subsequently used to index the vector of 200 colors stored in rbow when plotting.

With that, you can enjoy the fruits of your labor! The following code plots the bivariate kernel density estimate of the Chorley-Ribble observations as a perspective plot using facet coloring to reflect the relative height of the surface along the z-axis. Border lines are suppressed to show off the color clearly, the z-axis is scaled down slightly, and a color legend is inserted on the right side (ensure the shape package has been loaded for that) after manipulating the default figure margins via mar in a call to par to create extra space for it. You can find the result in Figure 25-21.

```
R> par(mar=c(0,1,0,7))
R> persp(chor.dens.WIN$x,chor.dens.WIN$y,chor.dens.WIN$z,border=NA,
        col=rbow[zf.colors],theta=-30,phi=30,scale=FALSE,expand=750,
        xlab="Eastings (km)",ylab="Northings (km)",zlab="Kernel estimate")
R> colorlegend(col=rbow,zlim=range(chor.dens.WIN$z,na.rm=TRUE),
              zval=seq(0,0.02,0.0025),main="KDE",digit=4,
              posx=c(0.85,0.87),posy=c(0.2,0.8))
```

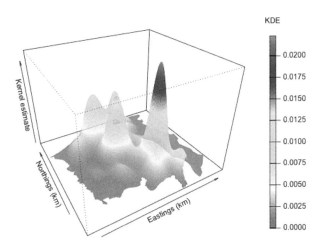

Figure 25-21: A perspective plot of the Chorley-Ribble density estimate, demonstrating facet coloring that changes according to the z-value of the surface. See Plate VIII in the color insert.

I've included the optional argument scale=FALSE in the execution of persp. This retains a one-to-one aspect ratio in the x- and y-coordinate directions; this is useful since you're looking at geographical data. Unfortunately, this also forces the density estimate values on the z-axis to be scaled in the same way, which makes no sense in the context of the current plot. To avoid the small scale resulting in a supremely flat appearance of the surface itself, you need to use expand to artificially amplify the surface along the third axis.

In this instance, multiplying it by a factor of around 750 provides a visually pleasing result. Note that this would not be necessary if you left the scale argument at its default TRUE value (since, in that case, R internally scales all three axes for a one-to-one-to-one aspect ratio).

### 25.6.3    Rotating with Loops

There's one last bit of fun you can have with perspective plots if you want to get an overall impression of the plotted surface. Using a simple for loop (Section 10.2.1) to increment either theta or phi, you can perform a series of repeated calls to persp, each one at a slightly new angle. Doing this in sequence results in an animation—essentially a cartoon—of a rotating surface, allowing you to see it from all different sides.

Consider the following basic function in the R editor:

```
persprot <- function(skip=1,...){
    for(i in seq(90,20,by=-skip)){
        persp(phi=i,theta=0,...)
    }
    for(i in seq(0,360,by=skip)){
        persp(phi=20,theta=i,...)
    }
}
```

Using an ellipsis (see Section 11.2.4), persprot simply takes all the arguments you'd usually supply to a call to persp, barring theta and phi. Then comes a for loop, which immediately calls persp with theta=0 and the content of the ellipsis. The for loop alters the vertical viewing angle, starting with phi=90 (birds-eye view) and moving down to a mildly elevated phi=20. A second for loop then completes a full 360-degree horizontal rotation by altering theta.

The only formally tagged argument is skip, which determines the amount phi and theta increment by at each iteration. The default, skip=1, simply moves through the integer-valued angles. Increasing skip will reduce the time it takes to complete the rotation, though it makes for a more jagged animation.

Depending on the type of graphics device you're using, you may want to experiment with skip. Note that not all graphics device types will be well-suited to the animation effect sought by running this rather crude function (for example, it's not appropriate if you're using RStudio—see Appendix B). That said, when running the base R GUI applications on OS X or Windows, I find persprot works well under default graphics settings.

Import the function to try it; let's do so here for a perspective plot of a kernel estimate of the probability density function of the spatial quakes locations you first examined as Example 3, Section 25.4.1. With the MASS package already loaded, produce the density estimate on a 50 × 50 evaluation grid with the following line.

```
R> quak.dens <- kde2d(x=quakes$long,y=quakes$lat,n=50)
```

Then you use persprot just as you'd use persp, without needing to specify either theta or phi.

```
R> persprot(x=quak.dens$x,y=quak.dens$y,z=quak.dens$z,border="red3",shade=0.4,
           ticktype="detailed",xlab="Longitude",ylab="Latitude",
           zlab="Kernel estimate")
```

Figure 25-22 shows a series of screenshots of the rotating plot.

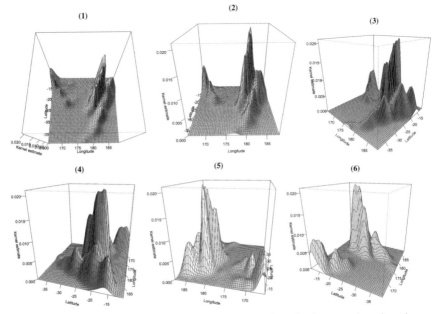

Figure 25-22: A rotating perspective plot of a KDE surface for the spatial earthquake locations, after a call to the custom persprot function

## Exercise 25.5

In Exercise 25.3 (a), you revisited the nuclear data set from the boot package and fitted two multiple linear regression models aiming to model mean construction cost by permit date issue and plant capacity—one with main effects only and the other including an interaction term between the two continuous predictors.

a.  Refit the two versions of the model and produce perspective plots of the response surfaces based again on a $50 \times 50$ evaluation grid, taking the following into account:

- Use mfrow in a call to par to display the two perspective plots next to each other. In the same call to par, override the default figure margins to have only one line of space on each side (par is explored in this role in Chapter 23).
- Use zlim to ensure both plots are displayed on the same scale of vertical axis, spin each one horizontally 25 degrees, and ensure detailed axis markings and tidy titles.
- Is there any visual indication that the presence of the interaction term has had any meaningful impact on modeling the response?

b.  Start a fresh plot. To get a better idea of the difference between the two surfaces, produce a perspective plot of the $z$-matrix obtained as the elementwise difference between the two individual $z$-matrices for the two fitted models in (a). What, in general, is the effect of including the interaction term?

Turn your attention back to the topographical information on the Auckland volcano, as the built-in R object volcano: an $87 \times 61$ matrix of elevation values (in meters). You first looked at this in Section 25.4.1 as a contour plot.

c.  Produce the most basic, default perspective plot of the volcano, using simple integer sequences for the $x$- and $y$-coordinates.

d.  The plot in (c) is decidedly unappealing for a number of reasons. Produce a more realistic depiction of the volcano as per the following:

- Use a new graphics device with the margin widths reset to one, one, one, and four lines on the bottom, left, top, and right, respectively.
- The help file ?volcano reveals the $x$- and $y$-coordinates to which the volcano $z$-matrix corresponds is in 10-meter units. Using scale and altering expand, replot the surface with the correct aspect ratio in all three axes.
- Suppress all axis tick marks and notation using axes.
- The facets should be colored according to 50 colors generated from the built-in terrain.colors palette, and the facet border lines should be suppressed.

- Find your choice of visually appealing viewing angle.
- Use colorlegend from the shape package to place a color legend referencing elevation in meters in the space to the right of the plot. Experiment with the arguments to find appropriate placement and tick mark labels.

Here's my version of the improved plot:

In Exercise 25.4, you looked at the spatial distribution of intentionally lit fires in a region of Spain. Ensure the spatstat package is loaded, and then rerun the following lines to obtain the relevant data objects:

```
R> fire <- split(clmfires)$intentional
R> firewin <- clmfires$window
```

e. Borrow the code from Exercise 25.4 (d) and (e) to reproduce the kernel density estimate of this dispersion of observations, based on a 256 × 256 evaluation grid, truncated to the study region. Then, display it as a perspective plot according to the following:
- Just as with the pixel image, use 50 colors from the built-in heat.colors palette to color the facets by $z$ value. Note the truncated $z$-matrix for this function contains NA values.
- Border lines on the surface should be suppressed, and you should find your preferred choice of viewing angle.
- Use scale to ensure the correct spatial aspect ratio. In doing so, you'll also need to adjust the $z$-axis expansion by a factor of around 5,000,000 for the density surface to be visible along the vertical, given the natural scaling of the density estimate on the specified evaluation grid.
- Employ detailed axis labeling and simply entitle the axes "X", "Y", and "Z" as appropriate.

My product is given here.

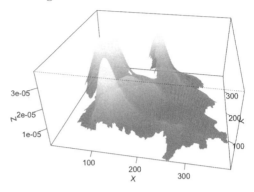

f.   Use the persprot function defined in Section 25.6.3 to view the surface from (e), setting skip=10.

## Important Code in This Chapter

| Function/operator | Brief description | First occurrence |
|---|---|---|
| palette | List integer colors | Section 25.1.1, p. 632 |
| col2rgb | Named color to RGB | Section 25.1.1, p. 632 |
| rgb | RGB to hex code | Section 25.1.1, p. 633 |
| rainbow, heat.colors, gray, terrain.colors, cm.colors, topo.colors, gray.colors | Built-in palettes | Section 25.1.2, p. 635 |
| colorRampPalette | Custom palette (integer) | Section 25.1.3, p. 636 |
| colorRamp | Custom palette ([0,1] interval) | Section 25.1.4, p. 640 |
| colorlegend | Color legend (shape) | Section 25.1.5, p. 641 |
| scatterplot3d | 3D scatterplot (scatterplot3d) | Section 25.2.1, p. 649 |
| expand.grid | All unique evaluation coords. | Section 25.3.1, p. 654 |
| letters | Alphabet characters | Section 25.3.1, p. 655 |
| contour | Contour plot | Section 25.4.1, p. 657 |
| kde2D | Bivariate KDE (MASS) | Section 25.4.1, p. 660 |
| filled.contour | Color-filled contour plot | Section 25.4.2, p. 664 |
| image | Pixel images | Section 25.5.1, p. 668 |
| inside.owin | Test inside region (spatstat) | Section 25.5.2, p. 674 |
| persp | Perspective plot | Section 25.6.1, p. 680 |

# 26

## INTERACTIVE 3D PLOTS

When it comes to 3D plots, it's important to be able to view them from different angles to interpret the function or surface that's been displayed. The rgl package by Adler et al. (2015) offers some fantastic, simple-to-use R functions that allow you to rotate and zoom in on three-dimensional plots with your mouse. In this chapter, you'll look at a few examples that show off the possibilities of rgl.

Under the hood, rgl utilizes OpenGL—a standard application programming interface—to render the graphics on your computer screen. Install rgl (for example, by calling install.packages("rgl") at the prompt) and then call library("rgl") to load it.

## 26.1 Point Clouds

Let's begin with the most basic of 3D plots—*point clouds*. In statistics, this tool is typically used to provide scatterplots of three continuous variables, as you saw when you created static 3D scatterplots.

### 26.1.1 Basic 3D Cloud

Return to the built-in iris data, composed of four measurements taken on three species of flower. Create the following four vectors in your workspace for accessibility, as you did in Section 25.2.1:

```
R> pwid <- iris$Petal.Width
R> plen <- iris$Petal.Length
R> swid <- iris$Sepal.Width
R> slen <- iris$Sepal.Length
```

You use the plot3d function of rgl to display an interactive 3D cloud of points. It's called in the way familiar for scatterplots—by supplying the *x*-, *y*-, and *z*-coordinates to the x, y, and z arguments, respectively. The following line opens an RGL device and produces a scatterplot of petal width, length, and sepal width from the iris data:

```
R> plot3d(x=pwid,y=plen,z=swid)
```

You'll probably want to increase the size of the device with your mouse to see the data better. Then, by left-clicking the plot and holding the button down, you can move the mouse to rotate the plot in any direction you like. If you right-click the plot and hold, you control the zoom. Specifically, right-clicking and holding while moving the mouse upward will zoom out, and right-clicking and holding while moving the mouse downward will zoom in. The axis tick marks and labels automatically appear on different sides based on your viewing angle. Figure 26-1 shows this plot.

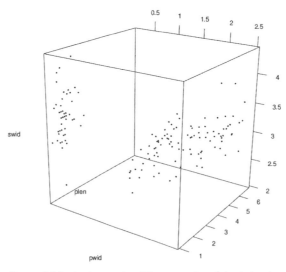

*Figure 26-1: An interactive 3D scatterplot of the iris data using the plot3d function of rgl. This is the default appearance of plotting the petal width, petal length, and sepal width.*

## 26.1.2   Visual Enhancements and Legends

You can alter the appearance of a plot3d scatterplot in new ways as well as some familiar ways. For example, the optional type argument, defaulting to "p" for "points," plots the points as dots as in the most recent scatterplot. To draw points as visible 3D spheres, use type="s". You can control the size of any plotted points or spheres by using size, and you can control the color (or colors) by using col. The legend3d function is the rgl analog of legend and is also useful; it works by changing the background image upon which the interactive plot sits.

To illustrate these modifications, let's replot the same iris observations. First, close any currently open RGL graphics devices. Then, execute the following:

```
R> plot3d(x=pwid,y=plen,z=swid,size=1.5,type="s",
          col=c(1,2,3)[as.numeric(iris$Species)])
```

This will start a new RGL device, coloring spheres according to flower species. As usual, you pass the col argument a vector of the same length as the plotted coordinates, and it will assign the color to the corresponding point in an element-wise fashion. You specify the size parameter on a slightly different scale than the traditional R graphics parameter cex, and it changes according to the value of type—inspect the help file ?plot3d for details. With a little experimentation, it's not difficult to find a size value that suits the plot.

To add a legend, first resize the RGL device with your mouse to your preferred display size and then execute the following line:

```
R> legend3d("topright",col=1:3,legend=levels(iris$Species),pch=16,cex=2)
```

This inserts a static, unmovable legend referencing the plotted species by color. The legend3d function actually calls the base R legend function, so they are conveniently used in the same way. With the static legend in place, the scatterplot remains fully interactive, and you can continue to rotate and zoom. Figure 26-2 shows all this.

The legend3d function changes the background canvas, which is why you have to open a new device and resize it manually before you add the legend. If you now produced a new rgl plot in the same device without closing it first or resetting the background, the flower species legend would still be there. If you're making multiple rgl plots, you can reset the background to its default white canvas at any time by calling the following:

```
R> bg3d(color="white")
```

If you try that with the most recent plot still active, you'll see that the flower species legend disappears and the scatterplot remains. Alternatively, you can just close your RGL device when you're done so that a new device will be used for any subsequent plot.

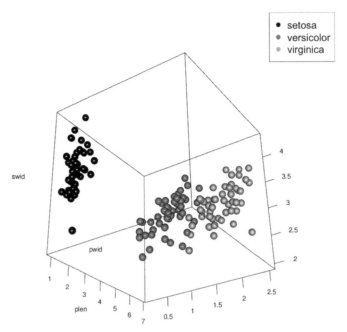

Figure 26-2: Replotting the iris petal width, length, and sepal width data with plot3d. Observations are plotted as spheres, increased in size, and colored according to species type; a legend is added via legend3d.

### 26.1.3 Adding Further 3D Components

You can also add new observations and lines to a current 3D plot. The rgl package includes the functions points3d, lines3d, and segments3d, reminiscent of points, lines, and segments from base R graphics. As an example, in Section 25.2.2, you used an optional argument to add vertical lines from the base of the *x-y* plane to each plotted point in scatterplot3d. In a plot3d scatterplot, you would use segments3d to the same effect. In addition, you can add the grid that's drawn by default on the same plane in a scatterplot3d plot by using the grid3d function for rgl graphics.

Let's put that into practice. Take a look back at Figure 25-8 on page 652 (Plate V in the color insert). To create a similar plot using rgl functionality, where color is used to reference the fourth continuous variable, sepal length, first re-create the palette and set up the colors for each observation. This is done here with a categorization of 50 colors (see Section 25.1.4).

```
R> slen.pal <- colorRampPalette(c("purple","yellow2","blue"))
R> cols <- slen.pal(50)
R> slen.cols <- cut(slen,breaks=seq(min(slen),max(slen),length=51),
                    include.lowest=TRUE)
```

Then, either close any currently active RGL devices or clear the background of the one in focus. A call to plot3d starts the plot with appropriately colored spheres.

```
R> plot3d(x=pwid,y=plen,z=swid,type="s",size=1.5,col=cols[slen.cols],
        aspect=c(1,1.75,1),xlab="Petal width",ylab="Petal length",
        zlab="Sepal width")
```

You supply a vector of length 3 to the aspect argument, describing the relative lengths of the *x*-, *y*-, and *z*-axes, in that order. By changing the second entry to 1.75, you're lengthening the *y*-axis by that multiplicative factor relative to the others. This creates the stretched-out effect along the *y*-axis. The colors are assigned by vector indexing using the slen.cols factor vector, and xlab, ylab, and zlab are used to tidy up the axis titles.

Now, to add a vertical line from the *x-y* plane to each observation, you need to understand how to use the segments3d function. Unlike its base R counterpart, segments3d doesn't separate the "from" and "to" coordinates into different arguments (recall the use of x0, y0, x1, and y1 in segments and arrows). Instead, it takes each sequential pair of observations provided to the x, y, and z arguments to be the beginning and end of each line segment, in that order.

So, to draw the vertical lines on the existing RGL device, you first need to set up these vectors containing a "from" location and a "to" location in the 3D space. Consider the following code:

```
R> xfromto <- rep(pwid,each=2)
R> yfromto <- rep(plen,each=2)
R> zfromto <- rep(min(swid),times=2*nrow(iris))
R> zfromto[seq(2,length(zfromto),2)] <- swid
```

The first two lines set up the vectors for the *x*- and *y*-components, xfromto and yfromto, respectively, by simply replicating each observation twice. These are easy, since the "from–to" values don't change in these coordinate directions. The *z*-component does change, however. You first create the zfromto vector by replicating the smallest sepal width value, min(swid), by two times the size of the data set, so you have a vector that matches xfromto and yfromto in length. Then, every second position of zfromto is overwritten using the elements of the sepal width vector. This gives you "from" *z* values for all observations, namely min(swid), matched (in the pairwise fashion as required for segments3d) with the "to" *z* values in swid itself. Together with xfromto and yfromto, you'll therefore end up with lines that go from the bottom *x-y* plane of the plot (the vertical position of which is automatically level at min(swid)) up to the actual swid value (which is of course the corresponding *z* value of each sphere).

To help understand the way they've been set up, print the coordinate vectors to your console screen so you can see what they hold. Then a call to segments3d places the lines on the plot.

```
R> segments3d(x=xfromto,y=yfromto,z=zfromto,col=rep(cols[slen.cols],each=2))
```

To ensure the color of each line matches its corresponding sphere, you also need to replicate each entry of the vector-indexed collection of colors provided by cols[slen.cols] twice, which implies a constant "from–to" color.

Then, executing the following line places a reference grid over the lower *x-y* plane:

```
R> grid3d(side="z-")
```

To the side argument you specify the axis you want held constant (in this case, *z*) and at which end to place the grid (in this case, because you want the grid at the lower end of the *z*-axis, you specify with a minus symbol). To place the grid at the upper end of the vertical axis, which is on the top side of the rectangular prism, you would specify side="z+".

Lastly, you can add a custom, continuous-color legend to the plot to reference the sepal length. The bgplot3d function is a more general version of legend3d; it allows you to specify any plotting commands you like to define the RGL device background. Let's do so using the colorlegend function of the shape package, first explored in Section 25.1.5. Make sure you have the shape package loaded and that your RGL device of the scatterplot is sized to your liking. On my machine, I execute the following:

```
R> bgplot3d({plot.new();colorlegend(slen.pal(50),zlim=range(slen),
                        zval=seq(4.5,7.5,0.5),digit=1,
                        posx=c(0.91,0.93),posy=c(0.1,0.9),
                        main="Sepal length")})
```

The bgplot3d function can take multiple plotting commands, which you need to provide as a code chunk within braces, {}, with each command separated by a semicolon (;). In this example, the initial call to plot.new() silently initializes the background of the RGL device so that you can add the continuous-color legend. Without that call, colorlegend will still work, but a warning will be issued. Figure 26-3 shows the final result, with the scatterplot still spinnable and zoomable with your mouse.

The ability to rotate a 3D scatterplot and any of the plots you'll see over the next few sections with simple mouse commands is especially handy when you're exploring visuals of higher-dimensional data. You're not restricted to a single viewpoint, and you don't need to manually decide on a viewing angle before actually producing the plot. The rgl functionality also makes it easy to add extra elements to an existing plot—something that's harder to do with `scatterplot3d` or `persp` plots. That said, certain features you might take for granted in more traditional plotting can be difficult to mirror in interactive plots. For example, no equivalent of the `pch` graphical parameter is readily available in `rgl`. To plot different symbols, you would need to design, render, and place new 3D shapes.

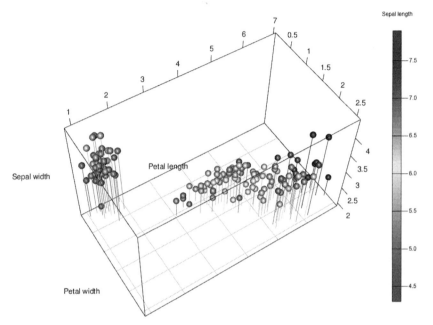

*Figure 26-3: This demonstrates the addition of lines and a plane grid to a `plot3d` 3D scatterplot of the `iris` data to mimic the earlier `scatterplot3d` example of the same data. See Plate IX in the color insert.*

## Exercise 26.1

Turn back to the survey data frame in the MASS package, checking the description of the present variables in the help file `?survey` if you need to. Create a copy of survey containing only the writing handspan, nonwriting handspan, left- or right-handedness, sex, and height columns. Then use `na.omit` to remove any rows of this subsetted data frame that contain missing values.

a.  Produce a basic interactive 3D point cloud of student height on the *z*-axis, writing handspan on the *x*-axis, and nonwriting handspan on the *y*-axis.

b. Create a more informative version of the scatterplot in (a) that uses color to distinguish between sexes and uses point size to distinguish between left- and right-handed individuals, following these guidelines:

- Start by plotting only those points that correspond to right-handed individuals. Set the color via vector indexing using the numeric version of sex for right-handed individuals—females should be black, males red.
- Set the plotted point size as 4 for the right-handed individuals and ensure tidy axis labels.
- Using `points3d`, add the points for left-handed individuals to the existing plot. Colors are to be assigned according to sex in the same way as for the right-handed students, but this time, the point size should be set at 10.
- Resize the RGL device to your liking and add a legend to the top-left corner that references the four types of points: `"Male RH"`, `"Female RH"`, `"Male LH"`, and `"Female LH"`. In setting the legend, use a `pch` value of `19` and use `pt.cex` values of `0.8` and `1.5` for right- and left-handed individuals, respectively.

For reference, my version of the rotatable 3D scatterplot is shown here:

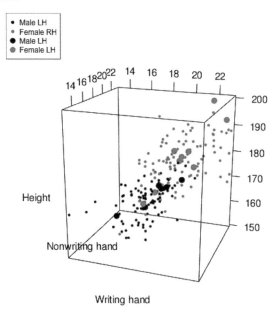

In Exercise 25.2 on page 652, you looked at a static 3D scatterplot of the built-in `airquality` data. Again, create a copy of the data frame, omitting any rows with `NA` entries.

c. Create a similar version of the plot from the earlier exercise using rgl functionality, displaying wind speed, solar radiation, and temperature on the *x*-, *y*-, and *z*-axes, respectively, according to the following guidelines:

  – Set up 50 colors from the built-in topo.colors palette. Set up the appropriate color index vector for the ozone values, based on the categorization approach.
  – Plot the observations as size 1 spheres, colored as before, and modify the aspect ratio so that the *y*-axis is 1.5 times the length of the other two axes. Provide neat axis titles.
  – Add correspondingly colored lines, one for each observation, stretching vertically upward from the *x*-*y* plane to meet the plotted spheres. Also, place a grid on the lower *x*-*y* plane.
  – Modify the background of the RGL device to include a color legend referencing the ozone level; use a sequence of values between 60 and 95, in steps of 5, to label it.

Here's my result:

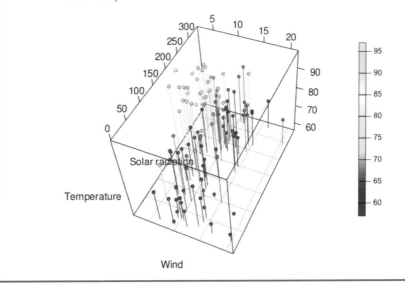

## 26.2  Bivariate Surfaces

Next you'll look at plotting bivariate surfaces—a continuous surface calculated with respect to a 2D *x*-*y* evaluation grid—with rgl. In Chapter 25, you plotted these using contour, filled.contour, image, and persp in base R graphics. Anything that you're able to plot using those functions can also be plotted as an interactive perspective plot with the persp3d function of rgl.

### 26.2.1  Basic Perspective Surface

Take the `mtcars` response surface of mean MPG as a function of horsepower and weight (first used in Section 25.4.1) as an easy initial example. In Section 25.6.1, you plotted static, base R perspective plots of this surface. The next few lines will refit the multiple linear regression model and re-create the $20 \times 20$ evaluation grid $x$- and $y$-sequences:

```
R> car.fit <- lm(mpg~hp*wt,data=mtcars)
R> len <- 20
R> hp.seq <- seq(min(mtcars$hp),max(mtcars$hp),length=len)
R> wt.seq <- seq(min(mtcars$wt),max(mtcars$wt),length=len)
R> hp.wt <- expand.grid(hp=hp.seq,wt=wt.seq)
```

To create the surface, predict using the evaluation grid in `hp.wt` as you've done previously, but this time, include the calculation of a prediction interval for the raw observations.

```
R> car.pred <- predict(car.fit,newdata=hp.wt,interval="prediction",level=0.99)
```

(You'll use the interval in a later example.) Then, construct the $z$-matrix and draw a green `persp3d` surface with the following two lines:

```
R> car.pred.mat <- matrix(car.pred[,1],nrow=len,ncol=len)
R> persp3d(x=hp.seq,y=wt.seq,z=car.pred.mat,col="green")
```

The result is shown on the left of Figure 26-4. If you compare it to Figure 25-19 on page 681, you can see that it shows the same surface. The default lighting and shadowing effect produced by the `persp3d` surface helps with depth perception, similar to the `shade` argument to `persp`. The main benefit of this version is the mouse-based rotation and zoom interactivity.

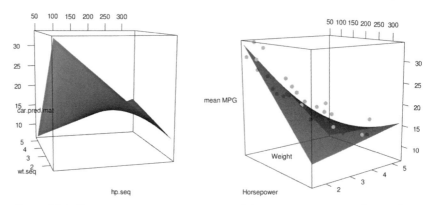

Figure 26-4: Two interactive `persp3d` versions of the `mtcars` response surface. Left: Default appearance in green. Right: Red, 70 percent opacity surface, with original data superimposed in the 3D space. Both plots can be rotated and zoomed with the mouse.

## 26.2.2 Additional Components

Another useful attribute of a persp3d plotted surface is the ability to add further components easily—something that's nowhere near as straightforward in base R functionality. You'll continue using the objects just created for the mtcars response surface.

### Adding Points

As this response surface is based on a model fitted to data on the three variables of horsepower, weight, and MPG, it would be useful to view the raw observations alongside the fitted model. For this, you can use points3d, which works just like points in base R graphics. Execute the following:

```
R> persp3d(x=hp.seq,y=wt.seq,z=car.pred.mat,col="red",alpha=0.7,
            xlab="Horsepower",ylab="Weight",zlab="mean MPG")
R> points3d(mtcars$hp,mtcars$wt,mtcars$mpg,col="green3",size=10)
```

Resize the RGL device to your liking and keep the device open. These two commands plotted the predicted mean MPG response surface, this time in red at 70 percent opacity using the optional alpha argument, and then added the raw observations to the same image in green, slightly enlarged from their default size. You can see this plot on the right of Figure 26-4; you can now compare the fit of the response surface to the raw data and view it from any angle.

### Adding Surfaces

You can also add more perspective surfaces! Let's continue to add to the current plot using the car.pred object you created for Figure 26-4. The response surface is stored as the first column in car.pred; the corresponding lower and upper prediction limits are stored as the second and third columns—flip back to Section 20.4.2 for a discussion of predict for linear regression models. To add these prediction bounds to the response surface displayed on the right of Figure 26-4, you first need to store each bounding surface as a z-matrix corresponding to the x-y evaluation grid.

```
R> car.pred.lo <- matrix(car.pred[,2],nrow=len,ncol=len)
R> car.pred.up <- matrix(car.pred[,3],nrow=len,ncol=len)
```

Then, simply call persp3d for each of these z-matrices and use the optional add argument set to TRUE—this instructs the persp3d function to add to the existing graphic without refreshing the plot.

```
R> persp3d(x=hp.seq,y=wt.seq,z=car.pred.up,col="cyan",add=TRUE,alpha=0.5)
R> persp3d(x=hp.seq,y=wt.seq,z=car.pred.lo,col="cyan",add=TRUE,alpha=0.5)
```

Here you've also set the color for each additional surface to cyan and set the opacity at 50 percent. You can see the result on the left of Figure 26-5.

After rotating it with your mouse, you'll be able to see that the observations all fall between the 3D 99 percent prediction interval bounds for this particular model.

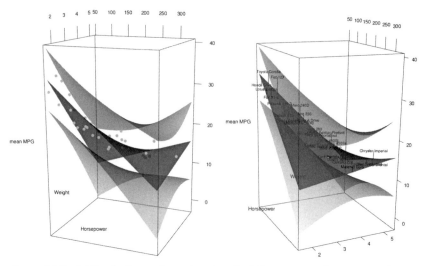

Figure 26-5: Adding further surfaces denoting the 99 percent prediction interval to an existing persp3d plot of the fitted mtcars model. Left: Green points make up the raw observations. Right: Raw observations are labeled with added text, and corresponding line segments mark the corresponding residuals. See Plate X in the color insert.

Alternatively, you could label raw observations with the row names attribute of the original mtcars data frame as added text so that you could identify which car is which in the plot. In this case, the names are obtained as a vector of character strings using the built-in rownames function. To add text to an existing 3D graphic, rgl has its own analog of the traditional text function, text3d. Executing the following four lines replots the translucent red response surface, adds the appropriate text at the $(x, y, z)$ coordinates corresponding to that car, and again adds the cyan prediction interval:

```
R> persp3d(x=hp.seq,y=wt.seq,z=car.pred.mat,col="red",alpha=0.7,
           xlab="Horsepower",ylab="Weight",zlab="mean MPG")
R> text3d(x=mtcars$hp,y=mtcars$wt,z=mtcars$mpg,texts=rownames(mtcars),cex=0.75)
R> persp3d(x=hp.seq,y=wt.seq,z=car.pred.up,col="cyan",add=TRUE,alpha=0.5)
R> persp3d(x=hp.seq,y=wt.seq,z=car.pred.lo,col="cyan",add=TRUE,alpha=0.5)
```

The text is harder to visually locate than the green dots, so it makes sense to point their locations out on the fitted surface—and what better way to do so than using the fitted model residuals? The segments3d function is ideal for this purpose, as you know from the 3D scatterplots of the iris data. First, you need to set up the "from–to" vectors in the three coordinates (refer to Section 26.1 for an explanation of segments3d).

```
R> xfromto <- rep(mtcars$hp,each=2)
R> yfromto <- rep(mtcars$wt,each=2)
R> zfromto <- rep(car.fit$fitted.values,each=2)
R> zfromto[seq(2,2*nrow(mtcars),2)] <- mtcars$mpg
```

Here, again, the *x*- and *y*-axis values don't change when moving from the "from" location to the "to" location, so these are simply double-replicates of each horsepower and weight entry of the original data frame. You need to instruct the *z*-axis "from" values to remain as the fitted values of the model (in other words, the actual vertical location of the response surface), and the "to" values are the raw data *z* values. Then, a final call to segments3d draws on the residuals, as standard black line segments, for each text3d-labeled car.

```
R> segments3d(x=xfromto,y=yfromto,z=zfromto)
```

Take a moment to interact with the final product, shown on the right of Figure 26-5.

### 26.2.3   Coloring by z Value

One advantage of persp3d plots is that you can color the surface according to the *z* values without needing to do anything special. Recall that if you were using the base R persp function, coloring by *z* value would require a minor workaround, because you'd need to calculate the relevant vertical position as the average of four adjacent *z*-matrix entries that make up each facet (see Section 25.6.2).

Fortunately, this isn't necessary with persp3d. Continuing one last time with the mtcars response surface, you can set up your desired color palette and assign colors to the entries of the *z*-matrix themselves without having to average out each set of four adjacent values first.

```
R> blues <- colorRampPalette(c("cyan","navyblue"))
R> blues200 <- blues(200)
R> zm <- car.pred.mat
R> zm.breaks <- seq(min(zm),max(zm),length=201)
R> zm.colors <- cut(zm,breaks=zm.breaks,include.lowest=TRUE)
```

Then, using categorization to assign color to values on a continuum, you just need to index blues200 by zm.colors when specifying the col value in persp3d.

```
R> persp3d(x=hp.seq,y=wt.seq,z=car.pred.mat,col=blues200[zm.colors],
           alpha=0.6,xlab="Horsepower",ylab="Weight",zlab="mean MPG")
```

Figure 26-6 shows the result.

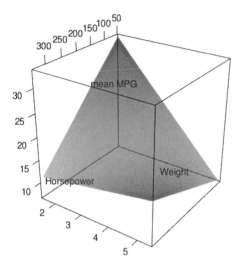

*Figure 26-6: Showing a result of direct color assignment corresponding to the z-matrix value in use of* `persp3d` *for the* `mtcars` *response surface*

## 26.2.4 Dealing with the Aspect Ratio

Taking a break from the `mtcars` model, you'll now return to the bivariate kernel density estimate of the Chorley-Ribble data, used in Sections 25.5.2 and 25.6.2. Load the `spatstat` package to access the `chorley` data and use the `MASS` package to access the `kde2D` function. Repeated from earlier for convenience, the following code calculates the KDE surface using `kde2D`, stored as the `$z` component of the `chor.dens.WIN` object:

```
R> chor.WIN <- chorley$window
R> chor.dens.WIN <- kde2d(chorley$x,chorley$y,n=256,
                        lims=c(chor.WIN$xrange,chor.WIN$yrange))
R> chor.xy <- expand.grid(chor.dens.WIN$x,chor.dens.WIN$y)
R> chor.out.mat <- matrix(!inside.owin(x=chor.xy[,1],y=chor.xy[,2],
                                    w=chor.WIN),
                        256,256)
R> chor.dens.WIN$z[chor.out.mat] <- NA
```

It also truncates the surface to fall within the polygon that represents the geographical study region by setting all elements of the z-matrix outside that polygon to `NA` (you studied in detail how to do this in Section 25.5.2).

Then, executing the next few lines of code generates 200 colors from the built-in `rainbow` palette that you've used previously for this KDE plot and categorizes the entries of the truncated z-matrix appropriately:

```
R> zm <- chor.dens.WIN$z
R> rbow <- rainbow(200,start=0,end=5/6)
```

```
R> zm.breaks <- seq(min(zm,na.rm=TRUE),max(zm,na.rm=TRUE),length=201)
R> zm.colors <- cut(zm,breaks=zm.breaks,include.lowest=TRUE)
```

Note again that the difference here is you don't need to calculate facet averages as you did in Section 25.6.2—cut is applied directly to zm.

Before calling persp3d, it's worth remembering that since you're dealing with a geographical area, you should consider the aspect ratio in the $x$- and $y$-coordinate directions. As you saw in Section 26.1, the aspect argument in rgl functions operates a little differently than the asp argument in image or the scale/expand arguments in persp. In rgl plots, including persp3d, aspect requests a numeric vector of length 3, which defines the relative scale of the $x$-, $y$-, and $z$-axes, in that order.

To determine the appropriate relative scales for the Chorley-Ribble data, you need to calculate the total $x$-axis and $y$-axis widths that the study region is defined upon and find their ratio.

```
R> xd <- chor.WIN$xrange[2]-chor.WIN$xrange[1]
R> xd
[1] 23
R> yd <- chor.WIN$yrange[2]-chor.WIN$yrange[1]
R> yd
[1] 21.38
R> xd/yd
[1] 1.075772
```

This was done using the $xrange and $yrange components of the spatstat polygon, subtracting the lower limit from the upper in each case. The final ratio of xd/yd reveals that you almost have a one-to-one scale, though technically the region is physically wider in the $x$-axis, by a factor of around 1.076, than it is in the $y$-axis.

Factoring that in, you can call persp3d to plot the KDE surface correctly.

```
R> persp3d(chor.dens.WIN$x,chor.dens.WIN$y,chor.dens.WIN$z,
        col=rbow[zm.colors],aspect=c(xd/yd,1,0.75),
        xlab="Eastings (km)",ylab="Northings (km)",
        zlab="Kernel estimate")
```

You use aspect to stipulate that the $x$-axis should be scaled according to a factor of xd/yd relative to the $y$-axis, that the $y$-axis is taken as the reference scale of 1, and that the $z$-axis should be squashed by a factor of 0.75 relative to the $y$-axis. This is arbitrarily set so that the graphic is similar to the original persp plot in Figure 25-21 on page 685 (Plate VIII in the color insert).

Let's finish off the plot by adding a color legend. Ensure you have the shape package loaded, resize your RGL device containing the result of the most recent call to persp3d, and execute the following command:

```
bgplot3d({plot.new();
        colorlegend(col=rbow,zlim=range(chor.dens.WIN$z,na.rm=TRUE),
```

```
zval=seq(0,0.02,0.0025),main="KDE",digit=4,
posx=c(0.87,0.9),posy=c(0.2,0.8))})
```

Remember that bgplot3d must be used to change the background of the current RGL device—refer to the end of Section 26.1. You might want to experiment a little with posx and posy to find your preferred placement of the color legend. Figure 26-7 shows the result on my machine.

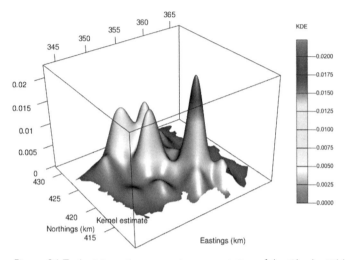

Figure 26-7: An interactive persp3d representation of the Chorley-Ribble kernel density estimate, colored according to z-axis values, with a static color legend

## Exercise 26.2

Return to the measurements in the built-in airquality data frame. Create a copy of the data frame containing the variables pertaining to temperature, wind speed, ozone level, and month; delete all rows with any missing values. You're now going to experiment with an rgl visualization of an earlier regression model for mean temperature.

a. Refit the multiple linear model from Exercise 25.4 on page 676, which regressed temperature against the main effects and an interactive effect of wind speed and ozone. Use expand.grid and predict to construct a z-matrix of the response surface; include the estimation of a 95 percent confidence interval for the fitted mean. Then, use rgl functionality to produce an interactive 3D plot of the response surface and color it yellow.

b. Using the built-in topo.colors palette, replot the response surface, assigning the colors according to the z value and setting the opacity at 80 percent. Tidy up the axis titles, resize the RGL device, and leave the plot open.

c. Enhance the plot from (b) as follows:

i. Generate exactly five colors from a custom palette that goes from "red4" to "pink" and add the raw wind, ozone, and temperature observations as points to the plot of the response surface. The points should be colored according to month (May through September) using these five colors in order. Set the size of the added points to 10.

ii. Add vertical lines that denote the residuals of the fitted model to the plot; in other words, each observation should have a vertical line connecting it to the corresponding fitted value of the response surface. These added lines should use the custom palette from earlier to match the color of each data point.

iii. Add the upper and lower 95 percent confidence limits you stored as part of the model prediction in (a). The added surfaces should both be gray with 50 percent opacity.

iv. Add a legend to the top-right corner of the interactive plot referencing the five colors of points/lines according to month. Use a pch value of 19 and a cex value of 2.

The result should look like this:

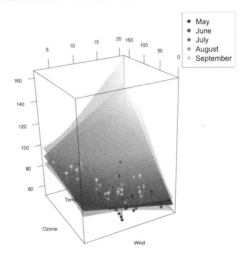

Next, load the spatstat package and revisit the clmfires data set. Execute the following lines to restrict attention to only the intentionally lit fires and to obtain the geographical study region:

```
R> fire <- split(clmfires)$intentional
R> firewin <- clmfires$window
```

d. Reproduce the static perspective plot from Exercise 25.5 (e) on page 689 as an interactive perspective plot, based on the following guidelines. Then keep the plot open.

- Calculate the KDE surface of the $x- and $y-coordinates of fire, truncated to the study region in firewin, using a $256 \times 256$ evaluation grid.
- Use the built-in color palette heat.colors to color the surface according to the z value. Set the opacity to 70 percent.
- Ensure the x-y axes have the correct ratio. Then reduce the vertical aspect ratio to be 0.6 relative to the y-axis.
- Suppress the z-axis title but add neat "X" and "Y" titles to the other two axes.

e. Make the following enhancements to the plot:

i. Add the raw observations so they lie underneath the surface itself. To do this, set a constant z value for each data point as the minimum (and non-NA) value of the z-matrix.

ii. You can obtain the vectors of the x- and y-coordinates of the irregular polygon that forms the study region by using the vertices function of spatstat as follows:

```
R> firepoly <- vertices(firewin)
R> fwx <- firepoly$x
R> fwy <- firepoly$y
```

By supplying these two vectors to the appropriate x and y arguments of the lines3d function, add the study region to surround the superimposed observations lying flush to the x-y plane underneath the plotted surface. Again, you'll need to specify the z value as the minimum z-matrix value for all drawn lines. Set lwd=2 for a slightly thicker line than the one drawn by default.

Your production should look something like this:

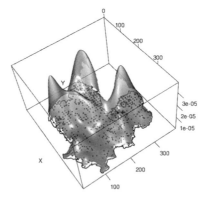

## 26.3　Trivariate Surfaces

So far you've looked at bivariate functions of the form $z = f(x, y)$, where your evaluation grid is two-dimensional. In other words, you evaluate the function $f$ from an $x$ value and a $y$ value; the $x$ and $y$ values are plotted on the first two axes, and the values of $f$ are used to plot a third dimension. Next you'll plot *trivariate* functions, which can be thought of as $w = f(x, y, z)$. That is, the evaluation grid is itself three-dimensional, and $f$ gives you a fourth value, $w$, to use for plotting the surface.

### 26.3.1　Evaluation Coordinates in 3D

When dealing with a trivariate mathematical function, you need an $x$, a $y$, and a $z$ value to evaluate the result. Rather than a flat evaluation grid, you'll have an evaluation *lattice* that sits in a cube or some other 3D prism.

As a perfect first example of a trivariate function, you'll create a "color cube" of RGB colors, where each point is the result of three values in red, green, and blue—refer to Section 25.1.1 for details. You'll use the three physical axes to reflect the evaluation lattice in the red, green, and blue values, and the result will be a point plotted with that color in that position in the 3D space.

The following code sets up the evaluation lattice in the three coordinate directions:

```
R> reds <- seq(0,255,25)
R> reds
 [1]   0  25  50  75 100 125 150 175 200 225 250
R> greens <- seq(0,255,25)
R> blues <- seq(0,255,25)
R> full.rgb <- expand.grid(reds,greens,blues)
R> nrow(full.rgb)
[1] 1331
```

The first four lines generate equally spaced increasing sequences in the three colors spanning the standard 0 to 255 RGB integer range. Then you use the built-in expand.grid function to generate the data frame of all unique color triplets according to these three sequences, resulting in an evaluation lattice of exactly $11^3 = 1331$ specific coordinates. Note that expand.grid works in the same way for higher-dimensional evaluation grids as it does for bivariate $x$-$y$ grids (refer to Section 25.3.1).

Finally, a call to plot3d places spheres at each 3D evaluation coordinate (recall the use of the rgb command from Section 25.1.1):

```
R> plot3d(x=full.rgb[,1],y=full.rgb[,2],z=full.rgb[,3],
          col=rgb(full.rgb,maxColorValue=255),type="s",
          size=1.5,xlab="Red",ylab="Green",zlab="Blue")
```

Figure 26-8 shows the result from two different angles so you can see how the intensities of the red, green, and blue components of an RGB triplet control the color of each point.

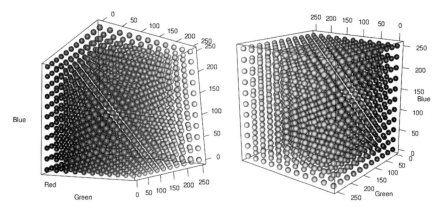

Figure 26-8: The `rgl` "color cube," created as spheres whose fourth-dimension result (color itself) is the product of evaluating the trivariate RGB function. See Plate XI in the color insert.

## 26.3.2 Isosurfaces

One of the problems with plotting the individual spheres of points at each 3D evaluation coordinate is revealed by Figure 26-8—it's difficult to see spheres "inside" the 3D prism. This same issue complicates the more general case of visualizing a continuous trivariate function.

To remedy this, you could instead produce an *isosurface*, which can be thought of as a kind of trivariate analog of a contour plot.

With an isosurface, you select a certain level of the values $w = f(x, y, z)$ and join up all the entries of $w$ at that level inside the 3D space to form a shape or "blob." These blobs show where in the 3D space the trivariate function takes on the chosen value. If you then plot these blobs at various levels, you get a 3D version of the contour plots you created in Section 25.4.1, showing which levels hold the highest densities of observations.

### Higher-Dimensional Probability Densities

Cast your mind back to the univariate normal probability density function detailed in Section 16.2.2. First, I'll introduce the idea of higher-dimensional density functions with the bivariate version of the normal distribution, and then I'll go one step further and use the trivariate version to illustrate isosurface plotting.

To work with multivariate normal distributions, you can use the mvtnorm package, installed with a call to install.packages("mvtnorm"). Just like the rnorm function for the univariate normal, the rmvnorm function is used to generate random variates from a specified multivariate normal. Once you've installed mvtnorm, execute the following code:

```
R> library("mvtnorm")
R> rand2d.norm <- rmvnorm(n=500,mean=c(0,0))
R> plot(rand2d.norm,xlab="x",ylab="y")
```

This produces the plot on the left of Figure 26-9. The rmvnorm function is used to generate 500 independent variates from the standard bivariate normal distribution. You center the variates around the coordinate (0,0) by passing a numeric vector to mean. By default, independent standard deviation components of 1 are used in both *x-y* coordinate directions.

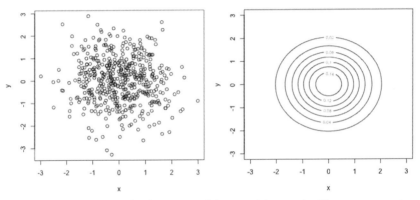

*Figure 26-9: Viewing randomly generated data, and the standard bivariate normal density they came from, using* mvtnorm *functionality*

To actually view the bivariate density function, you need to decide on the *x-y* evaluation grid and construct the *z*-matrix as usual using expand.grid. The following code sets up an evenly spaced sequence to use in both coordinate directions and uses the dmvnorm function (this is the multivariate version of dnorm and gives you the density function value at specified coordinates) to fill the *z*-matrix:

```
R> vals <- seq(-3,3,length=50)
R> xy <- expand.grid(vals,vals)
R> z <- matrix(dmvnorm(xy),50,50)
```

Then, you can use contour (or persp or persp3d) to view the density from which the data in rand2d.norm were generated for comparison (truncated to the limits −3 to 3 in both axes). The following line produces the plot on the right of Figure 26-9:

```
R> contour(vals,vals,z,xlab="x",ylab="y")
```

## Basic One-Level Isosurface

Now let's increase the dimension once more—what does a *trivariate* normal density function look like?

First, let's take a look at some data generated from this density. The following code generates 500 random variates again:

```
R> rand3d.norm <- rmvnorm(n=500,mean=c(0,0,0))
R> plot3d(rand3d.norm,xlab="x",ylab="y",zlab="z")
```

However, since you supplied a vector of length 3 as the mean argument to rmvnorm, the function knows you have three dimensions to work with. You're telling it that you want the data to come from a trivariate normal, with means of 0, 0, and 0 in each coordinate direction. You can see the rgl point cloud of the data produced via plot3d on the left of Figure 26-10.

To calculate and display the actual trivariate density function that generated these data, you'll need a 3D evaluation lattice, as noted at the beginning of Section 26.3.

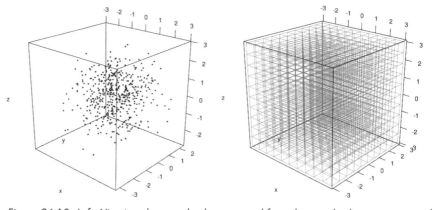

Figure 26-10: Left: Viewing data randomly generated from the standard trivariate normal distribution. Right: Concept of the 3D evaluation lattice upon which the trivariate density function itself will be plotted.

Take a look at the plot on the right of Figure 26-10. It shows a 3D $11 \times 11 \times 11$ evaluation lattice based on sequences spanning $[-3,3]$ in $x$, $y$, and $z$. This should give you a clear idea of how increasing the dimension of a continuous function works. Each intersection in the $11 \times 11 \times 11$ grid is the 3D equivalent of each intersection of the solid lines in the 2D $6 \times 4$ evaluation grid in Figure 25-9 on page 656, and each of the $10^3$ mini-3D cubes in this 3D lattice is the 3D equivalent of a 2D facet, as noted in the discussion of Figure 25-20 on page 683. (As with Figure 25-9, you can find the code to plot this 3D lattice on the book's website.)

To plot the result of the trivariate function, you need the unique evaluation coordinates of the evaluation lattice. Using vals, the sequence of values between $-3$ and 3 created earlier, the following code produces a data frame of all $50^3 = 125,000$ unique 3D evaluation lattice coordinates.

```
R> xyz <- expand.grid(vals,vals,vals)
R> nrow(xyz)
[1] 125000
```

Then you use dmvnorm to get the numeric value of the standard trivariate normal just as you did in the bivariate setting. The function automatically knows you're requesting the trivariate density because your data argument xyz has three columns.

```
w <- array(dmvnorm(xyz),c(50,50,50))
```

Note that the result is stored appropriately as a $50 \times 50 \times 50$ 3D array—refer to Section 3.4 for details on array. Look at the conceptual diagram of a 3D array (Figure 3-3 on page 53) and compare it to the 3D lattice on the right of Figure 26-10. The trivariate normal values in the object w are clearly represented by a 3D block of numbers sitting at each corresponding unique evaluation coordinate in the defined 3D space.

An isosurface can be produced using the contour3d function, part of the misc3d package (Feng and Tierney, 2008), which works closely with rgl. To use it, you need to decide on the level (or levels) at which to plot the surface. For densities, you typically make this choice with respect to what's called the $\alpha$-level contours; for more details, see the authoritative text on the theory of multivariate densities by Scott (1992). In brief, for some density $f$, these levels delineate the $(1 - \alpha) \times 100$ percent "most dense" observations by setting the isosurface to be drawn at positions in the multivariate evaluation lattice that correspond to the density value given by $\alpha \times \max(f)$.

For the trivariate standard normal, the maximum value of the density is located at the mean at the coordinate $(0, 0, 0)$.

```
R> max3d.norm <- dmvnorm(c(0,0,0),mean=c(0,0,0))
R> max3d.norm
[1] 0.06349364
```

You'll use this when producing the next couple of plots. Next, install misc3d, load it with library("misc3d"), and then call contour3d.

```
R> contour3d(x=vals,y=vals,z=vals,f=w,level=0.05*max3d.norm)
```

This produces an isosurface in an RGL device that you can rotate and zoom as you desire; you can see the result on the left of Figure 26-11. You supply contour3d with the arguments x, y, and z as evenly spaced sequences in the x-, y-, and z-coordinate directions, respectively (all are defined by the vector vals in this case). You supply the corresponding 3D array, defining the entire result of the trivariate function to f, and pass the level (or levels) at which you want to draw the isosurface itself to level. Here, I've chosen the $\alpha$-level to leave only 5 percent of the probability in the tails of the distribution, meaning that 95 percent of the total mass is held within the "blob."

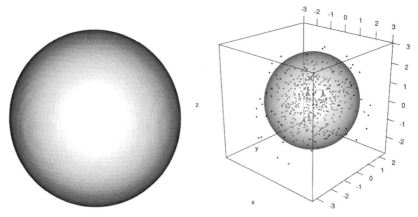

Figure 26-11: An isosurface of the trivariate standard normal density function produced using contour3d of the misc3d package. Left: Stand-alone plot drawn at an α level of 0.05. Right: Adding the same surface to an existing rgl plot of randomly generated trivariate normal observations, at 50 percent opacity.

The plot matches what you might expect—the shape of the trivariate density is relatively clear based on the plot of the randomly generated data you produced earlier. However, without scale, it's little more than a statistical golf ball. It's often more helpful to view the data alongside the density from which they came, which is also easy to do. The following code replots the data in rand3d.norm using plot3d and calls contour3d again to draw at the α level of 0.05:

```
R> plot3d(rand3d.norm,xlab="x",ylab="y",zlab="z")
R> contour3d(x=vals,y=vals,z=vals,f=w,level=0.05*max3d.norm,add=TRUE,alpha=0.5)
```

Just as with the traditional R contour function, if you want to use contour3d to add to an existing rgl plot (as is the case here), you need to explicitly specify add=TRUE. You can also use the optional alpha argument to adjust opacity, reduced to 50 percent in this example, to "see inside" the density isosurface.

## Controlling Multiple Levels with Color and Opacity

Playing with opacity is especially useful when you want to plot the isosurface at multiple α levels at once. Color is also useful in this way, as a variable that can represent a fourth dimension without adding an additional physical axis to the graph.

To view the trivariate normal density at multiple levels, consider the plot produced by executing the following code:

```
R> plot3d(rand3d.norm,xlab="x",ylab="y",zlab="z")
R> contour3d(x=vals,y=vals,z=vals,f=w,
            level=c(0.05,0.2,0.6,0.95)*max3d.norm,
```

```
color=c("pink","green","blue","red"),
alpha=c(0.1,0.2,0.4,0.9),add=TRUE)
```

Figure 26-12 shows the result. Here, you replot the 500 randomly generated trivariate normal observations, and another call to contour3d now draws contours at four specific $\alpha$-levels of the trivariate density—0.05, 0.2, 0.6, and 0.95. You use the optional color argument to render these in pink, green, blue, and red, respectively, and progressively increase the opacity of each level with the alpha argument.

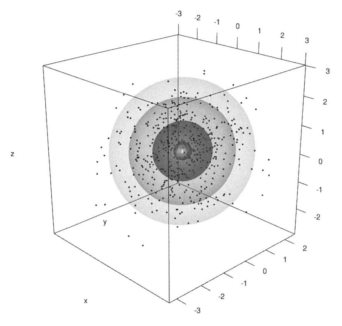

Figure 26-12: An isosurface of the trivariate normal density plotted at four levels, over the randomly generated observations. Color and opacity are used to distinguish among the different numeric levels of the plotted function. See Plate XII in the color insert.

You should be able to see that you can gauge the increase in the denseness of the points in the 3D space for this distribution in a similar way as you'd use standard 2D contours to appraise the distribution of bivariate observations.

### 26.3.3   Example: Nonparametric Trivariate Density

For an extended example using real data, look once more at the built-in quakes data frame, which includes the spatial location, magnitude, and depth of 1000 seismic events.

In Section 25.4.1, you constructed bivariate kernel density estimates of the 2D longitude-latitude spatial coordinates, using the MASS function kde2D. As noted there, KDE extends naturally to higher dimensions. The goal now is to calculate and visualize a density estimate of the same spatial earthquake data, but this time to do so based on the trivariate coordinates of longitude, latitude, *and* depth, in 3D space.

## Raw Data

First, let's look at the raw observations. The following code creates a copy of the quakes data, extracting those three variables and rendering depth negative. I do this so that, when plotted, earthquake depth corresponds to moving *down* the vertical axis to give the impression of depth below sea level.

```
R> quak <- quakes[,c("long","lat","depth")]
R> quak$depth <- -quak$depth
```

In the usual rgl fashion, you create a point cloud of the raw data with the following:

```
R> plot3d(x=quak$long,y=quak$lat,z=quak$depth,
          xlab="Longitude",ylab="Latitude",zlab="Depth")
```

Figure 26-13 shows the result. If you spin the plot so that you're looking down directly from the top with a bird's-eye view, you'll recognize the 2D spatial patterning that you've plotted already; see, for example, Figure 13-1 (page 265), Figure 23-1 (page 578), or Figure 25-12 (page 662).

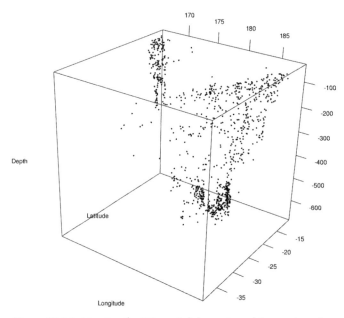

Figure 26-13: Viewing the 3D spatial dispersion of the earthquake occurrences—latitude, longitude, and depth

## Calculating the 3D Estimate

The evaluation lattice for this kernel estimate will be defined by the entire 3D space in which the longitude-latitude-depth data reside, exactly as illustrated in Section 26.3.2.

To actually calculate the 3D KDE surface for the quak data, you'll use the impressive functionality of the contributed package ks (Duong, 2007). Install the package and load it with a call to library("ks"). The kde function within the ks package allows you to use kernel smoothing to estimate the probability density of 1D through 6D data.

The first argument you supply to kde is your data, in the form of a matrix or data frame, with the tag x. Note that the order of the columns in your data object matters when using kde. When called as follows with quak, and taking into account the extraction order of the three variables in the earlier code that created quak, the *x*-, *y*-, and *z*-coordinate axes in the resulting 3D kernel estimate will correspond to longitude, latitude, and depth, respectively.

```
R> quak.dens3d <- kde(x=quak,gridsize=c(64,64,64),compute.cont=TRUE)
```

This matches the way the data are displayed in Figure 26-13. The gridsize argument specifies the lattice resolution in each axis. In this example, I've settled on a 64 × 64 × 64 lattice; by default, kde chooses the range of evaluation in each coordinate direction so that it's slightly wider than the observed data. Finally, to plot the result, it's useful to also specify the argument compute.cont=TRUE; I'll go into the reason for this in a moment.

The returned object has several components. The 3D estimate is provided as an appropriately sized array as the $estimate member; if you want to check, execution of the following line confirms it matches the desired lattice resolution:

```
R> dim(quak.dens3d$estimate)
[1] 64 64 64
```

The $eval.points component holds a list whose members are the specific evaluation coordinates, which are equally spaced sequences in each of the three axes. The number of members reflects the dimension of the problem, and their order corresponds to the specific axis. You can extract them with the following lines:

```
R> x.latt <- quak.dens3d$eval.points[[1]]
R> y.latt <- quak.dens3d$eval.points[[2]]
R> z.latt <- quak.dens3d$eval.points[[3]]
```

If you print these vectors to your console screen, you'll see that each is a vector of length 64, with x.latt, y.latt, and z.latt corresponding to the variables matching the order of the columns in the data frame quak.

## Isosurface Level Selection

The selection of the level to display depends on the range of values that make up the result of the trivariate function itself. When you choose `compute.cont=TRUE` in the call to `kde`, you're automatically provided with a collection of appropriate levels. These are returned in the component `$cont` as a numeric vector with a length of exactly 99, representing each integer between 1 percent and 99 percent.

Internally, these levels are calculated by working out the result of the trivariate function at the location of each of the originally observed data points and then using `quantile` to obtain all the integer-valued percentiles (from 99 percent to 1 percent) of these density values (for a refresher on quantiles, refer to Section 13.2.3). These are returned in decreasing order; in other words, `quak.dens3d$cont[1]` corresponds to the 99th percentile, and `quak.dens3d$cont[99]` is the 1st percentile.

Though these values are obtained in a different way from the $\alpha$-levels you experimented with when plotting the trivariate normal density, you essentially end up with same interpretation when visualizing the result— these values allow you to draw the isosurfaces at the level of estimated observation "denseness" that you want. For example, the lower quartile (aka the 25th percentile) is extracted with the following:

```
R> quak.dens3d$cont[75]
      25%
2.002741e-05
```

This provides the value of the KDE trivariate function that is estimated to separate the most spatially diffuse 25 percent of the observations from the rest (in other words, so that the resulting blobs encapsulate the most spatially dense 75 percent of the data).

**NOTE**    *At the time of writing, both the `rgl` and `misc3d` packages are dependencies of `ks`. This means they are loaded automatically when you load `ks`, so you don't need to call `library("rgl")` or `library("misc3d")` explicitly in this case, and `plot3d` and `contour3d` are already available to you. This may change as the developers update their packages over time.*

When you execute the following code, it first replots the quak data that the density estimate is based on and then adds the corresponding isosurface using the lower point-wise density quartile as the desired level. You can see the result on the left of Figure 26-14.

```
R> plot3d(x=quak$long,y=quak$lat,z=quak$depth,
          xlab="Longitude",ylab="Latitude",zlab="Depth")
R> contour3d(x=x.latt,y=y.latt,z=z.latt,f=quak.dens3d$estimate,
             color="blue",level=quak.dens3d$cont[75],add=TRUE)
```

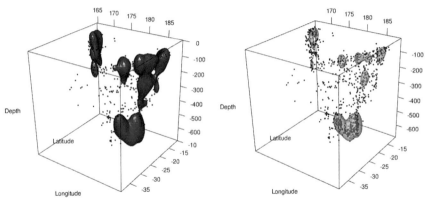

*Figure 26-14: Isosurfaces (3D contour plots) of a trivariate kernel density estimate drawn at point-specific density quantiles, based on use of the contributed* kde *and* contour3d *functions. Left: Solid blue delineation of the lower quartile—the 25 percent most diffuse points. Right: Green delineation of the median—the 50 percent most diffuse from the 50 percent most dense—with opacity reduced by half.*

Looking at the image, the blue blobs representing the 3D contour at the specified level are clear to see. Higher levels of the trivariate function, that is, more densely grouped points, are "inside" these blobs. In other words, the blue shapes encapsulate the observations associated with the highest 75 percent of estimated density with respect to latitude, longitude, and depth. To see inside the isosurfaces, you can adjust the opacity with alpha.

Let's take the level that delineates the observations associated with the lower and upper 50 percent of estimated density values.

```
R> quak.dens3d$cont[50]
       50%
3.649565e-05
```

Then, rerun the call to plot3d to replot the raw quak data. After that, a call to contour3d produces the result on the right of Figure 26-14, allowing you to see through the green blobs.

```
R> contour3d(x=x.latt,y=y.latt,z=z.latt,f=quak.dens3d$estimate,
            color="green",level=quak.dens3d$cont[50],add=TRUE,alpha=0.5)
```

Lastly, you'll highlight the top 80 percent of the mostly densely clustered observations using multiple levels. Execute the following:

```
R> qlevels <- quak.dens3d$cont[c(80,60,40,20)]
R> qlevels
         20%          40%          60%          80%
1.771214e-05 2.964305e-05 4.249407e-05 9.543976e-05
```

This obtains four levels—the quantiles above which the 80 percent, 60 percent, 40 percent, and 20 percent most densely clustered observations are identified. Then set up a couple of vectors to control the color and opacity of each increasing level of denseness accordingly.

```
R> qcols <- c("yellow","orange","red","red4")
R> qalpha <- c(0.2,0.3,0.4,0.5)
```

The range of colors and alpha levels means the isosurface will darken in color and become more opaque as the density increases.

One final time, replot the raw quak data using plot3d as earlier. Then it's simply a matter of supplying your vectors of length 4 to each appropriate argument in contour3d.

```
R> contour3d(x=x.latt,y=y.latt,z=z.latt,f=quak.dens3d$estimate,
             color=qcols,level=qlevels,add=TRUE,alpha=qalpha)
```

Figure 26-15 shows the results. You can see that the tightest grouping of earthquakes occurs quite deep and toward the eastern edge of the 3D spatial prism (the visible "three-chamber" density blob is a well-known feature of these particular data).

## 26.4 Handling Parametric Equations

In most of the examples in the chapter so far, the surfaces are directly defined by the coordinates of a regular evaluation grid or lattice, but there are situations where the final axis you want to visualize is *not* a function of some evaluation grid. This occurs quite naturally when you simply want to draw familiar geometric shapes but also extends to more complicated situations in mathematics.

In this section, you'll plot from a collection of parametric equations, which together define the shape or surface of interest. This section will assume you're familiar with the fundamental trigonometric functions *sine* and *cosine*, as well as the conversion of angles from *degrees* to *radians*, since by default R deals exclusively with the latter. That said, I'll walk you through the relevant calculations and R code as needed.

### 26.4.1 Simple Loci

Using mathematical terminology, a *locus* (plural *loci*) is a set of points that satisfy, and are defined by, a particular set of parametric equations. In R, these equations govern how individual numeric elements of the resulting objects are calculated, which you can then easily plot using familiar functions.

**NOTE** *When discussing loci, any reference to 2D or 3D space refers to Euclidian space, which is the standard way in which you've dealt with coordinates in the x-, y-, and z-axes so far.*

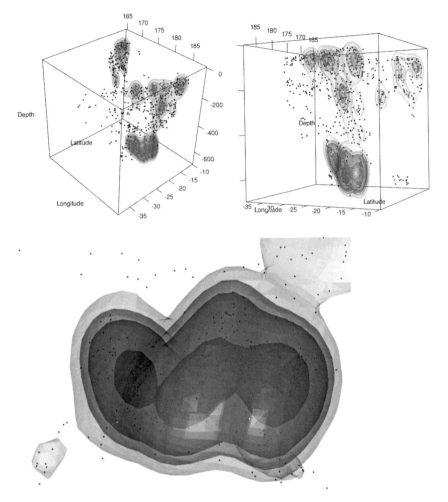

*Figure 26-15: Three screenshots of the trivariate kernel density estimate of the earthquake observations, taken from varying angles and with different levels of zoom. Increasing levels of denseness are reflected by isosurfaces of darkening yellow-to-red color and increased opacity. See Plate XIII in the color insert.*

## 2D Circle

Let's start with a simple example. One of the most immediately recognizable shapes defined in this way is a 2D circle. To find any point on a circle, you need to know the circle's center and its radius, and you need to provide a specific angle at which to look (typically taken to be relative to a perfectly horizontal line). Any planar 2D point $(x, y)$ that lies on a circle can be expressed with the following equations if you take the center to be at the coordinate $(a, b)$, with a fixed radius of $r > 0$ and looking at the angle $\theta$:

$$x = a + r\cos(\theta) \quad \text{and} \quad y = b + r\sin(\theta) \tag{26.1}$$

If you're working in degrees, then technically $0 \le \theta < 360$; to convert to radians, you must multiply by $\pi/180$ such that $0 \le \theta < 2\pi$.

To draw a circle based on the equations in (26.1), first decide on a radius, then decide on a center point, and then generate the corresponding values of $x$ and $y$. Consider the following code:

```
R> radius <- 3
R> a <- 1
R> b <- -4.4
R> angle <- 0:360*(pi/180)
R> x <- a+radius*cos(angle)
R> y <- b+radius*sin(angle)
R> plot(x,y,ann=FALSE)
R> abline(v=a)
R> abline(h=b)
```

The circle will have a radius of 3 and be centered at $(1, -4.4)$. Given the sequence defined as angle, note the plot will place a point at each integer angle from 0 to 360 degrees—I've allowed the upper limit to be equal to exactly 360 to fully complete the rotation—after which you convert to radians (with multiplication by $\pi/180$) in order to use the built-in R functions cos and sin. The geometric value of pi ($\pi = 3.1415...$) is held within the ready-to-use R object pi (see the help file ?Constants). The last three lines execute the plot, shown in Figure 26-16.

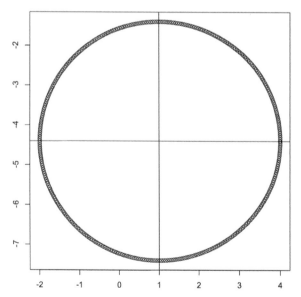

Figure 26-16: Drawing a 2D circle in R with center $(1, -4.4)$ and radius 3, following the relevant parametric equations of the locus

The key takeaway here is that $y$ isn't calculated as a direct result of $x$ in the same way as you might obtain a fine, evenly spaced increasing sequence in $x$ and then evaluate $y$ when plotting, for example, a linear regression model. Rather, the equations in (26.1) jointly define the rules of the locus in the 2D space.

## 3D Cylinder

Plotting surfaces with three dimensions is done in much the same way, only now your equations set up the rules for all satisfying points in the $x$-, $y$-, and $z$-axes.

For example, points lying on a hollow cylinder can be defined by the following equations:

$$x = r\cos(\theta), \quad y = r\sin(\theta), \quad \text{and} \quad z = z \qquad (26.2)$$

To actually plot points that satisfy these rules, you need to decide on a fixed radius $r$, recognize that $0 \le \theta < 360$ (in degrees), and define a fixed maximum height $h$ so that you can ensure $0 \le z \le h$. With that information, to generate vectors for $x$, $y$, and $z$, you need to first set up numeric sequences spanning the possible values of $\theta$ and $z$. Consider the following code:

```
R> r <- 3
R> h <- 10
R> zseq <- 0:h
R> theta <- 0:360*(pi/180)
```

These lines show a radius of 3 set as r and a maximum height of 10 as h. The sequence in $z$ is set up as the 11 integer values from 0 to 10 in zseq—this will allow you to place points on the locus at each of these defined $z$ values. The sequence for $\theta$ is set up as $0 \le \theta < 2\pi$ in theta (note the necessary conversion to radians). Then, you need all unique combinations of these parameter values to get all relevant $(x, y, z)$ coordinates for plotting. You know how to do that from Section 25.3.1, using expand.grid.

```
R> ztheta <- expand.grid(zseq,theta)
R> nrow(ztheta)
[1] 3971
```

Calling nrow on the result shows that you now have $11 \times 361 = 3971$ unique height-angle values. Now you're able to generate the values for $x$, $y$, and $z$ as defined by (26.2). You could use a for loop (Section 10.2.1), cycling through each row of ztheta, but a neater way would be to use implicit looping in apply (refer to Section 10.2.3 for details).

```
R> x <- apply(ztheta,1,function(vec) r*cos(vec[2]))
R> y <- apply(ztheta,1,function(vec) r*sin(vec[2]))
R> z <- apply(ztheta,1,function(vec) vec[1])
```

Note that disposable functions (see Section 11.3.2) are used to operate on the two-element height-angle (in that order) vectors that make up each row of ztheta.

You can use persp3d from rgl to plot this kind of parametrically defined surface, but in a slightly different way than in earlier sections of this chapter. The calculated x, y, and z coordinates must now *all* be supplied as identically sized, appropriately arranged matrices. This is because there's no longer an evenly spaced evaluation grid in the *x*- and *y*-coordinate directions—along with the *z* values, the *x* and *y* values have all been defined through an application of (26.2). In these types of plots, you effectively have a *latent* evaluation grid defined by the unique combinations of parameter values (height and angle in this case).

The matrices in all of the *x*-, *y*-, and *z*-coordinates are given by the three $11 \times 361$ matrices filled with x, y, and z in the typical column-wise fashion.

```
R> xm <- matrix(x,length(zseq),length(theta))
R> ym <- matrix(y,length(zseq),length(theta))
R> zm <- matrix(z,length(zseq),length(theta))
```

At this point, it's worth introducing the built-in outer function, which takes a sequence of values in two variables and produces all unique combinations of values, computes the result at each combination, and then returns the results as a matrix—doing the three tasks just done by expand.grid, apply, and matrix in one go. With this approach, you could create xm, ym, and zm indentically by simply calling the following:

```
R> xm <- outer(zseq,theta,function(z,t) r*cos(t))
R> ym <- outer(zseq,theta,function(z,t) r*sin(t))
R> zm <- outer(zseq,theta,function(z,t) z)
```

The only difference here is that the anonymous function provided as the third argument must be explicitly defined in terms of two separate arguments that represent the values of the necessary height and angle parameters.

However you obtain xm, ym, and zm, it's now just a matter of calling persp3d with these coordinate matrices. Make an additional call to points3d to emphasize the precise evaluation points that are returned in those matrices. The result of the next two lines is shown on the left of Figure 26-17.

```
R> persp3d(x=xm,y=ym,z=zm,col="red")
R> points3d(x=xm,y=ym,z=zm)
```

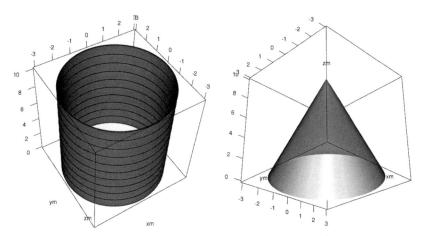

*Figure 26-17: Drawing a cylinder and a cone using* persp3d, *with matrix arguments in all three coordinate directions. The loci are defined by the corresponding parametric equations. The black rings visible on the cylinder represent the actual evaluation points stored in the required matrices* xm, ym, *and* zm.

### 3D Cone

This next example will show that once you understand the procedure involved with setting up the *x*-, *y*-, and *z*-coordinate matrices, you can display virtually any 3D shape or surface with ease. Taking *r*, *h*, and *θ* to be the base radius, maximum height, and angle, respectively, a cone follows these equations:

$$x = \frac{h-z}{h} r \cos(\theta), \quad y = \frac{h-z}{h} r \sin(\theta), \quad \text{and} \quad z = z \qquad (26.3)$$

Using the same objects r, h, zseq, and theta from earlier, the following code alters the disposable functions in outer to reflect (26.3). The right of Figure 26-17 shows the result.

```
R> xm <- outer(zseq,theta,function(z,t) (h-z)/h*r*cos(t))
R> ym <- outer(zseq,theta,function(z,t) (h-z)/h*r*sin(t))
R> zm <- outer(zseq,theta,function(z,t) z)
R> persp3d(x=xm,y=ym,z=zm,col="green")
```

### 26.4.2   *Mathematical Abstractions*

Many areas of mathematics, applied mathematical modeling, and statistics utilize high-dimensional shapes. To round off this chapter, and indeed the book, let's employ rgl to take a look at a couple of famous abstractions using skills from Section 26.4.1.

## Möbius Strip

A classic example is the *Möbius strip*—a continuous surface that has only one side and one edge. It can be expressed using the parametric equations

$$x = F(v,\theta)\cos\theta, \quad x = F(v,\theta)\sin\theta, \quad \text{and} \quad z = \frac{v}{2}\sin\left(\frac{\theta}{2}\right) \qquad (26.4)$$

where

$$F(v,\theta) = 1 + \frac{v}{2}\cos\left(\frac{\theta}{2}\right)$$

with $-1 \leq v \leq 1$ and $0 \leq \theta < 2\pi$ (assuming angles measured in radians). The parameter $v$ controls the position of the point along the width of the strip, and $\theta$ controls the rotation angle.

You can draw the strip in the same way as the cylinder and cone from earlier. First, set up the sequences over the possible values of $v$ and $\theta$, done here at a resolution of 200 each:

```
R> res <- 200
R> vseq <- seq(-1,1,length=res)
R> theta <- seq(0,2*pi,length=res)
```

Next, use outer to obtain the $200 \times 200$ matrices in each of the $x$-, $y$-, and $z$-coordinates as per (26.4).

```
R> xm <- outer(vseq,theta,function(v,t) (1+v/2*cos(t/2))*cos(t))
R> ym <- outer(vseq,theta,function(v,t) (1+v/2*cos(t/2))*sin(t))
R> zm <- outer(vseq,theta,function(v,t) v/2*sin(t/2))
```

Then, a quick call to plot3d from the rgl package will show you the 40,000 locations, based on the defined vseq and theta sequences, that lie on the Möbius strip. The result of the following line is shown on the left of Figure 26-18:

```
R> plot3d(x=xm,y=ym,z=zm)
```

Let's display the strip as a continuous surface using persp3d to fully appreciate the one-side/one-edge phenomenon. The image on the right of Figure 26-18 shows the result of the following code:

```
R> persp3d(x=xm,y=ym,z=zm,col="orange",axes=FALSE,xlab="",ylab="",zlab="")
```

Note the use of axes to suppress the default box and axes and the use of empty strings to remove the default axis titles denoting xm, ym, and zm.

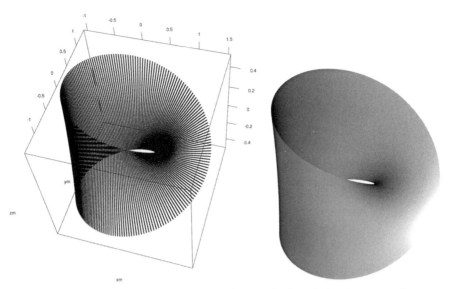

*Figure 26-18: Drawing a Möbius strip in R. Left: Specifically calculated points on the strip, visualized with* `plot3d`. *Right: Surface formed by joining up the points on the left via* `persp3d`.

You could also use color in a more interesting way to emphasize the wraparound nature of a Möbius strip. Taking inspiration from a similar collection of colors defined in Section 25.1.3, create the following custom palette:

```
R> patriot.colors <- colorRampPalette(c("red4","red","white","blue",
                                         "white","red","red4"))
```

This palette has been specifically generated to go from a dark red to white to blue but also to wrap around from the blue back to white and the dark red. This is needed for the way in which colors from `patriot.colors` will be assigned, in a point-wise fashion, to the plotted strip.

The color vector for plotting the surface will need to be of length $200^2 = 40,000$ given the preset value of res (governing the length of both vseq and theta). To fill the vector, execute the following:

```
R> patcols <- patriot.colors(2*res-1)
R> stripcols <- rep(NA,res^2)
R> for(i in 0:(res-1)){
+    stripcols[1:res+res*i] <- patcols[1:res+i]
+ }
```

The first line generates exactly 399 colors from `patriot.colors`, and the second sets up a vector of the required length, which will store the assigned colors (stripcols). The for loop ensures that at the first iteration, the elements 1 through 200 in stripcols will be assigned colors 1 through 200 from patcols; at the second iteration, the elements 201 through 400 will be assigned colors 2 through 201 from patcols, and so on. This gives the colors their wraparound appearance.

To properly understand the for loop, first look at the order of the arguments as supplied to the calls to outer. Having vseq first and theta second implies that each column of 200 in the resulting matrices corresponds to a span from $-1$ to $1$ in $v$, which refers to moving from one end of one of the lines of points to the other, that is, along the width of the strip. By using an index variable i from 0 to 199 (inclusive), the loop assigns each consecutive block of 200 elements in stripcols (increased at every iteration via +res*i) the 200 elements from the 399 patcols by collectively moving forward exactly one element (increased at every iteration via +i). What that does is change the color from red to white to blue in the plotted lines of points in the early stages of the loop, but as it progresses, working its way around the strip, the specific span of that palette is incrementally shifted until it goes from blue to white to red as you rotate around to the last few plotted lines of points. The effect is a smooth change in color as you alter both $v$ and $\theta$. You can see the result of the following in Figure 26-19:

```
R> persp3d(x=xm,y=ym,z=zm,col=stripcols,aspect=c(2,2.5,1.5),axes=FALSE,
          xlab="",ylab="",zlab="")
```

Figure 26-19: A patriotic Möbius strip, created with the careful construction of an appropriate color vector. See Plate XIV in the color insert.

The plot can still be rotated and zoomed as usual on your computer. You can experiment with aspect to alter the specific axis aspect ratios purely to enhance the appearance of the final product; here I've widened both the x- and y-axes relative to the z-axis.

## Torus

Another shape often expressed in 3D space is a *ring torus* (plural *tori*). This is the classic topological "shape with one hole" and resembles, for lack of a better word, a doughnut. The mathematical properties of tori are quite useful in many fields.

Parameterization of a torus may be achieved with these equations:

$$x = F(\theta_2; \alpha, \beta) \cos\theta_1, \quad x = F(\theta_2; \alpha, \beta) \sin\theta_1, \quad \text{and} \quad z = \alpha \sin\theta_2 \quad (26.5)$$

where
$$F(\theta_2; \alpha, \beta) = \beta + \alpha \cos\theta_2$$

with $0 \le \theta_1 < 2\pi$ and the same for $\theta_2$ (assuming angles measured in radians). The fixed values $\alpha$ and $\beta$ control the radius of the "tube" (in other words, the relative thickness of the doughnut) and the overall size of the torus in terms of distance from the middle of the hole to the middle of the tube. Provided $\alpha < \beta$, the equations in (26.5) give you the classic ring torus shape; you can get different kinds of tori by relaxing that condition on $\alpha$ and $\beta$.

Setting $\alpha = 1$ and $\beta = 2$, the following code uses the theta object defined earlier for the Möbius strip to compute the matrices in the x-, y-, and z-coordinate directions as per (26.5):

```
R> alpha <- 1
R> beta <- 2
R> xm <- outer(theta,theta,function(t1,t2) (beta+alpha*cos(t2))*cos(t1))
R> ym <- outer(theta,theta,function(t1,t2) (beta+alpha*cos(t2))*sin(t1))
R> zm <- outer(theta,theta,function(t1,t2) alpha*sin(t2))
```

Refer to Section 26.4.1 to remind yourself of the usage of outer if you need to do so.

Then, this line reveals the calculated points of the torus:

```
R> plot3d(x=xm,y=ym,z=zm)
```

And this gives you the final appearance of the continuous surface:

```
R> persp3d(x=xm,y=ym,z=zm,col="seagreen4",axes=FALSE,xlab="",ylab="",zlab="")
```

Figure 26-20 shows the results of both.

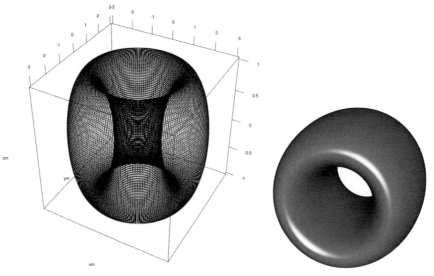

Figure 26-20: Drawing a ring torus in R. Left: Specifically calculated points on the surface, visualized with plot3d. Right: Shape formed by joining up the points on the left using persp3d.

Earlier, you used a specifically constructed color vector to color the Möbius strip, but you can assign color on any such surface by identifying the specific points in the defined matrices that you want to exert control over. Since it's the last example in the book, let's have a little fun with the current mathematical doughnut to illustrate this point-wise indexing.

First, the dough needs to look authentic. The following line sets up the vector of length $200^2 = 40,000$ to store the colors you're going to use. Initially, every element is set as a dough-colored tan.

```
R> donutcols <- rep("tan",res^2)
```

Next, add some icing. If you look at the distribution of the points, displayed in the plot on the left of Figure 26-20, you can see that the "top half" of plotted locations on the surface of this torus are at $z$-coordinates greater than zero. With this, you can overwrite the relevant elements of donutcols with the following line:

```
R> donutcols[as.vector(zm)>0] <- "pink"
```

Lastly, any premium doughnut should have sprinkles. You need a mechanism to identify random locations on the top half of the surface and color them appropriately. To do this, you can use the built-in sample function to randomly select a subset of elements from an existing vector. If you have the integers 1 through 10, for example, and you want to randomly select four, you can execute the following:

```
R> sample(x=1:10,size=4)
[1] 8 9 2 6
```

The argument x takes the vector from which to select a sample, and size takes the number of elements you want to draw from that vector. Note that you're likely to get a different set of four random numbers when executing this line.

With that knowledge, you can use this code to make sprinkles:

```
R> sprinkles <- c("blue","green","red","violet","yellow")
R> donutcols[sample(x=which(as.vector(zm)>0),size=300)] <- sprinkles
```

This sets up five distinct sprinkle colors; then randomly selects 300 locations, strictly from locations on the iced surface area; and finally assigns to them the five colors. The nature of vector recycling means there will be exactly 60 of each color sprinkle, randomly placed on the top half of the torus. You can add more sprinkles by increasing size, though given the number of colors, you should ensure size remains evenly divisible by 5.

The following call completes the visual treat, shown in Figure 26-21:

```
R> persp3d(xm,ym,zm,col=donutcols,aspect=c(1,1,0.4),axes=FALSE,
           xlab="",ylab="",zlab="")
```

*Figure 26-21: A delicious mathematical doughnut. Coloring of the torus surface is achieved by identification of corresponding positions in a color vector and subsequent element replacement. See Plate XV in the color insert.*

You can find a more serious (and technical) visualization of a ring torus, in its role as a convenient computational structure used for generating specially defined high-dimensional random normal variates, in Davies and Bryant (2013).

---

### Exercise 26.3

Ensure you have the functionality of the mvtnorm, rgl, misc3d, and ks packages available in your current R session. By specifying different *covariance matrices*, you can control how the different components of a multivariate normal random variable relate to one another, which affects the appearance of the distribution itself. In the standard trivariate normal density, for example, the three elements $(x, y, z)$ are independent of one another. Executing the following code will generate 1000 observations from a nonstandard trivariate normal distribution where the three elements are related to one another in a specific way:

```
R> covmat <- matrix(c(1,0.8,0.4,0.8,1,0.6,0.4,0.6,1),3,3)
R> rand3d.norm <- rmvnorm(1000,mean=c(0,0,0),sigma=covmat)
```

Note that the covariance matrix covmat is supplied to the optional sigma and that the mean of this collection of points remains centered at $(0, 0, 0)$.

a.  Plot the generated data as an interactive 3D point cloud, with simple axis titles "x", "y", and "z". You should see how the points form an elliptical shape, in contrast to the spherical shape in the standard trivariate normal in Figures 26-11 and 26-12. Keep the plot open.

b.  Using a $50 \times 50 \times 50$ evaluation lattice between −3 and 3 in each of the three axes, calculate this particular trivariate normal density function using dmvnorm and store it as an appropriately sized array. Note that you'll also need to set sigma=covmat in any use of dmvnorm. Calculate the maximum value of the density and use this to superimpose upon the point cloud isosurfaces at three specific $\alpha$ levels—0.1, 0.5, and 0.9. Color the three isosurfaces "yellow", "seagreen4", and "navyblue" and set them at 20 percent, 40 percent, and 60 percent opacity, respectively.

c.  Now, use ks functionality to calculate a 3D kernel estimate of the density based on the 1000 generated observations. Ensure that the returned object contains the vector of 99 sensible contour levels. Replot the point cloud of (a) in a new RGL device and then make two separate calls to contour3d as follows.

i.  The first should superimpose the theoretical contour at an $\alpha$ level of 0.5 from (b) only. Use the same color and opacity as you did in (b).

ii. The second should draw the isosurface at the 50th percentile as estimated from the point-specific KDE surface. Make it red and reduce the opacity to 20 percent.

Here are my results from (b) on the left and (c) on the right. Note that the appearance of your KDE isosurface will vary because of the random generation of the 1000 data points, which dictate the final estimate.

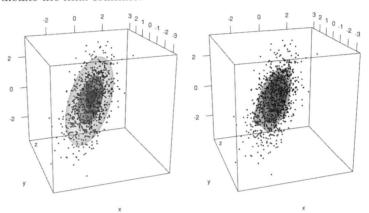

The MASS package has another data set you've not yet met. The Boston data frame object contains a number of descriptive observations concerning house prices in suburbs of Boston, Massachusetts, in the 1970s (Harrison and Rubinfeld, 1978). Load the MASS package and inspect the help file ?Boston to learn about the present variables.

d.  Focus on the variables for average number of rooms, percentage of lower-socioeconomic-status dwellings, and median value—you're going to experiment with the visualization of a 3D scatterplot as follows:

i.  Use rgl functionality to plot the three variables, with rooms, status, and value on the $x$-, $y$-, and $z$-axes, respectively; supply tidy axis titles. The data points should be plotted as gray spheres, with a size of 0.5. Keep the plot open.

ii. Use ks functionality to estimate the trivariate density function of these data. Base the estimate on a $64 \times 64 \times 64$ evaluation lattice; ensure the 99 integer percentiles of the observation-specific density levels are returned. Superimpose isosurface contours delineating the 75 percent, 50 percent, and 10 percent "most dense" observations using green, yellow, and blue. Set the opacity at 10 percent, 40 percent, and 50 percent, respectively.

iii. Finally, add reference grids along the three planes identified at the lower *z*-, upper *x*-, and upper *y*-axis locations.

e. Interpret the final plot from (d). For example, what values of the present variables tend to characterize the most common types of houses in Boston's suburbs?

Here's the result:

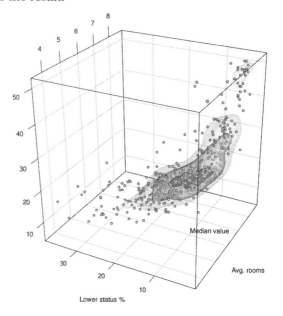

The *umbilic torus* is another interesting 3D shape in mathematics and can be defined by the following parametric equations:

$$x = \sin(\theta)F(\theta, \phi)$$
$$y = \cos(\theta)F(\theta, \phi)$$
$$z = \sin(\theta/3 - 2\phi) + 2\sin(\theta/3 + \phi)$$

In these equations, $F(\theta, \phi) = 7 + \cos(\theta/3 - 2\phi) + 2\cos(\theta/3 + \phi)$, and you allow for both $-\pi \le \theta \le \pi$ and $-\pi \le \phi \le \pi$.

f. Using a sequence of length 1000 for both $\theta$ and $\phi$, as well as 1000 colors generated from the built-in `rainbow` palette assigned directly to the `col` argument, produce an interactive 3D plot of the umbilic torus. Suppress the box, axes, and axis titles. In viewing the object from different perspectives, note that, much like the Möbius strip you plotted earlier, this shape has only one edge.

Here's the result:

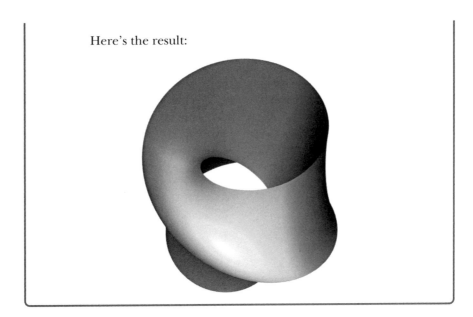

## Important Code in This Chapter

| Function/operator | Brief description | First occurrence |
| --- | --- | --- |
| plot3d | Interactive 3D point cloud | Section 26.1.1, p. 692 |
| legend3d | Add RGL device legend | Section 26.1.2, p. 693 |
| bg3d | Reset RGL device background | Section 26.1.2, p. 693 |
| segments3d | Add 3D line segments | Section 26.1.3, p. 696 |
| grid3d | Add plane grid | Section 26.1.3, p. 696 |
| bgplot3d | Alter/replot RGL device background | Section 26.1.3, p. 696 |
| persp3d | Interactive 3D perspective surface | Section 26.2.1, p. 700 |
| points3d | Add 3D points | Section 26.2.2, p. 701 |
| text3d | Add 3D text | Section 26.2.2, p. 702 |
| rmvnorm | Random multivariate normal variates | Section 26.3.2, p. 711 |
| dmvnorm | Multivariate normal density | Section 26.3.2, p. 711 |
| contour3d | Draw isosurface | Section 26.3.2, p. 713 |
| kde | Multivariate kernel estimation | Section 26.3.3, p. 717 |
| pi | Geometric value $\pi$ | Section 26.4.1, p. 722 |
| sin, cos | Sine and cosine | Section 26.4.1, p. 722 |
| outer | Outer array product | Section 26.4.1, p. 724 |
| sample | Random sample from vector | Section 26.4.2, p. 731 |

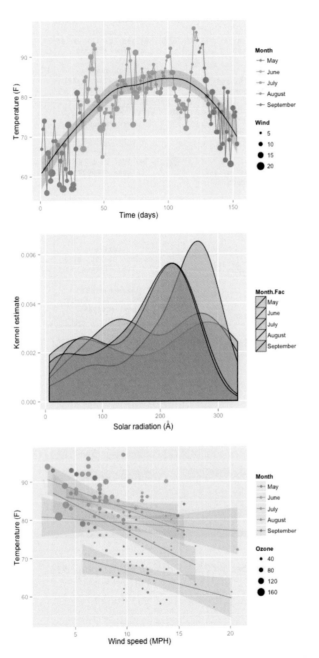

*Plate I: Produced in Section 24.3.1 as Figure 24-4; page 617*

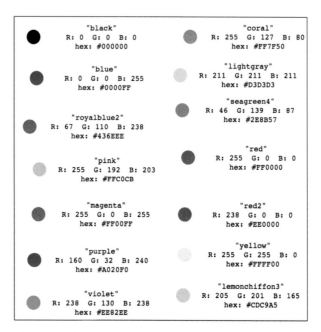

*Plate II: Produced in Section 25.1.1 as Figure 25-1; page 634*

*Plate III: Produced in Section 25.1.2 as Figure 25-2; page 636*

*Plate IV: Produced in Section 25.1.3 as Figure 25-3; page 637*

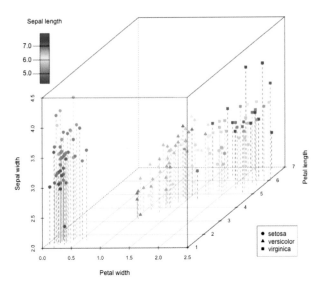

*Plate V: Produced in Section 25.2.2 as Figure 25-8; page 652*

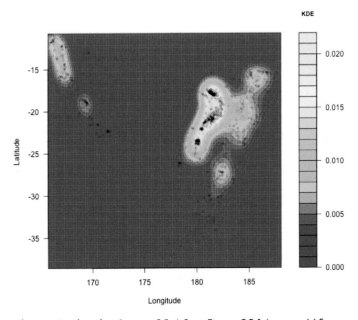

*Plate VI: Produced in Section 25.4.2 as Figure 25-14; page 665*

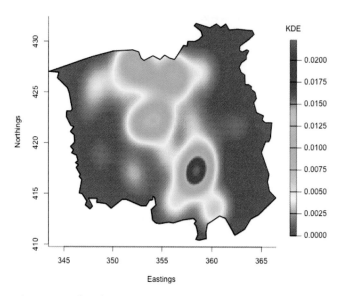

*Plate VII: Produced in Section 25.5.2 as Figure 25-18; page 676*

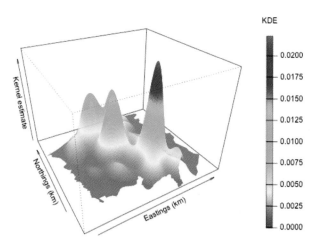

*Plate VIII: Produced in Section 25.6.2 as Figure 25-21; page 685*

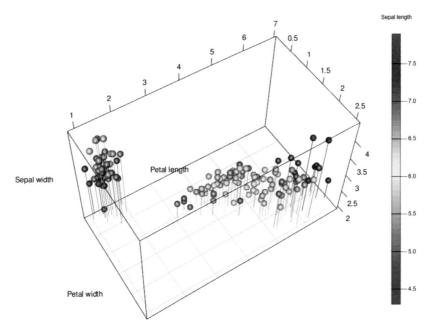

Plate IX: Produced in Section 26.1.3 as Figure 26-3; page 697

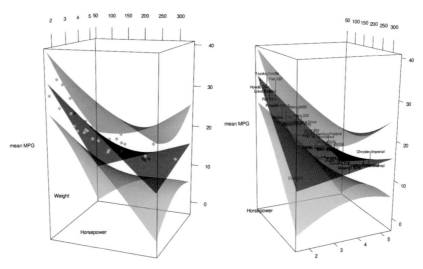

Plate X: Produced in Section 26.2.2 as Figure 26-5; page 702

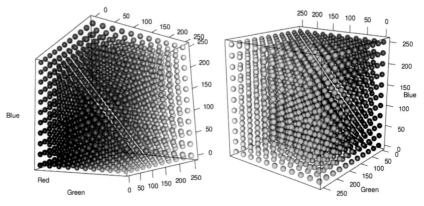

*Plate XI: Produced in Section 26.3.1 as Figure 26-8; page 710*

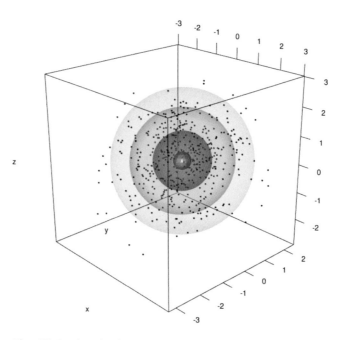

*Plate XII: Produced in Section 26.3.2 as Figure 26-12; page 715*

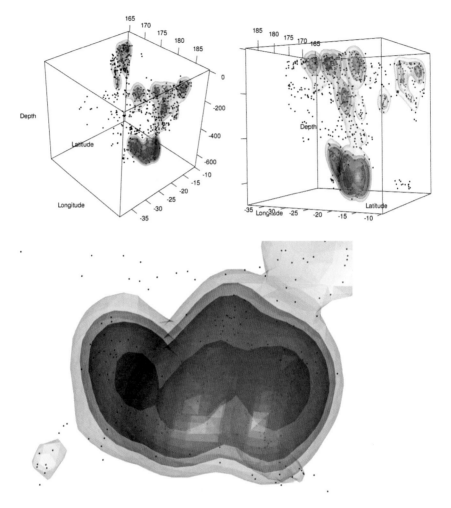

Plate XIII: Produced in Section 26.3.3 as Figure 26-15; page 721

Plate XIV: Produced in Section 26.4.2 as Figure 26-19; page 728

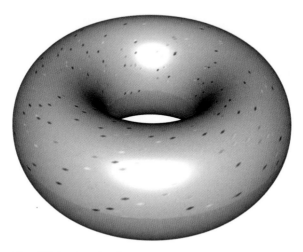

Plate XV: Produced in Section 26.4.2 as Figure 26-21; page 731

# A

## INSTALLING R AND CONTRIBUTED PACKAGES

 This appendix provides more detail on where to find R and how to install it and its contributed packages. R is available through CRAN—the Comprehensive R Archive Network—accessed through the R website at *https://www.r-project.org/*. I'll cover only the fundamentals here, but you'll find a substantial amount of information in the R FAQ by Hornik (2015) at *http://CRAN.R-project.org/doc/FAQ/R-FAQ.html*. This should be your first port of call if you need help with installing R and its packages. Installation of R and its contributed packages is dealt with in Sections 2 and 5 of the FAQ, respectively.

## A.1 Downloading and Installing R

Once at the R website, click the **CRAN mirror** link in the welcoming text or the **CRAN** link under the Download heading on the left, as shown in Figure A-1, and a page will load that asks you to select a CRAN mirror.

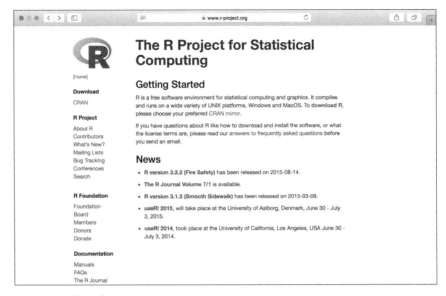

Figure A-1: The R home page

Pick one close to your geographical location and click the link. Figure A-2 shows my local mirror, which is at the University of Auckland; yours will look the same.

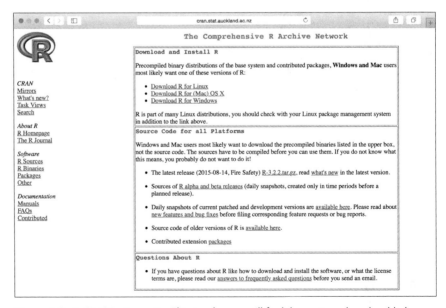

Figure A-2: A CRAN mirror site. This is where you'll find the various download links.

Then click the link for your operating system.

- If you're a Windows user, click the Windows link, and from that page choose the installation file (the binary executable) for the base distribution. Double-click the executable file and follow the instructions in the installation wizard. You'll want either the 32- or 64-bit version depending on your current Windows installation—you can find out which you have by going to Control Panel → System.

- If you're a Mac user, click the Mac OS X link and you'll be taken to a page with the packaged binary files. At the time of writing, there are still two versions available: one for OS X 10.9 (Mavericks) and later, and another for OS X 10.6 to 10.8 (this file has the *snowleopard* designation), though support for Snow Leopard is being phased out. Download the correct file for your operating system. Once the download is complete, double-clicking it will immediately launch the installer; follow the instructions therein. I also recommend getting the XQuartz window system, freely downloadable from *http://xquartz.macosforge.org/*, which provides support for additional devices for graphics.

- Linux users will be taken to a subdirectory with folders named after operating systems such as Debian or Ubuntu. Click the link relevant to you and you'll be taken to a page providing step-by-step command line instructions for installing R.

## A.2   Using Packages

R *packages* (or *libraries*) are collections of code that hold data and functionality used in R. It's essential to get comfortable with loading these libraries to access certain features and commands.

Packages come in three flavors. Those that make up the core functionality of the software are included with the installation and are automatically loaded when you open R. Also, a handful of *recommended* packages are included with a typical R installation but aren't automatically loaded. Finally, a huge collection of user-contributed packages—more than 7000 at the time of writing—extend R's applications vastly.

### A.2.1   Base Packages

The *base packages* provide the essential syntax and commands for programming, computing, and graphics production, as well as the built-in data sets, basic arithmetic, and statistical functionality, and they are immediately available when you start R. At the time of writing, there are 14.

```
base       compiler   datasets   grDevices   graphics   grid    methods
parallel   splines    stats      stats4      tlctk      tools   utils
```

You'll find a brief description each of these base packages in Section 5.1.1 of Kurt Hornik's R FAQ.

### A.2.2   Recommended Packages

At the time of writing, there are 15 *recommended packages*, as noted in Section 5.1.2 of the R FAQ. These packages are included with any standard R installation, and they extend the functionality of the base packages to include slightly more specialized (yet still ubiquitous) statistical methods and computational tools. In this book, you'll only be using MASS and boot from this list.

| | | | | | |
|---|---|---|---|---|---|
| KernSmooth | MASS | Matrix | boot | class | cluster |
| codetools | foreign | lattice | mgcv | nlme | nnet |
| rpart | spatial | survival | | | |

These recommended packages *aren't* loaded automatically. If you want to access functions or data sets from these packages, load them manually with a call to library. For example, to access the data sets available as part of MASS, execute the following at the prompt:

```
R> library("MASS")
```

Some packages provide a short welcome message upon loading, and R always informs you of any masking (see Section 12.3.1) that has occurred.

When you close your current R session, the package will close too, so you need to reload it if you open another instance of R and want to use it again. If you decide you no longer need a package in any given session and want to unload it to, for example, avoid any potential masking issues, use detach as follows:

```
R> detach("package:MASS",unload=TRUE)
```

You can find topics and technical details concerning package loading and unloading in Sections 9.1 and 12.3.1 of this book.

### A.2.3   Contributed Packages

On top of these built-in and recommended packages, there's a massive collection of user-contributed packages available through CRAN, serving all kinds of purposes and applications in statistics and mathematics, computing, and graphics. If you navigate to your local CRAN mirror site, the Packages link on the left of the page (see Figure A-2) will take you to a page with further links that provide up-to-date lists of all the available packages on CRAN. You'll also find the useful CRAN Task Views web page, which is a collection of subject-specific articles giving an overview of relevant packages, as shown in Figure A-3. This is a great way to get familiar with the types of specialized analyses that are possible in R.

Because of the sheer number of available packages, naturally R doesn't include them all upon its installation, and as a researcher you'll only ever be interested in a relatively small subset of methods at any one time.

*Figure A-3: The CRAN Task Views web page. Each article discusses prominent CRAN packages in use in the field.*

In this book, you'll make use of a handful of contributed packages. Some are used to access certain data sets or objects, and others are used for their unique functionality or to illustrate statistical methodology. They are listed here:

| | | | | |
|---|---|---|---|---|
| car | faraway | GGally | ggplot2 | ggvis |
| gridExtra | ks | misc3d | mvtnorm | rgl |
| scatterplot3d | shape | spatstat | tseries | |

When you want access to any contributed package, you need an Internet connection to first download and install it. Packages are generally less than a few megabytes in size. Once a package is installed, you then load it with the usual call to `library` to access the relevant functionality. For the packages listed above, you'll be prompted to do this when necessary in the relevant parts of the book.

You'll now look at a few ways to perform a package download and installation for R, using the `ks` package (Duong, 2007) as an example.

**NOTE** *Contributed R packages tend to be of good quality in terms of correctness, speed and efficiency, and user-friendliness. Although there are fundamental compatibility checks that must be passed before a submitted package is made available on CRAN, passing them isn't a sign of the overall quality and usability of a given package. You can gauge that only by using the package, studying its documentation, and seeking out any related publications.*

## Finding Packages on CRAN

Every R package on CRAN has its own standard web page, providing the direct links to the downloadable files and important information about the package. Locate the package name on one of the lists on CRAN and click it, or do a quick Google search for, in our case, *ks r cran*. Figure A-4 shows the top of the web page for ks.

| | |
|---|---|
| **ks: Kernel Smoothing** | |
| Kernel smoothers for univariate and multivariate data. | |
| Version: | 1.9.5 |
| Depends: | R (≥ 1.4.0), KernSmooth (≥ 2.22), misc3d (≥ 0.4-0), mvtnorm (≥ 1.0-0), rgl (≥ 0.66) |
| Imports: | grDevices, graphics, multicool, stats, utils |
| Suggests: | MASS |
| Published: | 2015-10-07 |
| Author: | Tarn Duong |
| Maintainer: | Tarn Duong <tarn.duong at gmail.com> |
| License: | GPL-2 ǀ GPL-3 |
| URL: | http://www.mvstat.net/tduong |
| NeedsCompilation: | yes |
| Materials: | ChangeLog |
| In views: | Multivariate |
| CRAN checks: | ks results |

Figure A-4: Descriptive information on the CRAN web page for the ks package

Along with basic information such as the version number and the maintainer's name and contact information, you'll see the Depends field. This is important for installation; if the R package you're interested in is dependent on other contributed packages (not all are), then you also need to install those packages for your package to install successfully.

Looking at Figure A-4, you can see that your R version needs to be later than 1.4, that ks requires KernSmooth (already installed—it's one of the recommended packages noted in Section A.2.2), and that it also needs misc3d, mvtnorm, and rgl. Fortunately, if you install an R package directly from R, the dependencies are also installed automatically.

## Installing Packages at the Prompt

The quickest way to download and install a contributed R package is to use the install.packages command directly from the R prompt. Starting with a fresh installation of R, I see the following on my iMac:

```
R> install.packages("ks")
--- Please select a CRAN mirror for use in this session ---
also installing the dependencies 'misc3d', 'mvtnorm', 'rgl'

--snip--
```

You may first be asked to choose a CRAN mirror. The list pops up and defaults to the secure HTTPS servers only; choosing HTTP will switch to the unsecured ones. I chose HTTP, found the New Zealand mirror, and clicked OK, as shown in Figure A-5. Once you've done this, your selected mirror will remain set as the go-to site until you reset it; see Section A.4.1.

Figure A-5: Pop-up windows for selecting a CRAN mirror site to use for downloading contributed packages. Left: Optional selection of HTTP servers (as opposed to HTTPS). Right: Selection of my local HTTP mirror.

After you click **OK**, R lists any dependencies that will also be downloaded and installed and then presents download notifications for each package (in the snipped output).

Some additional notes are warranted:

- You need to install a package only once, and it will save to your hard drive to be loaded with a call to library as usual.

- You might be prompted to use or create a local folder on your computer to store installed packages. This is done to ensure R knows where to get packages from when they're requested with library. Agreeing to this means you have a user-specific library of packages, which is generally a good idea.

- There are a number of optional arguments for install.packages; see the help file given by a call to ?install.packages at the prompt. For example, if you want to specify a CRAN mirror in the console, supply the relevant URL as a character string to the repos argument, or if you want to prevent dependencies from installing, use the dependencies argument.

- You can also install R packages from source, that is, from the uncompiled code, which may have more recent updates than the precompiled binary version of the package. If you're an OS X user, recent versions of R will ask you whether you'd like to download packages from source for packages with versions available that are more recent than the precompiled binary version. To do so, you need certain command line tools installed on your system; if the download fails, you can stick with the binary version by answering n ("no") at the download from source prompt.

### Installing Packages with the GUI

The basic R graphical user interface (GUI) used in this book gives you the option to download and install contributed packages from the console using menu items. Here, you'll look briefly at the Windows and OS X versions.

On Windows, click the **Packages → Install package(s)...** menu item, shown on the left of Figure A-6. Select a CRAN mirror and a tall window will open, listing all available packages in alphabetical order. Scroll to select the package you're interested in. You can see my selection of ks on the right of Figure A-6. Click **OK** to download and install the packages and dependencies.

Figure A-6: Initiating download and installation of a contributed R package (and any missing dependencies automatically) via GUI menus in Windows

For OS X R, click the **Packages & Data → Package Installer** item in the OS X menu bar, as shown on the top of Figure A-7. When the package installer opens, click the **Get List** button to bring up the table of available packages. Select the package you want, being sure to select the **Install Dependencies** box near the bottom of the installer before clicking **Install Selected**. R will then download and install everything it needs, including any dependencies, as shown on the bottom of Figure A-7. You can select more than one package. Note that the options on the bottom left of the installer allow you to choose exactly where the installed packages will be stored; if you're a nonadmin user, you may need to create a user-specific library, as mentioned earlier.

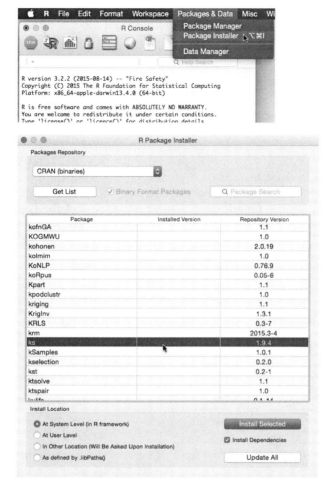

Figure A-7: Initiating download and installation of a contributed R package (and any missing dependencies) using the GUI-based package installer in OS X

## Installing Packages Using Local Files

Lastly, you can download the required package files from CRAN through your Internet browser as you'd download anything else, store them on your local drive, and then direct R to those local files.

On the CRAN web page for ks, you'll find a Downloads section as shown in Figure A-8. For Linux, choose the Package source file. For Windows or OS X, choose the corresponding *.zip* or *.pkg* file labeled *r-release*. The *r-oldrel* and *r-devel* versions should be used only if you're experiencing compatibility issues. The Old sources link holds archived source files for old versions.

On Windows, select **Packages → Install package(s) from local zip files...** (shown on the left of Figure A-6). This will open a file browser so you can navigate to the downloaded *.zip* file for the package; R will do the rest.

| Downloads: | |
|---|---|
| Reference manual: | ks.pdf |
| Vignettes: | kde |
| Package source: | ks_1.9.5.tar.gz |
| Windows binaries: | r-devel: ks_1.9.5.zip, r-release: ks_1.9.5.zip, r-oldrel: ks_1.9.5.zip |
| OS X Snow Leopard binaries: | r-release: ks_1.9.4.tgz, r-oldrel: ks_1.9.2.tgz |
| OS X Mavericks binaries: | r-release: ks_1.9.4.tgz |
| Old sources: | ks archive |

Figure A-8: The Downloads section of the CRAN web page for ks

On OS X, once you've downloaded the *.pkg* file, select **Packages & Data** → **Package Installer**. Use the drop-down menu at the top of the installer to select **Local Binary Package**, as shown in Figure A-9. To open a file browser to find the local file, you need to click the **Install...** button at the bottom of the installer. R will take it from there.

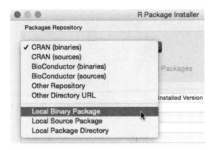

Figure A-9: Using the package installer on OS X to install an R package from a local file

It's important to know that this method *does not* automatically install any dependencies. You'll also need to install any packages your package depends on, any dependencies of those, and so on—so be sure to check the Depends field on the CRAN package web page, as noted earlier.

It's far easier to use install.packages directly from the R prompt, or through the GUI, to automate this process. You need to do local file installations only if the automatic methods fail for some reason, if you're installing a package that isn't on CRAN (or any other readily accessible repository—see Section A.4.2), or if the package isn't directly available for your operating system.

## A.3   Updating R and Installed Packages

There are roughly four new releases of R a year that address functionality, compatibility issues, and bug fixes. It's a good idea to keep up-to-date with these new releases. The R project home page and any CRAN mirror site will tell you the latest release, and you can execute news() at the R prompt for details on what's new.

Contributed R packages are also periodically updated and new package files uploaded to CRAN. Packages you've installed won't auto-update, so it's up to you to update them. It's difficult to predict how often such updates will be released, since it's completely up to the maintainers, but it's worth checking for newer versions of the packages you've installed every few months or, at the very least, when you upgrade your version of R.

Checking for package updates is easy. A simple call to update.packages(), with no arguments, will systematically look through your installed packages and flag any that have more recent versions available.

For example, on my current installation, executing the following tells me that a later version of MASS is available (along with several other packages not shown in this snipped output).

```
R> update.packages()
MASS :
 Version 7.3-43 installed in /Library/Frameworks/R.framework/Versions/3.2/
                            Resources/library
 Version 7.3-44 available at http://cran.stat.auckland.ac.nz
Update (y/N/c)? y

--snip--
```

Enter y to download the updated package from CRAN. If there are several packages with updates available, R will ask you one at a time whether you'd like to update, and you have to enter y (or N or c) for each.

You can also perform package updates using the R GUI menus (or manually with local file installation). In Windows, select **Packages** → **Update packages...** to open a list of available updates to your current packages. On OS X, a column in the populated table in the package installer provides information on the version of every package you currently have installed, as well as the version currently on CRAN, giving you the option to install the more recent version. There's also the Update All button, which is what you'd typically use.

## A.4 Using Other Mirrors and Repositories

Sometimes you might want to change the CRAN mirror associated with your typical package installation process or, indeed, change the target repository itself to one other than CRAN—there are several options.

### A.4.1 Switching CRAN Mirror

You'll rarely need to change your CRAN mirror, but you might want to if, say, your usual mirror site is inaccessible for some reason or you want to use R from a different location. To query your currently set repository, call getOption with "repos".

```
R> getOption("repos")
                          CRAN
"http://cran.stat.auckland.ac.nz"
```

To change this to, say, the mirror at the University of Melbourne, simply assign the new URL to the repos component in a call to options as follows:

```
R> options(repos="http://cran.ms.unimelb.edu.au/")
```

Any subsequent usage of install.packages or update.packages will now use this Australian mirror for downloads.

### A.4.2   Other Package Repositories

CRAN isn't the only repository of R packages. Other repositories include Bioconductor (see *https://www.bioconductor.org/*), Omegahat (see *http://www.omegahat.org/*), and R-Forge (see *https://r-forge.r-project.org/*), and there are several more. These repositories tend to deal with different subjects. Bioconductor, for example, hosts packages that deal with DNA microarray and other genomic analysis methodologies; Omegahat hosts packages that focus on web- and Java-based applications.

In terms of general statistical analyses, CRAN is the go-to repository for most users. To learn more about the other repositories, you can visit the associated websites.

## A.5   Citing and Writing Packages

It's important that the work put into R and its packages is recognized in the appropriate way when the software is used in, for example, data analysis as part of your research projects. Indeed, when you get to the stage where you're thinking about writing your own packages, it pays to be aware of the following.

### A.5.1   Citing R and Contributed Packages

For citing R and/or its packages, the citation command produces the relevant output.

```
R> citation()

To cite R in publications use:

  R Core Team (2016). R: A language and environment for statistical computing.
  R Foundation for Statistical Computing, Vienna, Austria. URL
  https://www.R-project.org/.

A BibTeX entry for LaTeX users is
```

```
@Manual{,
  title = {R: A Language and Environment for Statistical Computing},
  author = {{R Core Team}},
  organization = {R Foundation for Statistical Computing},
  address = {Vienna, Austria},
  year = {2016},
  url = {https://www.R-project.org/},
}
```

```
We have invested a lot of time and effort in creating R, please cite it when
using it for data analysis. See also 'citation("pkgname")' for citing R
packages.
```

Note that LaTeX users are conveniently catered to via automatically generated BibTeX entries.

You can also cite individual packages if these have been instrumental in
completing a particular piece of work. Here's an example:

```
R> citation("MASS")
```

```
To cite the MASS package in publications use:

  Venables, W. N. & Ripley, B. D. (2002) Modern Applied Statistics with S.
  Fourth Edition. Springer, New York. ISBN 0-387-95457-0

A BibTeX entry for LaTeX users is

  @Book{,
    title = {Modern Applied Statistics with S},
    author = {W. N. Venables and B. D. Ripley},
    publisher = {Springer},
    edition = {Fourth},
    address = {New York},
    year = {2002},
    note = {ISBN 0-387-95457-0},
    url = {http://www.stats.ox.ac.uk/pub/MASS4},
  }
```

## A.5.2  Writing Your Own Packages

Once you become an R expert, you may find yourself with a suite of functions, data sets, and objects that others might find useful or that you use
often enough to warrant packaging them up in a standardized, easily loadable format. There's certainly no obligation to submit your packages to
CRAN or any other repository, but if you do aim to do so, note that there

are rather strict requirements to assure users of the stability and compatibility of your package.

If you're interested in constructing your own installable R package, see the official *Writing R Extensions* manual, accessible on any CRAN mirror site by clicking the Manuals link under Documentation on the left of the home page; you can see this link in Figure A-2. If you're interested, you may also want to seek out the book by Wickham (2015*b*), which provides useful instruction on the R package-writing process and associated dos and don'ts.

# B

# WORKING WITH RSTUDIO

Although the base R application and GUI are all you need to unleash the full suite of available functionality, the bare-bones appearance of the console and code editor can be off-putting to some, especially beginners. One of the best integrated development environments (IDEs) designed specifically to enhance day-to-day use of the R language is RStudio.

Like R, the desktop version of RStudio (RStudio Team, 2015) is free and can be used on Windows, OS X, and Linux systems. Before installing RStudio, you must first have R installed, as described in Appendix A (OS X users will also want XQuartz; see Section A.1). Then, you can download RStudio from the official website at *https://www.rstudio.com/products/rstudio/download/*.

The RStudio website also hosts a variety of useful support articles and links, as well as instructions for various special enhancements, some of which are noted in Section B.2. If you need help with RStudio, see *https://support.rstudio.com/hc/en-us/*. In particular, you should take time to click the Documentation link; you can also view this through RStudio by selecting Help → RStudio Docs.

In this appendix, you'll get an overview of RStudio and its most commonly used tools.

# B.1   Basic Layout and Usage

The RStudio IDE is split into four panes, and you can customize the content and layout to suit your preferences. Figure B-1 shows my setup. In it, I'm playing with the ggvis code from Section 24.4.

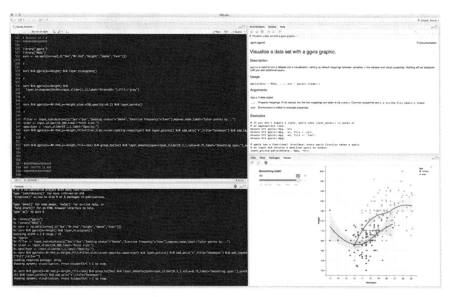

*Figure B-1: RStudio in action. The four panes can be arranged and suppressed as you like; here, you can see the code editor (top left), the console (bottom left), the help pages (top right), and the graphics viewer (bottom right). The panes on the right also have additional tabs to choose from.*

You write R code in the built-in editor and execute it in the console; the shortcut "send code to console" keystrokes are CTRL-ENTER or CTRL-R in Windows and ⌘-RETURN on a Mac. The textual output appears in the console as usual.

## B.1.1   Editor Features and Appearance Options

One of the most useful features of RStudio editor is the color-themed code highlighting and bracket matching. This makes for an easier coding experience than in the base R editor, particularly when writing long chunks of code. There are also autocomplete options that pop up as you're typing in either the editor or the console. You can see an example of this in Figure B-2.

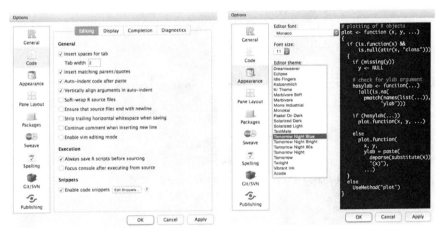

Figure B-2: RStudio's autocomplete feature includes hints about each option.

These features and more can be enabled, disabled, and customized using the RStudio options (select **Tools → Global Options...** on both Windows and OS X; for OS X, you can also select **RStudio → Preferences...**); you can see the Code and Appearance options in Figure B-3.

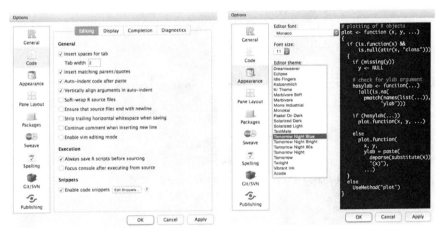

Figure B-3: The options panes for code editing (left) and appearance (right)

## B.1.2   Customizing Panes

Next you'll probably want to sort out the arrangement and content of the four RStudio panes. Two panes will always be the editor and the console, but you can set up a number of additional tabs to be displayed on the two utility panes. These include a file browser that you can use to search for and open R scripts on your local machine, plot and document viewers, standard R function help files, and a package installer.

You can configure your utility panes with the drop-down menus and checkboxes in the Pane Layout section of the RStudio options. Figure B-4 shows my current settings; one change I've made from the default arrangement is to make the help files appear in the topmost utility pane, with the graphical displays on the bottom, since I often want to refer to function documentation while experimenting with my plots.

Figure B-4: Pane layout and arrangement options

## B.2   Auxiliary Tools

RStudio gives you access to a number of handy tools to use in conjunction with R, which I'll briefly highlight here. For more information on a particular feature, check the supporting documentation at *https://support.rstudio. com/hc/en-us/*.

### B.2.1   Projects

RStudio *projects* assist with development and file management when you're working on more complicated endeavors. In these situations, you're typically working with multiple script files, you might want to save separate R workspaces, or you might have certain R options set to specific or non-default values. RStudio facilitates this process so you don't have to set it up manually.

At the top right of the RStudio window, you'll see a Project: (None) button. Click it and you'll see a short menu, as shown in Figure B-5; set up a basic project folder by clicking **New Project** and selecting the **New Directory** → **Empty Project** items.

Essentially, creating a new project does the following:

- Sets the working directory as the project folder

- Saves, by default, the R workspace, history, and all *.R* source files in said folder

- Creates a *.Rproj* file, which can be used to open a saved project at a later date, and stores RStudio options set specifically for that project

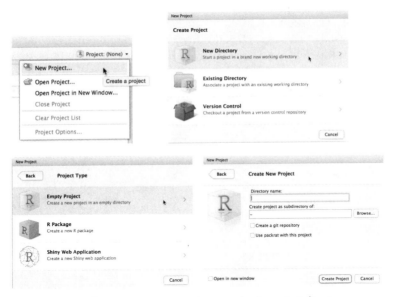

Figure B-5: RStudio project menus; setting up a basic project directory

When you're working on a specific project, its name will appear in place of (None) on the Project: (None) button.

### B.2.2 Package Installer and Updater

RStudio provides a package installer to manage the downloads and installations of contributed packages. You'll find the package manager in the Packages tab on your chosen utility pane. It lists only those packages that you currently have installed, along with their version numbers, and you can use the checkboxes next to each package name to load it (instead of using library at the R console prompt).

Mine appears in Figure B-6. I've just selected the box for the car package, which automatically executes the relevant call to library in the console.

| Files | Plots | Packages | Viewer | | |
| --- | --- | --- | --- | --- | --- |

| ☐ Install | ⟳ Update | Packrat | | | |
| --- | --- | --- | --- | --- | --- |
| Name | | Description | | Version | |
| **System Library** | | | | | |
| ☐ abind | | Combine Multidimensional Arrays | | 1.4–3 | ⊗ |
| ☐ acepack | | ace() and avas() for selecting regression transformations | | 1.3–3.3 | ⊗ |
| ☐ alphashape3d | | Implementation of the 3D alpha–shape for the reconstruction of 3D sets from a point cloud | | 1.1 | ⊗ |
| ☐ alr4 | | Data to accompany Applied Linear Regression 4rd edition | | 1.0.5 | ⊗ |
| ☐ assertthat | | Easy pre and post assertions. | | 0.1 | ⊗ |
| ☐ BH | | Boost C++ Header Files | | 1.58.0–1 | ⊗ |
| ☐ bitops | | Bitwise Operations | | 1.0–6 | ⊗ |
| ☐ boot | | Bootstrap Functions (Originally by Angelo Canty for S) | | 1.3–17 | ⊗ |
| ☐ brainR | | Helper functions to misc3d and rgl packages for brain imaging | | 1.2 | ⊗ |
| ☑ car | | Companion to Applied Regression | | 2.1–0 | ⊗ |
| ☐ CHsharp | | Choi and Hall Clustering in 3d | | 0.3 | ⊗ |

Figure B-6: The RStudio package installer, showing car being loaded by clicking its checkbox

The figure also shows the Install and Update buttons for packages. To install a package, click the **Install** button and enter the package you want in the field. RStudio will give you options as you type, as shown for the ks package on the left of Figure B-7. Ensure "Install dependencies" is checked in order to automatically install any additionally required packages.

To update packages, click the **Update** button to bring up the dialog box on the right of Figure B-7; here you can choose to update either individual packages or all of them by clicking the **Select All** button.

Figure B-7: Package installation and update features in RStudio

Naturally, you can still use the install.packages, update.packages, and library commands directly from the console prompt within RStudio if you prefer.

### B.2.3   Support for Debugging

Another nice feature of RStudio is its built-in tools for code debugging. Debugging strategies usually involve being able to "pause" your code at a specific point to inspect your objects and function values in a given "live" state. Specific techniques are best left to more advanced texts such as *The Art of Debugging* (Matloff and Salzman, 2008) and *The Art of R Programming* (Matloff, 2011); see also Chapter 9 of *Advanced R* (Wickham, 2015a). But I mention it here since the tools available in RStudio provide more convenient, higher-level support for debugging than do base R commands alone.

Once you're at the stage where you're starting to write programs comprised of multiple interlinked R functions, you might like to learn more. With respect to R and RStudio in particular, there's a good introductory article by Jonathan McPherson on the support website at *https://support.rstudio.com/hc/en-us/articles/205612627-Debugging-with-RStudio/*.

### B.2.4   Markup, Document, and Graphics Tools

When writing up reports on a project or tutorials for particular analyses, researchers often use a *markup* language. One of the best-known markup languages, particularly in the sciences, is LaTeX; it facilitates a unified approach to typesetting, formatting, and layout of technical documents.

There are specialized packages that incorporate R code into the compilation of these documents. These are in turn incorporated into RStudio, allowing you to create dynamic documents that make use of R code and graphics without switching between different applications.

You'll need a TeX installation on your computer to use these tools, which you can find at *https://www.latex-project.org/*. In this section, I'll briefly discuss the most widely used enhancements.

### Sweave

Sweave (Leisch, 2002) was arguably the first markup language to become popular with R; its functionality is included with any standard R installation. Sweave follows typical LaTeX markup rules; in your document, you declare special fields called *chunks*, in which you write R code and instruct any corresponding output to be displayed; the output can include both console text and graphics. When you compile the Sweave file (which has a *.Rnw* extension), the R code fields are sent to R for live evaluation, with the results appearing in the specified places of the finished product. To start a new document, choose **File → New File → R Sweave**, as shown in Figure B-8. For some examples and resources, visit the Sweave home page at *https://www.statistik.lmu.de/~leisch/Sweave/*.

*Figure B-8: Starting a new Sweave document in RStudio. The editor is used for markup and live code fields, and you use the Compile PDF button to render the result.*

### knitr

knitr (Xie, 2015) is an R package designed as an extension to Sweave, with some additional features that make document creation easier and more flexible. You can select knitr as the document "weaver" in the Sweave tab of the RStudio options, found by selecting **Tools → Global Options...** (see Figure B-9). To learn more about Sweave and knitr with respect to RStudio, consult Josh Paulson's article at *https://support.rstudio.com/hc/en-us/articles/200552056-Using-Sweave-and-knitr/*.

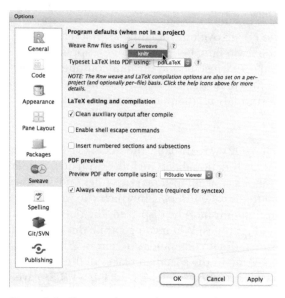

Figure B-9: Choosing knitr as the weaver of your markup document in the Sweave tab of the RStudio options

## R Markdown

R Markdown (Allaire et al., 2015) is another dynamic document creation tool, downloadable from CRAN as the rmarkdown package. Like both Sweave and knitr, its goal is to produce polished documents that can include R code and output automatically. Unlike Sweave and knitr, however, one of the objectives of R Markdown is to minimize the need to learn complicated markup languages like LaTeX, and as a result, its syntax is considerably simpler. Working from a *.Rmd* source file, you can create a variety of output document types, such as PDF, HTML, and Word.

To start a new R Markdown document, click the **R Markdown...** menu item under **File → New File**, as shown on the left of Figure B-8; this opens the New R Markdown dialog shown at the top of Figure B-10. Here, you can choose the appropriate document type for your project, and then you're provided with a basic template in the RStudio editor; one such template is shown on the bottom of Figure B-10. The template even points you toward the R Markdown home page at *http://rmarkdown.rstudio.com/*, which you should certainly investigate if you're interested in learning more. Garrett Grolemund also provides a useful collection of links for using R Markdown at *https://support.rstudio.com/hc/en-us/articles/205368677-R-Markdown-Dynamic-Documents-for-R/*.

*Figure B-10: Starting a new R Markdown file in RStudio. Templates relevant to your chosen output file type are automatically provided.*

## Shiny

Shiny is a framework for creating interactive web applications developed by the RStudio team. If you're interested in sharing your data, statistical models and analyses, and graphics, you can craft a Shiny app. The R package shiny (Chang et al., 2015) provides the required functionality. Shiny apps require you to have an R session running behind the scenes, which is what drives the plots as the user interacts with the application in a web browser.

Like other tools associated with RStudio, Shiny is a high-level framework intended to be friendly for both users and developers. Its emphasis is on creating interactive visuals not unlike the graphics you produced using ggvis in Section 24.4, which you can then deploy online for anyone to use.

You can find the Shiny website at *http://shiny.rstudio.com/*. The development team has put a tremendous amount of work into creating comprehensive tutorials, as well as a host of examples. Once you're comfortable with the app, you can even use Shiny to create interactive documents via R Markdown—notice the Shiny document option in the top image of Figure B-10.

# REFERENCE LIST

Adler, D., Murdoch, D., and others (2015). *rgl: 3D Visualization Using OpenGL.* *http://CRAN.R-project.org/package=rgl*, R package version 0.95.1247.

Allaire, J.J., Cheng, J., Xie, Y., McPherson, J., Chang, W., Allen, J., Wickham, H., Atkins, A., and Hyndman, R. (2015). *rmarkdown: Dynamic Documents for R.* *http://CRAN.R-project.org/package=rmarkdown*, R package version 0.8.1.

Anderson, E. (1935). "The irises of the Gaspe Peninsula." *Bulletin of the American Iris Society,* **59**, 2–5.

Atkinson, A.C. (1985). *Plots, Transformations and Regression.* Oxford University Press, UK.

Auguie, B. (2012). *gridExtra: functions in Grid graphics.* *http://CRAN.R-project. org/package=gridExtra*, R package version 0.9.1.

Baddeley, A.J. and Turner, R. (2005). "spatstat: An R package for analyzing spatial point patterns." *Journal of Statistical Software,* **12**(6), 1–42.

Becker, R.A., Chambers, J.M., and Wilks, A.R. (1988). *The New S Language.* Chapman & Hall, UK.

Bollen, K.A. and Jackman, R.W. (1990). "Regression diagnostics: An expository treatment of outliers and influential cases." In J. Fox and J.S. Long (eds.), "Modern Methods of Data Analysis," Sage, USA.

Canty, A. and Ripley, B.D. (2015). *boot: Bootstrap R (S-Plus) Functions.* *http://CRAN.R-project.org/package=boot*, R package version 1.3-15.

Chang, W., Cheng, J., Allaire, J., Xie, Y., and McPherson, J. (2015). *shiny: Web Application Framework for R.* *http://CRAN.R-project.org/package=shiny*, R package version 0.12.2.

Chang, W. and Wickham, H. (2015). *ggvis: Interactive Grammar of Graphics.* *http://CRAN.R-project.org/package=ggvis*, R package version 0.4.2.

Chatterjee, S., Hadi, A.S., and Price, B. (2000). *Regression Analysis by Example.* Wiley, USA, 3rd edition.

Chu, S. (2001). "Pricing the C's of diamond stones." *Journal of Statistics Education,* **9**(2). *http://www.amstat.org/publications/jse/v9n2/datasets.chu. html*.

Cox, D.R. and Snell, E.J. (1981). *Applied Statistics: Principles and Examples.* Chapman & Hall, USA.

Davies, T.M. and Bryant, D.J. (2013). "On circulant embedding for Gaussian random fields in R." *Journal of Statistical Software*, **55**(9), 1–21.

Davison, A.C. and Hinkley, D.V. (1997). *Bootstrap Methods and Their Applications.* Cambridge University Press, UK.

Dickey, D.A. and Arnold, J.T. (1995). "Teaching statistics with data of historic significance: Galileo's gravity and motion experiments." *Journal of Statistics Education*, **3**(1). *http://www.amstat.org/publications/jse/v3n1/datasets.dickey.html.*

Diggle, P.J. (1990). "A point process modelling approach to raised incidence of a rare phenomenon in the vicinity of a prespecified point." *Journal of the Royal Statistical Society Series A*, **153**, 349–362.

Dobson, A.J. (1983). *An Introduction to Statistical Modelling.* Chapman & Hall, UK.

Duong, T. (2007). "ks: Kernel density estimation and kernel discriminant analysis for multivariate data in R." *Journal of Statistical Software*, **21**(7), 1–16.

Faraway, J.J. (2005). *Linear Models with R.* Chapman & Hall, USA.

Feng, D. and Tierney, L. (2008). "Computing and displaying isosurfaces in R." *Journal of Statistical Software*, **28**(1), 1–24.

Fisher, J.C. (1976). "Homicide in Detroit: The role of firearms." *Criminology*, **14**, 387–400.

Fisher, R.A. (1936). "The use of multiple measurements in taxonomic problems." *Annals of Eugenics*, **7**(2), 179–188.

Fox, J. and Weisberg, S. (2011). *An R Companion to Applied Regression.* Sage, USA, 2nd edition.

Golub, G.H. and Van Loan, C.F. (1989). *Matrix Computations.* Johns Hopkins University Press, UK, 2nd edition.

Gupta, S. (2012). "How R Searches and Finds Stuff." *http://blog.obeautifulcode.com/R/How-R-Searches-And-Finds-Stuff/.*

Hand, D.J., Daly, F., McConway, K., Lunn, D., and Ostrowski, E. (eds.) (1994). *A Handbook of Small Data Sets.* Chapman & Hall, USA.

Härdle, W. (1990). *Applied Nonparametric Regression.* Cambridge University Press, USA.

Harraway, J.A. (1995). *Regression Methods Applied.* University of Otago Press, New Zealand.

Harrison, D. and Rubinfeld, D.L. (1978). "Hedonic prices and the demand for clean air." *Journal of Environmental Economics and Management*, **5**, 81–102.

Henderson, H.V. and Velleman, P.E. (1981). "Building multiple regression models interactively." *Biometrics*, **37**, 391–411.

Hornik, K. (2015). "The R FAQ." *http://CRAN.R-project.org/doc/FAQ/R-FAQ. html*.

Ihaka, R., Murrell, P., Hornik, K., Fisher, J.C., and Zeileis, A. (2015). *colorspace: Color Space Manipulation. http://CRAN.R-project.org/package= colorspace*, R package version 1.2-6.

James, D.A. and DebRoy, S. (2012). *RMySQL: R interface to the MySQL database. http://CRAN.R-project.org/package=RMySQL*, R package version 0.9-3.

Kruschke, J.K. (2010). *Doing Bayesian Data Analysis: A Tutorial with R and BUGS.* Academic Press/Elsevier, USA.

Kuhn, M. and Johnson, K. (2013). *Applied Predictive Modeling.* Springer, USA.

Kusnierczyk, W. (2012). *rbenchmark: Benchmarking routine for R. http://CRAN. R-project.org/package=rbenchmark*, R package version 1.0.0.

Leisch, F. (2002). "Dynamic generation of statistical reports using literate data analysis." In W. Härdle and B. Rönz (eds.), "Compstat 2002 – Proceedings in Computational Statistics," Physika Verlag, Germany.

Lemon, J. (2006). "Plotrix: a package in the red light district of R." *R-News,* **6**(4), 8–12.

Ligges, U. and Fox, J. (2008). "R Help Desk: How can I avoid this loop or make it faster?" *R News,* **8**(1), 46–50.

Ligges, U. and Mächler, M. (2003). "Scatterplot3d—an R package for visualizing multivariate data." *Journal of Statistical Software,* **8**(11), 1–20.

Lock, R.H. (1993). "1993 new car data." *Journal of Statistics Education,* **1**(1). *http://www.amstat.org/publications/jse/v1n1/datasets.lock.html*.

Matloff, N. (2011). *The Art of R Programming: A Tour of Statistical Software Design.* No Starch Press, USA.

Matloff, N. and Salzman, P.J. (2008). *The Art of Debugging: With GDB, DDD, and Eclipse.* No Starch Press, USA.

McNeil, D.R. (1977). *Interactive Data Analysis.* Wiley, USA.

Mirai Solutions GmbH (2014). *XLConnect: Excel Connector for R. http://CRAN. R-project.org/package=XLConnect*, R package version 0.2-7.

Montgomery, D.C., Peck, E.A., and Vining, G.G. (2012). *Introduction to Linear Regression Analysis.* Wiley, USA, 5th edition.

Murrell, P. (2011). *R Graphics.* Chapman & Hall, USA, 2nd edition.

Murrell, P. and Ihaka, R. (2000). "An approach to providing mathematical annotation in plots." *Journal of Computational and Graphical Statistics,* **9**, 582–599.

Neuwirth, E. (2014). *RColorBrewer: ColorBrewer Palettes. http://CRAN.R-project. org/package=RColorBrewer*, R package version 1.1-2.

R Core Team (2015). *foreign: Read Data Stored by Minitab, S, SAS, SPSS, ... https://CRAN.R-project.org/package=foreign*, R package version 0.8-66.

Reinhart, A. (2015). *Statistics Done Wrong: The Woefully Complete Guide.* No Starch Press, USA.

Ripley, B. and Lapsley, M. (2013). *RODBC: ODBC Database Access. http://CRAN.R-project.org/package=RODBC*, R package version 1.3-10.

Royston, P. (1982). "Algorithm AS 181: The W test for Normality." *Applied Statistics*, **31**, 176–180.

RStudio Team (2015). *RStudio: Integrated Development Environment for R.* RStudio, Inc., USA. *http://www.rstudio.com/*.

Schloerke, B., Crowley, J., Cook, D., Hofmann, H., Wickham, H., Briatte, F., and Marbach, M. (2014). *GGally: Extension to ggplot2. http://CRAN.R-project.org/package=GGally*, R package version 1.0.1.

Schorling, J.B., Roach, J., Siegel, M., Baturka, N., Hunt, D.E., Guterbock, T.M., and Stewart, H.L. (1997). "A trial of church-based smoking cessation interventions for rural African Americans." *Preventive Medicine*, **26**, 92–101.

Scott, D.W. (1992). *Multivariate Density Estimation: Theory, Practice, and Visualization.* Wiley, USA.

Snee, R.D. (1974). "Graphical display of two-way contingency tables." *The American Statistician*, **28**, 9–12.

Soetaert, K. (2014). *shape: Functions for plotting graphical shapes, colors. http://CRAN.R-project.org/package=shape*, R package version 1.4.2.

Sterne, J.A.C. and Smith, G.D. (2001). "Sifting the evidence—what's wrong with significance tests?" *British Medical Journal*, **322**(7280), 226–231.

Takezawa, K. (2006). *Introduction to Nonparametric Regression.* Wiley, USA.

Tippett, L.H.C. (1950). *Technological Applications of Statistics.* Wiley, USA.

Tong, H. (1990). *Non-Linear Time Series: A Dynamical System Approach.* Oxford University Press, UK.

Trapletti, A. and Hornik, K. (2013). *tseries: Time Series Analysis and Computational Finance. http://CRAN.R-project.org/package=tseries*, R package version 0.10-32.

Urbanek, S. (2013). *RJDBC: Provides access to databases through the JDBC interface. http://CRAN.R-project.org/package=RJDBC*, R package version 0.2-3.

Venables, W.N. and Ripley, B.D. (2002). *Modern Applied Statistics with S.* Springer, USA, 4th edition.

Wand, M.P. and Jones, M.C. (1995). *Kernel Smoothing.* Chapman & Hall, USA.

Warnes, G.R., Bolker, B., Gorjanc, G., Grothendieck, G., Korosec, A., Lumley, T., MacQueen, D., Magnusson, A., Rogers, J., and others (2014). *gdata: Various R programming tools for data manipulation. http://CRAN.R-project.org/package=gdata*, R package version 2.13.3.

Wickham, H. (2009). *ggplot2: Elegant Graphics for Data Analysis*. Springer, USA.

Wickham, H. (2015*a*). *Advanced R*. CRC Press/Taylor & Francis, USA.

Wickham, H. (2015*b*). *R Packages*. O'Reilly Media, USA.

Wilkinson, L. (2005). *The Grammar of Graphics*. Springer, USA, 2nd edition.

Willems, J.P., Saunders, J.T., Hunt, D.E., and Schorling, J.B. (1997). "Prevalence of coronary heart disease risk factors among rural blacks: A community-based study." *Southern Medical Journal*, **90**, 814–820.

Woosley, A.I. and Mcintyre, A.J. (1996). *Mimbres Mogollon Archaeology*. University of New Mexico Press, USA.

Xie, Y. (2015). *Dynamic Documents with R and knitr*. Chapman & Hall/CRC, USA, 2nd edition.

# INDEX

## A

α (significance level), 387
abline function, 134, 137, 456, 525
add1 function, 533, 571
add_axis function, 626, 628, 629, 630
addition, 17, 18
    of matrices, 49–50
additive effect, 468
add_legend function, 626, 628,
    629, 630
add.smooth argument, 552
adjust argument, 628
adjustcolor command, 612–613,
    643–645, 665
adjusted measure, 460
Adjusted R-squared (coefficient of
    determination), 460, 548
advanced plot customization.
    *See* customizing plots;
    plotting
aesthetic mapping, with geoms,
    143–146
aggregate function, 444,
    446–447, 450
AIC (Akaike's Information
    Criterion), 541–548
AIC function, 542
airquality data frame, 614, 615, 617,
    653, 676, 698
Akaike's Information Criterion
    (AIC), 541–548
algorithms, model selection,
    529–548
    backward selection, 537–541
    forward selection, 533–537
    nested comparisons, 529–532
    stepwise AIC selection, 541–548
all function, 63, 64
alpha (significance level), 426
alpha argument, 643, 644, 701, 714
alpha.f argument, 613, 643
alternative argument, 391, 405
alternative hypothesis, 386
American Standard Code for
    Information Interchange
    (ASCII), 160

analysis of variance. *See* ANOVA
AND operator, 65
Anderson-Darling test, 438
angle argument, 370
anonymous functions, 236
anorexia data set, 401
ANOVA (analysis of variance),
    435–450
    Kruskal-Wallis test, 447–450
    one-way, 435–442
        ANOVA table construction,
        439–440
        building ANOVA tables with
        aov function, 440–442
        equivalence with, 481–483
        hypotheses and diagnostic
        checking, 436–439
    two-way, 443–447
        main effects and inter-
        actions, 444–447
        suite of hypotheses, 443–444
anova function, 531, 571
any value, 63, 64
aov function, 440–442, 481
appearance constants, setting with
    geoms, 141–143
apply function, 204–209, 236, 723
args.legend argument, 292
arguments, 8, 222–232
    defaults, setting, 225–227
    ellipses and, 176–177, 228–233
    lazy evaluation, 222–225
    matching, 172–177
        ellipses and, 176–177
        exact, 172–173
        mixed, 175–176
        partial, 173–174
        positional, 174–175
    missing, checking for, 227–228
Arguments section, of help files, 10
arithmetic, 18–19. *See also* math
arithmetic operators, displaying in
    plots, 601
array function, 53, 58
arrays, multidimensional, 52–58
arr.ind argument, 71
arrows function, 134, 138, 353, 585

# E

e (Euler's number), 19
each argument, 25–26, 417
earthquake data example, 660–663
editor, in RStudio, 752–753
editor pane, 5–6
element-wise checks, 184–186
ellipses, arguments and, 176–177, 228–233
else statements, 183–184
empty parentheses, 216
empty pixels, 671–679
end argument, 635, 672
e-notation, 20–21
environments, 166–168
   global, 166
   local, 167–168
   package environments and namespaces, 166–167
equal sign (=), 21–22
errors. *See also* exception handling
   overview, 420–421
   Type I errors, 421–423
     Bonferroni correction, 423
     simulating, 421–423
   Type II errors, 424–428
     other influences on error rate, 426–428
     simulating, 425–426
escape sequences, 76–77
Euclidian space, 720
Euler's number (e), 19
evaluation grid, constructing, 654–655
events, probability and, 310
   complement of an event, 312–313
   intersection of two events, 311–312
   union of two events, 312
exact argument matching, 172–173
Examples section, of help files, 10
Excel, file format for, 153
exception handling, 241–248
   errors and warnings, 242–244
   try statements, 244–248
     suppressing warning messages, 246–248

using in body of function, 245–246
exp function, 19
expand argument, 682, 688, 705
expand.grid function, 654–655, 659, 675, 678, 706, 709, 724
explanatory variables, 451, 453, 485, 490, 528, 589
explicit attributes, 114, 115
exponential distribution, 359–362
   dexp function, 359–360
   vs. exponential function, 361
   pexp function, 360–361
   qexp function, 361–362
exponential function, 19–20
exponents, 17
expression function, 598–599
externally defined helper functions, 234–235
extractAIC function, 542
extraction
   of elements in vectors
     using indexes, 28–32
     using logicals, 68–72
   from matrices, 43–44
extraneous variables, 486
extrapolation, 466

# F

F (abbreviation for FALSE), 60
F (cumulative distribution function), 323
facet_grid command, 619–620
facets
   coloring, 682–686
   multiple plots using, 616–623
     facets mapped to categorical variable, 619–623
     independent plots, 616–618
facet_wrap command, 619–620
factor function, 80
factor variable, 435
factors, 79–87
   combining and cutting, 83–86
   defining and ordering levels, 82–83
   identifying categories, 79–81

RStudio, *continued*
  basic layout and usage, 752–754
    customizing panes, 753–754
    editor features and
      appearance options,
      752–753
rt function, 357
runif function, 347

# S

S3 classing structure, 116
Salaries data frame, 622–623,
  628, 646
sample function, 730–731
sampling distributions, 367–377
  distribution for sample mean,
    368–373
    Dunedin temperatures
      example, 369–373
    standard deviation
      known, 369
    standard deviation
      unknown, 369
  distribution for sample
    proportion, 373–377
sapply function, 223
saturations, 632
save.image command, 11
saving
  scripts, 12
  workspace image files, 11–12
scalar multiple, of matrices, 49
scale_fill_grey function, 292–293
scale_linetype_manual function, 297
scale_shape_manual function, 304
scale_x_discrete function, 292–293
scale_y_continuous function,
  292–293
scatterplot3d function, 649, 652
scatterplot3d package, 741
scatterplots, 300–308
  matrix of plots, 303–308
  single plot, 301–302
scatter.smooth function, 612
scope argument, 533, 539, 542

scoping, 165–171
  environments, 166–168
    global environment, 166
    local environments, 167–168
    package environments and
      namespaces, 166–167
    reserved and protected names,
      170–172
    search path, 168–170
scripts
  overview, 5
  saving, 12
scrolling through commands, 5
sd (standard deviation), 277, 278
search function, 168–169, 252
search path, 168–170
segments function, 134, 137, 345
segments3d function, 694–696, 702
semicolon, 666
seq function, 24–26, 169, 350,
  638, 654
sequences, creating numeric, 24–26
setTxtProgressBar function, 249–250
setwd function, 7, 151
shade argument, 682, 700
shading. *See* smoothing and shading
shape package, 644, 647, 648, 652,
  653, 669, 705, 741
shapiro.test function, 554, 571
Shapiro-Wilk test, 438, 554
shiny package, 759
short-form logical operator, 452
side-by-side boxplots, 299–300
sigma argument, 350, 422, 560
significant digits, 643
silent argument, 244, 246
simple linear regression, 451–483
  categorical predictors, 468–483
    binary variables, 468–472
    changing reference level,
      477–478
    equivalence with one-way
      ANOVA, 481–483
    multilevel variables, 472–477
    treating categorical variables
      as numeric, 478–480

## About the Author

Tilman M. Davies is a lecturer at the University of
Otago in New Zealand, where he teaches statistics
at all university levels. He has been programming
in R for 10 years and teaches R programming in
all his courses. His research on spatial point
pattern modeling has been recognized by the
New Zealand Statistical Association's Worsley
Award, and he has been awarded a prestigious
Marsden Fast-Start grant by the Royal Society of
New Zealand to work on related problems. He
organizes an annual three-day Introduction to R
workshop, which inspired him to write this book
as a guide for beginners.

## About the Technical Reviewer

Debbie Leader has been an R user for many
years. She is passionate about teaching university
students the fundamentals of statistics. In particular, she enjoys guiding students to develop
expertise with R so that they come to appreciate
it as a valuable component in their statistical tool-
box. Debbie joined Massey University as a Senior
Tutor in Statistics on completion of her PhD in
Statistics from the University of Auckland in 2010.

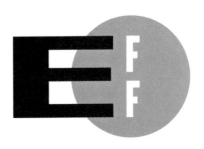

**The Electronic Frontier Foundation** (EFF) is the leading organization defending civil liberties in the digital world. We defend free speech on the Internet, fight illegal surveillance, promote the rights of innovators to develop new digital technologies, and work to ensure that the rights and freedoms we enjoy are enhanced — rather than eroded — as our use of technology grows.

# EFF.ORG
## ELECTRONIC FRONTIER FOUNDATION
Protecting Rights and Promoting Freedom on the Electronic Frontier

*The Book of R* is set in New Baskerville, Futura, Dogma, and TheSansMono Condensed, using the LaTeX $2_\varepsilon$ package nostarch by Boris Veytsman. The book was printed and bound by Sheridan Books, Inc. in Chelsea, Michigan. The papers are 50# Finch Offset and 60# Sappi Matte, which are certified by the Forest Stewardship Council (FSC).

The book uses a layflat binding, in which the pages are bound together with a cold-set, flexible glue and the first and last pages of the resulting book block are attached to the cover. The cover is not actually glued to the book's spine, and when open, the book lies flat and the spine doesn't crack.

# RESOURCES

Visit *https://www.nostarch.com/bookofr/* for resources, errata, and more information.

*More no-nonsense books from*  **NO STARCH PRESS**

## DATA VISUALIZATION WITH JAVASCRIPT

*by* STEPHEN A. THOMAS
MARCH 2015, 384 PP., $39.95
ISBN 978-1-59327-605-8
*full color*

## THE ART OF R PROGRAMMING
**A Tour of Statistical Software Design**

*by* NORMAN MATLOFF
OCTOBER 2011, 400 PP., $39.95
ISBN 978-1-59327-384-2

## STATISTICS DONE WRONG
**The Woefully Complete Guide**

*by* ALEX REINHART
MARCH 2015, 176 PP., $24.95
ISBN 978-1-59327-620-1

## THE MANGA GUIDE TO REGRESSION ANALYSIS

*by* SHIN TAKAHASHI, IROHA INOUE,
*and* TREND-PRO CO., LTD.
MAY 2016, 232 PP., $24.95
ISBN 978-1-59327-728-4

## THE MANGA GUIDE TO STATISTICS

*by* SHIN TAKAHASHI *and*
TREND-PRO CO., LTD.
NOVEMBER 2008, 232 PP., $19.95
ISBN 978-1-59327-189-3

## DOING MATH WITH PYTHON
**Use Programming to Explore Algebra,
Statistics, Calculus, and More!**

*by* AMIT SAHA
AUGUST 2015, 264 PP., $29.95
ISBN 978-1-59327-640-9

**PHONE:**
800.420.7240 OR
415.863.9900

**EMAIL:**
SALES@NOSTARCH.COM

**WEB:**
WWW.NOSTARCH.COM